T0296270

INTELLIGENT IMAGE AND VIDEO COMPRESSION

INTELLIGENT IMAGE AND VIDEO COMPRESSION

Communicating Pictures

SECOND EDITION

DAVID R. BULL

FAN ZHANG

Academic Press is an imprint of Elsevier
125 London Wall, London EC2Y 5AS, United Kingdom
525 B Street, Suite 1650, San Diego, CA 92101, United States
50 Hampshire Street, 5th Floor, Cambridge, MA 02139, United States
The Boulevard, Langford Lane, Kidlington, Oxford OX5 1GB, United Kingdom

Library of Congress Cataloging-in-Publication Data
A catalog record for this book is available from the Library of Congress

British Library Cataloguing-in-Publication Data
A catalogue record for this book is available from the British Library

ISBN: 978-0-12-820353-8

For information on all Academic Press publications
visit our website at https://www.elsevier.com/books-and-journals

Publisher: Mara Conner
Acquisitions Editor: Tim Pitts
Editorial Project Manager: Isabella C.Silva
Production Project Manager: Kamesh Ramajogi
Designer: Mark Rogers

Typeset by VTeX

Contents

List of figures

List of tables

List of algorithms

About the authors

Professor David R. Bull PhD, FIET, FIEEE, CEng obtained his PhD from the University of Cardiff in 1988. He currently holds the Chair in Signal Processing at the University of Bristol, where he is head of the Visual Information Laboratory and Director of the Bristol Vision Institute, a group of some 150 researchers in vision science, spanning engineering, psychology, biology, medicine, and the creative arts. In 1996 David helped to establish the UK DTI Virtual Centre of Excellence in Digital Broadcasting and Multimedia Technology and was one of its Directors from 1997 to 2000. He has also advised Government through membership of the UK Foresight Panel, DSAC, and the HEFCE Research Evaluation Framework. He is now also Director of the UK Government's new MyWorld Strength in Places programme.

David has worked widely across image and video processing focused on streaming, broadcast, and wireless applications. He has published over 600 academic papers, various articles, and four books and has given numerous invited/keynote lectures and tutorials. He has also received awards including the IEE Ambrose Fleming Premium for his work on primitive operator digital filters and a Best Paper Award for his work on link adaptation for video transmission. David's work has been exploited commercially and he has acted as a consultant for companies and governments across the globe. In 2001, he cofounded ProVision Communication Technologies Ltd., who launched the world's first robust multisource wireless HD sender for consumer use. His recent award-winning and pioneering work on perceptual video compression using deep learning has produced world-leading rate-quality performance.

Dr. Fan Zhang PhD received his BSc (Hons) and MSc degrees from Shanghai Jiao Tong University (2005 and 2008, respectively), and his PhD from the University of Bristol (2012). He is currently a Research Fellow in the Visual Information Laboratory at the University of Bristol, working on video compression and immersive video processing. His research interests include perceptual video compression, video quality assessment, and immersive video formats. Aaron has published over 30 academic papers and has contributed to two previous books on video compression. His work on superresolution-based video compression has contributed to international standardization processes and he was a cowinner of the 2017 IEEE Grand Challenge on Video Compression.

Preface

Throughout history, pictures have played a major role in communicating information, emotion, and entertainment. Images are pervasive and the way we exchange and interpret them is central to everyday life.

A very large proportion, approximately 50%, of the human brain's cortical matter is dedicated to vision and related tasks. Comparing this with around 20% for motor functions and only 10% for problem solving helps to explain why we can build computers that beat us at chess but cannot yet, despite recent advances in AI, design machines that come close to replicating the entire human vision system. It also reflects the importance of vision for function and survival, and explains its capacity to entertain, challenge, and inform.

Optical images were unknown before the 10th century and it was not until the 17th century that it was realized that "images" existed in the eyes – via light projected onto our retinas. The eye is a relatively simple organ which captures two inverted and distorted impressions of the outside world, created by light reflected from external objects. So how does the eye create such a rich interface to the outside world? The brain processes the crude images on the retina to detect, assimilate, and predict threats, to produce perceptions of beauty, and to perceive other means for survival and reproduction. These perceptions are not just derived from instantaneous images, but by the temporal and spatial relationships which they represent, placed in the context of a wealth of knowledge obtained from prior experiences. This combination of sensing, stored knowledge, and dynamic neural processing uniquely enables animals, to varying degrees, to perform recognition, prediction, and planning tasks.

Although many contemporary applications use computer vision to aid image understanding (for example in autonomous vehicles, manufacturing, medical diagnosis, robotics, and surveillance), in most cases, the final consumer and interpreter is a human being. In many ways, this captures the essence of communicating pictures – the convergence of information engineering and mathematics with psychology, biology, and the creative arts.

Early visual communications

Visual communication has a very long history, from early cave paintings to Egyptian hieroglyphs, through the inventions of paper, moveable type, photography, and film, to modern art, television, and "new digital media." To give some feel for its beginnings, Fig. 0.1 shows a simple piece of ochre which was discovered in a cave in South Africa. While there is no evidence of drawings in the cave, the markings on the ochre strongly indicate that it was used for drawing – some 200,000 years ago. This is the earliest evidence of humans using tools for visual communication.

Exploration and exploitation of different media have grown ever more sophisticated over the ensuing 200,000 years. The first evidence for humans engraving patterns dates from the mid stone age (South Africa, c.60,000 BCE) and exists in the form of hatched bands on ostrich

FIGURE 0.1 Drawing Ochre: c. 200,000 BCE, South Africa (Courtesy Professor Alice Roberts).

eggshells. Another equally famous example of prehistoric art, known as Venus of Willendorf, is a stone-carved sculpture representing the female form and dates from the Paleolithic period (c.24,000 BCE). Later examples of paintings on cave walls, presumed to be a means of recording significant events (but surely also admired for their beauty) can be traced back as far as c.15,000 BCE. Particularly fine examples can be found in Lascaux, France.

Mobility and print

Mobility in image communications started with the need to exchange information. Early Mesopotamian writing systems (c.4000 BCE) used a stylus and clay for recording numbers whereas Chinese tortoise-shell carvings date back even further, to c.6000 BCE. Papyrus was first manufactured in Egypt as far back as the third millennium BCE and this was rivaled by parchment from c.600 BCE, which was prepared from animal skins. Paper, made from wood pulp, is attributed to the Chinese c.200 BCE, and provided a cheaper and more flexible alternative to silk. The availability of this new medium and its impact on cheap mass communication was unimaginable at the time.

Printing (leading to the mass reproduction of text and images, typically with ink on paper) was clearly one of the most significant developments in image and textual communications, supporting widespread and consistent dissemination of art and information. Over time, print provided knowledge to the masses and enabled an increased accessibility to what was previously only in the domain of the privileged. The origins of printing can be found in the cylinder seals used in Mesopotamia around 3500 BCE. The invention of woodblock printing is attributed to the Chinese c.200 CE, as is the introduction of moveable type in 1040.

Perhaps the most significant step forward was taken by Johannes Gutenberg in 1454, with the development of the movable-type printing press. This high-quality process led to the production of the Gutenberg Bible (1455), which rapidly spread across Europe, triggering the Renaissance. Lithography – the use of a flat stone or metal plate to transfer ink to paper –

was introduced in 1796 by Alois Senefelder, followed by chromolithography in 1837. It was not until 1907 that screen printing was invented, with phototypesetting in 1960. The printing process has developed dramatically over the past 50 years, with the introduction of the laser printer in 1969, 3D printing in 1983, and the digital press in 1993.

Artistic impression, painting, and perspective

Byzantine and Gothic art from the Early Modern Era and Middle Ages primarily reflects the dominance of the church through the expression of religious stories and icons – depiction of the material world was deemed unnecessary at the time. There was extensive use of gold in paintings for religious emphasis and this has been shown through contemporary studies to influence the way eyes are drawn toward or away from key areas (especially when viewed under the prevailing candle-lit conditions).

Figures were normally idealized and drawn without perspective. This, however, changed with the advent of the European Renaissance (between the 14th and 17th centuries). During this period, art returned to the material world, with painters such as Holbein depicting real people with emotions, and Filippo Brunelleschi and Leon Battista Alberti formalizing linear perspective to create depth. Leonardo Da Vinci (scientist, engineer, musician, anatomist, cartographer, and painter) was perhaps the most famous contributor to Renaissance art. His painting Mona Lisa (completed 1519) is worthy of special mention – a half-length portrait of a woman whose facial expression possessed an enigmatic quality that still fascinates today.

The separation of the Church of England from Rome was used, first by Henry VIII in his Great Bible (1539 – based on Tyndale's earlier incomplete version) and then by James I in the King James Bible of 1611, to bring an English version to the population, reinforcing the message of an independent church. Two thousand five hundred copies of the Great Bible were produced and the King James Bible is widely acknowledged as the best selling book in history.

In the 18th century (the so-called Enlightenment Period), inspiration in art was derived more from our expanding knowledge of the universe and an associated deterministic interpretation of the world. This was then progressively transformed and challenged in the 19th and 20th centuries, leading to more abstract forms of art such as impressionism, expressionism, cubism, and surrealism, all influenced by experiments with perception, the distortion of reality, and uncertainty. The systematic use of complementary colors in art was introduced by impressionist painters such as Monet, Seurat, Van Gogh, Renoir, and Cézanne. Claude Monet's powerful painting *Impression, Sunrise* (which gave its name to the movement) depicted a small orange sun and its reflections onto clouds and water in a blue landscape.

Photographic film and motion pictures

The process of creating art on canvas was transformed in 1840 when Louis Daguerre (France) and William Fox Talbot (UK) independently invented the photographic process. Like the printing press 400 years before, this revolutionized the way we see and communicate pictures. At once, it opened up new opportunities for artistic creation as well as, for the first time enabling communication of the "true" representation of people and events – "the photograph never lies."

A number of other key inventions emerged in the late 19th century that facilitated the creation and public display of stored moving images. Firstly, US inventor Thomas Edison is credited with inventing the electric light bulb (although this was the culmination of many years of work by him and others) in New Jersey in 1879. It was probably Edison who was the first to realize that light was the key to displaying moving images. Around the same time, George Eastman in Rochester, USA, founded the Eastman Kodak Company and invented roll film, bringing photography into mainstream use. This was used in 1888 by the world's first filmmaker Louis Le Prince, followed in 1895 by the first public motion picture presentation created by the Lumière brothers in Lyon, France. Louis Lumière was the first to realize that the grab-advance mechanism of a sewing machine could be used to advance, pause, and expose Eastman's film, illuminating it with one of Edison's light bulbs and projecting the light via a lens (i.e., a back to front camera) onto a screen in front of an audience.

Leon Bouly and Thomas Edison himself were also amongst the world's first film makers. One of the earliest films by the Lumière brothers (1896) was called *Arrival of a Train at Ciotat*. This is said to have unnerved the audience due to the realism of the train coming toward them "through" the screen, an impact perhaps more profound or "immersive" at the time than some of today's 3D movies! We think of widescreen formats as a recent innovation, used in cinema since 1953, but only recently in TV. However, the first use of widescreen was in 1897, when Enoch Rector replaced conventional 35-mm film with a 63-mm alternative to form a broader image of a boxing match, in order to show more of the action.

Movies in the early 20th century moved from using technology for capturing reality to story telling and innovations such as special effects, which both mirrored and distorted reality. The development of cinematography followed, exploring close-ups, camera angles, and narrative flow. Sound arrived in the 1920s and color movies in the 1930s.

As Howard Hughes found when making the film *Hell's Angels* (1930), communicating pictures can become very expensive. With production costs of about $3 million, the film (which told the story of WWI aviators) and its director were ridiculed by other Hollywood producers, such as Louis Mayer, for its lavish use of real aircraft, props, and cameras. Hughes was vindicated, however, by taking nearly $8 million at the box office, a record at the time.

The Cohen brothers pioneered digital color grading of feature films in *O Brother, Where Art Thou* (2000), the first to be entirely processed via a digital intermediary. Martin Scorcese's film *The Aviator* (2004), itself about the life of Howard Hughes, took this to another level, using grading techniques to emulate historical film stock. For the first 50 minutes, scenes are rendered only in shades of red and cyan-blue – green objects are rendered as blue. This emulated the early bipack color movies, in particular the multicolor process, which Hughes himself owned. In turn, scenes in the film depicting events after 1935 emulate the saturated appearance of three-strip Technicolor. It is also of interest that, unlike *Hell's Angels*, almost all the flying scenes in *The Aviator* were created using scale models and computer graphics. The film won five Academy Awards. Since then the use of color grading, particularly the exploitation of blue-yellow (or amber-teal) complementarity, has become almost universal in filmmaking.

Television

Television was perhaps the technology that brought the most dramatic change to the way we acquired information and used our leisure time in the 20th century. Television has its roots in the discovery of selenium's photoconductivity (Willoughby Smith, 1873). Early mechanical versions started with the German student Paul Nipkow, who invented the scanning disk image rasterizer in 1884 (although he never made a working model). This scanning method solved the key problem of how to map a 2D image onto a 1D radio signal. Constantin Perskyi is accredited with being the first to use the word "television" in his presentation to the International Electricity Congress in Paris in 1900.

After contributions by DeForest, Korn, and others in the US and Russia, a crude still image was transmitted by Rignoux and Fournier in Paris in 1909. Their scanner employed a rotating mirror with a receiver based on 64 selenium cells. This technology was developed through the early 20th century, culminating in the first public demonstration of TV in 1925 by John Logie Baird in London. AT&T's Bell Telephone Laboratories transmitted halftone still images later the same year. None of these, however, really achieved psychovisually acceptable moving images since they were only transmitted at around five frames per second.

The first demonstration of motion television is attributed to Baird in January 1926 in London, when he presented his 30-line electromechanical system to members of the Royal Institution. Baird's work continued apace, with transmission over fixed and wireless channels. The first transatlantic transmission, between London and New York, was achieved in 1928. It is interesting to note that Baird also worked on low-light imaging and 3D TV at that time!

In the USA, Philo Farnsworth transmitted the first fully electronic television picture in 1927 – an image of a straight line – later that year followed by his *puff of smoke* sequence. The first experimental TV service was established in Germany in 1929 and later that year, a similar operation was established in France. Although Baird's system provided only 240 scanning lines, it did form the basis of the world's first public service, launched by the BBC from Alexandra Palace in 1936. Baird's method was, however, rapidly replaced by Marconi-EMI's "high-definition" 405-line all-electronic system.

Six hundred twenty-five line monochrome broadcasts were introduced in 1964 and the BBC launched its BBC2 service on UHF using a 625-line system with a PAL color system in 1967. Four hundred five-line transmissions continued until 1985 when the frequencies were reused for digital audio broadcasting (DAB) and private mobile radio (PMR) services. Color TV was actually introduced much earlier in the US (1953) but was slow to take off due to a lack of content and high prices. The US networks converted to color in 1965, coincident with the introduction of GE's Porta-Color TV set, and it gained popularity rapidly. From the earliest days, operators realized that television broadcasting was expensive in terms of radio bandwidth, especially as frame rates and resolution (or the numbers of scanning lines) increased. This was addressed in a number of ingenious ways – through the use of interlaced fields and through the use of color subcarriers.

The digital video revolution really took off when television (both satellite and terrestrial) went digital (in the UK) in 1998. This was only made possible by extensive worldwide effort in digital video broadcasting, in particular by the Motion Picture Experts Group (MPEG), chaired by Leonardo Chiariglione. This led to the now ubiquitous MPEG-2 Audiovisual

Coding framework, which still forms the basis for most digital TV in the world. Analog transmission has now been discontinued in much of the world (change-over was completed in the UK in 2012).

The first public high-definition television (HDTV) broadcast in the United States occurred in July 1996. Although HDTV broadcasts had been demonstrated in Europe since the early 1990s, the first regular broadcasts were started in 2004 by the Belgian company Euro1080. UHDTV (4K) terrestrial broadcasting is not yet (in 2020) commonplace but 4K on-demand services are growing rapidly – Netflix first streamed House of Cards in 4K format in 2014. UHD Blu-ray discs and players have been available since 2016.

Pervasive media and the internet

The advancement of visual media through film and television has been accompanied by numerous other technologies that now allow us to see what used to be too far away, too small, or simply invisible to the human eye. We can now visualize almost everything using advanced telescopes, microscopes, and computer-based imaging. But... this explosion of visual information has brought with it new challenges – in particular how to store and transmit the captured information in an efficient manner over bandwidth-limited networks without compromising quality. To do this effectively, we need to exploit our knowledge of the human visual system and the mathematical properties of signals, identify redundancies in the pictures, and hence compress them efficiently.

Over the past two decades, video has been one of the primary drivers for advances in communication and consumer technologies. Real Networks were one of the pioneers of streaming media introducing RealPlayer in 1997. Streaming was first incorporated into Windows Media Player 6.4 in 1999 and Apple introduced streaming in Quicktime 4 in 1999. The demand for a common streaming format was settled through widespread adoption of Adobe Flash Player, although this is now no longer widely supported.

Internet video has been enabled by advances in network technology and video compression. In the same way that the MPEG-2 video compression standard transformed digital TV, H.264/AVC (jointly developed by the International Telecommunications Union [ITU-T] Video Coding Experts Group [VCEG] and the International Organization for Standardization [ISO]/International Electrotechnical Commission [IEC] Moving Picture Experts Group [MPEG]) enabled widespread accessibility to internet video.

Sites like YouTube changed how we access and share video. Created in 2005 by three former PayPal employees, it enabled users to upload, view, and share videos. It used Adobe Flash Video and HTML5 technology with advanced codecs such as H.264/AVC to display a wide variety of content such as movie clips, TV clips, music videos, amateur footage, and blogs. Google bought YouTube in 2006 for $1.65 billion. Every day people watch over a billion hours of YouTube video, with more than 70% of it viewed on mobile devices.[1] Similarly companies such as Netflix have disrupted conventional TV broadcasting through on-demand streaming of high-quality series and films with significant investment in original content. Netflix reported some 170 million subscribers worldwide in 2020. In 2019 Netflix consumed around 13% of total internet download bandwidth and YouTube consumed some

[1] YouTube Statistics: https://www.youtube.com/intl/en-GB/about/press/.

9%. YouTube and Netflix together account for approximately half of North America's peak hours download capacity traffic.[2]

More recently video sharing apps linked to social networking services have boomed. Instagram, now owned by Facebook, started in 2010 with simple photo sharing, providing users with a range of filters to "enhance" their content. It now supports short video clips with geotagging and hashtags to aid searching. Instagram exceeded 1 billion registered users in 2019. Other video platforms such as TikTok, set up by ByteDance in 2017, offer similar services, but focus on very short clips with extensive user support for content creation with trends including memes, lip-synced songs, and comedy videos. It is estimated that TikTok has 800 million users. WeChat by Tencent is a more generic app but also with extensive service offerings including video and group chat. The use of video conferencing services for small- and large-scale meetings have boomed recently, driven in large part by the global lockdowns in place due to the COVID-19 pandemic. The verb *to Zoom* is now in common parlance...

It is therefore a very exciting yet challenging time for video compression. The predicted growth in demand for bandwidth, driven largely by video applications, is probably greater now than it has ever been.[3] There are several primary drivers for this:

1. **Increased numbers of users consuming video**: By 2022 there will be 4.8 billion users accessing the internet with 28.5 billion connected devices.
2. **More of these users are consuming video**: There will be 4.8 zettabytes (4.8×10^{21} bytes) of global annual internet traffic by 2022, equivalent to all movies ever made crossing global IP networks in 53 seconds. Video will account for 82% of all internet traffic by 2022.
3. **An extended video parameter space**: Trends toward increased dynamic range, spatial resolution, and frame rate, coupled with increased traffic due to VR and AR applications and increased use of distributed gaming platforms – all requiring increased bit rates to deliver improved levels of immersion or interactivity. There will be 800 million connected 4K TV sets with 22% of all IP traffic being UHD.
4. **Increased use of mobile devices**: New services, in particular mobile delivery through 5G to smartphones. TVs will account for 24% of all video traffic in 2022, while smartphone usage will rise to 44%.
5. **Increased use of streaming services**: These dominate the internet through social networking and on-demand movies and TV.

Recent and ongoing standards activities, *high-efficiency video coding* (HEVC) and *versatile video coding* (VVC), chaired by Jens-Rainer Ohm and Gary Sullivan, alongside parallel activities driven by the Alliance for Open Media (AOM), have driven the video community to challenge existing methods, delivering significant improvements over their predecessors. HEVC delivered impressive rate-quality gains over H.264/AVC and likewise H.266/VVC outstrips HEVC by over 40%. In the longer term, alternative approaches – perhaps merging conventional compression techniques with machine learning – have the potential to create a new content-driven rate-quality optimization framework for video compression.

[2] Sandvine Global Internet Phenomena Report September 2019: https://www.sandvine.com/global-internet-phenomena-report-2019.

[3] Cisco Visual Networking Index (VNI) Complete Forecast Update, 2017–2022, December 2018.

This book

This book builds on its first edition – *Communicating Pictures* – which first appeared in 2014 (revised in 2017). The importance of video compression remains as important as it was in 2014, with more and more products and services being designed around video content. And, with the emergence of new immersive services, coupled with increased user numbers and expectations, the tension between available bandwidth and picture quality has never been greater. Since its first publication, we have seen significant further developments in video formats and compression standards. High dynamic range production and delivery methods are now mainstream and standards have evolved to support higher spatial and temporal resolutions and wider color gamuts. Immersive formats and associated coding profiles that support volumetric representations and 360-degree video have also emerged. New, more reliable, perceptual metrics have been developed that are now used in mainstream production by major streaming companies. Artificial intelligence and computer vision methods are being increasingly applied that offer significant potential for enhanced coding gains coupled with content understanding. There has also been an increasing tension between royalty bearing and royalty-free standards and between MPEG and the newer AOM. There is also an increasing awareness of the computational demands of compression processes – especially in the context of energy consumption in data centers.

In this book we attempt to place the topic in a practical framework. We provide sufficient mathematical background to enable a proper understanding of the material, but without detracting from the design and implementation challenges that are faced. The linkage between coding and assessment methods and our understanding of visual perception is also key and is emphasized throughout. An algorithmic approach is adopted, where appropriate, to capture the essence of the methods presented. While, on the whole, we keep algorithmic descriptions independent of specific standards, we consistently cross-reference them to current standards such as H.264/AVC, H.265/HEVC, and H.266/VVC.

After an introduction to the motivation for compression and the description of a generic coding framework in Chapter 1, Chapters 2, 3, and 4 describe the basic concepts that underpin the topic. Chapter 2 looks at the perceptual aspects of visual information, characterizing the limits and attributes of our visual system and identifying the attributes that can be exploited to improve rate-quality performance. Chapter 3 provides a grounding in some of the fundamental mathematical concepts that underpin discrete-time signal and image processing. These include signal statistics, sampling, filters, transforms, information theory, quantization, and linear prediction. It also includes a new introduction to machine learning methods. Chapter 4 looks, in more detail, at the practical basics, covering acquisition formats, coding structures, and color spaces, as well as providing an introduction to quality assessment and rate-distortion theory.

The most common approach to spatially decorrelating images is the block transform. This topic is covered in Chapter 5, where decorrelating transforms are defined and derived, and their properties are investigated. The primary focus is on the work horse of compression, the discrete cosine transform (DCT), including its associated algorithmic and implementation issues. Wavelets are an important alternative to DCT-based coding, and these are used in JPEG2000. Chapter 6 develops a framework for wavelet-based compression and explains the benefits and practical issues of this approach such as boundary extension, filter selection, and quantization. Coverage of lossless coding is provided in Chapter 7 – primarily in the context of entropy coding in lossy encoders. This culminates in an operational description of a full image encoder-decoder (or codec) using JPEG as an example together with a detailed overview of CABAC.

Chapter 8 begins the journey into coding moving pictures, with a detailed description of temporal decorrelation and motion estimation. The focus here is on the trade-offs between accuracy and complexity; a range of fast algorithms are described algorithmically and compared in terms of their efficiencies. This leads on to Chapter 9, which integrates the concepts and methods from previous chapters into a description of the ubiquitous hybrid motion-compensated block-based codec. This chapter introduces many of the codec features and refinements that have facilitated significant coding gains in recent years, including subpixel and multiple reference frame estimation, variable block sizes, loop filtering, and intra-prediction.

Chapter 10 addresses the important and challenging topic of measuring and managing the quality of digital pictures. This is related, throughout, to human perception, starting with an analysis of the advantages and disadvantages of the mean squared error, then continuing to describe various perceptually inspired objective metrics. This chapter also includes a detailed description of how to conduct subjective assessment trials, including aspects related to subject screening, test environments, test conditions, and posttest analysis. Rate-quality optimization (RQO) is also covered here. It is only through effective RQO that modern codecs can make informed local mode decisions during coding.

After all the components of the video codec have been described in Chapters 1 to 10, Chapter 11 offers insight into the problems associated with the transmission of coded video in real-time over lossy and bandwidth-limited networks (wireless and fixed). It explains the conflicts between conventional video compression, based on variable-length coding and spatio-temporal prediction, and the requirements for error-resilient transmission. It describes how the presence of packet and bit errors influence a codec's performance and how we can mitigate their impact through packetization strategies, cross-layer optimization, resilient entropy coding, channel coding, and adaptive streaming over HTTP.

The final two chapters, Chapters 12 and 13, provide a view of the current state-of-the-art in standards and coding efficiency. Chapter 12 explains how the methods described throughout the book have been integrated into today's standards, specifically H.264/AVC, H.265/HEVC, AV1, and the most recent H.266/VVC standard. The book concludes in Chapter 13 with some insight into how we might code and communicate pictures in the future, including examples of synthesis-based parametric and region-based compression methods. A particular focus in this second edition is on the rapidly developing application of machine learning to video compression – so-called deep video compression. These areas are the topic of ongoing research and are candidates for future standardization.

Unique aspects of the second edition

While the basic underpinnings described in the first edition remain relevant and important today, some changes have nonetheless impacted the requirements, architectures, performance, and emerging standards. In recognition of this, *Intelligent image and video compression: communicating pictures, 2nd edition* will incorporate significant revisions and updates alongside new innovations, further distinguishing it from its competition and making it more accessible and relevant to both students and professionals. Bull's concept of the creative continuum, introduced in the first edition, is more relevant now than ever. This emphasizes the interactions and interdependencies between content creation, processing, delivery, and consumption, and between the disciplines of engineering, computer science, psychology, and the creative arts. This is brought to the fore – we fully explain the dependencies within the video parameter space and their impact on compression.

The primary advances that will be addressed in the second edition include:

- **More immersive applications and format extensions**: Since the publication of *Communicating Pictures*, the video parameter space has been extended to support the capture, processing, and distribution of formats such as HDR and UHDTV. Virtual and augmented reality applications are also increasing, albeit perhaps not as rapidly as predicted a few years ago. All of these provide increased challenges for video compression with increased sensitivities to coding artifacts that demand even greater rate-quality performance from the video codec. The need for compressing higher spatial and temporal resolutions, higher dynamic ranges, multiple views, and 360-degree content hence has become increasingly important. The second edition introduces some of the video compression requirements associated with these more immersive formats and their applications.
- **New approaches to perceptual quality assessment and measuring engagement**: The importance of perceptually aligned video quality metrics has become prominent in recent years and several new innovations have emerged. The second edition has revised Chapter 10 on visual quality assessment to include the latest advances in perceptual metrics such as VMAF (itself based on machine learning) and also enhance the description of codec comparison methods, test sequence selection, and crowdsourced subjective testing. New immersive formats claim higher levels of engagement, but can introduce new viewing problems that interact with compression artifacts (e.g., vergence-accommodation issues) and sensory misalignment can cause viewing discomfort or even nausea. These issues will be reviewed in the chapter on the human visual system. New methods that are capable of continuously measuring subject immersion (e.g., based on dual task methodologies) will also be described in Chapter 10.
- **New and extended databases**: In recent years, an increased need has emerged for larger and more comprehensive datasets for training and comparing compression technologies. This is particularly true when deep learning methods are employed. The second edition includes a definition of the requirements for these databases and provides an updated review and description of recently available databases, including the use of synthesis and augmentation to enhance coverage.
- **The pervasion of machine (deep) learning**: Artificial intelligence in the form of deep learning has taken the computer vision community by storm over the past few years, continuously delivering state-of-the-art performance in scene understanding. Its potential impact on video compression is very clear, as demonstrated by the work of the authors and others. The importance of machine learning algorithms and how they can be used to optimize compression tools, create new coding architectures, and enhance quality metric performance are explored in the second edition and a review of recent innovations is included. We explain how intelligent resampling based on deep learning can exploit content characteristics and human perception and we review the work ongoing in the wider research community.

- **The dominance of adaptive video streaming over content delivery networks**: Approximately 80% of all internet traffic is now video, with about half of it delivered by streaming companies such as Netflix, YouTube, and others mentioned above. How this content is packaged and delivered to achieve the best rate-quality performance is important to both corporates and consumers. Major innovations and standards have emerged in recent years to support dynamic adaptive streaming and these are described herein.
- **The emergence of new video coding standardization initiatives**: Since the publication of the first edition, several new coding standards have been published or initiated. These include those by the Alliance of Open Media (AV1 and AV2) and by MPEG/JVET (VVC). This second edition provides a comprehensive update of Chapter 12 on standards to include detailed coverage of the key attributes of these new standards.

Several good books have been written on image and video compression, but few take the holistic approach presented in *Communicating pictures, 2nd edition*. This book is targeted primarily at 4th (Senior) year and Masters courses on image and/or video compression or coding. We have tried to balance theory with practice, so we hope it will also be of significant value to practitioners and those wishing to update their skills in industry, as well as new students. Graduates with experience in video compression are highly sought-after in industry across the world and most leading EEE, ECE, and CS departments now include coverage of this topic in their programs.

We hope you find this book interesting, informative, clear, and concise. We have tried to make it relevant and to present concepts in practical context, in a style that is rigorous yet interesting and accessible. We include numerous worked examples together with many additional tutorial questions to reinforce the more important messages.

Picture processing is a popular topic that students enjoy, perhaps because it so intimately relates technology to our everyday personal experiences. This connection between the visual senses and technology, where psychology, biology, and esthetics meet mathematics, engineering, and computer science, is what makes the subject so fascinating. Video compression allows us to experience the "illusion" of high quality while throwing away most of the original data. We hope you will benefit from and have pleasure reading this book. Enjoy the magic!

Additional resources

For the solutions manual, a list of errata, and selected color images, please visit the accompanying resources website at https://fan-aaron-zhang.github.io/Intelligent-Image-and-Video-Compression.htm.

A software (MATLAB®)-based teaching aid has also been developed at the University of Bristol by Steve Ierodiaconou, Aaron Zhang, Alex Mackin, Paul Hill, and myself. This provides a range of demonstrations that will enable students to better understand some the issues associated with image and video compression. It is available at https://fan-aaron-zhang.github.io/Intelligent-Image-and-Video-Compression.htm.

Acknowledgments

This book would not have been possible without the support of many people over many years. We would both like to thank all our outstanding colleagues who currently work with us in the Visual Information Laboratory at Bristol, who have been supportive and professional. We would particularly like to highlight Angeliki Katsenou, Paul Hill, Pui Anantrasirichai, and Di Ma. We would like to acknowledge the amazing support given by Bridget Everett and, before her, Sarah Rogers and Jen Hawkins, who have kept our group functioning while we do the fun bits. We would also like to thank Danier Duolikun for his excellent proof reading – finding many errors in what we thought was a perfect manuscript! Finally we would like to thank all the colleagues from across the world who have provided valuable feedback on and corrections to the first edition.

David R. Bull: Much of my work on image communications has benefited from rich interactions with colleagues including Nishan Canagarajah, Andrew Nix, Angela Doufexi, Dimitris Agrafiotis, and Joe McGeehan. Alongside these academic colleagues there is a long list of PhD students, without whom much of my work would not have happened and I am grateful to all of them. Those who have directly contributed results include Alex Mackin, Mariana Afonso, Miltiadis Papadopoulos, Mohammed Al-Mualla, Mohsin Bokhari, Przemek Czerepinski, TK Chiew, James Chung-How, Pierre Ferre, Francois Martin, Robert O'Callaghan, David Redmill, Oliver Sohm, Richard Vigars, Dong Wang, Yang Zhang, and Fan Zhang. Particular thanks go to Aaron, who has made a massive contribution to this second edition. I would also like to acknowledge colleagues and collaborators whom I have worked with in other organizations across the world, particularly at BBC, Netflix, YouTube, the University of Cambridge, QMUL, the University of Aachen, HHI Fraunhofer Berlin, Thales, and NTT. These are too numerous to list, but they have, in different ways, been an inspiration. A special thanks to Professor Ed Delp form Purdue University, who has been a source of wisdom and advice throughout my career.

If this book has a unique feel, it almost certainly comes from the influences of the wide range of people we work with in the Visual Information Laboratory and in Bristol Vision Institute (BVI). It has been an honor to be the first Director of BVI, which has provided such rich interactions with colleagues in psychology, biology, engineering, computer science, mathematics, biochemistry, neuroscience, ophthalmology, medicine, and the arts. In particular I must highlight psychologists Tom Troscianko and Iain Gilchrist, biologist Innes Cuthill, and anatomist/mechanical engineer J. Burn. BVI, at least in part, owes its existence to the inspiration of Professor Richard Gregory FRS and Professor Tom Troscianko. Richard sadly died in 2010 and Tom passed away too early, in 2011.

As any teacher will know, a key contribution to our own education comes from our students, who take relish and pride, in equal proportion, from pointing out errors, inconsistencies, and ambiguities. I have been privileged to have taught (and be taught by!) many excellent students over the past 25 years.

Finally I would like to thank my family and in particular my partner, Janice, for her love, support, and patience throughout (fortunately she's an academic so she understands!). I dedicate this book to her.

Fan Zhang: First I would like to thank Professor David R. Bull for offering me this opportunity to contribute to the writing of this book. His direct and constant guidance have influenced me throughout my research journey. I am also grateful to all the talented researchers I have worked with in the past, including Alex Mackin, Mariana Afonso, Miltiadis Papadopoulos, and Angeliki Katsenou. Finally I wish to thank my wife for her understanding, support, and love, which encouraged me to complete this work.

CHAPTER

1
Introduction

Visual information is the primary consumer of communications bandwidth across all broadcast, internet, and surveillance networks. The demand for quality and quantity of visual content is increasing daily and this is creating a major tension between available network capacity and required video bit rate. Network operators, content creators, and service providers all need to transmit the highest-quality video at the lowest bit rate and this can only be achieved through the exploitation of perceptual redundancy to enable video compression.

This chapter provides a descriptive introduction to image and video compression. We first look at what compression means and why we need it. We then examine the primary drivers for video compression in today's world. Finally we consider the requirements of a video compression system and explain why standards are so important in supporting interoperability.

1.1 Communicating pictures: the need for compression

1.1.1 What is compression?

Before we look at why we need compression, let us consider what it actually does. For now we can define compression as the ratio of the size (usually in bits or bytes) of the original (input) image or video B_i to that of the (output) image or video used for storage or transmission B_o:

$$\mathrm{CR} = \frac{B_i}{B_o} \quad (\textit{unitless}). \tag{1.1}$$

This simple definition does not account for the resultant quality of the compressed pictures, but it will suffice for now.

Jumping ahead to Section 1.1.2, we will see that typical video compression ratio requirements are currently between 100:1 and 200:1, with this increasing to many hundreds or even thousands to one as new more demanding formats emerge. If we consider Fig. 1.1, we can see what that means in terms of a geometric analogy – in order to achieve a compression ratio of 256:1 we must replace the large square on the left by the small one at the bottom right of the figure. There is a common expression, "You can't squeeze a quart into a pint pot",[1] but that

[1] A quart is a unit of volume (in the imperial or United States customary system) equal to two pints, or one-quarter of a gallon.

https://doi.org/10.1016/B978-0-12-820353-8.00010-4

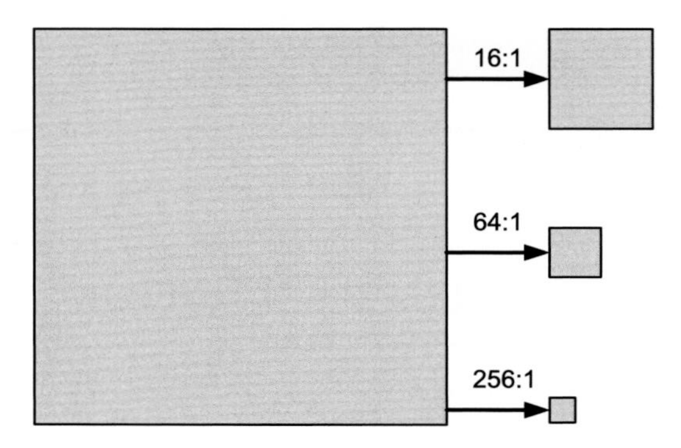

FIGURE 1.1 A geometric interpretation of compression.

is exactly what compression does. In fact, for the example of the bottom square in the figure, it must squeeze 128 quarts into a pint pot!

So what does this mean for pictures? Consider a small image, of dimension 10 × 10 samples (known as picture elements, *pels* or *pixels*) where each pixel is stored as a single byte to produce a 10 × 10 matrix of byte values. After compression (for a compression ratio of 100:1) we might expect that these 100 values will be replaced by just one single byte *and* that this is achieved without introducing any noticeable distortion in the image! At first this appears ridiculous, but that is what we need to achieve! Actually it is only what we need to achieve on average – as the only way we could reduce 100 numbers precisely to a single value was if they were all the same (e.g., the image was of constant luminance) or they fitted some other known model.

In reality most images are not constant-valued, but they are highly correlated spatially. Similarly, most video sequences are also correlated temporally. As we will see, it is the modeling and exploitation of this spatio-temporal correlation, primarily through prediction and transformation, that enables us to compress images and video so effectively. Consider your Blu-ray disc – most people think of these as extremely high-quality. However, they are encoded at around 20 Mbps – a compression ratio of around 60:1 compared to the original movie.[2]

So now let us look in a little more detail at why we need to compress pictures.

1.1.2 Why do we need compression?

Picture formats and bit rate requirements

We will consider picture formats in detail in Chapter 4, but introduce them briefly here in order to better understand the requirements of a picture compression system. Pictures are normally acquired as an array of color samples, usually based on combinations of the red, green, and blue primaries. They are then often converted to some other more convenient color

[2] Assuming a 1080p24 4:2:2 format original at 12 bits per pixel. See Chapter 4 for more details about formats and color subsampling.

TABLE 1.1 Typical parameters for common digital video formats and their (uncompressed) bit rate requirements (*indicates encoding at 10 bits). UHDTV, ultrahigh-definition television; HDTV, high-definition television; SDTV, standard-definition television; CIF, Common Intermediate Format; QCIF, Quarter CIF.

Format	Spatial sampling (V×H)	Temporal sampling (fps)	Raw bit rate (30 fps, 8/10 bits)
UHDTV (4:2:0) (ITU-R BT.2020)	Lum: 7680×4320 Chrom: 3840×2160	24, 25, 30, 50, 60,120	14,930 Mbps*
HDTV (4:2:0) (ITU-R BT.709)	Lum: 1920×1080 Chrom: 960×540	24, 25, 30, 50, 60	933.1 Mbps*
HDTV (4:2:2) (ITU-R BT.709)	Lum: 1920×1080 Chrom: 960×1080	24, 25, 30, 50, 60	1244.2 Mbps*
SDTV (ITU-R BT.601)	Lum: 720×576 Chrom: 360×288	25, 30	149.3 Mbps
CIF	Lum: 352×288 Chrom: 176×144	10–30	36.5 Mbps
QCIF	Lum: 176×144 Chrom: 88×72	5–30	9.1 Mbps

space that encodes luminance separately from two (possibly subsampled) color difference signals.

Table 1.1 shows typical sampling parameters for a range of common video formats. Without any compression, it can be seen, even for the lower resolution formats, that the bit rate requirements are high – often much higher than what is normally provided by today's communication channels. Note that the chrominance signals are encoded at a reduced resolution as indicated by the 4:2:2 and 4:2:0 labels.[3] Also note that two formats are included for the HDTV case (the same could be done for the other formats); for broadcast quality systems, the 4:2:2 format is actually more representative of the original bit rate as this is what is produced by most high-quality cameras. The 4:2:0 format, on the other hand, is what is normally employed for transmission after compression.

Finally, it is worth highlighting that the situation is actually worse than that shown in Table 1.1, especially for ultrahigh-definition (UHD) standards, where higher frame rates and longer word lengths will normally be used. For example at 120 frames per second (fps) with a 10 bit word length for each sample, the raw bit rate increases to 60 Gbps for a single video stream! This will increase even further if 3D or multiview formats are employed.

Available bandwidth

On the other side of the inequality are the bandwidths available in typical communication channels. Some common communication systems for broadcast and mobile applications are characterized in Table 1.2. This table must be read with significant caution as it provides the theoretical maximum bit rates under optimal operating conditions. These are rarely, if

[3] See Chapter 4 for more details on color spaces and subsampling.

TABLE 1.2 Theoretical download bandwidth characteristics for common communication systems.

Communication system	Maximum bandwidth
3G mobile (UMTS, basic)	384 kbps
4G mobile (LTE cat 4)	150 Mbps
4G+ LTE Advanced	450 Mbps
5G mobile	1–10 Gbps
Broadband (VSDL)	55 Mbps
Broadband (VSDL2)	200 Mbps
Broadband (VDSL2-VPlus)	300 Mbps
WiFi (IEEE 802.11n)	300 Mbps
WiFi 6 (IEEE 802.11ax)	10 Gbps
Terrestrial TV (DVB-T2 (8 MHz))	50 Mbps

ever, achieved in practice. Bandwidth limitations are more stringent in wireless environments because the usable radio spectrum is limited, the transmission conditions are variable and data loss is commonplace.

The bit rates available to an individual user at the application layer (which is after all what we are interested in) will normally be greatly reduced from the figures quoted in Table 1.2. The effective throughput (sometimes referred to as *goodput*) is influenced by a large range of internal and external factors. These include:

- overheads due to link layer and application layer protocols,
- network contention and flow control,
- network congestion and numbers of users,
- asymmetry between download and upload rates,
- network channel conditions,
- hardware and software implementations that do not support all functions needed to achieve optimal throughput.

In particular, as channel conditions deteriorate, modulation and coding schemes will need to be increasingly robust. This will create lower spectral efficiency with increased coding overhead needed in order to maintain a given quality. The number of retransmissions will also inevitably increase as the channel worsens. As an example, a DVB-T2 channel (which typically supports multiple TV programs) will reduce from 50 Mbps (256 QAM @ 5/6 code-rate) to around 7.5 Mbps when channel conditions dictate a change in modulation and coding mode down to 1/2 rate QPSK. Similarly, for standard WiFi (802.11n) routers, realistic bandwidths per user can easily reduce well below 300 Mbps even with channel bonding in use – speeds lower than 100 Mbps are not uncommon and can reduce to 10 Mbps or less in congested networks. Emerging WiFi 6 networks targeted at dense Internet of Things applications have demonstrated 10 Gbps, but can reduce to around 10 Mbps for the lowest channel bandwidth and most robust modulation and coding modes. Typical broadband download speeds depend on the VDSL technology used and on the distance from the cabinet (where the fiber terminates). A 100-Mbps VDSL2 link can fall to around 20 Mbps at a distance of 1 km from the cabinet. 3G, 4G, and 5G mobile download speeds also never reach their stated theoretical maxima. Typical useful bandwidths per user for 4G at the application layer will rarely exceed 50% of the theoretical optimum and are more likely to be around 10%. The actual speeds for each user will depend on factors such as location, the distance from the mast, and the amount of traffic. It is also important to note that, for broadband and cellular networks, the upload speeds are typically between 10% and 50% of those for downloading data.

On that basis, let us consider a simple example which relates the raw bit rates in Table 1.1 to the realistic bandwidth available. Consider a digital HDTV transmission at 30 fps using DVB-T2, where the average bit rate allowed in the multiplex (per channel) is 15 Mbps. The raw bit rate, assuming a 4:2:2 original at 10 bits, is approximately 1.244 Gbps, while the actual bandwidth available dictates a bit rate of 15 Mbps. This represents a compression ratio of approximately 83:1.

Download sites such as YouTube typically support up to 6 Mbps for HD 1080p[4] format, but more often video downloads will use 360p or 480p (640×480 pixels) formats at 30 fps, with a bit rate between 0.5 and 1 Mbps encoded using the H.264/AVC standard. In this case the raw bit rate, assuming color subsampling in 4:2:0 format, will be 110.6 Mbps. As we can see, this is between 100 and 200 times the bit rate supported for transmission.

Example 1.1. Compression ratio for UHDTV

Consider the case of 8K UHDTV with the original video in 4:2:0 format (a luminance signal of 7680×4320 and two chrominance signals of 3840×2160) at 10 bits per sample and a frame rate of 60 fps. Calculate the compression ratio if this video is to be transmitted over an internet link with an average bandwidth of 15 Mbps.

Solution. The 4:2:0 color subsampling method has, on average, the equivalent of 1.5 samples for each pixel (see Chapter 4). Thus, in its uncompressed form the bit rate is calculated as follows:

$$R = 7680\text{ (H)} \times 4320\text{ (V)} \times 1.5\text{ (samples/pixel)} \times 10\text{ (bits)} \times 60\text{ (fps)} = 29859840000\text{ bps},$$

i.e., a raw bit rate approaching 30 Gbps. Assuming this needs to be transmitted in a channel of bandwidth of 15 Mbps, then a compression ratio of 1991:1 would be required!

$$\mathrm{CR} = \frac{29859840000}{15000000} \approx 1991.$$

Hopefully this section has been convincing in terms of the need for compression. The tension between user expectations in terms of quality and ease of access on the one hand and available bandwidth on the other has existed since the first video transmissions, and this has promoted vigorous research in the fields of both coding and networks. Fortunately the advances in communications technology have mirrored those in video compression, enabling

[4] p stands for progressive transmission – see Chapter 4 for further details.

the transmission of high (and in most cases very high)-quality video that meets user expectations.

In the next section we examine the applications that are currently driving video compression performance as well as those that are likely do so in the future.

1.2 Applications and drivers

1.2.1 Generic drivers

By 2022 it is predicted that the number of network-connected devices will reach 28.5 billion, for around 5 billion internet users. Cisco predicts [1] that this will result in 4.8 zettabytes (1 zettabyte $= 10^{21}$ bytes) of global internet traffic, with 82% of this being video content and 22% being at UHD resolution. The generic drivers for this explosion in video technology are as follows:

- increased numbers of users with increased expectations of quality and accessibility,
- the ubiquity of user-generated image, video, and multimedia available through social networking and streaming sites,
- the emergence of new ways of working using distributed applications and environments such as the cloud,
- the demand for rapid and remote access to information, driven by the need for improved productivity, security, and responsiveness.
- emerging immersive and interactive entertainment formats for film, television gaming, and streaming.

In order to satisfy the emerging demands for accessibility and quality, video communication solutions require a high-performance core network, an adaptive wireless infrastructure, robust data transport protocols, and efficient content representations. The interactions between these components in the delivery chain can be thought of as a jigsaw puzzle (Fig. 1.2) where complex and dynamically changing pieces need to fit together. Only in combination can these provide a truly end-to-end flexible and robust solution to the delivery challenges of the future. Video compression is a key component in the solution.

1.2.2 Application drivers and markets

The range of applications which rely wholly or partially on video technology is enormous and increasing; as a consequence, the market for video compression technology is growing rapidly. Some examples are provided below.

Consumer video

Entertainment, personal communications, and social interaction provide the primary applications in consumer video, and these will dominate the video landscape of the future. There has been a massive increase in the consumption and sharing of content on mobile devices and this is likely to be the major driver over the coming years – some 1.6 billion

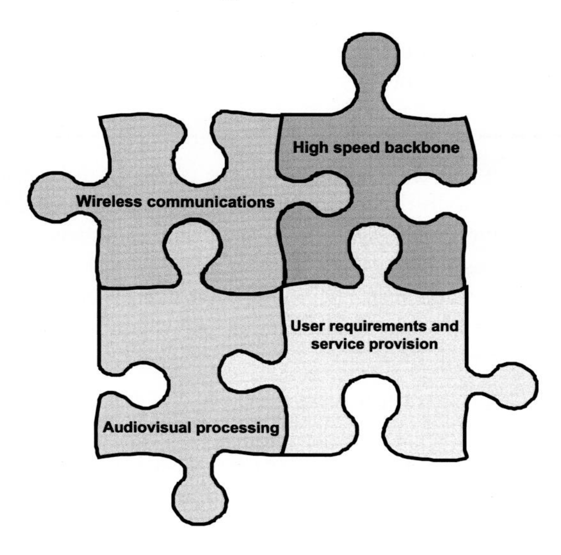

FIGURE 1.2 The multimedia communications jigsaw puzzle.

smartphones were sold in 2020 [1]. In contrast, the Quarterly Global TV Shipment and Forecast Report by NPD DisplaySearch [2] predicted global TV sales to reach 270 million in 2016. In reality the actual sales figure for 2019 was 214 million [3].

The key drivers for growth in this sector are:

- internet streaming, peer-to-peer distribution, and personal mobile communication systems,
- the demand for more immersive content (higher resolutions, frame rates, and dynamic range) including emerging virtual reality (VR) and augmented reality (AR) applications,
- social networking, user-generated short-form content, and content-based search and retrieval,
- in-home wireless content distribution systems,
- distributed gaming platforms.

Business, manufacturing, and automation

Computer vision and visual communications are playing an increasingly important role in business. Particularly, since the COVID-19 global pandemic and the desire to reduce business travel, the demand for higher-quality video conferencing and sharing of visual content have increased. Similarly, in the field of automation, vision-based systems are playing a key role in transportation systems and autonomous vehicle developments, and are now underpinning many manufacturing processes, often demanding the storage or distribution of compressed video content. The drivers in this case can be summarized as:

- videoconferencing, tele-working, and other interactive services,
- publicity, advertising, news, and journalism,
- design, modeling, and simulation,

- transport systems, including vehicle guidance, assistance, and protection,
- automated manufacturing and robotic systems.

Security and surveillance

In an uncertain world, we are increasingly aware of our safety and security, and video monitoring is playing an increasingly important role in this respect. In equal measure, the increased deployment of surveillance systems does raise concerns for many in terms of violations of personal privacy, especially when linked to powerful visual analytics software. This increase shows no signs of abating – Memoori reported recently that video surveillance product sales reached $19.15 billion in 2019 and predicted sales of $35.82 billion in 2024 [4]. The key drivers in this sector are:

- surveillance of public spaces and high-profile events,
- national security,
- battlefield situational awareness, threat detection, classification, and tracking,
- emergency services, including police, ambulance, and firefighters.

Healthcare

The healthcare market is becoming increasingly reliant on imaging methods to aid diagnoses and to monitor the progression of disease states. 3D imaging modalities such as CT and MRI produce enormous amounts of data for each scan and these need to be stored as efficiently as possible while retaining the highest quality. Video is also becoming increasingly important as a point-of-care technology for monitoring patients in their own homes to reduce costs and limit hospital admissions. The primary healthcare drivers for compression are:

- point-of-care monitoring,
- emergency services and remote diagnoses,
- tele-surgery,
- medical imaging.

It is clear that all of the above application areas require considerable trade-offs to be made between cost, complexity, robustness, and performance. These issues are addressed further in the following section.

1.3 Requirements and trade-offs in a compression system

1.3.1 The benefits of a digital solution

We live in a world where digital technology dominates. In the context of video compression, digital solutions offer many significant advantages over their analog counterparts. These include:

- ease of editing, enhancement, special effects, transcoding, and multigeneration processing,
- support for indexing, metadata extraction, annotation, search, and retrieval,
- ease of content protection, conditional access, rights management, and encryption,

- support for scalable and interactive delivery,
- separability of source and channel coding to provide better management of rate, distortion, and robustness to errors,
- flexible support for new formats and different content types such as multiview or 360.

1.3.2 Requirements

The basic requirement of a video compression system can be stated quite simply: we want the highest quality at the lowest bit rate. However, in addition to this, a number of other desirable features can be listed. These include:

- **Robustness to loss**: We want to maintain high quality even when signals are transmitted over error-prone channels.
- **Reconfigurability and flexibility**: To support delivery over time-varying channels or heterogeneous networks.
- **Low complexity**: Particularly for low-power portable implementations or large-scale data center applications.
- **Low delay**: To support interactivity and sensory alignment.
- **Authentication and rights management**: To support conditional access, for content ownership verification, or to detect tampering.
- **Standardization**: To support interoperability.

1.3.3 Trade-offs

In practice we cannot usually satisfy all of the above requirements, either due to cost or complexity constraints or because of limited bandwidth or lossy channels. We therefore must compromise and make design trade-offs that provide the best solution given the prevailing constraints. Areas of possible compromise include:

- **Lossy vs. lossless compression:** Ideally, we do not want the video to be distorted at all compared to the original version. In practice this is not usually possible as perfectly lossless (reversible) compression places severe limitations on the compression ratio achievable. We must therefore exploit any redundancy in the image or video signal in such a way that it delivers the desired compression with the minimum perceived distortion.
- **Rate vs. quality:** In order to compromise between bit rate and quality, we can adjust a number of parameters. These include: frame rate, spatial resolution (luma and chroma), dynamic range, prediction mode, and latency. The influence of these parameters on the final video quality is however highly content-dependent. For example, with a fixed camera and a completely static scene, frame rate will have little or no influence on the perceived quality. In contrast, for a high-motion scene, a low frame rate will result in unpleasant jerkiness.
- **Complexity and power consumption vs. cost and performance:** In general, a more complex video encoder will support more advanced features and thus produce a higher-quality result than a lower-complexity version. However, more complex architectures invariably are more expensive, consume more energy, and can introduce greater delay. In addition,

they may not be realizable using software implementation, possibly demanding custom hardware solutions.

- **Delay vs. performance:** Low latency is key in interactive and conversational applications. However, if an application is tolerant to latency in encoding or decoding, then performance gains can be made elsewhere. For example, increased error resilience features could be incorporated to retransmit data corrupted during transmission or, alternatively, more reference frames and more flexible coding structures could be used to improve motion prediction performance (see Chapters 8 and 9).

1.4 The basics of compression

A simplified block diagram of a video compression system is shown in Fig. 1.3. This shows an input digital signal (using for example one of the formats from Table 1.1) being encoded, transmitted, and decoded. Not shown are possible preprocessing stages such as filtering, (de)interlacing, segmenting, color subsampling, or gamma correction; these are covered in Chapter 4. A full description of the operation of this block-based hybrid motion-compensated video encoder is provided in Chapter 9.

1.4.1 Still image encoding

If we ignore the blocks labeled as motion compensation, the diagram in Fig. 1.3 describes a still image encoding system, such as that used in JPEG. The intraframe encoder performs (as its name suggests) coding of the picture without reference to any other frames. This is normally achieved by exploiting spatial redundancy through transform-based decorrelation (Chapters 5, 6, and 9) followed by variable-length symbol encoding (VLC) (Chapter 7). The image is then conditioned for transmission using some means of error-resilient coding that makes the encoded bitstream more robust to channel errors. Methods for achieving error resilience are described in Chapter 11.

At the decoder, the inverse operations are performed and the original image is reconstructed at the output.

1.4.2 Encoding video

A video signal is simply a sequence of still images, acquired typically at a rate of 24, 25, 30, 50, or 60 frames per second. Video can of course be encoded as a series of still images using intraframe methods as described above. However, we can achieve significant gains in coding efficiency if we also exploit the temporal redundancy that exists in most natural video sequences. This is achieved using interframe motion prediction, as represented by the motion compensation block in Fig. 1.3. This block predicts the structure of the incoming video frame based on the contents of previously encoded frames (Chapter 8). The encoding continues as for the intraframe case, except this time the intraframe encoder block processes the low-energy residual signal remaining after prediction, rather than the original frame.

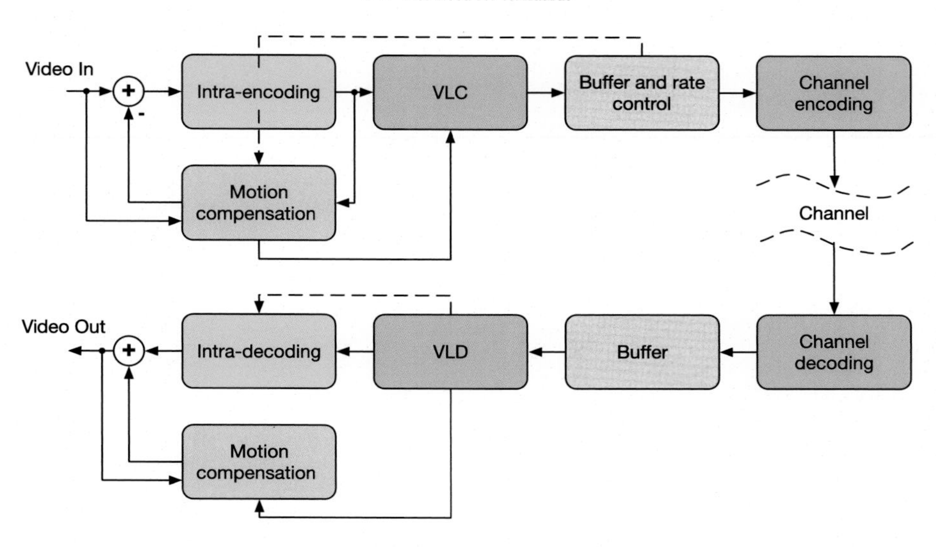

FIGURE 1.3 Simplified video compression architecture.

After variable-length encoding, the encoded signal will be buffered prior to transmission. The buffer serves to match the output bit rate to the instantaneous capacity of the channel and this is normally achieved using a rate control algorithm which adjusts coding parameters as described in Chapter 10.

Because of the reliance on both spatial and temporal prediction, compressed video bitstreams are more prone to channel errors than still images, suffering from temporal as well as spatial error propagation. Methods of mitigating this, making the bitstream more robust and correcting or concealing the resulting artifacts, are described in Chapter 11.

1.4.3 Measuring visual quality

In the final analysis, we need to assess how good our encoding system is, and the absolute arbiter of this is the human visual system. Because of this, subjective testing methodologies have become an important component in the design and optimization of new compression systems. However, such tests are very time consuming and cannot be used for real-time rate-distortion optimization. Hence there has been a significant body of work reported on the development of objective metrics that provide a realistic estimate of video quality. These are discussed alongside subjective evaluation methods in Chapter 10.

1.5 The need for standards

The standardization of image and video formats and compression methods has been instrumental in the success and universal adoption of video technology. We briefly introduce

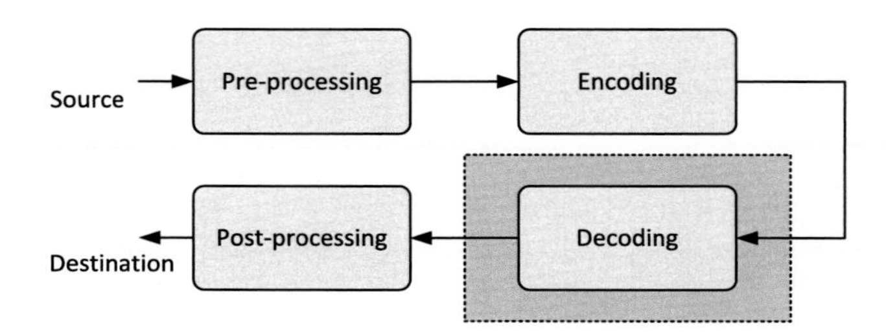

FIGURE 1.4 The scope of standardization.

the motivation for and history of standards here; a more detailed description of the primary features of recent standards is provided in Chapter 12.

1.5.1 Some basic facts about standards

A few basic facts about image and video compression standards follow.

1. Standards are essential for interoperability, enabling material from different sources to be processed and transmitted over a wide range of networks or stored on a wide range of devices. This interoperability opens up an enormous market for video equipment, which can exploit the advantages of volume manufacturing, while also providing the widest possible range of services for users.
2. Standards define the bitstream format and decoding process, not (for the most part) the encoding process. This is illustrated in Fig. 1.4. A standard-compliant encoder is thus one that produces a compliant bitstream and a standard-compliant decoder is one that can decode a standard-compliant bitstream, i.e., a standard specifies the syntax that a bitstream must follow and the corresponding decoding process, not explicitly the encoding process.
3. The main challenge lies in the bitstream generation, i.e., the encoding. This is where manufacturers can differentiate their products in terms of coding efficiency, complexity, or other attributes.
4. Many encoders are standard-compliant, but some are better than others! Compliance is no guarantee of absolute quality.

1.5.2 A brief history of video encoding standards

A chronology of video coding standards is represented in Fig. 1.5. This shows how the International Standards Organization (ISO) and the International Telecommunications Union (ITU-T) have worked both independently and in collaboration on various standards since 1988. In recent years, most coding standards have benefited from close collaboration between these organizations.

Study Group SG.XV of the CCITT (now ITU-T) actually produced the first international video coding standard, H.120, in 1984. H.120 addressed videoconferencing applications at 2.048 Mbps and 1.544 Mbps for 625/50 and 525/60 TV systems, respectively, but was never a commercial success and not widely deployed. H.261 [5] followed this in 1988 with a codec

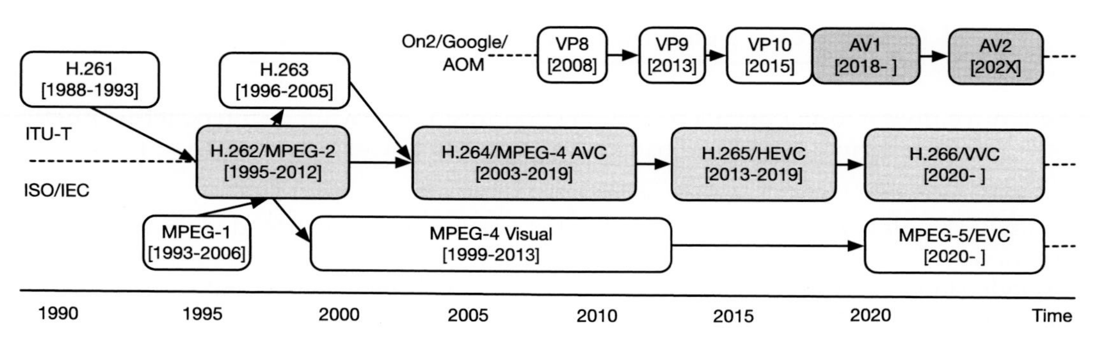

FIGURE 1.5 A chronology of video coding standards from 1988 to the present date (dates in parentheses indicate first published standard and last published revision).

based on a $p{\times}64$ kbps ($p{=}1\cdots30$) targeted at ISDN conferencing applications. This was the first block-based hybrid compression algorithm using a combination of transformation (the discrete cosine transform [DCT]), temporal differential pulse code modulation (DPCM), and motion compensation. This architecture has stood the test of time as all major video coding standards since have been based on it.

In 1988 the Moving Picture Experts Group (MPEG) was founded, delivering a video coding algorithm targeted at digital storage media at 1.5 Mbps in 1993. This was followed in 1995 by MPEG-2 [7], specifically targeted at the emerging digital video broadcasting market. MPEG-2 was instrumental, through its inclusion in all set-top boxes for more than a decade, in truly underpinning the digital broadcasting revolution. A little later, in 1996, ITU-T produced the H.263 standard [6]. This addressed the emerging mobile telephony, internet, and conferencing markets at the time. Although mobile applications were slower than expected to take off, H.263 had a significant impact in conferencing, surveillance, and applications based on the then-new Internet Protocol.

MPEG-4 [8] was a hugely ambitious project that sought to introduce new approaches based on object-based, as well as, or instead of, waveform-based methods. It was found to be too complex and only its Advanced Simple Profile (ASP) was used in practice. This formed the basis for the emerging digital camera technology of the time.

Around the same time ITU-T started its work on H.264. They delivered their standard, in partnership with ISO/IEC, in 2003 [9]. In the same way that MPEG-2 transformed the digital broadcasting landscape, so has H.264/AVC transformed the mobile communications and internet video domains; despite more recent developments, H.264/AVC remains the most prolific codec in use today. In 2013, the joint activities of ISO and ITU-T delivered the HEVC standard [10–12], offering bit rate reductions of up to 50% compared with H.264/AVC. In parallel with ISO/ITU activities, the Alliance for Open Media (AOM) was formed in 2015 to promote open royalty-free coding standards. It released its first standard (AV1) in 2018 [14], which achieves some 10% savings over HEVC. Most recently, ISO and ITU-T have again combined forces to develop the new H.266/VVC coding framework, released in 2020 [13]. VVC currently offers potential savings over HEVC of some 40%.

MPEG-2, H.264/AVC, and H.265/HEVC are in widespread use today, and AV1 implementations are increasing rapidly. Further details on coding standards are provided in Chapter 12.

1.6 The creative continuum: an interdisciplinary approach

Before we dive into the details of compression in the following chapters, let us briefly consider what is required to enable us to create an optimal viewing experience. In essence, we need:

- an understanding of the influence of, and interactions within, an extended visual parameter space;
- to understand the way humans view and engage with mediated visual content and to develop perceptually reliable measures of quality and immersion;
- to use of these measures to characterize the impact of distortion, dynamic range, color palette, spatial resolution, and temporal resolution;
- acquisition, delivery, and display processes supported by appropriate formats that preserve or enhance immersive properties while minimizing the bit rate;
- parameter optimization and adaptation methods that take account of the influences of content and use-context;
- an understanding of the impact of cinematographic methods, e.g., shot length, framing, camera placement, film-grain noise, and camera/subject motion.

These points intimately link production, cinematography, and acquisition to delivery, display, consumption, and quality assessment. So, rather than an end-to-end delivery system we should instead consider a set of continuous relationships within an extended parameter space as indicated in Fig. 1.6. We should no longer consider acquisition, production, compression, display, and assessment as independent processes, but should instead view them as a *creative continuum*, where:

- The **creation** processes must be matched to the acquisition formats to ensure optimal exploitation of the format in terms of sets, lighting, etc.
- The **capture** processes must employ formats and parameters that enable immersive experiences.
- The **experience** must provide maximum engagement with the displayed content. Quantitative measurements of experience are essential if we are to fully understand the influences of the delivery and display processes.
- Importantly, the **delivery** processes must ensure that the content is delivered in a manner that preserves the quality and immersive properties of the format.

We hope that the above discussion makes obvious the need for an interdisciplinary approach to this topic, one which pulls on strengths from psychology and the creative arts as well as from its traditional domains of electronic engineering and computer science.

1.7 Summary

This chapter has introduced the context and the primary drivers for image and video compression technology in today's world. We have seen that compression exists because of the

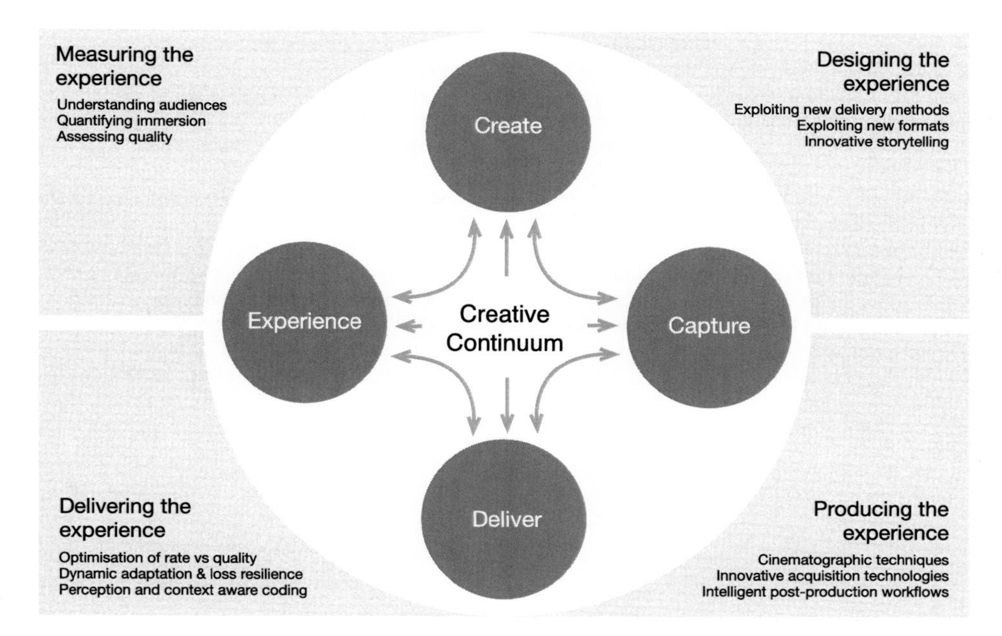

FIGURE 1.6 The creative continuum.

tensions between, on the one hand, the user demands and expectations for increased quality, mobility, and immersion, and on the other hand, the limited bandwidth and high traffic volumes that characterize most communications networks. Communicating pictures is a key part of modern life and compression is an essential element in ensuring that they are delivered faithfully, reliably, and in a timely manner.

We have examined the requirements of a compression system and have seen that these represent a compromise between coding rate, picture quality, and implementation complexity. Finally the justification for universal compression standards has been presented, as these are essential in order to enable interoperability across networks, terminals, and storage devices. The standardization process is also a primary driver for continued research in the field and creates a productive interaction between industry and academia.

References

[1] Cisco Visual Networking Index (VNI), Complete forecast update, 2017–2022, http://www.cisco.com/en/US/netsol/ns827/networking_solutions_sub_solution.html, December 2018.
[2] Global TV Shipment and Forecast Report, NPD DisplaySearch, 2013.
[3] https://www.statista.com/statistics/760281/global-tv-set-unit-sales/. (Accessed April 2020).
[4] https://memoori.com/portfolio/the-physical-security-business-2019-to-2024/. (Accessed April 2020).
[5] Int. Telecommun Union-Telecommun ITU-T, Video codec for audiovisual services at p×64 kbit/s (03/1993), 1994.
[6] ITU-T, Video Coding for Low Bitrate Communication, ITU-T Rec. H.263 (01/2005)), 2005.

[7] ITU-T and ISO/IEC JTC 1, Generic Coding of Moving Pictures and Associated Audio Information—Part 2: Video, ITU-T Rec. H.262 (02/2012) and ISO/IEC 13818-2 (MPEG-2 Video), 2013.
[8] ISO/IEC JTC 1, Coding of Audio-Visual Objects—Part 2: Visual, ISO/IEC 14496-2 (MPEG-4 Visual), version 1, 1999, version 2, 2000, version 3, 2004.
[9] ITU-T and ISO/IEC JTC 1, Advanced Video Coding for Generic Audiovisual Services, ITU-T Rec. H.264 and ISO/IEC 14496-10 (AVC) (06/2019), 2019.
[10] G. Sullivan, J-R. Ohm, W. Han, T. Wiegand, Overview of the high efficiency video coding (HEVC) standard, IEEE Trans. Circuits Syst. Video Technol. 22 (12) (2012) 1648–1667.
[11] Joint Collaborative Team on Video Coding (JCT-VC) of ITU-T SG 16 WP 3 and ISO/IEC JTC 1/SC 29/WG 11 ISO/IEC 23008-2 and ITU-T Recommendation H.265, High Efficiency Video Coding (HEVC) (11/2019), 2019.
[12] J-R. Ohm, G. Sullivan, H. Schwartz, T. Tan, T. Wiegand, Comparison of the coding efficiency of video coding standards—including high efficiency video coding (HEVC), IEEE Trans. Circuits Syst. Video Technol. 22 (12) (2012) 1669–1684.
[13] Joint Video Experts Team (JVET) of ITU-T SG 16 WP 3 and ISO/IEC JTC 1/SC 29/WG 11, Document: JVET-S2001: Versatile Video Coding (Draft 10), July 2020.
[14] Alliance for Open Media. AOMedia Video 1 (AV1). [Online]. Available: https://aomedia.org/. (Accessed April 2020).

CHAPTER

2

The human visual system

A study of the structure, function, and perceptual limitations of the human visual system (HVS) provides us with clues as to how we can exploit redundancy in visual information in order to compress it with minimum degradation in perceived quality. The way humans view and perceive digital images and video relates strongly to how we perceive the world in our everyday lives. While a full understanding of the HVS is still a long way off, we do know enough to be able to create the illusion of high quality in an image sequence where, perhaps, 199 out of every 200 bits from the original are discarded.

This chapter is intended to provide the reader with a basic understanding of the HVS and, where possible, to relate its characteristics to visual redundancy and ultimately to a means of compressing image and video signals. We first consider the physical architecture of the HVS and the constraints it imposes on the way we see the world. We then review perceptual aspects of vision related to brightness, contrast, texture, color, and motion, indicating the physical limits of key visual parameters. Visual masking is a key component in perception-based compression, so we end by looking at how certain spatio-temporal visual features can be masked by certain types of content.

This chapter is primarily a review of the work of others who are much more expert in the biological, neurological, and psychological aspects of the subject than the authors. For further information, the reader is directed to many excellent texts on this topic, including those by Snowden et al. [1], Mather [2], Wandell [3], Marr [4], and Cormack [5].

2.1 Principles and theories of human vision

The vision system is the most powerful and complex of our senses and we have only begun to understand it since the pioneering work of Helmholtz in the 1890s [6].

Between 40% and 50% of the neurons in the human brain are estimated to be associated with visual information processing. This compares to approximately 10% for hearing, 10% for the other senses, and 20% for motor functions, leaving 10–20% for everything else – things like playing chess [1]. This is not entirely surprising because visual activity is linked to almost everything we do. In basic terms we interpret colors, classify shapes, detect and assess motion, estimate distances, and do a pretty good job at creating a 3D world out of the 2D images that fall on our retinas.

https://doi.org/10.1016/B978-0-12-820353-8.00011-6

So what is the difference between what we see and what we perceive? Quite a lot actually. For example we fill in the blind spots where the optic nerve passes through the back of the eye, we correct the distorted imperfect images that fall on the retina, and ignore things that get in the way like our nose and the blood vessels in the eye. At a higher level, it is clear that the visual system does not exist to faithfully record images, but instead to interpret the world and survive in it – to recognize friends, to assess threats, to navigate, to forage, and to find mates. Humans have also used visual stimuli to entertain, communicate, and inform, probably over most of our 200,000-year history – from cave paintings to modern media.

A typical retina contains some 130 million light-sensitive rod and cone cells. Its neural processors are an extension of the visual cortex and it is the only part of the brain visible from outside the body. Each of the optic nerves, which carries signals from the eye to the brain, consists of around one million fibers and it is these that transmit the signals from the retina to the rest of the visual cortex. This contrasts starkly with other senses – for example, each auditory nerve comprises about 30,000 fibers.

We will examine the structure of the retina and the visual cortex in Sections 2.2 and 2.3, but first let us take a brief look at the theories that exist about how our visual system operates.

Theories of vision

Visual perception is highly specialized and complex. Historically, much of the research into the biological and anatomical aspects of the vision system has been based on primate brains, which appear to be very similar to the HVS. More recently however, technologies such as fMRI have revolutionized neuroscience and neuropsychology, providing ever increasing insight into the workings of the human system.

Until the late 1970s there were two main competing theories of vision – from the *empiricists* and the *rationalists*. Their views on the workings of the visual system differed significantly. Empiricists believed in innate, low-level, bottom-up processing, allocating less importance to higher-level functions. Julesz [7] for example showed that depth perception can arise from random dot stereograms with no high-level meaning, thus inferring that this could not be driven by a top-down mechanism. Empiricist models were largely based on the extraction of features which facilitated the generation of connected patterns. The pioneering work by Hubel and Wiesel in 1959 [8] showed the orientation and scale selectivity of the low-level visual system and Uhr and Vossler in 1961 demonstrated character recognition from such low-level features. Empiricists however did not consider context.

In contrast, rationalists focused on high-level control of vision and the construction of percepts by the brain. They hypothesized mental models that are used to influence our search for and verification of lower-level visual features [9]. Early work in this area developed *picture grammars* that described 2D properties using 2D structures [10]. These, however, described scene syntax rather than scene semantics and again did not show how 3D visual inputs could be represented. This in turn led to a *scene-grammar* approach, which attempted to map 3D scenes to 2D descriptors. This was achieved using either high-level shapes, such as cubes, or low-level structural elements, such as vertices and edges. Such models still failed to describe how humans can generate perceptions of unfamiliar objects since the models rely on prior knowledge of "grammars" to describe the visual input in a high-level fashion.

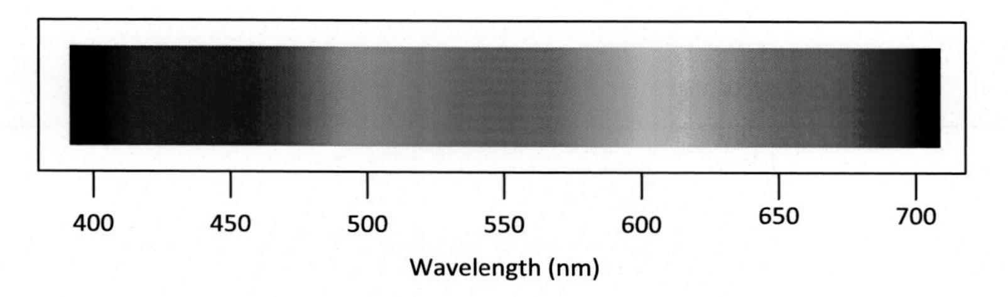

FIGURE 2.1 The visible spectrum.

From the rationalist approaches emerged a hybrid *connectionist* approach. It was not until Marr's contribution in 1982 [4] that this approach was formalized. Connectionist theories emphasized interconnections between processing units embodying both the bottom-up (data-driven) extraction of features combined with the top-down (interpretative) influence that generates perceptions based on this low-level information. This can be strongly modulated by processes such as attention. This connectionist distributed processing type of approach is still influential and draws heavily on biological evidence providing close correlations with experimental data.

2.2 Acquisition: the human eye

Photons from the sun or other sources of light reflect off physical objects, and some of these photons enter our eyes. Our visual system senses energy radiated in the part of the electromagnetic spectrum from about 380 to 720 nm (420 to 790 THz), as shown in Fig. 2.1. There is nothing particularly special about this range except that our physiology has adapted to it, and it has adapted to it because it is useful for things like finding food and detecting threats and potential mates. Some animals see other ranges; for example, many birds and insects see in the UV range, as this enables them to see certain features in, for example, flowers. The human eye has a peak sensitivity around 555 nm in the green region of the spectrum – perhaps not a coincidence since the Earth's surface mostly reflects light in the green band from 550 to 560 nm due to foliage cover [1].

The human eye senses brightness approximately logarithmically over a broad range of luminance values and can also see colors that are not present in the spectrum. Unsaturated colors, such as pink, are not present and neither are colors like magenta, which is formed from a combination of multiple wavelengths. A diagram of the human eye in cross-section is shown in Fig. 2.2. Let us now look in a little more detail at its main components.

2.2.1 Retinal tissue layers

The eye is an approximately spherical volume filled with fluid, comprising four layers of tissue – the sclera, the retina, the choroid, and the ciliary body.

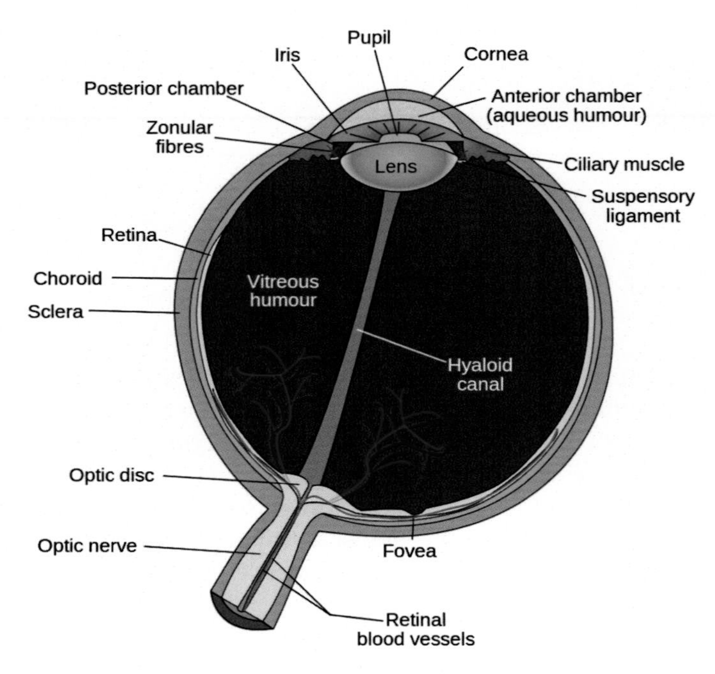

FIGURE 2.2 Cross-section of the human eye (public domain: http://commons.wikimedia.org/wiki/File:Schematic_diagram_of_the_human_eye_en.svg).

The sclera

The sclera is the layer of white tissue that provides the main strength of the eye's structure. It becomes transparent at the front of the eye, where light enters through the cornea.

The ciliary body

The ciliary body is a ring of tissue that surrounds the eye. It contains the ciliary muscles that adjust the refractive power of the lens by changing its shape.

The retina

The retina contains the photoreceptors that are sensitive to light as well as several other types of neuron that process and combine signals from the photoreceptors and transmit these to the main brain areas via the optic nerve. A fundus photograph of a healthy human retina is shown in Fig. 2.3. The macula is the darker area in the center and the optic disc, where the nerve fibers pass through the back of the eye, can be seen as the brighter area on the left. Interestingly, and despite its size, our vision system does a pretty good job of creating an impression of vision at this blind spot even though there are no photoreceptors present! Major nerve pathways are seen as white striped patterns radiating from the optic disk and blood vessels can also be clearly seen.

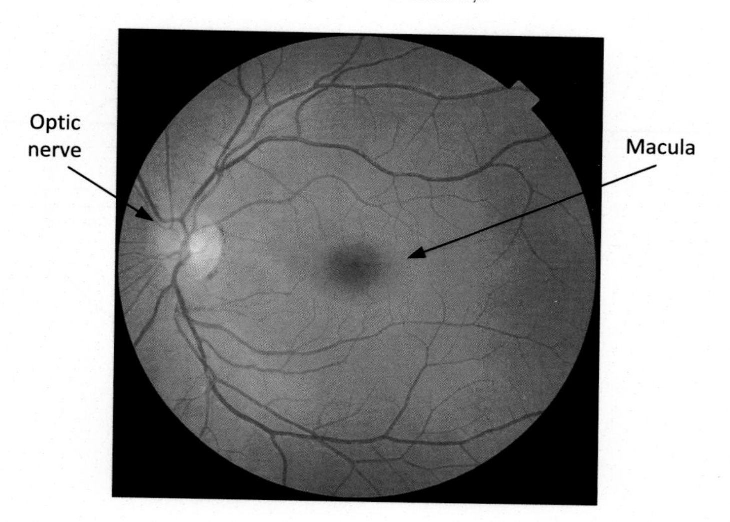

FIGURE 2.3 Fundus image of a healthy retina (public domain: http://commons.wikimedia.org/wiki/File:Fundus_ photograph_of_normal_right_eye.jpg).

The choroid

The choroid is the capillary bed that provides nourishment to the photoreceptors. It also contains the light sensitive, UV absorbing pigment – melanin.

Let us now look at the front of the eye, where the light enters, in a little more detail.

2.2.2 Optical processing

The cornea

The cornea is the sensitive transparent circular region at the front of the eye where light enters. It refracts the light onto the lens, which then focuses it onto the retina.

The lens

The lens is the transparent structure located behind the pupil that, in combination with the cornea, refracts the incident light to focus it on the retina. For distant objects, the ciliary muscles cause it to become thinner and for close-up objects it becomes fatter. Interestingly the lens is only responsible for around 25% of the refractive power of the eye's optics; the rest is provided by the cornea. Although the lens has much less refractive power than the cornea, it enables changes in focal length that enable us to accommodate objects of interest at various distances.

The relationship between object distance d_o, retinal distance d_r, and focal length f is illustrated in Fig. 2.4 and Eq. (2.1). The focal length of the human eye is about 17 mm and the optical power is about 60 dioptres. Note that

$$\frac{1}{f} = \frac{1}{d_o} + \frac{1}{d_r}. \tag{2.1}$$

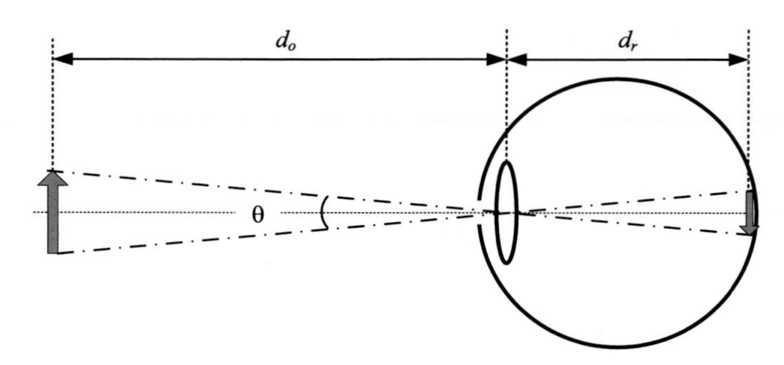

FIGURE 2.4 The focal length of the lens.

The iris

The iris is the individually colored circular region at the front of the eye containing the ciliary muscles that control the aperture of the pupil.

The pupil

The pupil is the aperture that controls how much light enters the eye and becomes smaller in brighter environments. The pupil however only adjusts by a factor of around 16:1 (dilation from 2 mm to 8 mm). Most of the compensation for varying light levels that allows us to adapt to some 8–9 orders of luminance magnitude is done in the retina by the photoreceptors and in other areas of the brain.

Now let us move to the transducer at the back of the eye – the retina.

2.2.3 Retinal photoreceptors and their distribution

When light hits the retina it is converted into electrical pulses by the responses of the photoreceptors. There are typically around 130 million sensor cells in the retina – 125 million rods and 6–7 million cones. The distribution of these is as shown in Fig. 2.5 [13]. Although the packing of cones is at its densest in the fovea, there are still several thousand per square millimeter in the periphery.

Rod cells

There are approximately 125 million rods in the retina. These are responsible for vision at low light levels (scotopic vision) and are so sensitive that they become saturated or bleached at mid light levels (mesopic vision). Rods do not mediate color vision and also provide lower spatial acuity than foveal cone cells. The photopigment, rhodopsin, is most sensitive to green light of wavelength around 498 nm. The central fovea – the region of highest visual acuity – has a very low density of rods, which is why, at night, we can see objects better by looking slightly away from them.

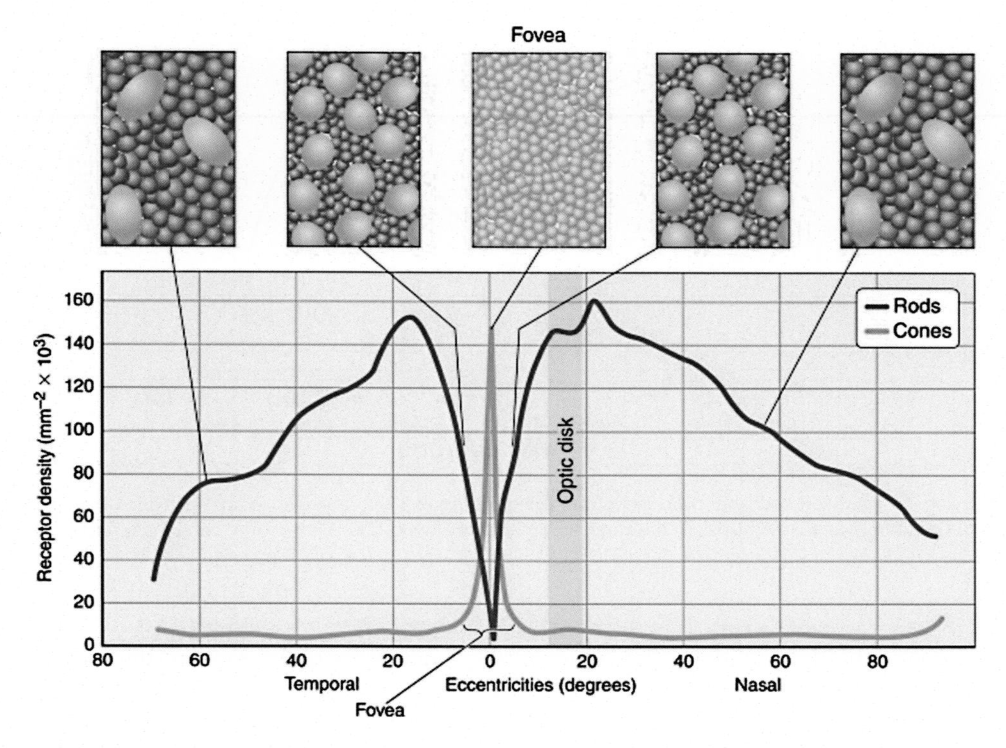

FIGURE 2.5 Photoreceptor distribution in the retina (reproduced with permission from [13]).

Cone cells

Cones operate at higher light levels than rods and provide what is known as photopic vision. In contrast to rods, they offer a range of different spectral sensitivity characteristics and they mediate color vision. Three types of cone exist and these are generally referred to as short-wavelength (S) or blue; medium-wavelength (M) or green; and long-wavelength (L) or red. These have broad overlapping spectral responses with peaks at around 420 nm, 534 nm, and 564 nm respectively. Light at blue wavelengths tends to be out of focus since it is refracted more than red and green light. The normalized frequency response characteristics of the rods and cones is shown in Fig. 2.6.

As shown in Fig. 2.5, the central fovea has the highest cone density and thus provides the region of highest visual acuity.

Macula

The macula is a depressed region of diameter 5.5 mm near the center of the retina which surrounds the fovea. It is in this region that the number of cone cells starts to increase dramatically. It has two or more layers of ganglion cells, and at its center lies the fovea. Because of its yellow color the macula absorbs excess blue and ultraviolet light and acts as a natural sunblock.

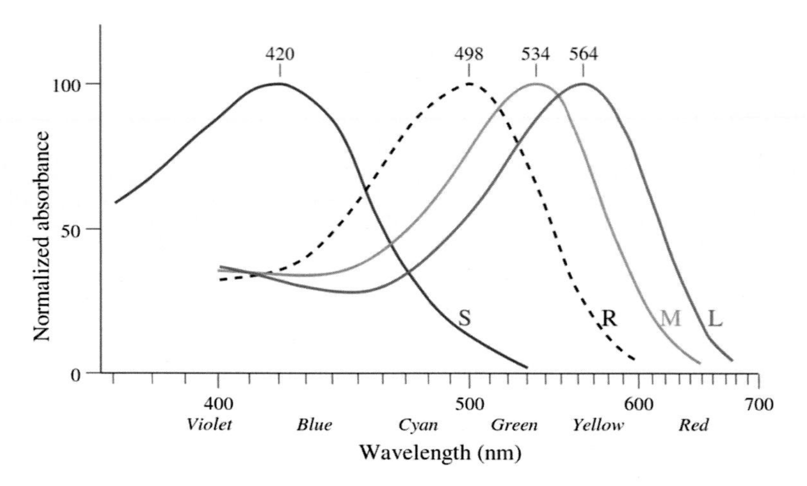

FIGURE 2.6 Normalized rod and cone responses for the human visual system (adapted from Bowmaker and Dartnall [14]; publicly available https://commons.wikimedia.org/wiki/File:1416_Color_Sensitivity.jpg).

Fovea

The fovea is the depression at the center of the macula about 1.5 mm in diameter that is responsible for central, high-resolution vision. As shown in Fig. 2.5, it has a very high spatial density of cones and very few rods. It also coincides with the region of the retina that is void of blood vessels and this further improves acuity as dispersion and loss are minimized. The center of the fovea, the foveola, is about 0.2 mm in diameter and is entirely formed of very compact, hexagonally packed, thin cones that are rod-like in structure. The fovea contains mostly M and L cones with around 5% of S cones.

The sampling resolution of the retinal image is highly nonuniform and falls off rapidly with increasing eccentricity. This creates a more blurred image in the peripheral vision which we do not perceive due to rapid eye movements and higher-level visual processing. The effective angle subtended at the fovea is around 2 degrees; hence we really do only "see" a small proportion of the visual scene sharply, yet we still perceive it all to be sharp. The average spacing of cone receptors in the fovea is 2–3 μm (in the periphery this increases to around 10 μm), and it is this that effectively limits our spatial resolution to approximately 1 arcminute. This is important when we assess the relative merits of high-resolution TV formats in terms of viewing distance. The spacing also gives us the maximum spatial frequency that can be perceived. In practice the eye cannot resolve spatial frequencies of more than about 60 cycles per degree (cpd).

Optic disc and nerve

The optic disc is the portion of the optic nerve that can be seen on the retina, where the signals from the ganglion cells in the retina leave the retina along axons on their way to the visual cortex. About half of the nerve fibers in the optic nerve originate from the fovea, with the remainder carrying information from the rest of the retina.

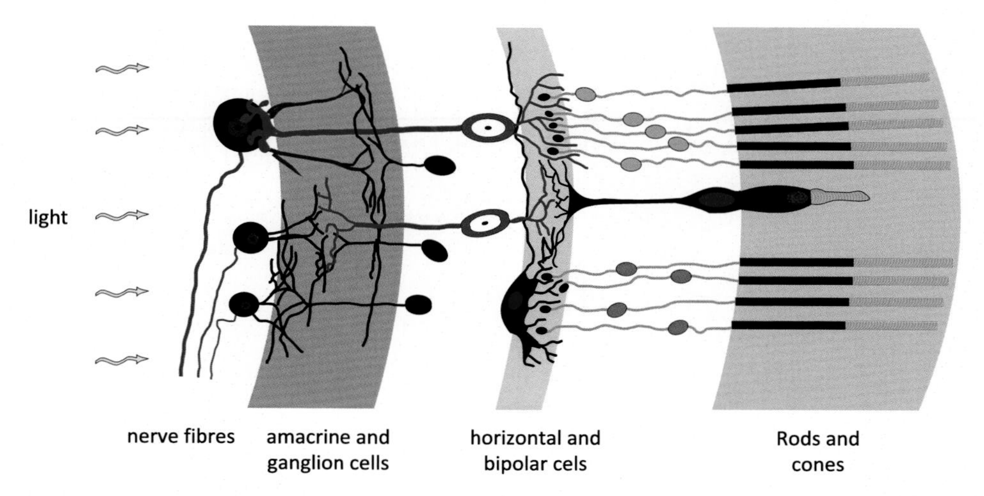

FIGURE 2.7 Retinal cell architecture (public domain image adapted from http://commons.wikimedia.org/wiki/File:Retina-diagram.svg).

The retina contains a number of other cell types, which provide early vision processing and reduce the amount of information that is transmitted to the main visual cortex. These cells are briefly explained in Section 2.2.4.

2.2.4 Visual processing in the retina

The responses from the retina's photoreceptors are processed by a complex (and far from perfectly understood) network of neurons. This network comprises bipolar, horizontal, amacrine, and retinal ganglion cells. Together, these provide early visual preprocessing, feature extraction, and detection, resulting in significantly reduced information flow through the optic nerve. This structure is shown in Fig. 2.7.

As depicted in Fig. 2.7, bipolar cells connect to either rods or cones and are activated by an increase in the photons incident on the associated photoreceptors. The horizontal cells connect laterally across the retina to a greater spatial extent and, because of their stimulation from a neighborhood of photoreceptors, enable lateral inhibition. Whereas the response of bipolar cells in isolation is rather crude, the influence of the horizontal cells is to add an opponent signal where the response of one or more photoreceptors can influence the response of surrounding receptors, thus shaping the receptive field. The horizontal cells are also known to modulate the photoreceptor signal under different illumination conditions [11].

Relatively little is known about the roles of amacrine cells, except that they can have extensive dendritic trees and, in a similar way to horizontal cells, contribute feedback to the receptive fields of both bipolar and ganglion cells.

There are approximately 1.6 million retinal ganglion cells and these create the earliest receptive fields in the vision system, providing a basis for low-level feature detection and the opponent coding of color (see Section 2.5). The axons from these form the optic nerve and transmit the electrochemical signals to the higher functions of the visual cortex. There are

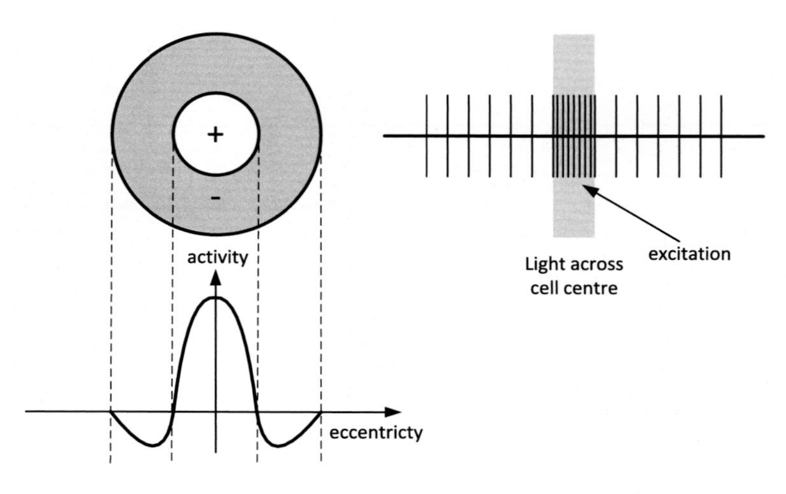

FIGURE 2.8 Spatial opponency, showing a center-surround cell and its firing pattern due to excitation.

approximately 100 times more photoreceptors than ganglion cells in the eye – providing evidence of the compression performed by the retina.

The receptive field of a cell is the region within which a stimulus must occur if it is to influence the response of that cell. Receptive fields are formed using spatial opponency, i.e., responses from one part of the receptive field are opposed or inhibited by other parts of the receptive field. The interactions between the bipolar and the horizontal cells provide an orientation-insensitive center-surround organization of the ganglion cell receptive field, where the center and surrounding areas act in opposition. This could comprise an excitatory center and an inhibitory surround, as shown in Fig. 2.8, or vice versa. Ganglion cells act as change detectors, responding only to differences in light intensities (or contrast) across their receptive field and not to absolute light levels. They are therefore sensitive to edges and can be model-led as a difference of Gaussian functions (DoG).

A scene is thus effectively encoded as intensity changes in the different color channels. However, it is clear that we see more than just edges, so this provides evidence that higher levels of the vision system perform infilling between receptive field responses with averages in intensity levels. This is interesting as it can be viewed as a simple form of image synthesis, providing a biological analog to perceptual redundancy removal. A particularly striking example of this is the visual in-filling of our blind spot, which contains no photoreceptors.

2.3 The visual cortex

A simple diagram showing the major visual pathways from the retinas to the primary visual cortex is shown in Fig. 2.9. The optic nerve axons from the retinal ganglion cells in each eye meet at the optic chiasm, where all the fibers from the nasal side of each eye cross to join those from the temporal side of the other eye and form two optic tracts. The left visual

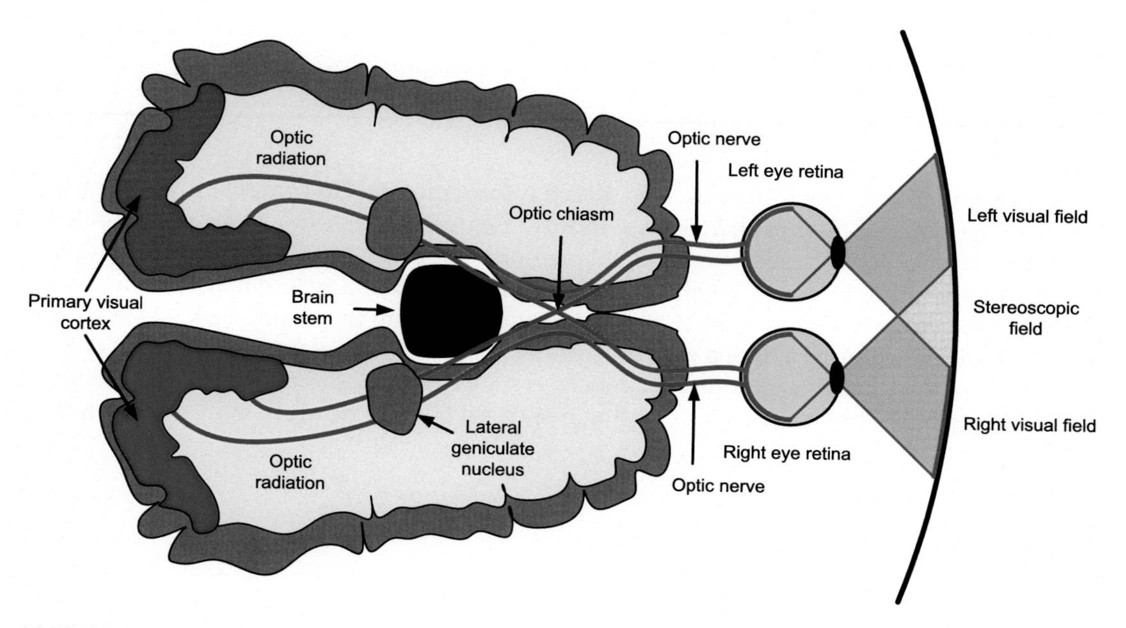

FIGURE 2.9 The visual cortex and visual pathways.

field is thus processed by the right hemisphere and vice versa. The fibers in the optic tract then synapse at the lateral geniculate nucleus (LGN), where they are distributed to other areas of the visual cortex. After the LGN, signals pass to the primary visual cortex, V1, located at the rear of the skull. They then pass to V2, and branch out to at least 20 other centers, each providing specialized functions such as detecting motion, recognizing faces, or interpreting color and shape.

The HVS provides its functionality largely through the combined processes of inhibition, excitation, biased competition, input races, opponent coding, and adaptation. These are considered in more detail below.

2.3.1 Opponent processes

As discussed above and depicted in Fig. 2.8, many neural functions in the vision system appear to encode data using opponent coding. Opponent coding is a process in which neurons encode opposite or opponent features in their single output. For example, a motion-sensitive cell has a base firing rate which reduces when motion in one direction is detected and which increases with motion in the opposite direction. The opponent processes of inhibition and excitation appear crucial to function, with inhibitory responses often being used to control a normally excitatory state.

We also interpret color by processing signals from cones and rods in opponency. The L, M, and S cone spectral responses overlap in terms of their wavelengths so it is more efficient for the visual system to use differences between their responses rather than individual responses.

This suggests that there are three opponent channels: red versus green, blue versus yellow, and black versus white. We will consider this further in Section 2.5.

2.3.2 Biased competition

Biased competition refers to the way in which information has to compete for limited resources in the visual pathways and cortices. Inputs from feature detectors in the visual system compete to be encoded into short-term visual memory. The competition between such inputs is weighted according to higher-level feedback, called attention, to ensure that those features relevant to the current task are prioritized. For example, during the Stroop task (subjects are asked to state the color of a series of words rather than read the words themselves) it has been observed that task-based attention causes enhancement of the relevant color feature-sensitive areas of the brain while suppressing responses from those areas that perform word processing.

2.3.3 Adaptation processes

Adaptation is the process by which a cell's response reduces over time when its input is constant. Adaptation to varying degrees is common in most cortical cells and, when coupled with opponent coding processes, is responsible for the after-effects we experience related to brightness, color, or motion. After an opponent-coding feature-tuned cell adapts to a particular stimulus and that stimulus is removed, the cell reacts – briefly signaling the opposite feature to which it has adapted. Examples of such after-effects include the sensation of slowness after traveling at high speeds, and the perception that static objects move in the opposite direction to that of a previously viewed moving object. Dramatic brightness and color after-effects are also frequently used as the basis for visual illusions. Even observing smiles for a prolonged period can make expressionless faces seem sadder!

2.3.4 V1 – the primary visual cortex

The primary visual or striate (layered) cortex (otherwise known as V1) is probably the best-understood area of the HVS. The role of V1 is to extract basic visual features from a scene and it has been shown that V1 contains receptive fields that are sensitive to line orientations, color, and spatial frequency. This effects visual data compaction through abstraction of visual stimuli into higher-level constructs. The majority of V1 connections are from the retina but it is also known to receive feedback connectivity from higher functional areas, which influence its cell's receptive fields.

Hubel and Wiesel [8] identified three primary detectors in V1:

- **Simple cells**: Tuned to specific orientations of edges.
- **Complex cells**: Phase-insensitive versions of simple cells, i.e., the response to an appropriately oriented stimulus is the same no matter where it falls within the receptive field.
- **Hypercomplex cells**: These show stimulus length sensitivity, i.e., the cell response increases as the stimulus length increases.

Neurons in V1 are generally grouped in columns that respond to a particular low-level feature such as line orientation, spatial frequency, or color. V1 is known to preserve the retinotopic mapping of the ganglion cells on the retina in its organization. This is however not a linear or uniform mapping since V1 exhibits cortical magnification, dedicating a disproportionate percentage of its cells to processing foveal information. This emphasizes the importance of the higher-resolution central part of the visual field.

2.3.5 V2 – the prestriate cortex

V2 is strongly linked to V1 with both feedforward and feedback connections and it also connects strongly to V3, V4, and V5. Although less well understood than V1, V2 cells appear to operate in a similar manner, combining V1 features tuned to orientation, spatial frequency, and color as well as responding to more complex patterns. Damage in V2 can lead to poor texture and shape discrimination. There is also evidence of V2 cell modulation by binocular disparity.

2.3.6 Dorsal and ventral streams

Connections from V1 to the V2 area are generally considered to diverge into two streams – dorsal and ventral (otherwise referred to as "where" and "what" or "action" and "perception"). The dorsal stream relates to motion and depth perception, originating in V2 and passing dorsally through the brain to the motion-sensitive parts of the visual cortex. The ventral stream on the other hand is associated with the perception of shapes and object recognition, again originating in V2, but in this case passing ventrally through V4 to the inferior temporal cortex (IT).

2.3.7 Extrastriate areas

The areas of the visual cortex beyond V1 are generally referred to as the *extrastriate* areas. These possess larger receptive fields and generally compute more complex and abstract features. They also generally exhibit strong attentional modulation. Although the function of these areas is not well understood, their connectivity is informative.

V3 is strongly connected to V2 with additional inputs from V1. It processes both dorsal and ventral streams and has been shown to respond to global motions.

V4 lies on the ventral stream and has been shown to respond to shapes of relatively high complexity. It also shows strong attentional modulation and a significant amount of color sensitivity. Damage to V4 can cause impairments to shape and texture discrimination.

V5 contains cells that respond to complex motion patterns and is believed to be the main area dedicated to the perception of motion.

The *medial superior temporal* (MST) area is fed primarily from the medial temporal (MT) area, and is also sensitive to motion. There is evidence that the MST computes optic flow especially in the context of global expansions and rotations.

The *visual short-term memory* (VSTM) can be considered to be a short-term buffer where prioritized data is stored for further processing. VSTM has been shown to persist across visual interruptions such as saccades and blinks and thus provides a link between the previously viewed scene and the new one after interruption.

The *inferior temporal cortex* (IT) provides complex shape recognition by parsing simpler shapes from lower levels in the HVS. It contains cells that trigger in response to specific shapes and patterns and also hosts cells that are known to respond to faces. It fuses information from both halves of the visual field (that up to this point was separated across the hemispheres of the brain). At higher stages of visual processing, the receptive fields become larger and more complex in their connectivity. For example, in the IT some receptive fields are thought to cover the entire visual field.

Many other important areas of the visual cortex have been identified and investigated, including the *fusiform face area* (FFA), largely responsible for recognition of faces and other complex entities. The *frontal eye field* (FEF) is a further important area that controls voluntary eye movements. It has been proposed that this is based on some form of visual saliency map, encoded with behaviorally significant features.

2.4 Visual fields and acuity

2.4.1 Field of view

The field of view of the human eye is approximately 95 degrees in the temporal direction (away from the nose), 60 degrees in the nasal direction (toward the nose), 75 degrees downward, and 60 degrees upward, enabling a horizontal field of view in excess of 180 degrees. Obviously, when rotation of the eye is included, the horizontal field of view increases still further, even without head movement.

2.4.2 Acuity

It is not too difficult to estimate the spatial acuity of the retina based simply on the distribution of cones in the fovea. We illustrate this in Example 2.1. The measured Nyquist frequency for the HVS is normally between 40 and 60 cpd. This upper value has been attributed to the photoreceptor sampling grid pattern, which is hexagonal rather than rectangular, and to other local enhancement techniques. No aliasing effects are normally experienced by the HVS and this is due to the fact that the optics of the eye perform low-pass filtering, acting as an antialiasing filter.

Example 2.1. Visual acuity of the retina

Assuming the photoreceptor densities indicated in Fig. 2.5 and that the distance from the lens to the retina is 16.5 mm, estimate the acuity of the human visual system in the fovea.

Solution. We can see from Fig. 2.5 that the cone density in the central fovea peaks at around 150,000 cones/mm^2. If we assume a very simple packing arrangement this gives around 400 cones per linear mm. To be conservative let us take the distance between cone centers as 0.003 mm. The optics of the eye dictates [2] that, for a lens-to-retina distance of 16.5 mm, there are 0.29 mm/deg subtended. The angular resolution is thus 0.003/0.29 = 0.0103 degrees between cone centers. Now, since we need at least 2 samples per cycle to satisfy the Nyquist criterion, this means that there are 0.0206 degrees per cycle. Hence the estimate for the maximum detectable spatial frequency is 49 cpd subtended.

2.4.3 Light, luminance, and brightness

Sensations of brightness are not direct analogs of photometric intensity (luminance). The number of photons that enter the eye and the perception of brightness are only related indirectly. For example, the apparent brightness of an achromatic target, under equiluminant conditions, will depend on the frequency content of the electromagnetic radiation reflected from its surroundings. Thus, a gray target on a relatively dark background looks brighter than the same target on a lighter background. As a result of this context-based perception, it is generally thought that the visual system computes brightness using luminance ratios across the contrast boundaries in a scene. The influence of color on brightness also provides evidence that perceptions of luminance are related to electromagnetic frequency [12].

The influence of local contrast can be seen, in combination with the effects of lateral inhibition, in the Mach bands shown in Fig. 2.10. Here the intensity is uniform over the width of each bar, yet the visual appearance is that each strip is darker at its right side than its left. Another striking example is provided by the Adelson grid, shown in Fig. 2.11, where squares A and B both appear very different due to shadowing, yet are actually identical.

There is significant confusion about the terms used in connection with light radiation, so let us define some terms related to light and how we perceive it [15].

Radiant intensity and radiance

The radiant energy (Joules) is the energy propagating from a source and the radiant flux Φ is the radiant energy per unit time (Joules/sec or Watts). Radiant intensity I is defined as the radiated light power in a given direction and is measured in terms of the amount of light passing through a solid angle (Watts/steridian). Radiance L is defined as the light intensity that falls on a unit projected area (Watts/steridian/m^2).

Luminance

None of the above measures take account of the composition of the light in terms of its wavelength or wavelengths. Luminance does this by weighting the radiance value according

FIGURE 2.10 Mach band effect.

FIGURE 2.11 Adelson's grid (reproduced with permission from http://web.mit.edu/persci/people/adelson/checkershadow_illusion.html).

to the human visual response. The luminance, Y, is thus normally represented by equation (2.2), where $L(\lambda)$ is the incident light intensity as a function of wavelength at a given position and time and $\bar{y}(\lambda)$ is a wavelength-dependent weighting (or relative luminous efficiency) that reflects the perceived luminance of the incident flux. Luminance is measured in candelas/m^2 (cd/m^2). We have

$$Y = K \int_{320}^{720} L(\lambda)\bar{y}(\lambda)d\lambda. \tag{2.2}$$

The constant K=685 lumens/watt. The function $\bar{y}(\lambda)$ will vary from person to person, but was measured by the CIE[1] in 1929 for a standard observer as shown in Fig. 2.12. If in the integral of Eq. (2.2), the term $L(\lambda)$ is replaced by radiant flux $\Phi(\lambda)$, then the result is termed luminous flux and has the units of lumens.

Brightness

Brightness is the perception elicited by the luminance of a visual target and, as such, represents a nonquantitative indication of the physiological sensation of light. It is normally assessed subjectively by an observer with reference to a white area adjusted to have the same brightness as the target viewed.

Luma

An image sensor normally produces output values proportional to radiance, with the incident light typically filtered into R, G, and B bands. However, before these emerge from the camera, they are usually nonlinearly mapped according to the response of the HVS, using a technique known as gamma correction. These outputs are normally labeled as R', G', and B'. Luma is the term used to denote the combination of these corrected components that

[1] The International Commission on Illumination.

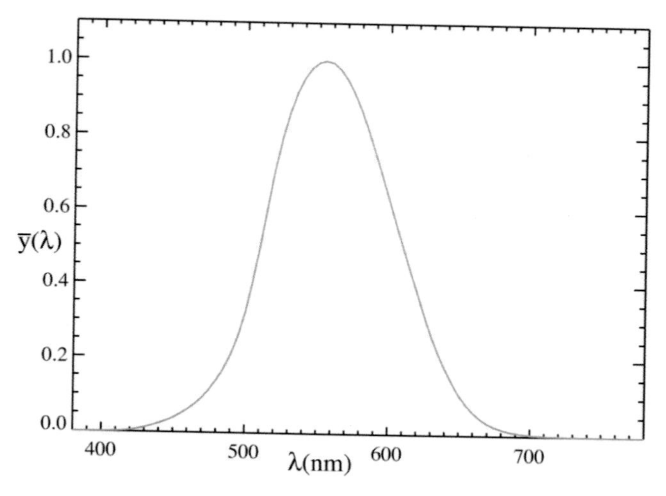

FIGURE 2.12 CIE luminous efficiency curve (publicly available: http://en.wikipedia.org/wiki/File:CIE_1931_Luminosity.png).

produces the gamma-corrected luminance signal, according to the specification of the color space used. Gamma correction is considered further in Chapter 4.

2.4.4 Light level adaptation

Our ability to adapt to a wide range of light levels is critical for function and survival. The photoreceptors, indeed the whole of the visual system, must remain sensitive as the ambient light intensity varies. This variation is quite dramatic as we are able to see in conditions from starlight (10^{-3} cd/m^2), through indoor lighting (10^2 cd/m^2), to bright sunlight (10^5 cd/m^2). The eye can thus function (at least to some degree) across some 8 orders of magnitude in luminance level – a ratio of 100,000,000:1.

The eye, of course, cannot cope with this dynamic range instantaneously. If we consider our eye's instantaneous dynamic range (where the pupil opening is fixed), then this is around 10 f-stops. If we include rapid pupil adaptation this increases to about 14 f-stops. The pupil diameter typically ranges from about 2 mm in bright conditions to around 8 mm in dark conditions, giving a 16:1 dynamic range or 4 f-stops.

Adaptation also takes place in both rods and cones. The cones adapt more quickly than the rods and this explains why it takes much longer to adapt to darkness (dark adaptation) than to brightness.

A typical dark adaptation characteristic is shown in Fig. 2.13. This illustrates the variation over time of an individual's threshold after adaptation to bright light. This figure shows that the dark adaptation process is quite slow, taking several minutes for complete adaptation, but also that the adaptation occurs in two stages. The first of these is due to the cones adapting and the second is due to the rods. The retinal photoreceptors adapt through pigment bleaching – but also by feedback from horizontal cells.

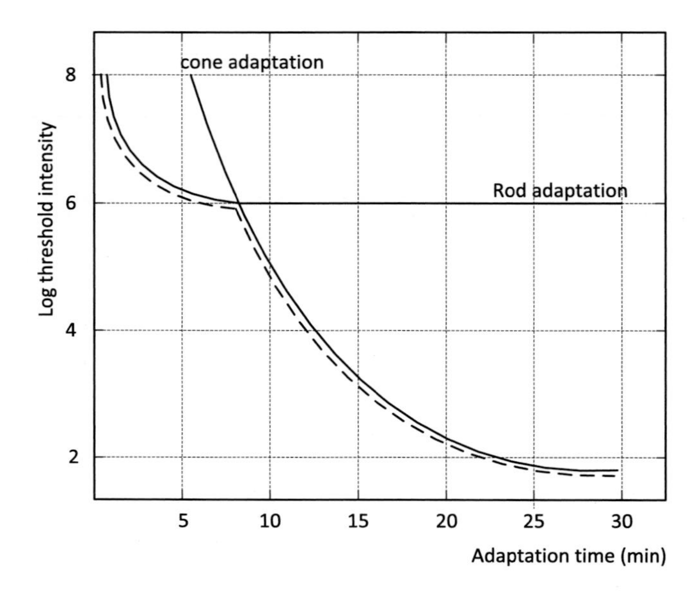

FIGURE 2.13 Dark adaptation of rods and cones.

2.5 Color processing

The sensations of color are the brain's response (qualia) to incident electromagnetic stimuli within a certain range of wavelengths:

- **Hue**: The sensation by which a region appears similar to one, or to proportions of two, of the perceived colors red, yellow, green, and blue.
- **Colorfulness**: The sensation by which a region appears to exhibit more or less of its hue.
- **Chroma**: The colorfulness of an area relative to the brightness of a reference white.
- **Saturation**: The colorfulness of an area relative to its brightness.

If you ask someone if they prefer to watch a color rather than a monochrome TV program or movie, invariably they would say yes on the basis that it is closer to reality. This is obvious – consider Fig. 2.14, which on the right shows a color image of Yellow Water in Kakadu, Australia and on the left a monochrome version of the same image. Most would agree that the color image is easier to interpret, providing improved differentiation between natural objects and features in the scene and also a better sense of depth and engagement.

Yet, despite few people seeing the world in black and white, we readily accept monochrome images and are perfectly happy to suspend disbelief, for the sake of entertainment or information. In some cases (for example in Steven Spielberg's *Schindler's List*) directors actually choose to use monochrome imagery to affect mood and increase impact in films. This acceptance of, and occasional preference for, luminance-only signaling may be associated with the way in which we process our trichromatic sensors.

FIGURE 2.14 Increased immersion from color images.

A final point worth highlighting here is that the choice of primary red, green, and blue colors in displays is not based on the sensitivities of the S, M, and L cones. Instead they are selected to be well spaced spectrally, thus allowing a large gamut of colors to be produced.

Following from Section 2.4 and Eq. (2.2) we can generate average cone responses R, G, B to incident radiation as follows, where $\bar{y}_r(\lambda)$, $\bar{y}_g(\lambda)$, and $\bar{y}_b(\lambda)$ are the luminous efficiencies for the red, green, and blue color channels, respectively:

$$R = K_r \int_{380}^{720} L(\lambda)\bar{y}_r(\lambda)d\lambda, \tag{2.3}$$

$$G = K_g \int_{380}^{720} L(\lambda)\bar{y}_g(\lambda)d\lambda,$$

$$B = K_b \int_{380}^{720} L(\lambda)\bar{y}_b(\lambda)d\lambda.$$

2.5.1 Opponent theories of color

We have three types of cone. However, most mammals only have two types – evolved from a single yellow–green cone centered around 550 nm (corresponding to the chlorophyll reflectivity maximum). A blue cone was then added centered around 430 nm – which assists with plant discrimination. Humans and other primates added a third cone type (actually the yellow–green cone probably split into two types) – at the red end of the spectrum. This enabled its owners to forage more effectively (red fruit can be more easily differentiated) and signal more dramatically. As a point of interest, some birds possess 4 channels and the mantis shrimp has 12!

As described earlier, color perception in the HVS employs an opponent-based approach. While the trichromatic theory explains how the retina enables the detection of color with three types of cones, at different wavelengths, opponent color theory explains how this information is processed. It is based on the observation that the HVS interprets color by processing signals from rods and cones in an antagonistic manner. The reason for this becomes clear if we consider color constancy requirements. Clearly if the light reflected from a certain object varies, we would ideally like to perceive it as a brighter or darker version of the same color. This can be achieved if we consider the ratio of the outputs from the photoreceptors rather than their absolute outputs. This was first identified by Herring in 1878, who realized that certain hues could not coexist in a single color sensation. For example we can experience red-yellow=orange and blue-green=cyan but cannot readily experience a reddish-green or a blueish-yellow. Hence the pairs red–green and blue–yellow are termed opponent pairs. Herring argued that we could not experience both colors in a pair simultaneously as they are encoded in the same pathway and that the excitation of one of these color channels inhibits activity in the other. This observation has been validated from psychophysics experiments and it is easy to convince yourself of it. For example if you stare at green long enough you get a red after-effect (and vice versa) and if you stare at blue long enough you get a yellow after-effect (and vice versa).

A diagram showing a possible opponent model is given in Fig. 2.15. This shows a luminance signal generated from the sum of the signals from green and red cones and two chrominance channels generated from ratios of red to green and yellow to blue. Three things should be noted here. Firstly, expert opinion varies on the nature of the combining processes and there is evidence for the chrominance channels being based on color differences (i.e., R-G and Y-B). However, because the sensitivities of the cone spectral responses overlap, most wavelengths will stimulate at least two types of cone. Some tristimulus values therefore cannot physically exist as an additive color space and imply negative values for at least one of the three primaries. Secondly, it is well known that color sensations vary significantly with context and spatial frequency. Theories to explain aspects of this have been proposed by authors such as Shapley and Hawken [21]. Finally, the observant reader will notice that there is no contribution from blue cones in the luminance channel. The main reason for this is that there are relatively few blue cones in the fovea and almost none at its center. This is probably because of the effects of chromatic aberration, which would significantly reduce acuity.

Our understanding of opponent processes suggests that the most contrasting color pairs are red–green and blue–yellow, and these are reflected in the RYB color model, where the complementary color pairs are red–green, yellow–purple, and blue–orange. The complementarity of colors is widely exploited in printing, art, and cinema color grading to create tension or emphasize difference. For example in movies, the exploitation of complementary colors, particularly blue–yellow (amber–teal) has become almost universal. All human flesh tones are clustered at various radii around the "orange" region of the color wheel. To focus attention on human subjects, by enhancing contrast and the perception of depth, color graders emphasize these flesh tones and introduce complementarity using teal in the shadows and background.

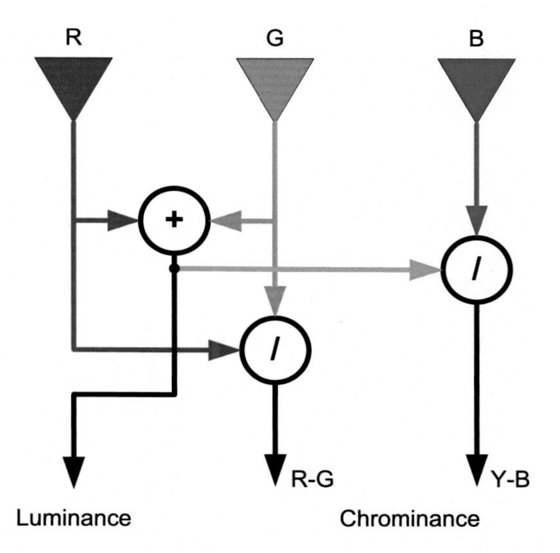

FIGURE 2.15 Opponent processing of color.

2.5.2 CIE 1931 chromaticity chart

The International Commission on Illumination (CIE) in 1931 defined a color mapping function based on a standard observer, representing an average human's chromatic response within a 2-degree arc, to primaries at R=435.8 nm, G=546.1 nm, and B=700 nm. Fig. 2.16 shows the CIE 1931 chromaticity chart. The boundary represents maximum saturation for the spectral colors, and the diagram forms the boundary of all perceivable hues. The colors that can be created through combinations of any three primary colors (such as RGB) can be represented on the chromaticity diagram by a triangle joining the coordinates for the three colors. Color spaces are discussed in more detail in Chapter 4.

2.6 Spatial processing

Spatial resolution is important as it influences how sharply we see objects. As discussed in Section 2.4, the key parameter is not the number of pixels in each row or column of the display, but the angle subtended, θ, by each of these pixels at the viewer's retina. We thus use the term *spatial* here to indicate things that are not moving as opposed to implying that there is some sense of spatial constancy, regardless of viewing distance.

2.6.1 Just noticeable difference, contrast, and Weber's law

Contrast is the visual property that makes an object distinguishable from other objects and from a background. It is related to the difference in the color and brightness of the object and

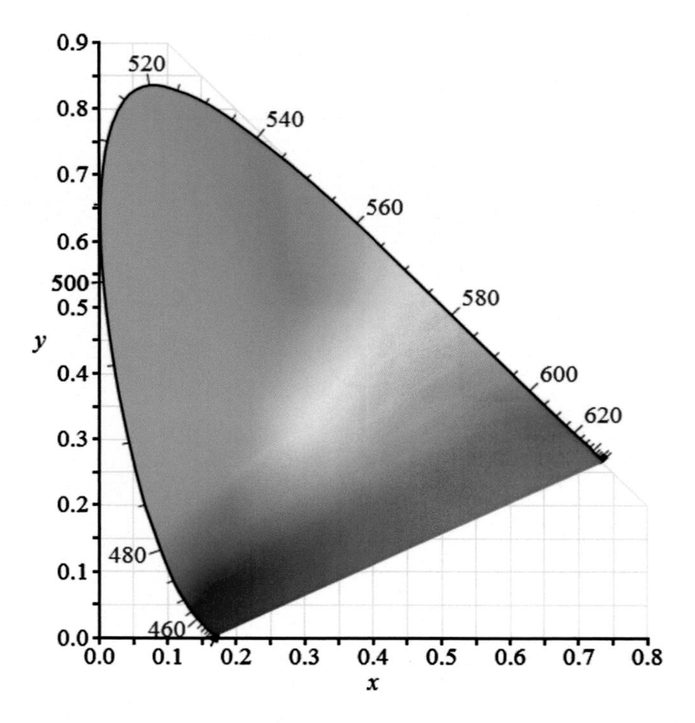

FIGURE 2.16 The CIE.1931 chromaticity chart (reproduced with permission from [22]).

other objects within the same field of view. Because the HVS is more sensitive to contrast than absolute luminance, we perceive the world similarly regardless of changes in illumination. It is well known that the perception of a constant increment in illumination is not uniform, but instead varies with illumination level. This is captured by Weber's law, which identifies that, over a reasonable amount of our visible range, the following expression for contrast, C, holds true, for a target T and a background B:

$$\Delta C = \frac{\Delta Y}{Y} = \frac{\text{JND}(Y_T - Y_B)}{Y_B} = constant. \tag{2.4}$$

Weber's law implies that the just-noticeable difference (JND) between two stimuli is proportional to the magnitude of the stimuli. This is sometimes confused with Fechner's law, which states that a subjective sensation is proportional to the logarithm of the stimulus magnitude. Fig. 2.17 shows an example test chart for JND testing. For most observers the JND value is between 0.01 and 0.03.

Weber's law fits well over 2 to 3 decades of $\log(Y)$ and is illustrated in Fig. 2.18. As background illumination is increased above or decreased below this region the slope will change. As we will see in later chapters, this subjective assessment of luminance sensitivity can be important in image and video quantization.

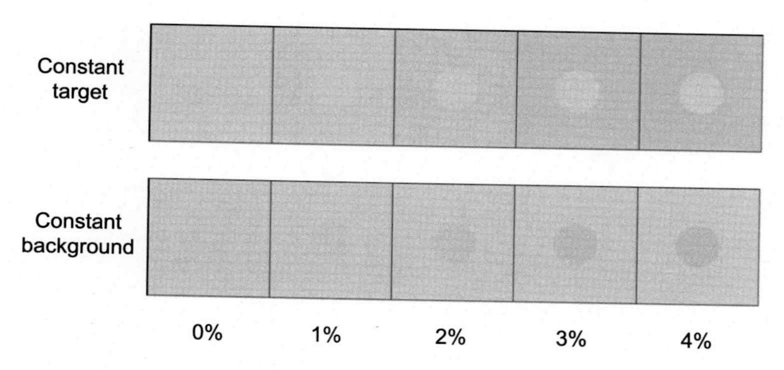

FIGURE 2.17 Just noticeable differences at different contrast increments.

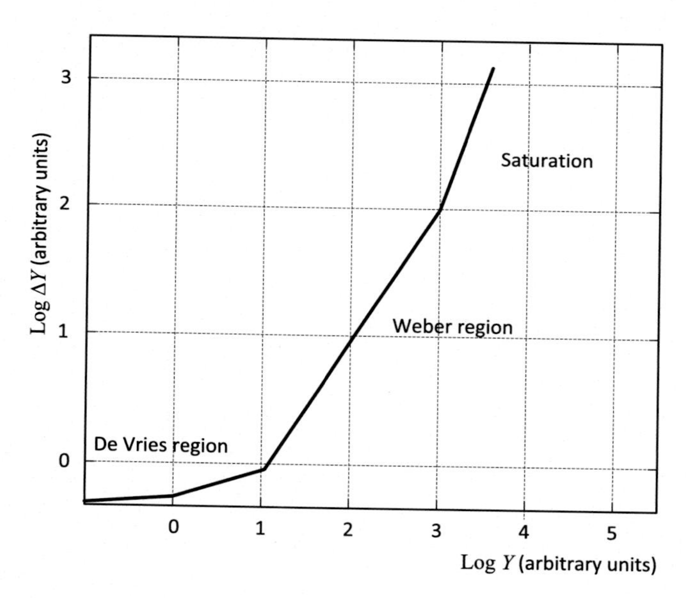

FIGURE 2.18 JND curve for human vision.

2.6.2 Frequency-dependent contrast sensitivity

One of the most important issues with HVS models concerns the relationship between contrast sensitivity and spatial frequency. This phenomenon is described by the contrast sensitivity function (CSF). The CSF model describes the capacity of the HVS to recognize differences in luminance and chrominance as a function of contrast and spatial frequency.

A spatial contrast sensitivity chart or sine wave grating is shown in Fig. 2.19. Typical contrast sensitivity responses for luminance and chrominance are shown in Fig. 2.20, where

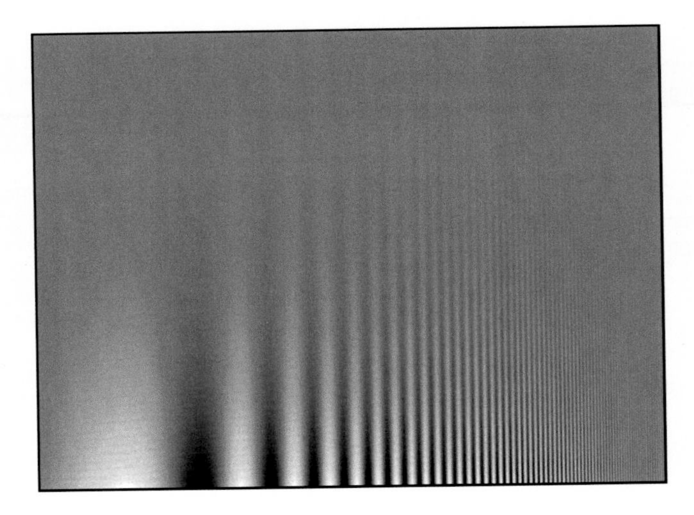

FIGURE 2.19 Contrast sensitivity chart.

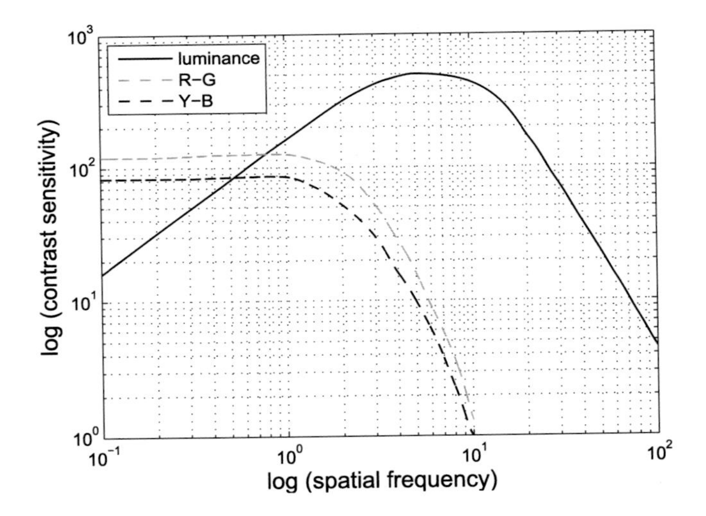

FIGURE 2.20 Luminance and chrominance CSF responses.

contrast sensitivity is defined as

$$C(f) = \frac{Y_B(f)}{\text{JND}(Y_T(f) - Y_B(f))}. \tag{2.5}$$

This illustrates the sensitivity of the visual system to spatial sinusoidal patterns at various frequencies and contrast levels. The band-pass characteristic is primarily due to the optical transfer function of the eye together with the retinal (lateral inhibition) processing and some properties of region V1. Its shape is influenced not only by the spacing and arrangement of photoreceptors, but also possibly by the limitation on the number of spatial patterns representable in V1.

Essentially, the HVS is more sensitive to lower spatial frequencies and less sensitive to high spatial frequencies. Mannos and Sakrison [23] presented a CSF model for gray-scale images as a nonlinear transformation followed by a modulation transfer function (Eq. (2.6)):

$$C(f) = 2.6\,(0.0192 + 0.114 f)\, e^{-(0.114 f)^{1.1}}, \tag{2.6}$$

where spatial frequency is

$$f = \sqrt{f_h^2 + f_v^2},$$

where f_h is the horizontal frequency component and f_v is the vertical frequency component. It is often preferable to work in terms of a spatial frequency normalized to the display characteristics in terms of cycles per pixel, thus

$$f_d = \frac{f}{f_e},$$

where f_e is the number of pixels per degree for the experiment being conducted, which depends on the resolution of the display, the dimensions of the display, and the viewing distance. This fits well with measured results, and its CSF response function (normalized) is shown in Fig. 2.21.

The contrast sensitivity of human vision is plotted here against spatial frequency. We can observe that:

1. The sensitivity to luminance information peaks at around 5–8 cycles/deg. This corresponds to a contrast grid with a stripe width of 1.8 mm at a distance of 1 m.
2. Luminance sensitivity falls off either side of this peak and has little sensitivity above 50 cycles/deg.
3. The peak of the chrominance sensitivity curves occurs at a lower spatial frequency than that for luminance and the response falls off rapidly beyond about 2 cycles/deg. It should also be noted that our sensitivity to luminance information is about three times that for R-G and that the R-G sensitivity is about twice that of B-Y.

The contrast sensitivities of the HVS lead us to our first basic means of compressing images – the perceptual CSF model can be used for reduction of imperceptible information and, if we use luminance and color difference signals as the basis of our representation, then we can sample the chrominance signals at around half the rate of the luminance signal, without any loss of perceived quality. Furthermore, both luminance and chrominance signals can be more coarsely quantized at higher spatial frequencies due to their reduced contrast sensitivity. These mechanisms will be explored in Chapter 4.

2.6.3 Multiscale edges

Edge localization models of early vision have generally featured some arrangement of tuned spatial filters at multiple image scales followed by a feature extraction process and

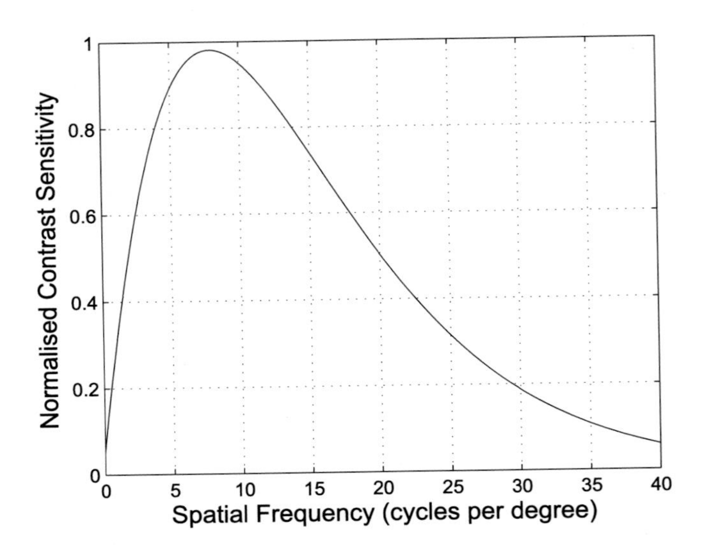

FIGURE 2.21 Luminance contrast sensitivity function.

finally integration of these features into a map (such as Marr's primal sketch [4]). The orientation and frequency selectivity of the visual system in many ways resembles multiscale transforms such as the wavelet or complex wavelet transform and these have been employed in various HVS models. As we will see later in the book, these tools are also used extensively in image and video compression, where the transformed image data in frequency, orientation, and scale can be exploited to remove invisible content by, for example, filtering with a CSF characteristic.

2.6.4 Perception of textures

Texture is an important visual cue that we exploit in tasks such as edge localization, depth perception, and general object recognition. The visual system appears to operate under the assumption of high entropy in a scene. For example, the distribution of pebbles on a beach provides good depth cues because they appear to get smaller the further they are from the viewer. It has been proposed that texture perception employs a filter–rectify–filter (FRF) process, which first applies a nonlinear transform to the output of the initial multiscale spatial filters, followed by a second filtering stage to provide a low-frequency surface on which to do edge localization. This model implies that humans should be less accurate at determining texture-defined edges compared to those associated with luminance gradients, and this has been confirmed experimentally.

This also suggests that the HVS would experience some degree of change blindness associated with textures and again this has been confirmed from subjective trials. Consider the two images in Fig. 2.22. They look identical but on closer inspection are actually significantly different. Again this observation reinforces the notion that an analysis–synthesis approach to compression may offer potential. More on this follows in Chapter 13.

FIGURE 2.22 Texture change blindness (images courtesy of Tom Troscianko).

2.6.5 Shape and object recognition

There is evidence from lesion studies that visual processing can be divided into three broad stages:

1. extracting local features,
2. building shapes and surfaces from these features,
3. creating representations of objects in the scene.

There is also some evidence that the HVS uses other properties from a scene to help group features, for example proximity, color and size similarity, common fate (regions with similar direction and speed properties are likely to be part of the same object), and continuity. Marr [4] proposed that features are grouped using average local intensity, average size, local density, local orientation, local distances between neighboring pairs of similar items, and local orientation of the line joining neighboring pairs of items.

To actually recognize objects, further integration of features over larger spatial distances must take place. A number of explanations for this exist, including feedback from higher functional areas of the visual cortex that modulate edge detector responses so as to enhance contours [2]. Surfaces are also likely to be important grouping cues as features on the same surface are likely to be part of the same object.

An object representation is thus likely to be formed from a combination of low-level image analysis with more symbolic higher-level abstract representations coupled through significant feedback between layers and regions. Object recognition is complex because objects have both intrinsic and extrinsic properties. The extrinsic factors, such as change of viewpoint, occlusion, or illumination changes, can have a major impact on recognizability. These extrinsic properties can provide context but can also confound recognition.

2.6.6 The importance of phase information

It has been widely reported that the phase spectrum is highly significant in determining the visual appearance of an image; distortions to phase information, especially over larger areas,

FIGURE 2.23 The importance of phase information in visual perception. Left: Original. Right: Phase-distorted version using the complex wavelet transform (reproduced from [16]).

can result in poor image quality. See for example Fig. 2.23. Equally, it has been shown that it is possible to obtain the phase indirectly from local magnitude information. There is evidence that complex cells in the human visual cortex V1 respond to local magnitude information rather than phase. It has thus been suggested that the HVS might use local magnitude information and local phase to determine an image's appearance. Vilankar et al. [16] conducted an experiment to quantify the contributions of local magnitude and local phase toward image appearance as a function of spatial frequency using images distorted using the dual-tree complex wavelet transform. They confirmed that both local magnitude and local phase do indeed play equally important roles and in some cases, local phase can dominate the image's appearance. While we are still at the early stages of understanding these processes, there is no doubt that they may in the future provide a better understanding of the effects of image compression on the HVS.

For the above reasons, it is widely accepted that signal processing operations on images should preserve phase information. Linear-phase finite impulse response (FIR) digital filters are therefore (almost) exclusively used in compression operations, both for pre- and postprocessing and for interpolation.

2.7 Perception of scale and depth

2.7.1 Size or scale

Size and scale are complex topics in vision. When a spatial pattern is viewed at increasing distance, the image it projects on the retina moves further from the retina, and the image it projects stimulates a response reflecting the higher spatial frequency. Thus, at this level

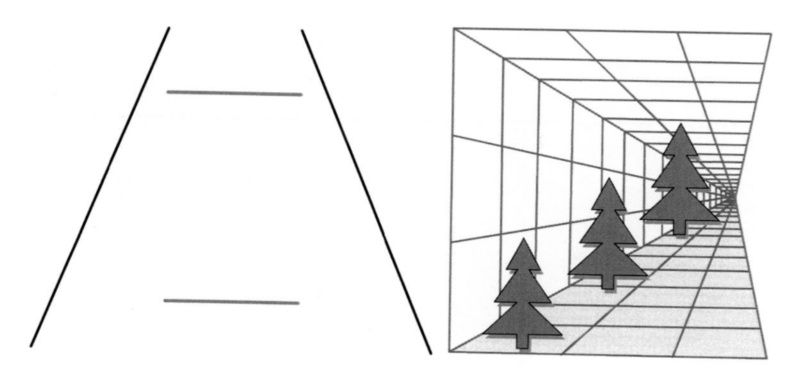

FIGURE 2.24 Perspective-based depth cues can be very compelling and misleading.

changes in scale simply relate to changes in spatial frequency. It has been shown that adaptation to a certain frequency at one distance does affect the perception of a different frequency at a different distance, even if both project the same spatial frequency at the retina. So, even though distance is not integrated into the frequency information from the scene, depth information must be incorporated at higher levels to compensate and allow humans to assess the size of an object at different distances. This ability to judge the size of objects regardless of distance is called size constancy and can lead to misleading and entertaining results, especially when combined with other depth cues such as perspective (Fig. 2.24).

2.7.2 Depth cues

There is no doubt that depth assessment is a dominant factor in our ability to interpret a scene; it also serves to increase our sense of engagement in displayed image and video content. Stereopsis, created through binocular human vision, is often credited as being the dominant depth cue, and has indeed been exploited, with varying degrees of success, in creating new entertainment formats in recent years. It is however only one of the many depth cues used by humans – and arguably it is not the strongest. A list of depth cues used in the human visual system is given below:

- **Our model of the 3D world**: Top-down familiarity with our environment enables us to make relative judgments about depth.
- **Motion parallax**: As an observer moves laterally, nearer objects move more quickly than distant ones.
- **Motion**: Rigid objects change size as they move away or toward the observer.
- **Perspective**: Parallel lines will converge at infinity – allowing us to assess relative depths of oriented planes.
- **Occlusion**: If one object partially blocks the view of another it appears closer.
- **Stereopsis**: Humans have two eyes and binocular disparity information, obtained from the different projections of an object onto each retina, enables us to judge depth.
- **Lighting, shading, and shadows**: The way that light falls on an object or scene tells us a lot about depth and orientation. See Fig. 2.25.

FIGURE 2.25 Pits and bumps – deceptive depth from lighting.

- **Elevation**: We perceive objects that are closer to the horizon as further away.
- **Texture gradients**: As objects recede into the distance, consistent texture detail will appear to be finer-scale and will eventually disappear due to limits on visual acuity.
- **Accommodation**: When we focus on an object our ciliary muscles will either stretch (a thinner lens for distant objects) or relax (a fatter lens for closer objects). This provides an oculomotor cue for depth perception.
- **Convergence**: For nearer objects, our eyes will converge as they focus. This stretches the extraocular muscles, giving rise to depth perception.

It should be noted that certain depth cues can be confusing and can cause interesting illusions. For example, we have an in-built model that tells us that light comes from above (i.e., the sun is above the horizon) and our expectation of shadows reflects this top-down knowledge. Fig. 2.25 shows exactly this. The top diagram clearly shows alternating rows of bumps and pits starting with a row of bumps at the top. The bottom diagram is similar except that it starts with a row of pits at the top. In reality, the only difference with these diagrams is that the bottom one is the top one flipped by 180 degrees. Try this by turning the page upside down. This provides an excellent example of how hard-wired certain visual cues are and how these can lead to interesting illusions. Another example of an illusion driven by top-down processes is the *hollow mask*. In Fig. 2.26, we can see a normal picture of Albert Einstein. In reality this is a photograph of a concave mask. Our visual system is so highly tuned to

FIGURE 2.26 The hollow mask illusion.

faces that, even when our disparity cues conflict with this, we cannot help but see this as a convex face. The right picture shows that even when we rotate the mask to an angle where the features are distorted and you can see it is clearly concave, it still looks like a convex face!

2.7.3 Depth cues and 3D entertainment

3D video technology exploits stereopsis generated through binocular vision. Stereopsis is produced using two cameras spaced at the interocular distance, to create two views which can be displayed in a manner so that the left view is only fed to the left eye and the right view is fed to the right eye. While this does indeed create a sensation of depth, and can be effective if used well for certain types of content, there are several issues that cause problems. These relate to conflict between the accommodation and the convergence of the eyes.

When we watch a 3D movie we view it on a 2D screen, so that is where we naturally focus. However 3D objects projecting out of the screen cause our eyes to converge because they appear nearer. This can lead to conflicting depth cues causing fatigue and, in some cases, nausea. In addition, it has been reported that up to 20% of the population do not see stereoscopically. Of the remainder, a large proportion do not like to watch movies in 3D and choose to watch them in 2D. There are a range of reasons given for this, including the wearing of glasses and the general feeling of forced, overemphasized, or inconsistent depth cues. In addition, current 3D LCD glasses cause attenuation of light levels, thus reducing dynamic range and contrast – both important factors in depth perception.

Equally, directors, cinematographers, and technologists are only gradually beginning to understand the complex interactions between their domains. For example, directors and cinematographers use shots to support narrative and the average shot length in a movie is surprisingly short (around 4 s). It is now known that the HVS takes longer to adapt to the transitions between shots in 3D than in 2D, so shot lengths should be correspondingly longer in the 3D case (more of this later). The rules of cinematography have had to change to cope with 3D.

2.8 Temporal and spatio-temporal response

The temporal response of the HVS is dependent on a number of internal and external factors. Essentially the impulse response of photoreceptors defines the maximum temporal sampling of the retinal image. However, external influences such as ambient lighting, display brightness, and viewing distance also cause variations. If an object is moving across the visual field and not tracked, its image will blur (motion blur) due to the fact that photoreceptors take more than about 100 ms to reach their peak response. These observations have significant impact, not only on camera and display design but also on cinematography (e.g., rules are often imposed to limit camera pan speeds) [17,20].

2.8.1 Temporal CSF

The visual system is sensitive to temporal variations in illumination. This characteristic is well known and exploited in the design of cameras and displays in determining their temporal update rates. The temporal limit is imposed by the rate of response of the photoreceptors, the retinal circuitry, and the response times in the visual cortex [2]. The eye retains the sensation of the incident image for a short time after it has been removed (persistence of vision). It has until recently been thought that an update rate of 60 Hz is sufficient to convey the perception of smooth motion. However, recent investigations are questioning this in the context of higher spatial resolution and larger displays. This is discussed in more detail in Chapter 13.

Kelly [19] showed that the visual sensitivity response to a temporally varying pattern at different frequencies was band-pass in nature, highly dependent on display brightness, and peaking at around 7 or 8 Hz for dimmer displays, becoming diminishingly small at around 24 Hz. However, for brighter displays, the peak rises to around 20 Hz, disappearing at around 80 Hz. The frequency at which the temporal variation becomes unnoticeable is important as it relates to the flickering we experience due to the refresh of a display. The variation with display brightness explains why we can (just about) get away with 24-fps sampling for cinema but need higher rates for TV screens and even higher rates for computer monitors (as we sit much closer to them).

2.8.2 Spatio-temporal CSF

A diagram showing the spatio-temporal characteristics of the HVS is shown in Fig. 2.27 [18]. This shows the band-pass temporal characteristic described above for low spatial frequencies. However, as the spatial frequency increases, the temporal response becomes more low-pass in nature and similarly, with higher temporal frequencies, the spatial response tends to a low-pass characteristic. Thus for faster moving objects, we cannot easily assimilate their

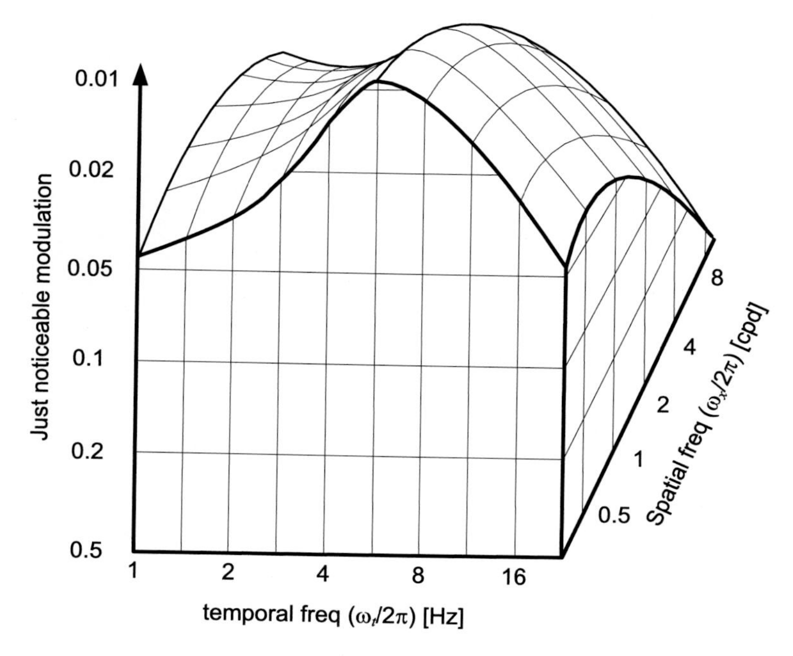

FIGURE 2.27 Spatio-temporal CSF (adapted from Kelly [18]).

spatial textures. This trade-off between spatial and temporal update rates has been exploited in interlaced scanning, as we will examine in Chapter 4.

2.8.3 Flicker and peripheral vision

The flicker fusion threshold is the frequency at which an intermittent light stimulus appears steady to the observer. This depends on a range of factors, including the frequency and depth of the modulation, the wavelength of the illumination, the position on the retina where the stimulation occurs, and the degree of light or dark adaptation. Other factors such as fatigue can also have an influence.

Tyler [24] demonstrated that the flicker threshold (or critical flicker frequency [CFF]) varies with brightness (higher for a brighter source) and with retinal location (rods respond faster than cones). As a consequence flicker can be sensed in peripheral vision at higher frequencies than in foveal vision. This is important as all of our assumptions about flicker are based on central vision and small experimental screens. This explains why, as screens have become brighter and their sizes have become larger, flicker has become more noticeable. This is why TV manufacturers have had to upsample to 300 or even 600 Hz.

Fig. 2.28 [24] shows the visual-field CFF contours for square-wave modulation (in cycles per second) as a function of eccentricity and meridian, with the field sizes scaled to stimulate a constant number of cones at each eccentricity, for two observers. This demonstrates flicker frequencies up to 90 Hz in the periphery.

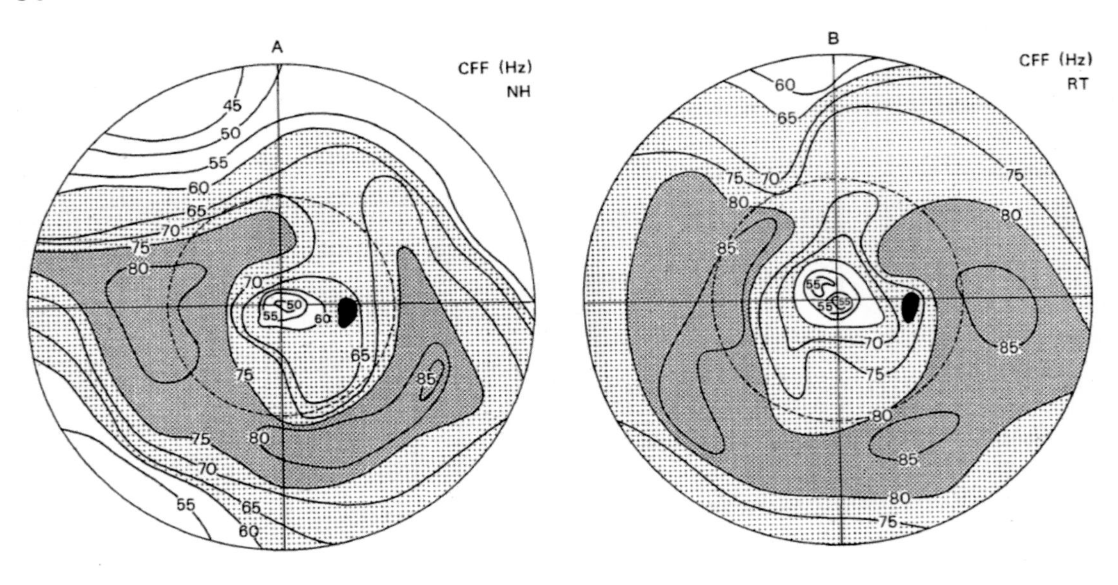

FIGURE 2.28 Variation of critical flicker frequency (reproduced from Tyler [24]).

It is also important to note that CFF generally refers to unstructured stimuli. If the interaction with spatial characteristics is of interest, the temporal CSF is more informative. Interestingly the temporal CSF has been observed to be independent of eccentricity up to 30 degrees on the nasal horizontal meridian [25]. In reality of course with natural scenes, complex interactions exist across different interacting patterns and motions in a scene.

2.9 Attention and eye movements

The HVS creates for us an illusion of fully perceiving a scene while in reality only a relatively small amount of information is passed to the higher brain levels. In this way we perceive only what is considered absolutely necessary, ignoring all other information, in order to compress the huge amount of visual data that continuously enters the visual system. The process that achieves this is referred to as attention and it supported at all cortical levels, but perhaps foremost through foveation and eye movements.

2.9.1 Saliency and attention

Attention modulates the responses of other parts of the visual system to optimize the viewing experience and to minimize the amount of additional information needed to achieve this. Attention can be bottom-up (exogenous), top-down (endogenous), or a combination of both. In the first case this happens when lower-level responses attract attention (e.g., through detection of motion in the periphery). This process is often referred to as saliency and is based on the strength of the low-level HVS response to the specific feature. Top-down or endogenous attention is guided (more) consciously by the task at hand.

There has been a substantial amount of research into models of saliency, perhaps most notably by Itti et al. [26], where top-down attention modulates bottom-up features. Eye movements are used to enable us to foveate on rapidly changing areas of interest and we will consider these briefly next. Attention and foveation have been proposed as a means of allocating bits optimally according to viewing in certain video compression scenarios [27,28].

2.9.2 Eye movements

Apart from when we compensate for head movements, our eyes move in two distinct ways:

1. **Saccades**: As rapid involuntary movements between fixation points. The visual system is blanked during saccading, so we do not experience rapid motion effects due to scanning. Saccades can take two forms:
   a. Microsaccades: Small involuntarily movements used to refresh cell responses that can fall off due to adaptation.
   b. Foveation: The eyes also move rapidly in photopic vision, to allow exploration of new areas of a scene providing an impression of increased acuity across the visual field.
2. **Smooth pursuit**: These are attention-guided smoother voluntary movements which happen, for example, when tracking moving objects. This process keeps the object of interest centered on the fovea so as to maintain high spatial acuity and reduce motion blur. Under the conditions of smooth pursuit, the spatio-temporal limits discussed earlier change dramatically. Girod [30] recomputed these characteristics and demonstrated that this type of eye movement has the effect of extending the temporal limit of our visual system, showing that frequencies of several hundred Hz can be perceived.

Eye tracking is used extensively in vision research [33] to assess the gaze and fixations of an observer in response to a stimulus. Some of the earliest and most famous research on this was published by Yarbus [29]. The results of Yarbus's experiment, where observers were provided with a range of viewing tasks associated with the Visitor painting, are shown in Fig. 2.29. It is interesting to observe how the saccade patterns relate to the specified task.

2.10 Visual masking

Visual masking is the reduction or elimination of the visibility of one brief stimulus, called the "target," by the presentation of a second stimulus, called the "mask." An overview of the influences of masking and other psychovisual effects is presented by Girod in [31].

2.10.1 Texture masking

The visibility threshold for a target increases when the background is textured rather than plain. Spatial (or texture) masking causes the contrast threshold for a given spatial

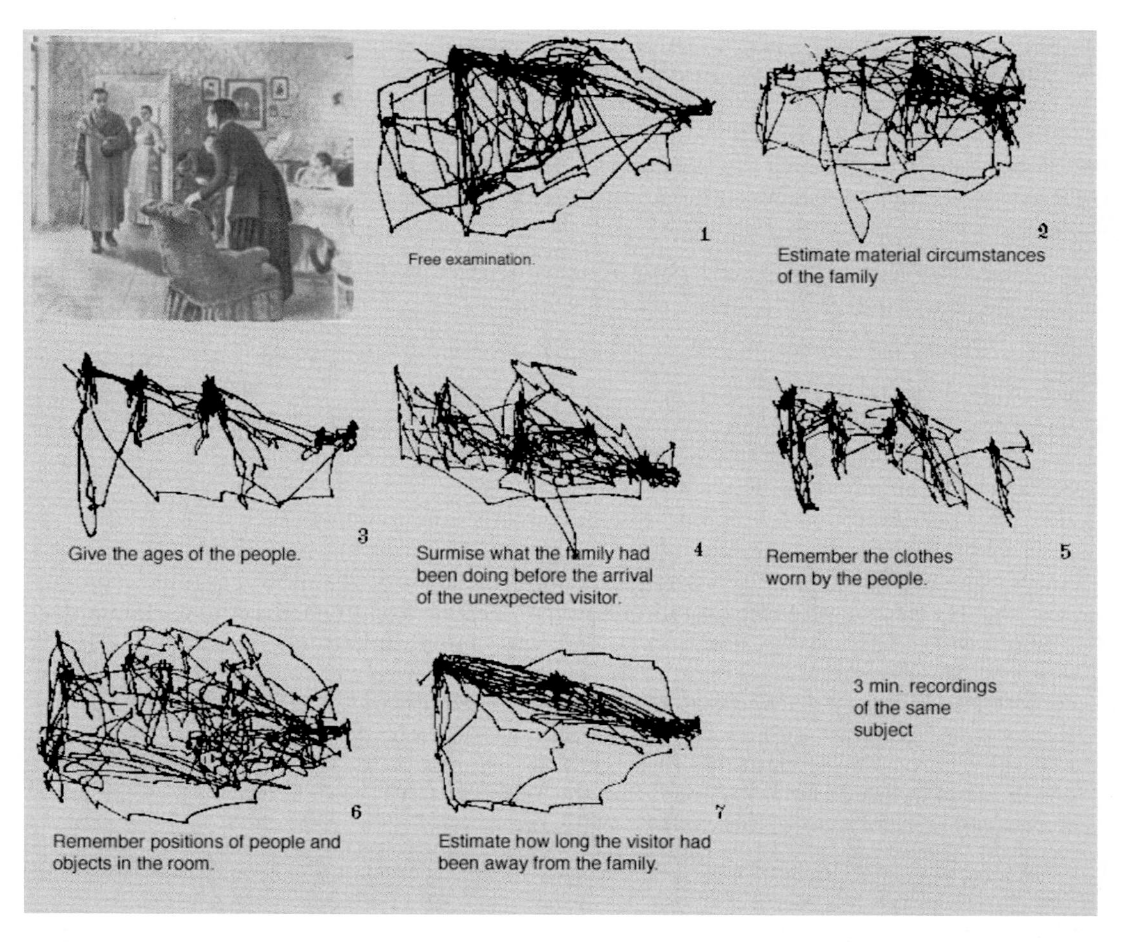

FIGURE 2.29 Eye movements in response to a task (from Yarbus [29]; publicly available from http://commons.wikimedia.org/wiki/File:Yarbus_The_Visitor.jpg).

frequency to rise when a high-contrast mask is present. The inhibitory nature of interacting receptive fields causes the HVS sensitivity to decrease to certain spatial patterns when they are viewed in the context of other patterns. The influence of the mask is related to the similarity between the spatial frequency content of the mask and the target. This effect can be explained by recognizing that the mask pattern elicits responses in the same spatial filters as stimulated by the target test pattern, making them indistinguishable at later visual processing stages [2]. An example of texture masking is shown in Fig. 2.30. This shows identical targets in four subfigures (a triangle, a square, and a circle). The targets are highly visible in the top left subfigure with an untextured gray background but they become increasingly difficult to detect as the noise level rises.

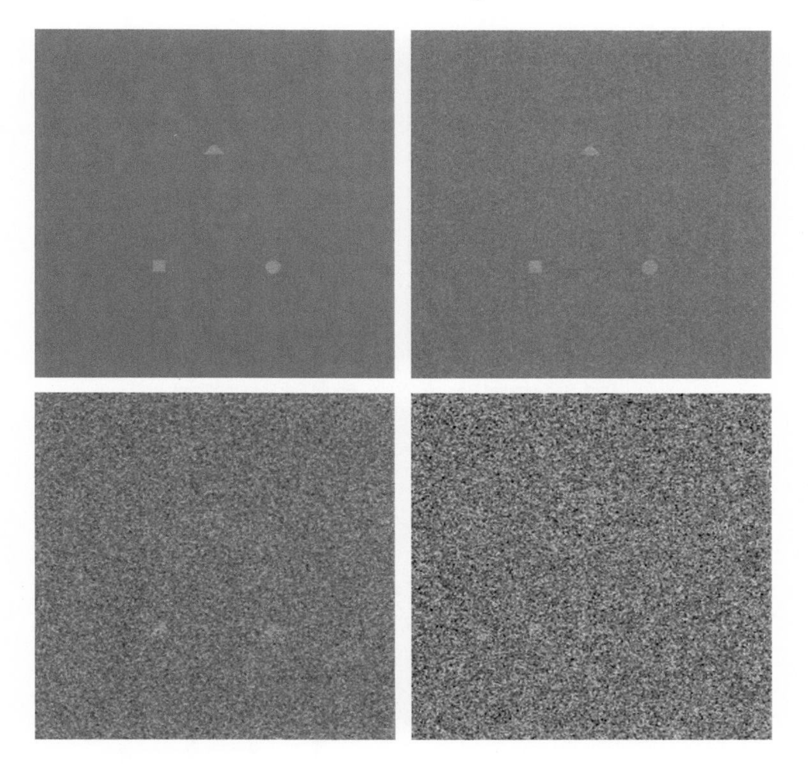

FIGURE 2.30 Example of texture masking. (Top left) Three cues are shown on a gray background. (Top right) With zero-mean, 0.001 variance Gaussian noise. (Bottom left) With 0.01 variance noise. (Bottom right) With 0.03 variance noise.

2.10.2 Edge masking

It has been observed that there is a distinct masking effect in the vicinity of a spatial edge [31]. This has also been demonstrated by Zhang et al. [32] for the case of higher dynamic range content. An example of this showing the experimental setup of the target and edge and the noise visibility variations according to edge contrast and distance of the target from the edge is shown in Fig. 2.31.

2.10.3 Temporal masking

Girod [31] demonstrated the masking effects of video in the presence of temporal discontinuities (such as those that occur at shot cuts or scene changes in a movie). This effect is shown in Fig. 2.32 for a gamma-predistorted video signal. It can be observed that as the extent of temporal discontinuity increases (shown here for an 8-bit edge), the temporal masking effect extends in duration over approximately 100 ms.

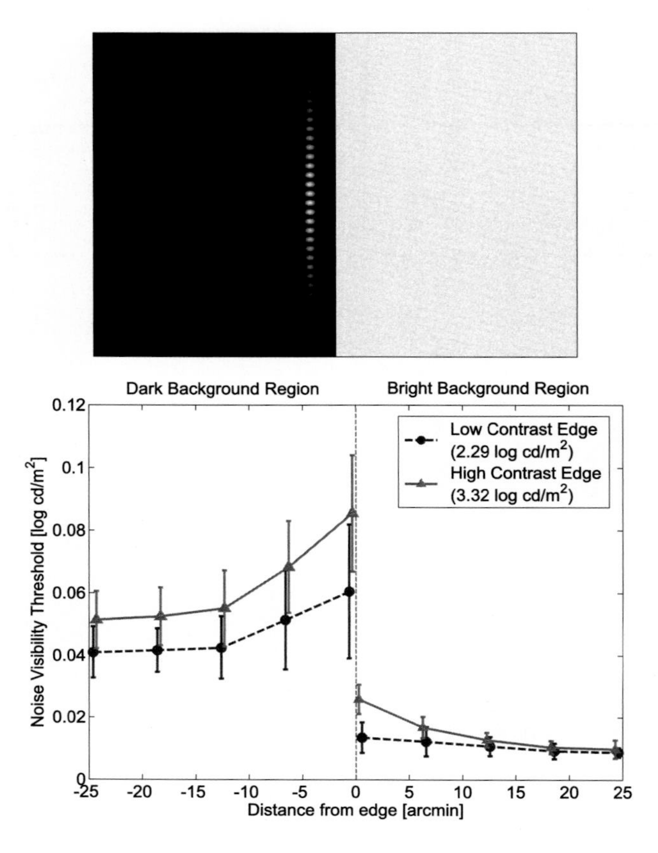

FIGURE 2.31 Edge masking for high and low dynamic range content.

2.11 A perceptual basis for image and video compression

As we have seen throughout this chapter, many factors influence the way we see things: ambient light, scene or display brightness, spatio-temporal content of the scene, viewing distance, attention, task, expectation, physiological variations between subjects, and many other subtle environmental factors.

Many researchers have tried to understand these influences on perception, using various abstract psychophysics experiments. While these provide very valuable insights into the limits and variations of our visual responses, few deal with real scenes and the complex interactions within them. However we can take away several important guidelines and some of these have been used in the design of acquisition and display devices and for optimizing signal representations and compression methods. Others hold potential for the future.

A summary of the observed characteristics of our visual system, alongside ways in which they can be exploited for picture processing and coding, is provided below. These are explored in much more detail throughout the remainder of this book.

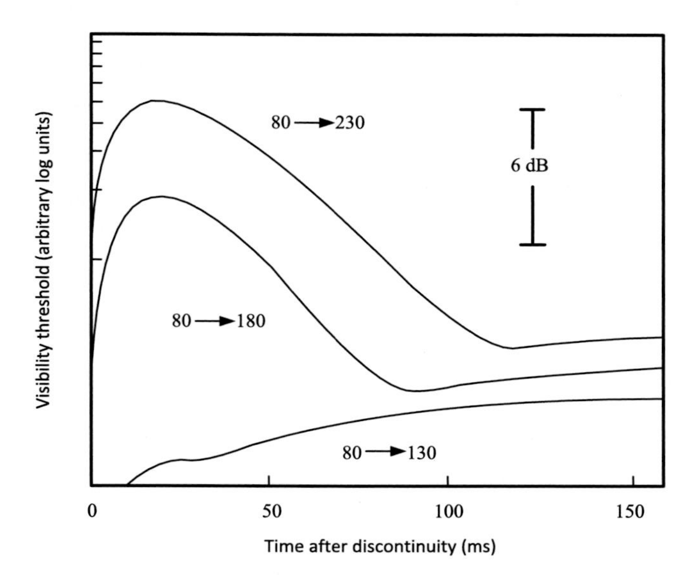

FIGURE 2.32 Temporal masking effects for various edge step sizes (adapted from Girod [31]).

Color processing

HVS characteristics	Implications
The HVS perceives color using three color channels in the visible part of the electromagnetic spectrum.	Acquisition and display technology and visual signal representations for transmission and storage can be based on a tristimulus framework.
Based on opponent theory, the HVS processes visual signals as a luminance channel and two chrominance channels.	Luminance-only content is readily accepted as natural. Visual signal representations can be based on color spaces that split luminance and chrominance.
The HVS has higher acuity in its luminance channel than its chrominance channels.	Chrominance channels can be allocated lower bandwidth than the luminance channel.

Spatial response

HVS characteristics	Implications
The HVS response to spatial frequencies typically peaks at 5–8 cpd and then falls off, becoming diminishingly small beyond 40–60 cpd.	Frequencies beyond the upper range need not be processed. Higher frequencies can be coded more coarsely with implications for quantization. The lower frequency luminance roll-off is not normally exploited.
Edges are an important aspect in human vision system processing. Sensitivity increases with contrast level.	Edges should be preserved. Artificial edge artifacts introduced by quantization are very noticeable and should be avoided.
The HVS can detect noise more readily in plain areas than in textured areas.	Spatial (contrast) masking of quantization noise can be exploited in textured regions.

Temporal response

HVS characteristics	Implications
The HVS temporal contrast sensitivity peaks at around 5–20 Hz and diminishes above 20–80 Hz (brightness-dependent).	Visual processing technology should be capable of sampling at twice the visual threshold and should take account of viewing conditions.
Our ability to see individual frames as a continuum (without flicker) occurs at approximately 50 Hz.	Frame rates for acquisition and display should normally be at least 50 Hz.
The critical flicker frequency is higher in the periphery than in the fovea.	Higher frame refresh rates are required for larger screens and closer viewing.

Spatio-temporal response

HVS characteristics	Implications
The eye can resolve spatial frequencies better at lower temporal frequencies, while at higher spatial frequencies, the temporal response becomes low-pass.	This trade-off between temporal and spatial acuity can be exploited in interlaced scanning to provide the perception of increased frame rates.

Processing textures

HVS characteristics	Implications
There is evidence that texture infilling occurs in the HVS.	Analysis–synthesis approaches to image and video coding hold potential.
Texture variations and changes are more difficult to detect than shape changes.	As above.

Depth cues

HVS characteristics	Implications
The HVS exploits many and varied depth cues at all stages of processing.	Stereoscopic 3D content may not be the only way to achieve more immersive content.

Dynamic range

HVS characteristics	Implications
Our visual response depends on the mean brightness of the display.	Cinema frame rates can be lower than those for TV or computer monitors. Higher dynamic range displays offer better perceptions of reality and depth.
There is a linear relationship between JND and background brightness over a wide range of brightness levels, with further increases at low and high values.	This can be exploited in applying intensity-dependent quantizations, allowing step sizes to increase more rapidly for low- and high-level signals.
There is a nonlinear relationship between luminance and perceived brightness.	Implications for signal coding in that a nonlinear (gamma) function is applied to camera outputs prior to coding.

Attention, foveation, and eye movements

HVS characteristics	Implications
Our high-acuity vision is limited to a 2-degree arc.	Bit allocation strategies could target those regions of an image where attention is focused (if known!).
The spatio-temporal frequency response of the HVS is altered significantly by eye movements. Under smooth pursuit conditions, much higher temporal acuity is possible.	Especially with larger and brighter screens, displays must be able to cope with temporal frequencies of many hundreds of hertz.

Physiological variations between subjects and environments

HVS characteristics	Implications
The responses of individuals will vary naturally and with age	Subjective quality assessment experiments should be based on results from a large number of subjects (typically > 20).
Human responses vary significantly according to viewing conditions.	Subjective tests should be conducted under tightly controlled viewing conditions, to enable cross-referencing of results.

References

[1] R. Snowden, P. Thompson, T. Troscianko, Basic Vision, Oxford University Press, 2006.
[2] G. Mather, Foundations of Sensation and Perception, 2e, Psychology Press, 2009.
[3] B. Wandell, Foundations of Vision, Sinauer Assoc., 1995.
[4] D. Marr, Vision, Freeman, 1982, reprinted, MIT Press, 2010.
[5] L. Cormack, Computational models of early human vision, in: A. Bovic (Ed.), Handbook of Image and Video Processing, Academic Press, 2000.
[6] H. von Helmholtz, Handbook of Physiological Optics, 1st ed., Hamburg and Leipzig, Voss, 1896.
[7] B. Julesz, Binocular depth perception of computer-generated patterns, Bell Syst. Tech. J. 39 (1960) 1125–1162.
[8] D. Hubel, T. Wiesel, Receptive fields of single neurones in the cat's striate cortex, J. Physiol. 148 (1959) 574–591.
[9] R. Gregory, Eye Brain, 5e, Princeton University Press, 1997.
[10] M. Clowes, Transformational grammars and the organization of pictures, in: A. Grasselli (Ed.), Automatic Interpretation and Classification of Images, Academic Press, 1969.
[11] H. Kolb, How the retina works, Am. Sci. 91 (2003) 28–35.
[12] R. Lotto, D. Purves, The effects of color on brightness, Nat. Neurosci. 2 (11) (1999) 1010–1014.
[13] D. Mustafia, A. Engela, K. Palczewskia, Structure of cone photoreceptors, Prog. Retin. Eye Res. 28 (4) (2009) 289–302.
[14] J. Bowmaker, H. Dartnall, Visual pigments of rods and cones in a human retina, J. Physiol. 298 (1) (1980) 501–511.
[15] A. Netravali, B. Haskell, Digital Pictures: Representation, Compression and Standards, 2e, Plenum Press, 1995.
[16] K. Vilankar, L. Vasu, D. Chandler, On the perception of band-limited phase distortion in natural scenes, in: Proc. SPIE 7865, Human Vision and Electronic Imaging XVI, 2011, 78650C.
[17] A. Watson (Ed.), Digital Images and Human Vision, MIT Press, 1998.
[18] D. Kelly, Adaptation effects on spatio-temporal sine wave thresholds, Vis. Res. 12 (1972) 89–101.
[19] D. Kelly, Visual responses to time-dependent stimuli, J. Opt. Soc. Am. 51 (1961) 422–429.
[20] A. Watson, A. Allhumada, J. Farell, Windows of visibility: psychophysical theory of fidelity in time sampled visual motion displays, J. Opt. Soc. Am. A 3 (3) (1986) 300–307.
[21] R. Shapley, M. Hawken, Neural mechanisms for color perception in the primary visual cortex, Curr. Opin. Neurobiol. 12 (4) (2002) 426–432.
[22] http://en.wikipedia.org/wiki/File:CIE1931xy_blank.svg.

[23] J. Mannos, D. Sakrison, The effects of visual error criteria on the encoding of images, IEEE Trans. Inf. Theory IT-20 (1974) 525–536.
[24] C. Tyler, Analysis of visual modulation sensitivity. III. Meridional variations in peripheral flicker sensitivity, JOSA A 4 (8) (1987) 1612–1619.
[25] V. Virsu, J. Rovamo, P. Laurinen, R. Nasanen, Temporal contrast sensitivity and cortical magnification, Vis. Res. 22 (1982) 1211–1217.
[26] L. Itti, C. Koch, E. Niebur, A model of saliency-based visual attention for rapid scene analysis, IEEE Trans. Pattern Anal. Mach. Intell. 20 (11) (1998) 1254–1259.
[27] D. Agrafiotis, C. Canagarajah, D. Bull, J. Kyle, H. Seers, M. Dye, A perceptually optimised video coding system for sign language communication at low bit rates, Signal Process. Image Commun. 21 (7) (2006) 531–549.
[28] S. Davies, D. Agrafiotis, C. Canagarajah, D. Bull, A multicue Bayesian state estimator for gaze prediction in open signed video, IEEE Trans. Multimed. 11 (1) (2009) 39–48.
[29] A. Yarbus, Eye Movements and Vision, Plenum, New York, 1967.
[30] B. Girod, Motion compensation; visual aspects, accuracy and fundamental limits, in: M. Sezan, Lagendijk (Eds.), Motion Analysis and Image Sequence Processing, Kluwer, 1993, pp. 126–152.
[31] B. Girod, Psychovisual aspects of image communication, Signal Process. 28 (3) (1992) 239–251.
[32] Y. Zhang, D. Agrafiotis, M. Naccari, M. Mrak, D. Bull, Visual masking phenomena with high dynamic range content, in: Proc IEEE Intl. Conf. on Image Processing, 2013, pp. 2284–2288.
[33] S. Liversedge, I. Gilchrist, S. Everling, Oxford Handbook of Eye Movements, Oxford University Press, 2011.

CHAPTER

3

Signal processing and information theory fundamentals

This chapter is not intended to be a complete course in discrete-time signal analysis or information theory – many excellent texts on this topic already exist covering both 1D [1] and 2D [2] signals. We assume here that the reader is familiar with the basics of sampling, linear systems analysis, digital filters, transforms, information theory, and statistical signal processing. Our purpose is to provide an overview of these important topics as they form the basis of many of the compression techniques described later in the book.

We firstly review the sampling theorem and consider the cases of 2D (image) and 2D+time (video) sampling. This provides the discrete time samples that represent our image or video and that are processed during the various stages of compression and coding. In Section 3.2, we introduce the means of describing signals using statistical representations. We focus on second order statistics and we use these to characterize the redundancy contained in an image or video signal. It is this redundancy that is exploited during the compression process.

Filters and transforms form the basic work horses of most compression systems and these are introduced in Section 3.3. The signal distortions introduced in most compression systems can be attributed to coefficient quantization after filtering or transformation; this important aspect is covered in Section 3.4. Prediction operators are also in common usage across many aspects of compression, from intra-prediction to motion estimation. The basic concepts of linear prediction are introduced in Section 3.5 and we pay particular attention to the mitigation of decoder drift. This architecture is important as it provides a framework for all current video compression standards.

As a precursor to coverage of lossless compression and entropy coding in Chapter 7, we review the basics of information theory: self-information, entropy, symbols, and their statistics. This topic is key to all symbol encoding methods in use today. Finally, in Section 3.7 we provide an introduction to the important topic of machine learning (ML), which is having an increasing impact on both compression methods and quality assessment.

3.1 Signal and picture sampling

The manner in which a signal is sampled has a profound effect on subsequent processing steps. Particularly in the context of sensory signals such as images or video, the samples must

Intelligent Image and Video Compression
https://doi.org/10.1016/B978-0-12-820353-8.00012-8

provide a suitable and realistic representation of the underlying (continuous) natural scene, when viewed on a screen. Sampling must thus be performed in a manner that avoids the introduction of perceptual spatio-temporal artifacts due to aliasing. The sampling process addresses three primary questions:

1. What spatial and temporal sampling rates are needed to avoid aliasing and to produce the desired signal fidelity?
2. How should the samples be distributed in space and time?
3. What preprocessing is required prior to sampling and what postprocessing is needed prior to reconstruction?

These issues are influenced by a number of factors, including: the frequency content of the signal being sampled, the perceptual thresholds of the observer as related to the prevailing viewing conditions, the characteristics of the capture and display devices, and the costs associated with storage and transmission. This section attempts to answer these questions.

3.1.1 The sampling theorem

In one dimension

Consider a continuous (let us assume time domain) signal $x_a(t)$ sampled using a sequence of delta functions $\delta(t-nT)$ to yield

$$x_s(t)=x_a(t)s(t), \tag{3.1}$$

where

$$s(t)=\sum_{n=-\infty}^{\infty}\delta(t-nT).$$

Using the Fourier transform we can obtain the frequency domain representation of these signals, thus

$$X_a(\omega)=\int_{-\infty}^{\infty}x_a(t)e^{-j\omega t}dt \tag{3.2}$$

and

$$S(\omega)=\frac{2\pi}{T}\sum_{k=-\infty}^{\infty}\delta(\omega-k\omega_s). \tag{3.3}$$

Invoking the modulation property of the Fourier transform gives

$$X_s(\omega)=\frac{1}{T}\sum_{k=-\infty}^{\infty}X_a(\omega-k\omega_s), \tag{3.4}$$

where ω_s represents the continuous frequency variable.

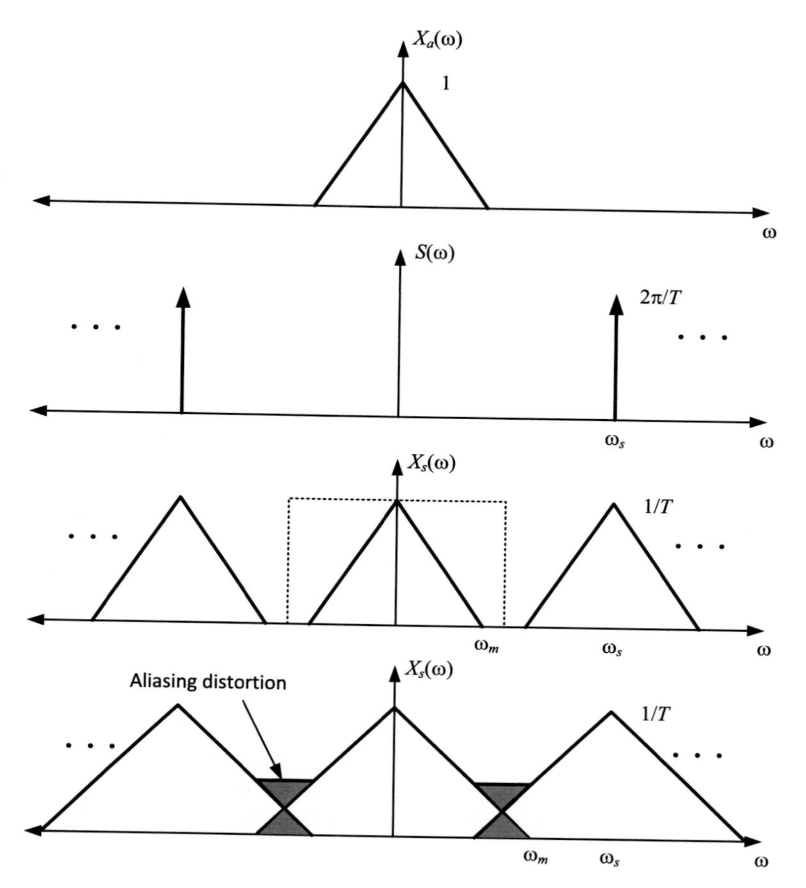

FIGURE 3.1 Spectral characteristics of sampling and aliasing.

The spectrum of a sampled signal thus comprises scaled replicated versions of the original (continuous-time) spectrum. In this context, Shannon's sampling theorem captures the requirements for sampling in order to avoid aliasing [3]:

> If a signal $x_a(t)$ is bandlimited with $X_a(\omega) = 0$ for $|\omega| > \omega_m$, then $x_a(t)$ is uniquely determined by its samples provided that $\omega_s \geq 2\omega_m$. The original signal may then be completely recovered by passing $x_s(t)$ through an ideal low-pass filter.

The spectral characteristics of the sampling process as described by Shannon's theorem are shown in Fig. 3.1. Paying particular attention to the bottom two subfigures, the penultimate subfigure shows the case where perfect reconstruction is possible with an ideal reconstruction filter. In contrast, the bottom subfigure shows the case where ω_m exceeds the frequency $\omega_s/2$ and aliasing distortion occurs. The frequency $\omega_s/2$ is often referred to as the Nyquist frequency.

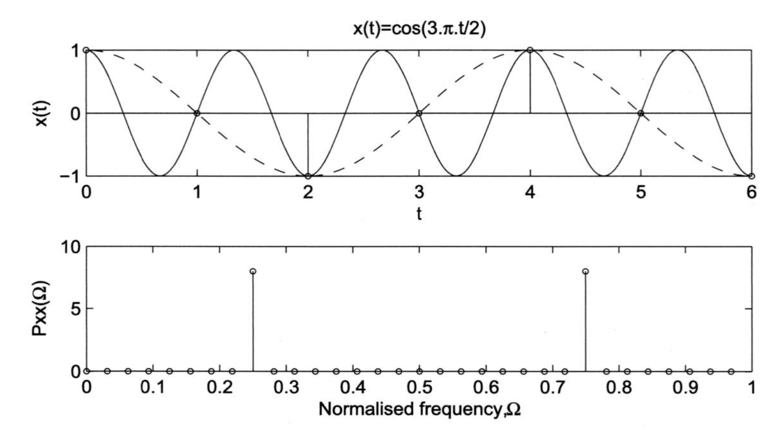

FIGURE 3.2 Demonstration of aliasing for a 1D signal: Top: Sinusoid sampled below Nyquist frequency. Bottom: Fourier plot showing spectral aliasing.

Example 3.1. 1D signal aliasing

Consider a sinusoidal signal $x(t) = \cos(3\pi t/2)$ sampled at $\omega_s = \frac{4}{3}\omega_m$ with $T = 1$. Show that this sampling violates the sampling theorem and characterize the impact of the alias in the time and frequency domains.

Solution. The result is shown in Fig. 3.2. The sinusoid $x(t) = \cos(3\pi t/2)$ is sampled at $\omega_s = \frac{4}{3}\omega_m$ with $T = 1$. As can be observed, an alias is created at a frequency of $\frac{\omega_s}{4}$ (to be accurate, this is actually an alias of the negative frequency component of the sinusoid at $-\omega_m$). This can be observed in both the signal domain plot (top) and the frequency domain plot (bottom). Here, as is conventional with discrete-time systems, we plot normalized frequency $\Omega = \omega T$, where the sampling is given by $\Omega = 2\pi$.

Extension to 2D

Assuming a rectangular (orthogonal) sampling pattern in 2D, we can extend our 1D approach to the case of sampling of a 2D signal. This gives

$$X_s(\omega_1, \omega_2) = \frac{1}{T_1 T_2} \sum_{k_1, k_2} X_a(\omega_1 - k_1\omega_s, \omega_2 - k_2\omega_s). \tag{3.5}$$

The effect of sampling and aliasing on the 2D spectrum is shown in Fig. 3.3. In 2D images, aliasing errors will take the appearance of ringing in a direction perpendicular to high frequencies or sharp edges. It should be emphasized that while Fig. 3.3 shows a circularly symmetric spectrum, in practice the spectrum could be skewed according to the prevailing horizontal and vertical frequencies or textures in the image.

Extension to 3D

In the case of spatio-temporal sampling, it is conventional to sample at a fixed frame rate using either progressive or interlaced sampling (see Chapter 4). As we saw in Chapter 2, the visual sensitivities in the spatial and temporal dimensions are quite different. For example, a conventional 1080p/50 4:4:4 format will capture 103,680,000 samples per second. This however comprises 2,073,600 samples per picture but only 50 pictures per second. Thus there is a large difference in the number of samples per unit dimension between the spatial and temporal domains. It appears that the temporal domain gets a raw deal and to some extent this is true. We will come back to this in Chapter 13.

As described in Chapter 2, visual sensitivity depends on the mean brightness of the display. For a bright TV signal, it is generally considered, based on the temporal contrast sensitivity function and critical flicker frequencies, that a temporal update rate of 50–70 Hz is sufficient, whereas for computer monitors, which are viewed more closely in brighter environments, a higher update rate is normally specified.

Example 3.2. How many spatial samples do we need?

Consider the case of a 16:9 screen with a width of 1 m, viewed at a distance of $3H$ (where H is the height of the screen). How many horizontal samples are required to satisfy the contrast sensitivity limits of the human visual system?

Solution. For the case of a 16:9 screen with a width of 1 m, viewed at $3H$, we can compute the number of horizontal samples required from simple trigonometry. The height of the screen is 0.5625 m and the half viewing angle is then

$$\theta = \tan^{-1}\left(\frac{0.5}{1.6875}\right).$$

Assuming a spatial resolution limit of 30 cycles per degree (see Chapter 2), this yields 1980 horizontal samples per line, which is pretty close to the 1920 that are used in practice.

Although this calculation is useful, as screen sizes increase and viewing distances decrease, there will be a demand to take a fresh look at spatio-temporal sampling, but more of that later in Chapter 13.

3.1.2 Multidimensional sampling lattices

A sampling lattice in real-valued K-dimensional space is the set of all possible vectors that can be formed from linear combinations of a set of K linearly independent basis vectors, $\mathbf{v}_k \in \mathcal{R}^K$, $k = \{1, 2, 3 \ldots K\}$. So the new vectors are formed from a weighted combination as follows:

$$\mathbf{w} = \sum_{k=1}^{K} c_k \mathbf{v}_k, \quad \forall c_k \in \mathcal{Z}. \tag{3.6}$$

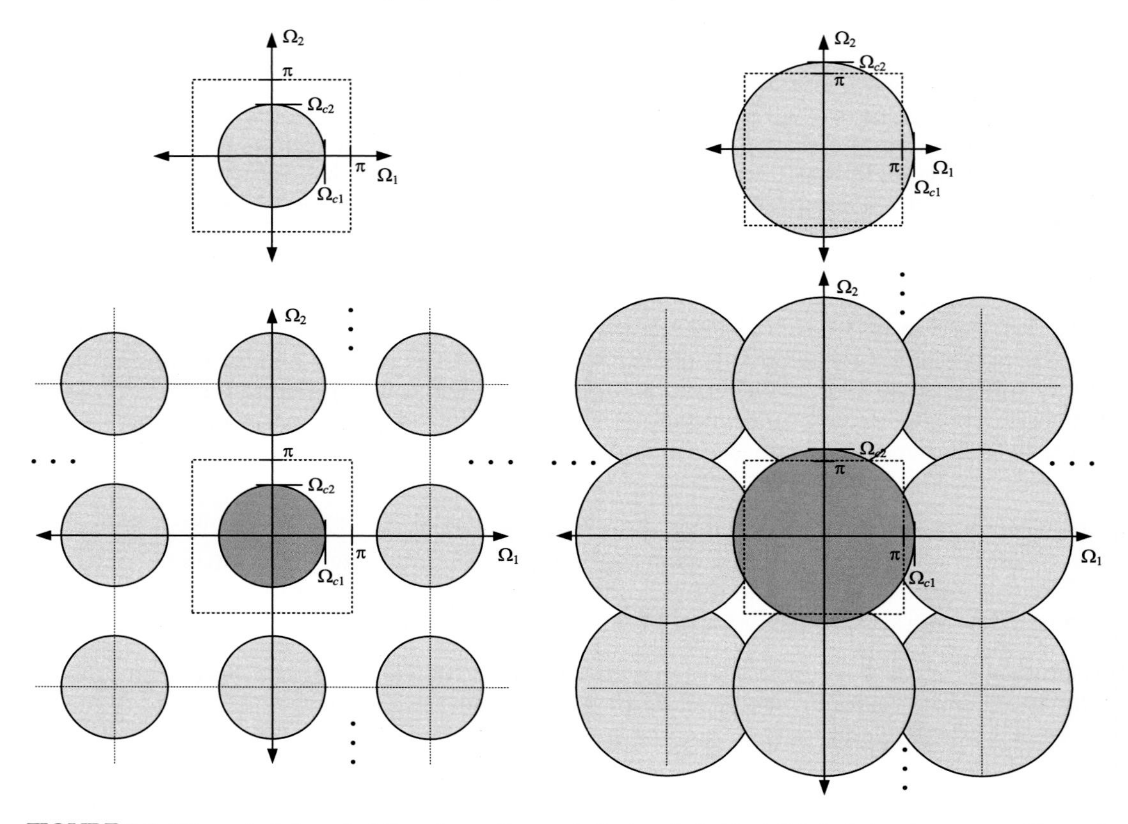

FIGURE 3.3 2D spectral characteristics of sampling and aliasing. Left: (Top) Original signal spectrum. (Bottom) Sampled signal spectrum with no aliasing. Right: (Top) Original signal spectrum. (Bottom) Sampled signal spectrum with aliasing due to sub-Nyquist sampling.

The matrix $\mathbf{V} = [\mathbf{v}_1, \mathbf{v}_2, \ldots, \mathbf{v}_K]$ that combines the K basis vectors is called the *sampling matrix* or the *generating matrix*. For example, the generating matrix for our familiar rectangular lattice is

$$\mathbf{V}_1 = \begin{bmatrix} 1 & 0 \\ 0 & 1 \end{bmatrix}. \tag{3.7}$$

However, we can explore more exotic sampling grids and their associated regions of support. Consider for example the generating matrix

$$\mathbf{V}_2 = \begin{bmatrix} \sqrt{3}/2 & 0 \\ 1/2 & 1 \end{bmatrix}. \tag{3.8}$$

The sampling lattice for this generating matrix is shown in Fig. 3.4. As can be seen, this represents an hexagonal sampling grid. The lattice structure represents a simple and convenient tool for generating and analyzing sampling structures that are not necessarily hypercubic in

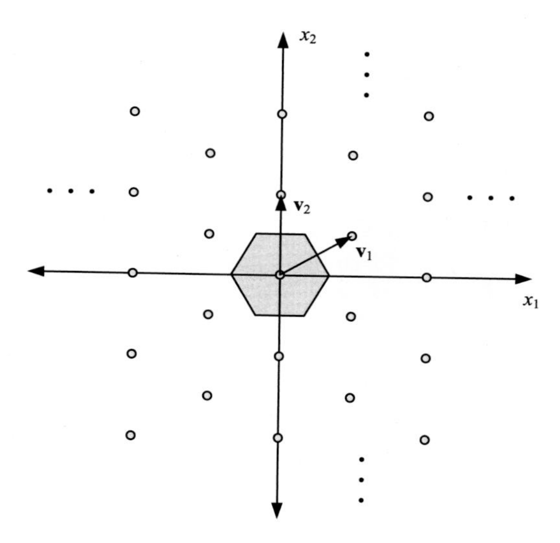

FIGURE 3.4 Hexagonal sampling lattice and its reciprocal as defined by Eq. (3.8).

structure. All the theorems for 1D and 2D sampling still apply and a generalized form of the Shannon sampling theory exists which dictates the structure and density of the sampling grid dependent on the signal spectrum. There is a significant body of work relating this more general approach to multidimensional sampling, which is relevant to image and video applications. The reader is referred to the work of Dubois [4] and Wang et al. [5] for excellent introductions to this topic.

3.2 Statistics of images

3.2.1 Histograms and distributions

A histogram is a discrete approximation of a probability density function based on the actual occurrences of quantized pixel values. It can provide important information about the distribution of sample amplitudes within the image, about the dynamic range of the signal as opposed to the word length of its representation, and about whether the image is under- or overexposed. An example histogram for the 256 × 256 image Stampe_SV4 is given in Fig. 3.5.

Spatial and subband distributions

An understanding of spatial domain and frequency domain distributions can be of significant benefit in compression systems. For example, the pyramid vector quantization method described in Chapter 11 exploits the Laplacian-like subband coefficient distribution, while the parametric video coding approaches outlined in Chapter 13 exploit the subband structure of dual-tree complex wavelet coefficients in providing segmentation into coherent texture regions.

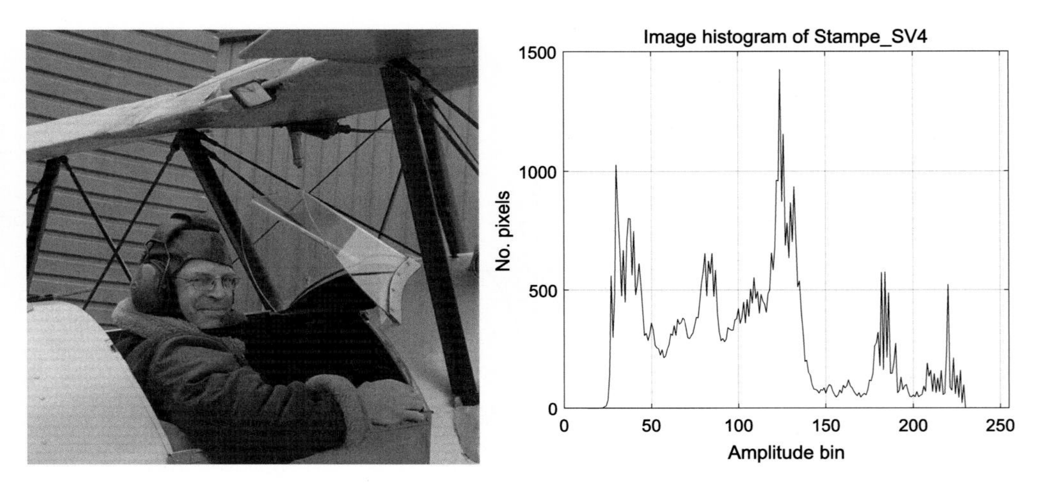

FIGURE 3.5 Example image histogram for 256 × 256 image Stampe_SV4.

3.2.2 Mean values

Mean and correlation provide a partial statistical characterization of a random process in terms of its averages and moments. They are useful as they offer tractable mathematical analysis, they are amenable to experimental evaluation, and they are well suited to the characterization of linear operations on random processes.

The mean for a stochastic process is given by

$$\begin{aligned}\mu_n &= E\{x[n]\} \\ &= \int_{-\infty}^{\infty} x[n]f(x[n])dx[n],\end{aligned} \tag{3.9}$$

where $f(\cdot)$ is the probability density function (pdf) of the process and $E\{\cdot\}$ is the expectation operator. In the case of a stationary process the pdf of the random variable is the same for all n and hence the mean is constant. Expectation can be interpreted as an average value obtained by repeating an experiment a number of times. Hence

$$\mu_n = E\{x[n]\} = \lim_{N\to\infty}\left[\frac{1}{N}\sum_{i=1}^{N} x_i[n]\right]. \tag{3.10}$$

3.2.3 Correlation in natural images

The autocorrelation (or autocovariance) of a sequence expresses the linear statistical dependencies between its samples. It is defined for a real-valued signal with a lag of m samples as

$$R_{xx}[m] = E\{x[n]x[n+m]\}, \tag{3.11}$$

where $x[n]$ is a stationary random process. In particular,

$$R_{xx}[0] = E\left\{|x[n]|^2\right\} \tag{3.12}$$

is the average power of the signal. In practice, the correlation is estimated based on a finite-length sequence $\mathbf{x}$ of N samples as follows:

$$R_{xx}[m] = \frac{1}{N}\sum_{n=0}^{N-|m|-1} x[n]x[n+m]. \tag{3.13}$$

Eq. (3.13) produces a biased estimate because of the attenuation for large lags. An unbiased estimate can also be used which is noisier, but preferable in certain situations:

$$R_{xx}[m] = \frac{1}{N-|m|}\sum_{n=0}^{N-|m|-1} x[n]x[n+m]. \tag{3.14}$$

The autocovariance is computed in the same manner as the autocorrelation, but with the signal means removed. When the autocorrelation or autocovariance functions are normalized by their maximum value, they are generally referred to as autocorrelation coefficients or autocovariance coefficients, respectively. These have values between -1 and $+1$. For example the autocorrelation coefficient is given by[1]

$$\rho_x(k) = \frac{r_x(k)}{r_x(0)}. \tag{3.15}$$

The autocorrelation matrix comprises autocorrelation values at various lags. The correlation matrix for a *wide-sense stationary* real-valued signal is the expectation of the outer product of the signal vector with itself. Thus,

$$\mathbf{R}_x = E\left\{\mathbf{x}\mathbf{x}^{\mathrm{T}}\right\}, \tag{3.16}$$

$$\mathbf{R}_x = \begin{bmatrix} r_x[0] & r_x[1] & \dots & r_x[M-1] \\ r_x[1] & r_x[0] & & r_x[M-2] \\ \vdots & & & \vdots \\ r_x[M-1] & r_x[M-2] & \dots & r_x[0] \end{bmatrix}. \tag{3.17}$$

The correlation matrix is Toeplitz and Hermitian. For the case of real-valued signals this means that it is transpose-invariant.

Spatial autocorrelation in natural images

Let us now consider the correlation properties of some images. Figs. 3.6 and 3.7 show the autocorrelation properties of two 512 × 512 images – *Acer* and *Stampe_SV4*. The first has a

[1] We have dropped the double subscript as it is redundant and where no confusion results we will use lowercase r to denote autocorrelation.

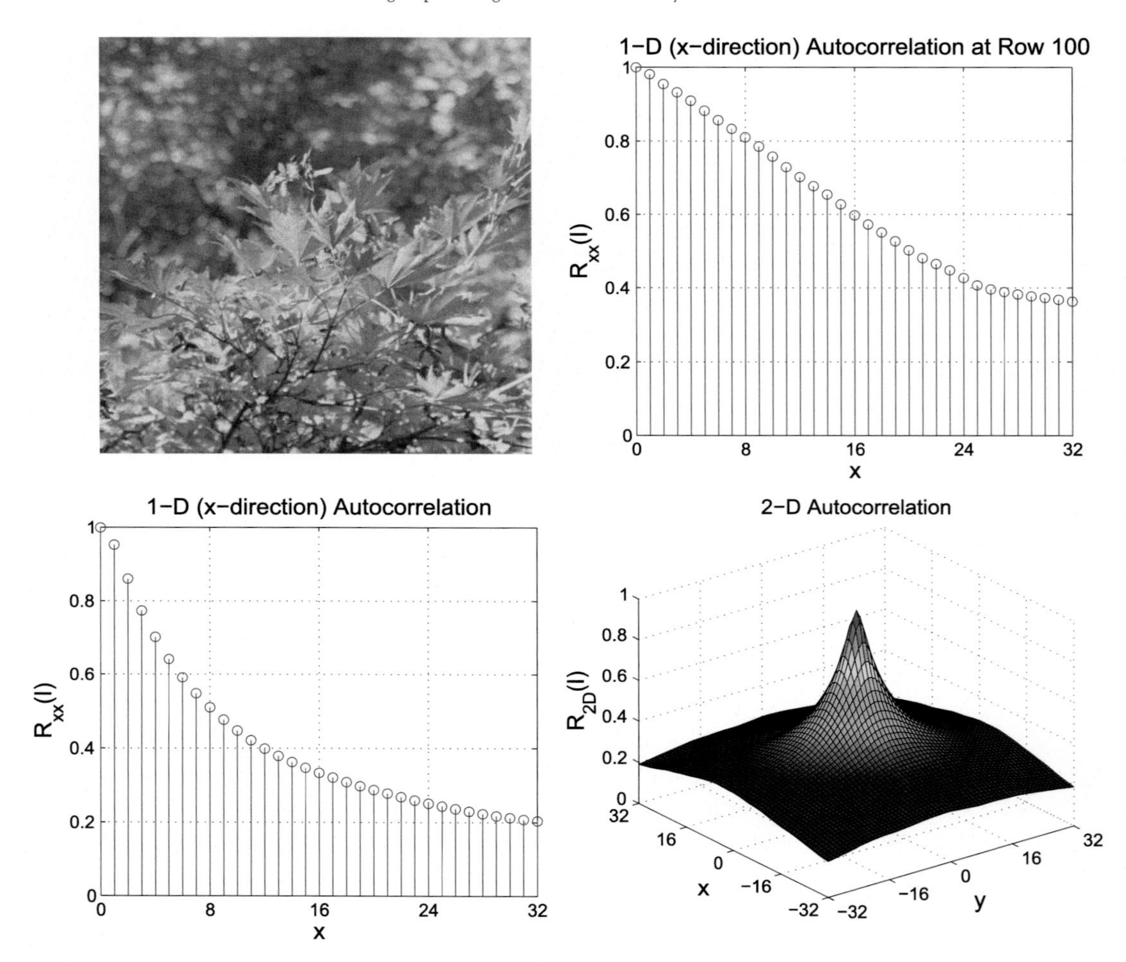

FIGURE 3.6 Autocorrelation plots for the Acer image (512 × 512). Top left to bottom right: Original image; autocorrelation function for row 100; autocorrelation function for the whole image; 2D autocorrelation surface.

highly textured, spatially active foreground with a blurred background and the second is highly structured with several plain areas. As well as the original image, each figure includes (i) an autocorrelation plot (unbiased with zero mean) for lags up to 32 pixels for a single row of pixels, (ii) a similar plot based on an average of all rows in the image, and (iii) a 2D version. It can be observed that in both cases the correlation falls off gradually with increasing lag and that the values for the Acer image fall off more quickly than those for the Stampe_SV4 image. This is because of the structure in the latter compared to the more stochastic nature of the former. Fig. 3.8 shows a similar series of plots, but this time for a 256 × 256 image, with only values up to a lag of 16 shown. As one would expect, the autocorrelation values fall off in a similar fashion to Fig. 3.7 except that the corresponding lag values are halved.

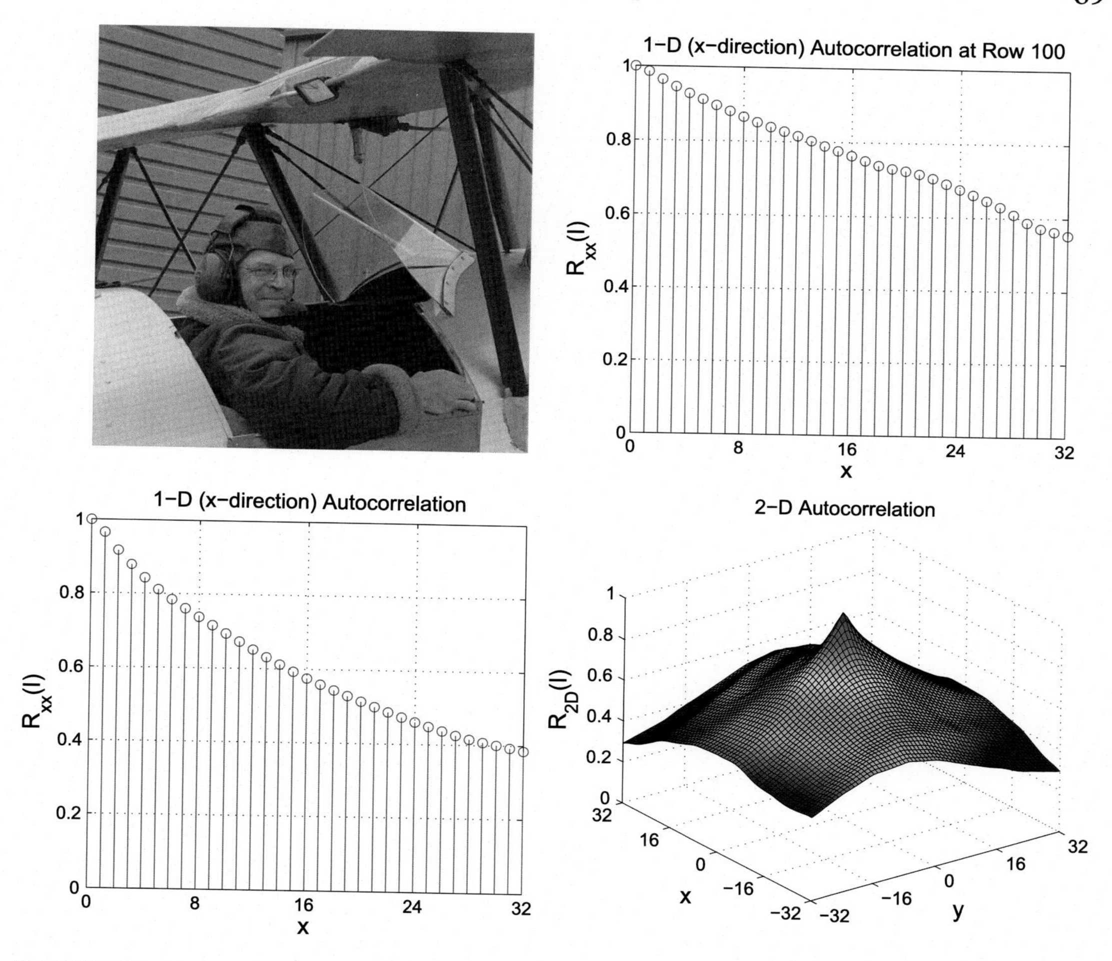

FIGURE 3.7 Autocorrelation plots for the Stampe_SV4 image (512 × 512). Top left to bottom right: Original image; autocorrelation function for row 100; autocorrelation function for the whole image; 2D autocorrelation surface.

Exploitation of this intersample correlation is the basis of prediction, transform, and filter-bank compression methods. In each case, the purpose of the transform or predictor is to decorrelate the spatial information to make it more amenable to frequency-related perceptual quantization. As we can see from the figures, the autocorrelation values for large lags become increasingly small, while those for smaller lags are high. Because decorrelating transforms and predictors work better on regions with higher correlation, they are normally applied to small regions (typically 8 × 8 blocks) rather than to the whole image. An 8 × 8 block has been conventionally chosen as a good compromise between computational efficiency and coding gain. It is important to note, however, that the image resolution will have a significant impact on this – as lower-resolution images will exhibit, like for like, lower correlations across the same size block.

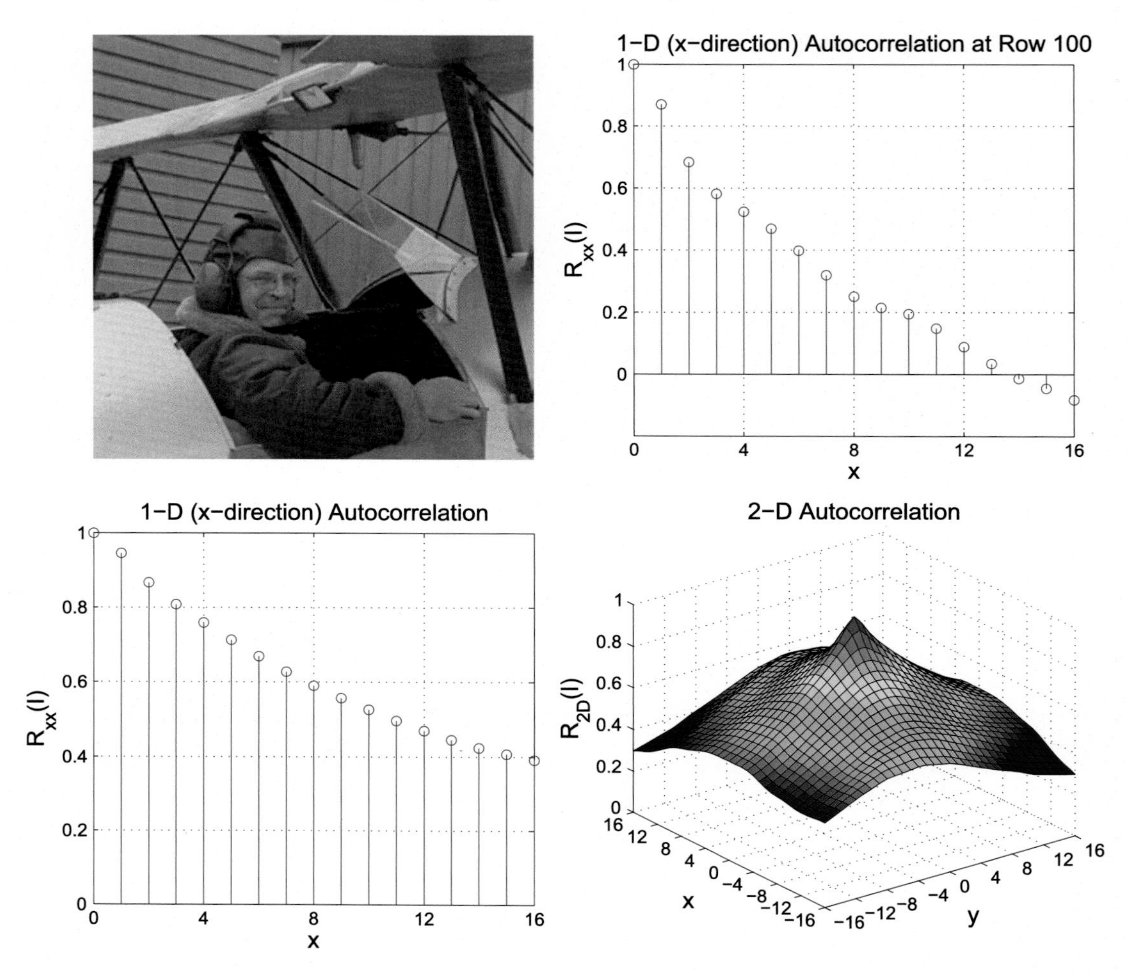

FIGURE 3.8 Autocorrelation plots for the Stampe_SV4 image (256 × 256). Top left to bottom right: Original image; autocorrelation function for row 100; autocorrelation function for the whole image; 2D autocorrelation surface.

Temporal autocorrelation in natural image sequences

Let us now consider the time axis. An example of temporal correlation, for the Foreman sequence, is shown in Fig. 3.9. This indicates a similar characteristic to the spatial case considered above. Two plots are shown. The first is the time correlation for a single pixel computed using 300 frames of the Foreman sequence. The second is the same calculation, but this time averaged over the 16 × 16 block of pixels indicated in the top subfigure. Again we can observe that significant temporal correlation exists between the adjacent video frames and indeed over 8 or so frames the correlation remains relatively high. This indicates that there is significant temporal redundancy present in most image sequences. As we will see later this is exploited in most compression methods, by applying motion-compensated prediction, prior to transformation.

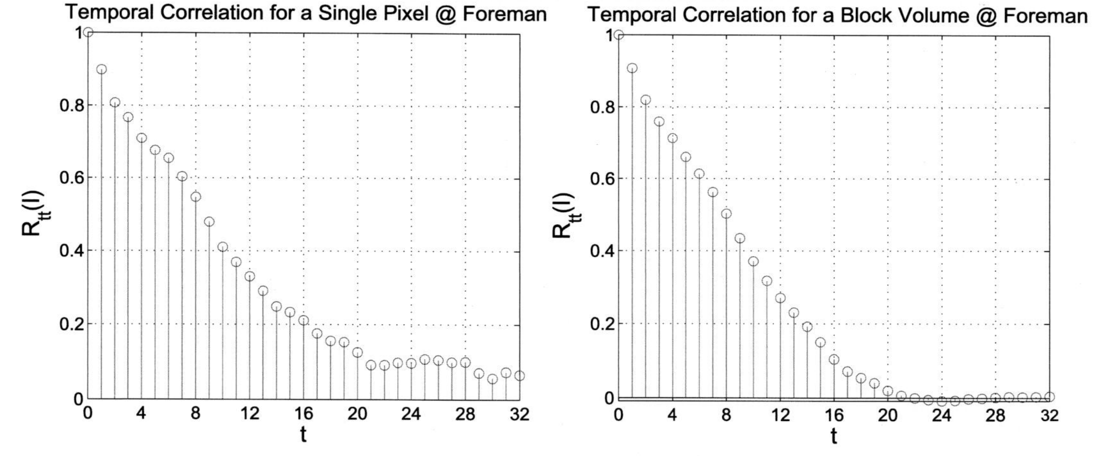

FIGURE 3.9 Temporal autocorrelation plots for Foreman (30 fps). Top to bottom right: Sample frame showing selected 16 × 16 block used; temporal correlation for a single pixel; temporal correlation for a 16 × 16 block.

3.3 Filtering and transforms

We assume here that the reader is familiar with basic linear system theory, filters, and transforms. For background reading, please refer to references such as [1,2,8,9].

3.3.1 Discrete-time linear systems

Many of the systems that we encounter in this book exhibit the properties of linearity and shift invariance.[2] Let us briefly describe these properties.

[2] Linear systems that are shift-invariant are often referred to as linear time-invariant (LTI). Although the independent variable is not always time, we will use this terminology here for convenience.

Shift invariance

A shift-invariant system is one where a shift in the independent variable of the input signal causes a corresponding shift in the output signal. So if the response of a system to an input $x_0[n]$ is $y_0[n]$, then the response to an input $x_0[n-n_0]$ is $y_0[n-n_0]$.

Linearity

If the response of a system to an input $x_0[n]$ is $y_0[n]$ and the response to an input $x_1[n]$ is $y_1[n]$, then a system is referred to as linear if:

1. The response to $x_0[n]+x_1[n]$ is $y_0[n]+y_1[n]$ (additivity).
2. The response to $ax_0[n]$ is $ay_0[n]$, where a is any complex constant (scaling).

A corollary of this is the principle of superposition, which states that if $x[n]=ax_0[n]+bx_1[n]$, then the output of a linear shift-invariant system will be $y[n]=ay_0[n]+by_1[n]$.

3.3.2 Convolution

If the response of an LTI system to a unit impulse, $\delta[n]$, is $h[n]$, then $h[n]$ is termed the *impulse response* of the system. The shift invariance, linearity, and sifting properties of LTI systems give

$$y[n]=\sum_{k=-\infty}^{\infty} x[k]h[n-k] \tag{3.18}$$

or

$$y[n]=x[n]\star h[n].$$

This is known as the convolution sum and can be used to determine the response of a discrete-time LTI system to an arbitrary input sequence.

Example 3.3. Convolution

Using the convolution Eq. (3.18), determine the response of a linear system with impulse response $h[n]=\{1,2,1\}$ to an input $x[n]=\{1,2,0,1,-1,2\}$.

Solution. The convolution sum embodies the principle of superposition. Hence we can form the overall system output from the sum of the responses to each input sample as follows:

n:	**0**	**1**	**2**	**3**	**4**	**5**	**6**	**7**
$y_0[n]$:	1	2	1					
$y_1[n]$:		2	4	2				
$y_2[n]$:			0	0	0			
$y_3[n]$:				1	2	1		
$y_4[n]$:					−1	−2	−1	
$y_5[n]$:						2	4	2
$y[n]$:	1	4	5	3	1	1	3	2

3.3.3 Linear filters

The transfer function of a discrete-time linear system is given by

$$H(z) = \frac{a_0 + a_1 z^{-1} + a_2 z^{-2} + \cdots + a_M z^{-M}}{b_0 + b_1 z^{-1} + b_2 z^{-2} + \cdots + b_N z^{-N}}. \tag{3.19}$$

Since $x[n-k] \stackrel{Z}{\longleftrightarrow} z^{-k}X(z)$, the equivalent difference equation is (with b_0 scaled to unity)

$$\begin{aligned} y[n] = {} & a_0 x[n] + a_1 x[n-1] + a_2 x[n-2] + \cdots + a_M x[n-M] \\ & - b_1 y[n-1] - b_2 y[n-2] - \cdots - b_N y[n-N] \end{aligned} \tag{3.20}$$

or

$$y[n] = \sum_{k=0}^{M} a_k x[n-k] - \sum_{k=1}^{N} b_k y[n-k]. \tag{3.21}$$

In cases where $\{b_k\} = 0$, the filter is described as having a *finite impulse response* (FIR). In other cases it is known as *infinite impulse response* (IIR). FIR filters are much more common in image and video processing than their IIR counterparts. The reason for this is primarily due to their phase response. We saw in Chapter 2 that phase distortion has an important influence on perceptual quality, and it is straightforward to design FIR filters with a linear phase characteristic. Such a characteristic only introduces a simple shift in the filter response with no phase distortion. We will provide some examples of practical FIR filters that are used in video compression a little later, but first let us examine the filter frequency response in a little more detail.

Extension to 2D

The 2D convolution of a signal $x[m,n]$ with an impulse response $h[m,n]$ is given by

$$y[m,n] = \sum_{k=-\infty}^{\infty} \sum_{l=-\infty}^{\infty} x[m-k, n-l]h[k,l] = x[m,n] \star h[m,n]. \tag{3.22}$$

2D digital filters are used extensively in image and video coding, for example, combined with sample rate changing, in the analysis or synthesis stages of filter-banks (Chapter 6), in interpolation filters for subpixel motion estimation (Chapters 8 and 9), or in more general pre- and postprocessing operations. However, for reasons of flexibility and complexity these are, wherever possible, implemented separably; i.e., as a cascade of 1D filtering stages.

Separability

Separability is an important property of digital filters and transforms that is exploited extensively in image and video processing. A system is separable when (for real-valued coefficients)

$$\mathbf{h} = \mathbf{h}_1 \mathbf{h}_2^{\mathrm{T}}, \tag{3.23}$$

where $h_1[m]$ is a 1D impulse response that operates across the rows of the signal and $h_2[n]$ is the corresponding impulse response that operates down the columns. We can now rewrite Eq. (3.22) as follows:

$$y[m,n] = \sum_{k=-\infty}^{\infty} h_1[k] \sum_{l=-\infty}^{\infty} x[m-k, n-1] h_2[l]. \tag{3.24}$$

If we define

$$y_2[m,n] = \sum_{l=-\infty}^{\infty} x[m, n-1] h_2[l],$$

then we can rewrite Eq. (3.24) as

$$y[m,n] = \sum_{k=-\infty}^{\infty} y_2[m-k, n] h_1[k]. \tag{3.25}$$

So this implies that if Eq. (3.23) holds, then we can achieve 2D filtering by first filtering the columns of the 2D signal with $h_2[n]$, followed by filtering the rows of the output from the first stage using $h_1[m]$. Due to the fact that we are cascading linear operations, these two stages can be interchanged if necessary.

It should be noted that not all filters are separable and that separability does impose some orientation constraints on the filter characteristic. In general, a filter is separable if all its rows and all its columns are linearly dependent. This property can be observed in the solution of Example 3.4.

Example 3.4. Filter separability

Consider the following two filter impulse responses:

$$\begin{aligned} \mathbf{h}_1 &= \begin{bmatrix} 0.5 & 0.3 & 0.2 \end{bmatrix}^{\mathrm{T}}, \\ \mathbf{h}_2 &= \begin{bmatrix} 0.6 & 0.1 & 0.3 \end{bmatrix}^{\mathrm{T}}. \end{aligned}$$

Form the equivalent 2D digital filter.

Solution. The 2D filter is given by

$$\mathbf{h} = \mathbf{h}_1 \mathbf{h}_2^{\mathrm{T}} = \begin{bmatrix} 0.30 & 0.05 & 0.15 \\ 0.18 & 0.03 & 0.09 \\ 0.12 & 0.02 & 0.06 \end{bmatrix}.$$

As a further exercise, satisfy yourself that the 2D filtering operation is identical to the use of separable filters.

3.3.4 Filter frequency response

From our understanding of convolution and digital filtering, we know that the response of a discrete linear system to a sampled sinusoidal input signal will be a number of weighted and shifted versions of the input signal. The output is thus also a sinusoidal signal at the same frequency, but with different amplitude and phase. We can therefore think of our filter in the frequency domain as a frequency-dependent modifier of amplitude and phase. If we generalize this with an input signal that is a complex exponential,

$$x[n] = e^{j\Omega n},$$

then from the convolution Eq. (3.18) we have

$$y[n] = \sum_k h[k]e^{j\Omega(n-k)} = e^{j\Omega n} \sum_k h[k]e^{-j\Omega k}.$$

Thus,

$$y[n] = x[n]H(\Omega),$$

where

$$H(\Omega) = \sum_{k=-\infty}^{\infty} h[k]e^{-j\Omega k}. \tag{3.26}$$

The term $H(\Omega)$ describes the change in amplitude and phase experienced by the complex exponential as it passes through the system and it is referred to as the *frequency response* of the system.

3.3.5 Examples of practical filters

LeGall wavelet analysis filters

A common set of separable filters that are employed to decompose a signal into low- and high-pass subbands prior to subsampling and quantization are the 5/3 LeGall filters. The impulse responses of the low- and high-pass filters are

$$\mathbf{h}_0 = [-1, 2, 6, 2, -1]/4,$$
$$\mathbf{h}_1 = [1, -2, 1]/4.$$

The frequency responses of these filters are shown in Fig. 3.10. It is interesting to note that the low-pass and high-pass filters are not symmetrical and that there appears to be significant spectral leakage between them. However, as we will see in Chapter 6, these filters do perform well and offer the property of perfect reconstruction when combined in a filter-bank.

Subpixel interpolation filters

A second important filtering operation is interpolation, for example where a region of a picture is upsampled prior to performing subpixel motion estimation (see Chapters 8 and 9).

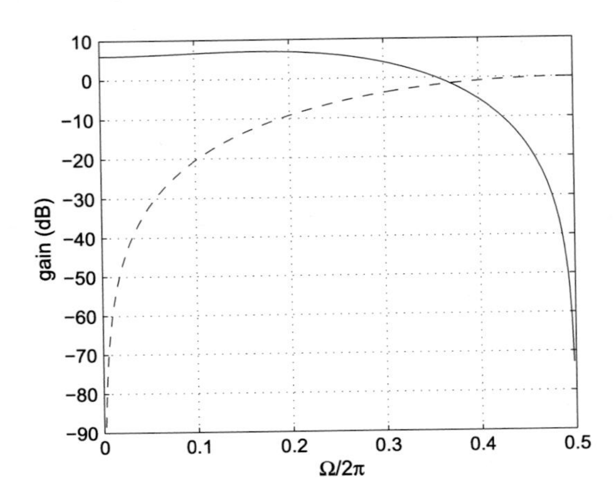

FIGURE 3.10 Filter-bank responses for the LeGall low-pass and high-pass analysis filters.

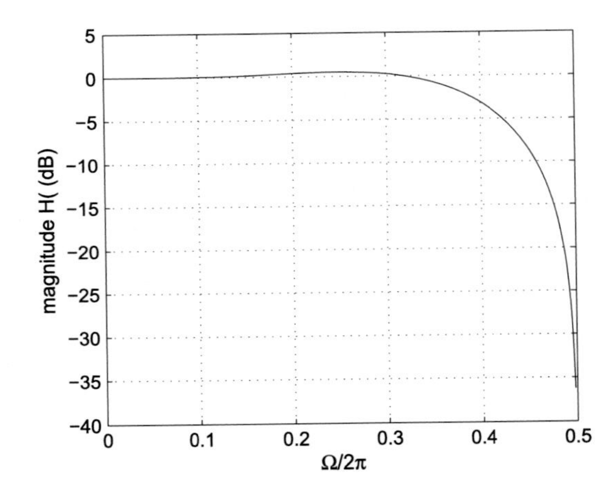

FIGURE 3.11 Filter response for the H.264 half-pixel interpolation filter.

For example, in the coding standard H.264/AVC, a 6-tap FIR filter is employed:

$$\mathbf{h} = [1, -5, 20, 20, -5, 1]/32. \tag{3.27}$$

The frequency response of this filter is shown in Fig. 3.11. It can be observed that the characteristic is not dissimilar to that of the LeGall low-pass analysis filter. However, in this case, the filter has no effect on the original integer-pixel resolution samples; it is simply used to generate the interpolated values. The filter characteristic is therefore a trade-off between obtaining a high cut-off frequency with rapid roll-off and low-pass band ripple. Such trade-offs are common in filter design [1].

3.3.6 Nonlinear filters

Nonlinear filters are not based on linear relationships between an input and an output via a system function. Instead they represent a much broader class of operations that do not have any explicit frequency domain transfer function. Nonlinear filters find application in a range of image processing and coding applications, such as denoising and edge preserving operations, and in some forms of prediction. The deblocking filters used in the HEVC, AVC, and VVC standards are content-adaptive nonlinear filters and these are described in Chapter 9.

A common class of filter that is used in some motion vector prediction applications is the rank order filter, in particular the median filter. We introduce these briefly below.

Rank order and median filters

A rank order filter is an operator that ranks the input samples and selects an output according to its position in the ranked list. The following are examples of rank order filters:

$$y[m,n] = \min_{\mathcal{N}(m,n)} (x(i,j)), \tag{3.28}$$

$$y[m,n] = \max_{\mathcal{N}(m,n)} (x(i,j)), \tag{3.29}$$

where $\mathcal{N}$ represents the neighborhood, related to location (m,n), over which the operation is applied. The min operator will attenuate isolated peaks or ridges in an image, whereas the max operator will fill isolated troughs or holes. It can be seen that repeated application of this type of filter will have a continued effect of eroding the isolated regions described.

The median operator will rank all the N samples within the region and output the ranked value in position[3] $(N+1)/2$; thus,

$$y[m,n] = \operatorname{med}_{\mathcal{N}(m,n)} (x(i,j)). \tag{3.30}$$

The median operator is better than a linear filter in removing outliers (e.g., salt and pepper noise). It will preserve straight edges, but will round corners and distort texture features. The extent of feature flattening will depend on the window size used.

We can of course combine these operations to produce nonlinear difference filters such as the following:

$$y[m,n] = \min_{\mathcal{N}(m,n)} (x(i,j)) - \max_{\mathcal{N}(m,n)} (x(i,j)), \tag{3.31}$$

which produces a nonlinear gradient of the image.

Morphological filters

The erosion and dilation operations referred to above can be generalized and combined to form a powerful toolbox of operators known as morphological filters. These shape-based operators support region erosion and dilation and geodesic reconstruction and have found

[3] Median filters are normally defined with an odd number of taps. In cases where the number of samples is even, it is conventional to take the average of the middle two terms.

applications in segmentation [6], contrast enhancement, and feature prioritization for low-bit rate video coding [7].

3.3.7 Linear transforms and the DFT

Linear transforms, in particular the *discrete cosine transform* (DCT), form the basis of most image intraframe and residual coding and are covered in detail in Chapter 5. We provide an introduction to these here, focusing on the *discrete Fourier transform* (DFT). Fourier transforms are used extensively across signal processing applications to perform spectral analysis based on discrete representations in both the signal domain and its associated frequency domain. An excellent introduction to Fourier transforms is provided in [8].

While the discrete-time Fourier series and the discrete-time Fourier transform provide a basis for analyzing periodic and aperiodic signals, respectively, the latter produces a continuous frequency domain function so it is not well suited for digital implementation. The ubiquitous DFT overcomes this and enables discrete-time processing in both the signal and its associated frequency domain.

The discrete Fourier transform

Let us consider the 1D version of the DFT initially. If $x[n]$ is a finite-duration signal defined over the range $0 \le n \le N_1 - 1$, a periodic signal $\tilde{x}[n]$ can be constructed which equals $x[n]$ over one cycle. Let $N \ge N_1$ and let $\tilde{x}[n]$ be periodic in N. We compute the discrete-time Fourier series of this new sequence:

$$c_k = \frac{1}{N} \sum_{n=\langle N \rangle} \tilde{x}[n] e^{-jk(2\pi/N)n}.$$

Letting $X(k) = N \cdot c_k$, this becomes the analysis equation of the DFT:

$$X(k) = \sum_{n=0}^{N-1} x[n] e^{-jk(2\pi/N)n}, \quad k = 0, 1, 2 \ldots N-1. \tag{3.32}$$

The inverse DFT is given by

$$x[n] = \frac{1}{N} \sum_{k=0}^{N-1} X(k) e^{jk(2\pi/N)n}, \quad n = 0, 1, 2 \ldots N-1. \tag{3.33}$$

The N terms of the DFT represent samples of the continuous function $X(\Omega)$, equally spaced over a 2π interval. These can be easily shown to equal the samples of the z transform on the unit circle in the z-plane.

Normally we use the DFT in matrix vector form:

$$\mathbf{X} = \mathbf{W}\mathbf{x}, \tag{3.34}$$

where

$$W_N = e^{-j(2\pi/N)}.$$

Hence we have in matrix form

$$\begin{bmatrix} X(0) \\ X(1) \\ X(2) \\ \vdots \\ X(N-1) \end{bmatrix} = \begin{bmatrix} W^0 & W^0 & W^0 & \dots & W^0 \\ W^0 & W^1 & W^2 & \dots & W^{N-1} \\ W^0 & W^2 & W^4 & & W^{2(N-1)} \\ \vdots & & & & \vdots \\ W^0 & W^{N-1} & W^{2(N-1)} & \dots & W^{(N-1)(N-1)} \end{bmatrix} \begin{bmatrix} x[0] \\ x[1] \\ x[2] \\ \vdots \\ x[N-1] \end{bmatrix}. \tag{3.35}$$

Similarly the inverse DFT is given by

$$\mathbf{x} = \frac{1}{N}\mathbf{W}^*\mathbf{X}. \tag{3.36}$$

The 2D DFT

In a similar manner to that for filters in Section 3.3.3, transform operations can be separable and this is the conventional way of applying 2D transforms, largely because of the associated complexity savings. We cover this in more detail in Chapter 5.

The DFT and compression

It may seem logical to attempt to exploit the ability of the DFT to analyze a signal in terms of its frequency domain components. There are however a number of significant issues with this, which mean that this is not done in practice. These are:

1. Although images are invariably real-valued, the DFT coefficients are complex-valued so we are doubling the amount of information.
2. For short sequences or small 2D regions, like those typical in compression, the discontinuities produced by the underlying assumptions of periodicity can have a major impact on performance as they produce ringing or spectral leakage in the frequency domain.
3. Ringing is normally addressed in spectral analysis applications by applying a nonrectangular window function (such as a Hamming window) to the input data prior to extension. These reduce spectral ringing but smear sharp frequency domain edges, distorting the characteristics of the underlying signal. Again, this is exactly what we do not want in a compression system!

Comparisons between the DFT and more appropriate transforms such as the DCT are made in Chapter 5.

3.4 Quantization

3.4.1 Basic theory of quantization

A quantizer maps an input value (or values in the context of a vector quantizer) to an output value. It can quantize either a continuous-amplitude discrete-time signal and produce a quantized digital signal (with finite word length) or a previously digitized sample and quantize it further to reduce its dynamic range. The former is common in acquisition devices such as A-to-D converters and the latter in compression systems where, for example, the coefficients of a transformed input signal might be mapped to a different range, with small values going to zero.

Once an image or video signal has been sampled to produce a discrete set of values with a given word length, the compression process normally involves some form of signal analysis (typically transformation) to decorrelate the content, followed by quantization to approximate the original signal in a way that retains as much perceptual quality as possible. Ignoring numerical imprecision, quantization is the stage in any compression process where loss is introduced. It is thus crucial to understand the impact of quantization on the perceptual quality of an image or video signal.

Uniform quantization

Uniform quantizers adopt a constant step size Δ between reconstruction levels. If the input to the quantizer is x, this will map to a reconstruction value, y_i, where y_i represents the center of the range as dictated by the step size parameter. We define a quantization error, $q[n] = x[n] - y[n]$, which is model-led as a zero-mean, uniformly distributed signal within the range $[-\Delta/2, \Delta/2]$. This signal has energy or variance given by

$$\sigma_q^2 = \frac{1}{\Delta} \int_{-\Delta/2}^{\Delta/2} q^2 dq = \frac{\Delta^2}{12}. \tag{3.37}$$

Two common uniform quantizer profiles are shown in Fig. 3.12. On the left is the conventional midtread quantizer for positive-valued signals, and on the right is a similar profile, but for two-sided input signals. Decision levels for x are indicated along with the corresponding reconstruction values for y.

For the case of the conventional midtread quantizer, the quantization operation on an input x, to produce a reconstruction level y_i, in terms of the decision levels x_i, is defined as

$$\dot{x} = Q(x) = y_i, \quad x_{i-1} < x \leq x_i. \tag{3.38}$$

This is typically achieved using a rounding operation, for example,

$$\dot{x} = \text{NINT}\left(\frac{x}{\Delta}\right), \tag{3.39}$$

where NINT is the operation of rounding to the nearest integer.

In order to reconstruct a quantized signal, it will normally have to be rescaled to its original range. This rescaling operation is defined as

$$\tilde{x} = \Delta \dot{x}. \tag{3.40}$$

3.4.2 Adaptation to signal statistics

Quantizers with non-uniform step sizes can be beneficial in cases when:

- the signal being quantized does not exhibit a uniform pdf;

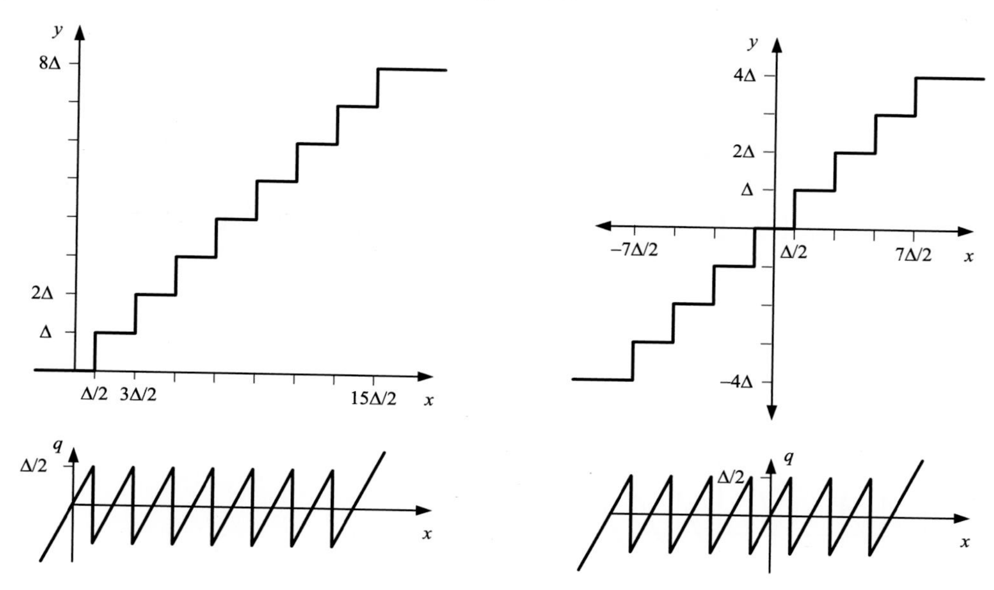

FIGURE 3.12 Common uniform quantizer characteristics.

- low-level signals or other signal ranges are known to contain excess noise;
- there is a desire to map a range of low-level signals to zero;
- there is a need to incorporate a nonlinear amplitude mapping as part of the quantization process.

Deadzone quantizer

The most common form of nonuniform quantizer is the *deadzone* quantizer. This has a broader decision range for the band of inputs close to zero. It has the benefit during compression of ensuring that noisy low-level signals are not allocated bits unnecessarily. The characteristic of a deadzone quantizer is shown in Fig. 3.13 (left).

Lloyd Max quantizer

The Lloyd Max algorithm [10] is a well-known approach to designing non-uniform quantizers optimized according to the prevailing pdf of the input signal. Rather than allocating a uniform step size, as would be optimal for a uniform pdf, the Lloyd Max approach identifies decision boundaries according to the mean values of equal area partitions of the pdf curve. The characteristic of a Lloyd Max quantizer is shown in Fig. 3.13 (right).

3.4.3 HVS weighting

In practice for many compression applications, the quantization step size is weighted according to some perceptual criterion, such as the spatial frequency of a coefficient produced by a decorrelating transform operation. This is considered in more detail in Chapter 5.

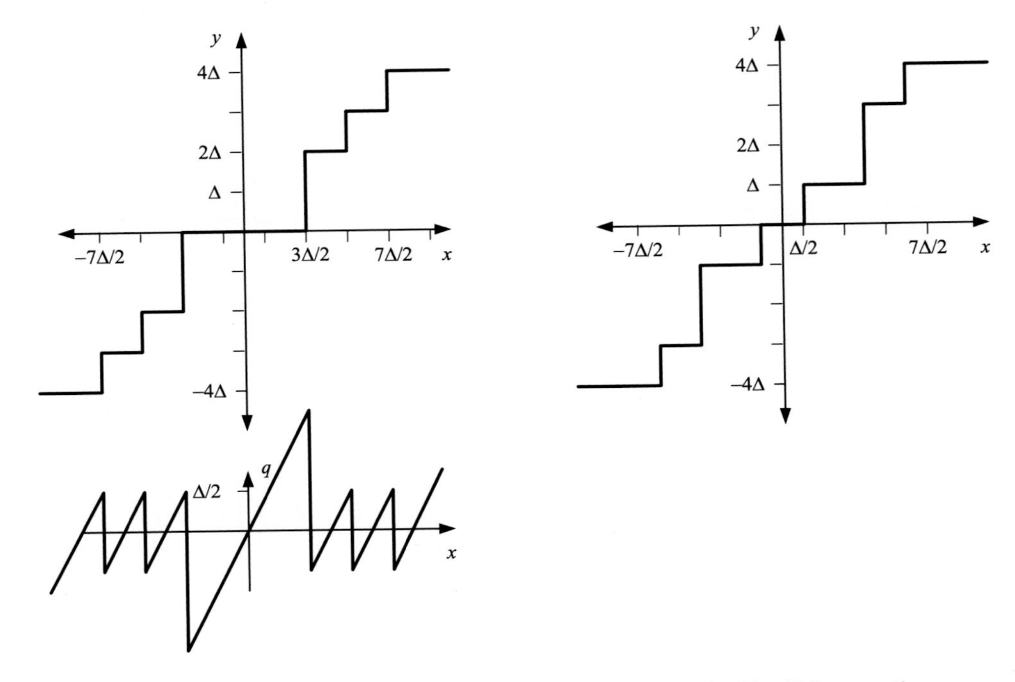

FIGURE 3.13 Common nonuniform quantizers. Left: Center deadzone. Right: Lloyd Max quantizer.

3.4.4 Vector quantization

Vector quantization is a method which maps a group of samples to a single quantization index. It is a method that has been adopted in highly robust codecs such as those based on pyramid vector quantization, as discussed in Chapter 11. We will not cover this in detail here but the interested reader is referred to [9,11].

3.5 Linear prediction

3.5.1 Basic feedforward linear predictive coding

The idea of linear prediction is to exploit statistical relationships in random data sequences in order to decorrelate them prior to coding.

A feedforward analysis filter for linear prediction is shown in Fig. 3.14 (top) and the corresponding synthesis filter is shown in Fig. 3.14 (bottom). Ignoring the quantized block for the time being, the transfer function of the analysis filter is given by

$$\begin{aligned} E(z) &= X(z) - X_p(z) \\ &= X(z)(1 - P(z)), \end{aligned} \tag{3.41}$$

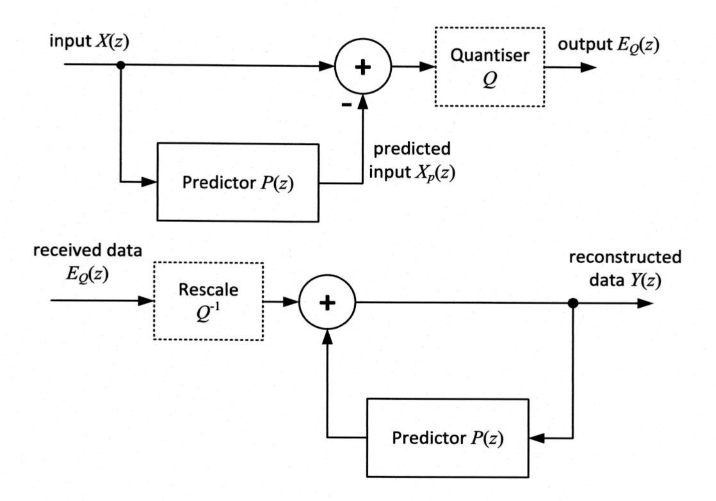

FIGURE 3.14 Feedforward linear prediction. Top: Encoder. Bottom: Decoder.

$$H(z) = \frac{E(z)}{X(z)} = 1 - P(z), \tag{3.42}$$

where the notation used is as shown in the figure. Assuming the predictor block in Fig. 3.14 is an FIR filter, then in the sample domain we have

$$e[n] = x[n] - \sum_{i=1}^{N} a_i x[n-i]. \tag{3.43}$$

If we now consider the decoder or synthesis filter,

$$Y(z) = E(z) + Y(z)P(z) \tag{3.44}$$

or:

$$Y(z) = \frac{E(z)}{1 - P(z)} = X(z). \tag{3.45}$$

Example 3.5. Linear predictive coding
Consider the case in Fig. 3.14 when $P(z) = z^{-1}$ and the input to the predictive coder is $x[n] = \{1, 2, 3, 2, 5, 4, 2, 4, 5, 6\}$. Assuming that no quantization takes place, confirm that the decoder output will be identical to the input sequence.

Solution. We have $e[n] = x[n] - x[n-1]$ and $y[n] = e[n] + y[n-1]$. So, assuming zero initial conditions and working through for each input sample,

$x[n]$	1	2	3	2	5	4	2	5	6
$e[n]$	1	1	1	−1	3	−1	−2	3	1
$y[n]$	1	2	3	2	5	4	2	5	6

Hence $y[n]=x[n]$.

Predictor dynamic range

The benefit of prediction filtering as a basis for coding is that although the dynamic range of the transmitted error signal can be double that of the original signal, the variance of the error signal is significantly reduced compared to the original. This is demonstrated in Fig. 3.15, where a sinusoidal signal with additive Gaussian noise ($\sigma_v^2 = 0.05$),

$$x[n] = \cos(0.01\pi n) + v[n], \tag{3.46}$$

is filtered using a predictor $P(z) = z^{-1}$.

Linear predictive coding with quantization

The problem with the feedforward architecture arises when quantization is introduced in the process. In this case we have

$$E_Q(z) = Q\left\{X(z) - X_p(z)\right\} \tag{3.47}$$

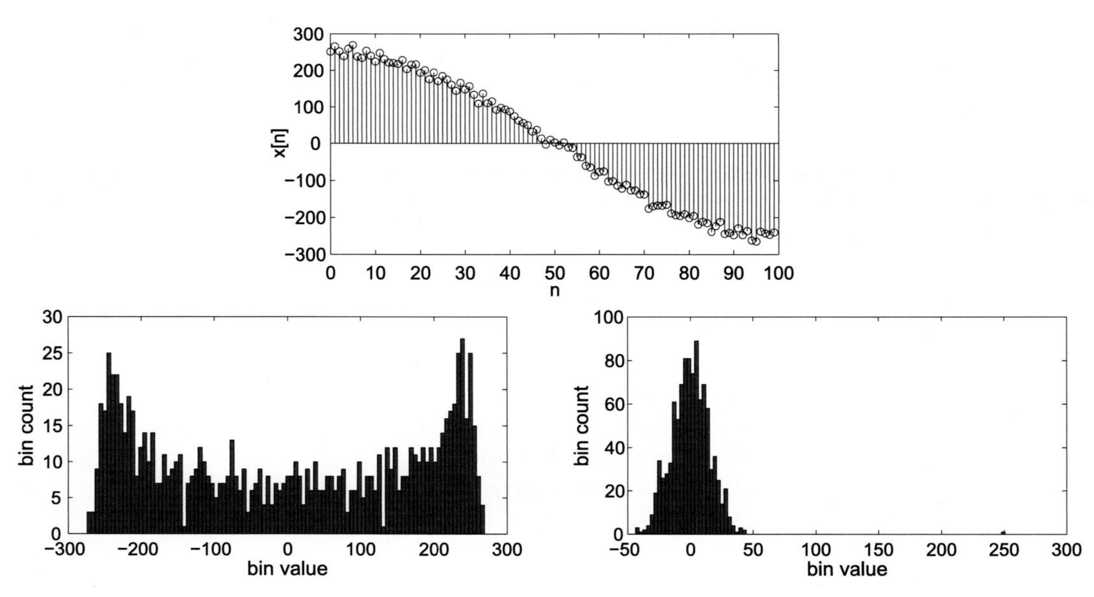

FIGURE 3.15 Prediction signal dynamic range. Top: Input signal. Bottom left: Distribution of 1000 samples of input signal. Bottom right: Distribution of 1000 samples of the prediction residual.

and

$$Y(z) = Q^{-1}\{E_Q(z)\} + Y(z)P(z) \tag{3.48}$$

or

$$Y(z) = \frac{Q^{-1}\{E_Q(z)\}}{1 - P(z)} \neq X(z). \tag{3.49}$$

As one would expect, the quantization process introduces errors in the reconstructed output. This is to be expected since the operation is a lossy one. However, the situation is worse because there is the potential for drift between the encoder and the decoder. This is illustrated in Example 3.6.

Example 3.6. Linear predictive coding with quantization

Consider the case in Fig. 3.14, where, as before, $P(z) = z^{-1}$ and the input to the predictive coder is $x[n] = \{1, 2, 3, 2, 5, 4, 2, 5, 6\}$. Assuming that quantization takes place, where the quantization and rescaling operations take the form

$$e_Q[n] = \text{rnd}\left(\frac{e[n]}{2}\right), \qquad e_R[n] = 2e_Q[n]$$

with rounding of 0.5 values toward 0, characterize the decoder output relative to the input signal.

Solution. Assuming 0 initial conditions and following the same approach as in Example 3.5, but this time including the quantization and rescaling operations,

$x[n]$	**1**	**2**	**3**	**2**	**5**	**4**	**2**	**5**	**6**
$e[n]$	1	1	1	−1	3	−1	−2	3	1
$e_Q[n]$	0	0	0	0	1	0	−1	1	0
$e_R[n]$	0	0	0	0	2	0	−2	2	0
$y[n]$	0	0	0	0	2	2	0	2	2

Hence $y[n]$ is not the same as $x[n]$. Furthermore significant drift has occurred due to the accumulation of rounding errors at the decoder.

3.5.2 Linear prediction with the predictor in the feedback loop

In order to avoid problems with drift, we must replicate the effects of the decoding operation within the encoder. This ensures that, in the absence of data corruption during transmission, both the encoder and the decoder process exactly the same data and operate in unison. This modified architecture is shown in Fig. 3.16. The expression for our encoded output (without quantization) is now given by

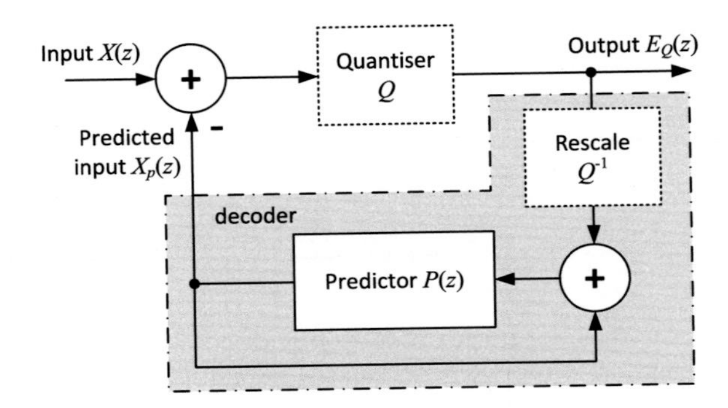

FIGURE 3.16 Feedback-based linear prediction.

$$E(z) = X(z) - X_p(z) \tag{3.50}$$
$$= X(z) - \frac{E(z)P(z)}{1 - P(z)}.$$

Therefore,

$$E(z) = X(z)(1 - P(z))$$

and, as before, at the decoder

$$Y(z) = E(z) + Y(z)P(z), \tag{3.51}$$

or

$$Y(z) = \frac{E(z)}{1 - P(z)} = X(z). \tag{3.52}$$

If we now include quantization in the process, the difference between the feedback and feedforward architectures will become clear. To ease analysis we model the quantization process as additive quantization noise, $V(z)$ with a variance of σ_N^2. Then we have, as illustrated in Fig. 3.17,

$$E(z) = X(z) - X_p(z) \tag{3.53}$$
$$= X(z) - \frac{[E(z) + V(z)]P(z)}{1 - P(z)}.$$

Therefore,

$$E(z) = X(z)(1 - P(z)) - V(z)P(z) \tag{3.54}$$

and at the decoder

$$Y(z) = E(z) + Y(z)P(z) + V(z), \tag{3.55}$$

or

$$Y(z) = \frac{E(z) + V(z)}{1 - P(z)}. \tag{3.56}$$

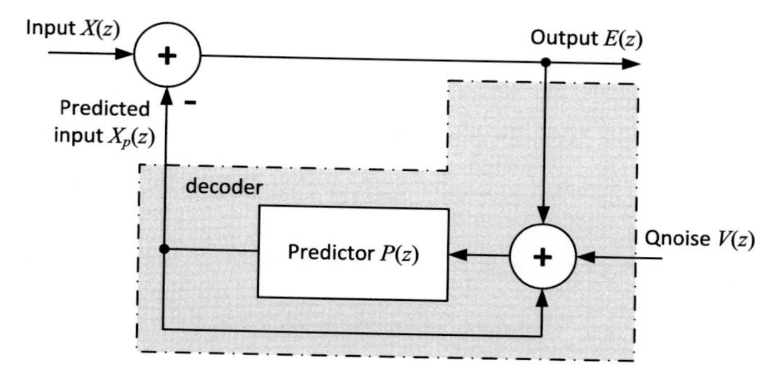

FIGURE 3.17 Feedback-based linear predictor with quantization noise modeling.

Substituting Eq. (3.54) for $E(z)$ in Eq. (3.56), it is straightforward to see that

$$Y(z) = X(z) + V(z). \tag{3.57}$$

So we no longer have any drift between encoder and decoder; we simply have a mismatch due to the additive noise of the quantization process. This is a well-known but very important result as this prediction architecture is used extensively in video compression.

Example 3.7. Feedback-based linear predictive coding with quantization

Consider the case in Fig. 3.16, where, as before, $P(z) = z^{-1}$ and the input to the predictive coder is $x[n] = \{1, 2, 3, 2, 5, 4, 2, 5, 6\}$. Assume that quantization takes place, where the quantization and rescaling operations take the form

$$e_Q[n] = \text{rnd}\left(\frac{e[n]}{2}\right), \qquad e_R[n] = 2e_Q[n]$$

and where 0.5 values are rounded off toward 0. Compute the decoder output and compare its characteristics with those of the input signal.

Solution. Assuming 0 initial conditions and following the same approach as in Example 3.6, but this time based on the architecture in Fig. 3.16,

$x[n]$	1	2	3	2	5	4	2	5	6
$e[n]$	1	2	1	0	3	0	−2	3	2
$e_Q[n]$	0	1	0	0	1	0	−1	1	1
$e_R[n]$	0	2	0	0	2	0	−2	2	2
$y[n]$	0	2	2	2	4	4	2	4	6

As expected, $y[n]$ is not identical to $x[n]$ because of the lossy quantization process. However, the variance of the predicted signal is significantly reduced, compared to the original, and also there is no drift between encoder and decoder.

3.5.3 Wiener Hopf equations and the Wiener filter

If we have a linear FIR predictor filter with impulse response, **h**, we wish the filter coefficients to be assigned values that force the error signal after prediction to be as small as possible. If we perform this optimization in terms of second order statistics (i.e., based on mean squared error [MSE]), then the result is referred to as a Wiener filter – an optimal nonrecursive estimator.

If we have a desired sequence (training sequence) **d**, then

$$e[n] = d[n] - y[n] \tag{3.58}$$

and we can define a cost function as

$$J = E\left\{|e[n]|^2\right\}, \tag{3.59}$$

where, when the gradient $\nabla(J) = 0$ with respect to all filter coefficients, the filter is said to be optimal in the MSE sense. It can be shown that a necessary and sufficient condition for J to attain a minimum is that the corresponding value of the estimation error is orthogonal to each sample that enters the estimation process at time n. This is known as the *principle of orthogonality*. Based on this observation, it is straightforward to show that the optimal filter coefficients can be computed from the autocorrelation of the filter input sequence, $r[k]$, and the cross-correlation of the filter input and the desired response $p[k]$ for a lag of k samples. This is known as the Wiener–Hopf equation:

$$\sum_{i=0}^{\infty} h_o[i]r[i-k] = p[k], \tag{3.60}$$

or in matrix vector form,

$$\mathbf{h}_o = \mathbf{R}^{-1}\mathbf{p}. \tag{3.61}$$

It is also possible to show, assuming stationarity, that the error surface is exactly a second order function of the filter coefficients. It will thus possess a unique minimum point corresponding to the optimal filter weights. For further details, the reader is referred to reference [12].

3.6 Information and entropy

Entropy quantifies the value of the information contained in a message – it is a measure of the uncertainty in a random variable. The concept of generating a quantitative measure of information was originated by Shannon in 1948 [13], building on the work of Nyquist (1924) and Hartley (1928).

Understanding the basics of information theory is important in the context of image and video compression, since it is used extensively for lossless symbol encoding after transformation and quantization. Typically, quantized transform coefficients are scanned and run-length encoded. Symbols representing runs of zeros and values of nonzero coefficients are coded using variable-length codewords, in a manner that reflects their information content. In this way a lossless reduction in bit rate can be achieved.

Before we look at symbol statistics, let us first review the basics of information theory. For a more in depth coverage, please refer to [14].

3.6.1 Self information

Information is used to reduce the uncertainty about an event or signal. Shannon [13] defined the *self-information* of an event A, which could for example comprise a set of outcomes from an experiment, as

$$i(A) = \log_2\left(\frac{1}{P(A)}\right) = -\log_2(P(A)). \tag{3.62}$$

The use of logarithms is intuitive, since $\log(1) = 0$ and $-\log(x)$ increases as x decreases from 1 to 0. Hence, if the probability of an event is low, then the information associated with it is high. For example, if $P(A) = 0.5$, then $-\log_2(0.5) = 1$, and if $P(A) = 0.25$, then $-\log_2(0.25) = 2$. Other logarithm bases can be used, but base 2 is the most common as it conveniently provides a measure of information in bits.

Independent events

The information obtained from two independent events is the sum of the information from each event. This can be seen as follows:

$$i(AB) = \log_2\left(\frac{1}{P(AB)}\right),$$

and if A and B are independent, then

$$P(AB) = P(A)P(B). \tag{3.63}$$

Therefore,

$$i(AB) = \log_2\left(\frac{1}{P(A)P(B)}\right) = \log_2\left(\frac{1}{P(A)}\right) + \log_2\left(\frac{1}{P(B)}\right),$$

or

$$i(AB) = i(A) + i(B). \tag{3.64}$$

3.6.2 Entropy

If we define a source that generates a series of random and independent events with outcomes from an alphabet $S = \{s_1, s_2, s_3, \ldots, s_N\}$, then the average self-information associated with an event is

$$H = \sum_{i=1}^{N} P(s_i)i(s_i) = -\sum_{i=1}^{N} P(s_i)\log_2(P(s_i)).$$

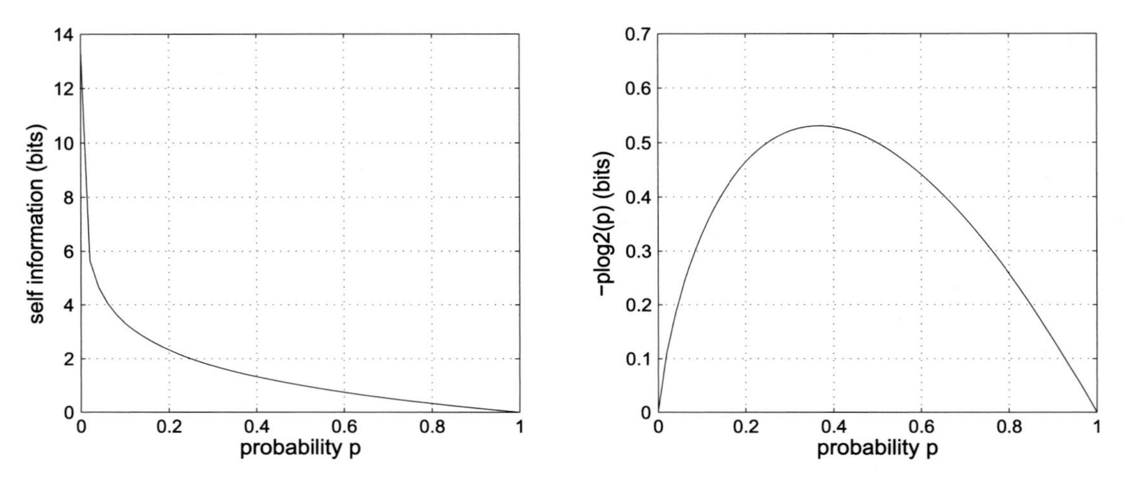

FIGURE 3.18 Self-information and probability. Left: Plot of self-information vs. probability for a single event. Right: Plot of the self-information of an event weighted by its probability.

The quantity H, the average self-information, is referred to as the entropy of the source and this tells us how many bits per symbol are required to code that source. Shannon demonstrated that no coding scheme can code the source losslessly at a rate lower than its entropy.

Entropy and first order entropy

The entropy of a source is given by

$$H(S) = \lim_{N\to\infty} \frac{1}{N} G_N, \tag{3.65}$$

where

$$G_N = -\sum_{i_1=1}^{N}\sum_{i_2=1}^{N}\cdots\sum_{i_M=1}^{N} P(d_1 = s_{i_1}, d_2 = s_{i_2}, \cdots, d_M = s_{i_M}) \log P\left(d_1 = s_{i_1}, d_2 = s_{i_2}, \cdots, d_M = s_{i_M}\right),$$

$D = \{d_1, d_2, d_3, \ldots, d_M\}$, and M is the length of the sequence. However, if each symbol is independent and identically distributed (i.i.d.), then it can be shown that

$$G_N = -M \sum_{i_1=1}^{N} P(d_1 = s_{i_1}) \log\left(P(d_1 = s_{i_1})\right), \tag{3.66}$$

and the entropy is then

$$H(S) = -\sum_{i=1}^{N} P(s_i) \log_2 (P(s_i)). \tag{3.67}$$

In most realistic cases Eqs. (3.65) and (3.67) do not give the same result. Hence we differentiate them by referring to Eq. (3.67) as the first order entropy.

The relationship between symbol probability and the self-information of that symbol is given in Fig. 3.18. The figure also illustrates the relationship between the probability of a symbol and the contribution that symbol makes to the overall entropy of the source. Fig. 3.18 indicates that symbols with probabilities at either end of the scale contribute little to the source entropy while those in the middle of the range contribute values around 0.5.

Example 3.8. Entropy

Consider an alphabet $S = \{s_1, s_2, s_3, s_4, s_5, s_6, s_7, s_8\}$. If the corresponding probabilities of occurrence of these symbols during transmission are $P(S) = \{0.06, 0.23, 0.3, 0.15, 0.08, 0.06, 0.06, 0.06\}$, calculate the first order entropy for the source.

Solution. First order entropy is given by

$$
\begin{aligned}
H &= -\sum_{S} P(s_i) \log_2 (P(s_i)) \\
&= -(0.06 \times \log_2 0.06 + 0.23 \times \log_2 0.23 + \ldots + 0.06 \times \log_2 0.06) \\
&= 2.6849 \text{ bits/symbol.}
\end{aligned}
$$

3.6.3 Symbols and statistics

In practice, since Eq. (3.65) is impossible to compute, first order entropy is almost always used instead. We therefore have to estimate the entropy of the source and this depends heavily on the statistics of the source symbols. This point is emphasized in Example 3.9.

Example 3.9. Source statistics and entropy

Consider the following sequence of symbols: $D = \{1, 2, 3, 2, 1, 2, 3, 4, 5, 6, 5, 6, 7, 8, 9, 10\}$. Compute the first order entropy for this sequence.

Solution. This seems like a straightforward question. We are given no information about the statistics of the source, so let us estimate the probabilities of each symbol, based on relative frequencies of occurrence. In this way we have

$$
\begin{aligned}
&P(4) = P(7) = P(8) = P(9) = P(10) = \frac{1}{16}, \\
&P(1) = P(3) = P(5) = P(6) = \frac{2}{16}, \\
&P(2) = \frac{3}{16}.
\end{aligned}
$$

Thus the first order entropy is

$$H = -\sum_{S} P(s_i)\log_2(P(s_i)) = 3.2028 \text{ bits.}$$

So this says that on average, we need just over 3.2 bits per symbol for this source.

However, if we now look closer at the sequence we can observe that the symbols are correlated. So what happens if we process this sequence with a simple first order linear predictor? Our residual sequence is E={1,1,1,−1,−1,1,1,1,1,1,−1,1,1,1,1,1}.

So now we have a new set of probability estimates:

$$P(1) = \frac{13}{16},$$

$$P(-1) = \frac{3}{16}.$$

In this case, we recalculate our entropy to be 0.6962 bits/symbol. So which is correct? In fact neither is right or wrong, it is just that in the latter case, we have better understood the structure of the source and provided a better estimate of its entropy.

For example, we can estimate the probabilities of occurrence of pixel values in an image, or a region of an image, through computing a histogram of its quantization levels and dividing the content of each quantization bin by the total number of pixels. Alternatively we can precompute them using a histogram of one or more similar training images. The latter is more common, although we rarely use actual pixel values as the input symbols in practice. More often we use some more efficient means based on prediction or run-length coding of quantized spatial or transform domain data. More of this later in Chapter 7.

3.7 Machine learning

3.7.1 An overview of AI and machine learning

Over the past decade, one of the most rapidly advancing scientific techniques likely to have a major impact on image and video compression has been artificial intelligence (AI), in particular ML methods based on deep neural networks (DNNs). With recent developments in high-performance computing and increased data storage capacities, these data-driven AI technologies have been empowered and are increasingly being adopted across various applications, ranging from intelligent assistants, marketing, and finance to command and control operations. Their success in solving image understanding and search tasks has been particularly relevant to consumer media applications, for example supporting search and retrieval operations and recommendation services. Although less extensively researched, it is evident that these methods are also highly relevant to image and video compression, in terms of both optimizing individual coding tools and perceptual quality assessment, but also potentially as a new end-to-end coding paradigm. In this section we introduce the basic concepts that underpin recent deep learning (DL) advances. This is a diverse, complex, and rapidly developing field with a huge literature base, so readers are referred to texts such as [15,16] and reviews [17,18] for further reading.

3.7.2 Neural networks and error backpropagation

AI embodies sets of codes, techniques, algorithms, and data that enable a computer system to make decisions similar to (or in some cases, better than) humans. When a machine exhibits broad human-like intelligence, this is often referred to as "general AI" or "strong AI." However, currently reported technologies are normally restricted to operation in a limited domain to work on specific tasks. This is referred to as "narrow AI" or "weak AI."

The main class of ML algorithms in use today are data-driven. These comprise a large number of highly interconnected but simple processing elements capable of "learning" directly from large amounts of example data without reliance on a predetermined equation or model. These algorithms adaptively converge to an optimal solution and generally improve their performance as the number of training samples and the number of training iterations increase. Several types of learning algorithm exist, including supervised learning, unsupervised learning, and reinforcement learning. Supervised learning algorithms build an internal model from a set of data that contains both the inputs and the desired outputs (each output usually representing a classification of the associated input vector), whilst unsupervised learning algorithms model the problems on unlabeled data. Reinforcement learning methods learn from trial and error and are effectively self-supervised. When these algorithms have converged, they are capable of generalizing to solve for data unseen during the training phase.

The model of a neuron

Modern ML methods have their roots in the early computational model of a neuron proposed by Warren MuCulloch (a neuroscientist) and Walter Pitts (a logician) in 1943 [19]. This so-called perceptron is shown in Fig. 3.19. The artificial neuron receives inputs that are independently weighted, summed, and then processed by a nonlinear activation function which represents the neuron's action potential. The output is then connected to other similar neurons. The structure resembles a biological neural structure where a neuron will process input signals from other neurons, transmitted along dendrites, and condition these through synaptic interfaces to produce an output signal on its axon that in turn feeds other dendrites.

The basic Pitts and McCulloch neuron takes the summation of the components of an input vector $\mathbf{x}$ to node j weighted by a vector $\mathbf{w}_j$ and biased with θ_j. It then passes this through a nonlinear activation function $\Phi(\cdot)$ to form an output o_j. Hence,

$$o_j = \Phi\left(\text{net}_j\right), \tag{3.68}$$

where

$$\text{net}_j = \sum_{i=0}^{I-1} w_{ji} x_i - \theta_j. \tag{3.69}$$

When the weighted summation exceeds a threshold (hard or soft) the neuron will fire, and

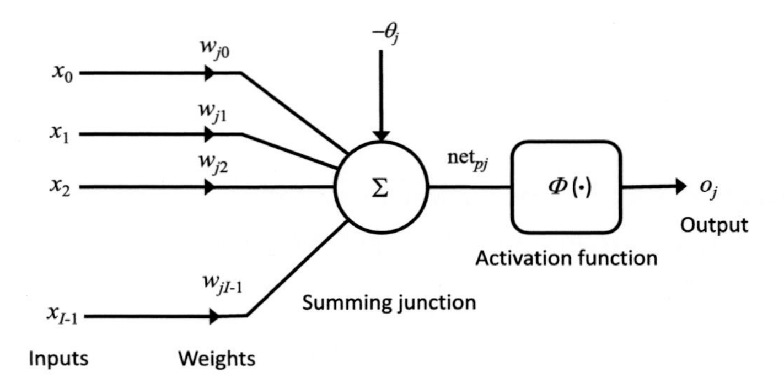

FIGURE 3.19 A basic neuron (perceptron) u_j (following the model of McCulloch and Pitts [19]).

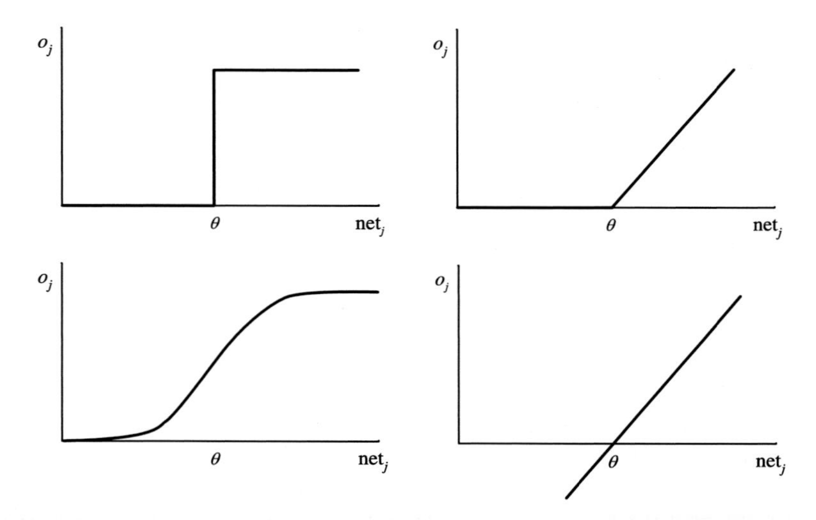

FIGURE 3.20 Example activation functions: (a) Step. (b) Rectified linear (ReLU). (c) Sigmoid (logistic) or *tanh*. (d) Linear.

this firing is dictated by the form of activation used. Typical activation functions for this type of system are shown in Fig. 3.20 and include a step function, a linear function, or a *tanh* function or a sigmoid function (Eq. (3.70)):

$$\Phi\left(\text{net}_j\right) = \frac{1}{1 + e^{-\text{net}_j}}. \tag{3.70}$$

Learning and the delta rule

In order to determine the optimal set of network weights, the network must be trained with a set of inputs $\{\mathbf{x}_p, p \in \mathbf{p}\}$ from a set of patterns $\mathbf{p}$ and a corresponding set of target outputs $\{\mathbf{t}_p, p \in \mathbf{p}\}$. The simplest form of associative learning is Hebb's law, which states that if a neuron u_j receives an input from an input x_{pi} and both are highly active, then the weight connecting them should be modified according to the relative strength of the activations between the input and the neuron's output. Considering a simple linear model, where the activation function is the identity function, Hebbian learning leads directly to the delta (or Widrow–Hoff) rule. For a given input pattern, the change that should be made to weight w_{ji} is thus

$$\Delta_p w_{ji} = \eta \left(t_{pj} - o_{pj}\right) x_{pi} = \eta \delta_{pj} x_{pi}, \tag{3.71}$$

where η is the learning rate and $\{\mathbf{o}_p, p \in \mathbf{p}\}$ are the actual outputs produced by the given input pattern. Although many cycles of pattern applications may be necessary, convergence will always be achieved for linearly independent pattern classes.

If we consider an error surface in weight space where the error measure is based on the sum (or mean) squared error, then, for a linear machine, the surface will be quadratic in the weights and hence the surface will contain a single unique minimum which can be reached via gradient descent. The error is given by

$$E_p = \frac{1}{2} \sum_j \left(t_{pj} - o_{pj}\right)^2, \tag{3.72}$$

and the total error is

$$E = \sum_{\mathbf{p}} E_p. \tag{3.73}$$

We can show that Eq. (3.71) implements gradient descent and thus minimizes the MSE over all pairs of input–output vectors. We wish to prove that

$$-\frac{\partial E_p}{\partial w_{ji}} = \delta_{pj} x_{pi}. \tag{3.74}$$

Using the chain rule,

$$\frac{\partial E_p}{\partial w_{ji}} = \frac{\partial E_p}{\partial o_{pj}} \cdot \frac{\partial o_{pj}}{\partial w_{ji}}. \tag{3.75}$$

From Eq. (3.72) we can see that

$$\frac{\partial E_p}{\partial o_{pj}} = -\left(t_{pj} - o_{pj}\right) = -\delta_{pj}. \tag{3.76}$$

Thus it can be seen that the contribution of unit j to the error surface is proportional to δ_{pj}. Since the units are linear,[4]

$$o_{pj} = \sum_i w_{ji} x_{pi}. \tag{3.77}$$

[4] We assume for convenience that the bias θ is treated as any other weight but permanently connected to -1.

Hence,

$$\frac{\partial o_{pj}}{\partial w_{ji}} = x_{pi}. \tag{3.78}$$

Substituting (3.76) and (3.78) in (3.75) gives Eq. (3.74), i.e.,

$$-\frac{\partial E_p}{\partial w_{ji}} = \delta_{pj} x_{pi},$$

but also,

$$\frac{\partial E}{\partial w_{ji}} = \sum_{\mathrm{p}} \frac{\partial E_p}{\partial w_{ji}}. \tag{3.79}$$

Hence the net change in weights after one complete cycle of pattern presentations is proportional to this derivative, thus showing that the delta rule implements gradient descent. This is known as the perceptron convergence theorem. Hence the weight update equation is

$$w_{ji,new} = w_{ji,old} + \eta \left(t_{pj} - o_{pj}\right) x_{pi}. \tag{3.80}$$

Multilayer networks

The basic neuron described above is capable of classification provided that the output classes are linearly separable. Many problems however do not satisfy this constraint, an often cited example being the Ex-Or problem. Multilayer networks were therefore developed to address these more complex types of problems. The multilayer perceptron (MLP) is a basic form of artificial neural network that connects its neural units in a multilayered (typically one input layer, one or more hidden layer(s) and one output layer) architecture (Fig. 3.21). These neural layers are generally fully connected to adjacent layers and must have nonlinear activation functions:

- The addition of hidden layers and nonlinear activation functions enables the network to create complex decision boundaries that in turn enable it to solve much more complex problems than the simple linear machine. However the resulting error surface is not quadratic in the weights and can have local minima.
- The nonlinear activation function is necessary as otherwise the network can be collapsed to an equivalent single-layer (linear) network and hence lose its ability to generate complex decision boundaries.
- This activation function must however be differentiable so that we can apply the delta rule and train the network.
- The fully connected nature of the MLP can lead to a large total number of parameters to optimize and make the network prone to overfitting.[5]

As for the simple perceptron, the MLP is trained in a supervised manner using a training set of data where input vectors (for example representing the pixels in an image) are explicitly paired with an output value – for example representing the class of the input image

[5] This overfitting problem has been addressed by more recent architectures and through careful dataset design.

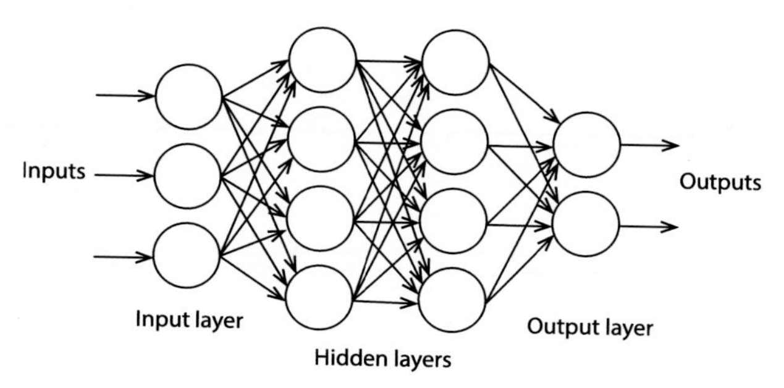

FIGURE 3.21 A basic multilayer perceptron with two hidden layers.

(e.g., "dog," "cat," etc.). MLPs are also trained by adapting the network weights in a way that minimizes the error between a target output value and that produced at a given time in the training cycle. This error is produced by a loss function that maps the event values from multiple inputs into one real number (for example using MSE) to represent the cost of that event. The goal of the training process is thus to minimize the loss function over multiple presentations of the input (training dataset). The performance of the trained network is then evaluated using an independent test set. Network training is generally achieved using a process referred to as error backpropagation, which is a form of gradient descent and an extension of the delta rule described above. The backpropagation algorithm was originally introduced in the 1970s, but was enhanced and popularized in the 1980s by Rumelhart, Hinton, and Williams [20], who proposed the generalized delta rule (error backpropagation) for training.

Error backpropagation

We have already described the delta rule for training simple two-layer networks. The problem with the MLP is how to adapt weights in its hidden layer(s), or more precisely, how to calculate δ_{pj} for the hidden units.

The backpropagation algorithm is a generalization of the delta rule, again based on gradient descent to minimize the sum squared difference between the target and the actual network outputs. The analysis is normally based on the assumption of a semilinear activation function (differentiable and monotonic) such as the sigmoid. The algorithm described below follows the original derivation in [20] and proceeds in two stages:

- *A feedforward phase*, where an input vector is applied and the signal propagates through the network layers, modified by the current weights and biases and by the nonlinear activation functions. Corresponding output values then emerge, and these can be compared with the target outputs for the given input vector using a loss function.
- *A feedback phase*: the error signal is then fed back (backpropagated) through the network layers to modify the weights in a way that minimizes the error across the entire training set, effectively minimizing the error surface in weight-space.

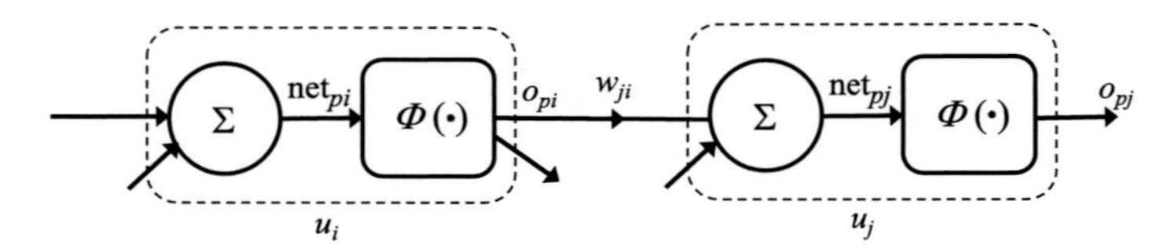

FIGURE 3.22 Error backpropagation notation for a three-layer system.

In a multilayer network (Fig. 3.21 and Fig. 3.22) we must be able to compute the derivative of the error with respect to any weight in the network and then change the weight according to

$$\triangle_p w_{ji} \propto -\frac{\partial E_p}{\partial w_{ji}}. \tag{3.81}$$

The solution to this problem again lies in the chain rule of partial differentiation:

$$\frac{\partial E_p}{\partial w_{ji}} = \frac{\partial E_p}{\partial \text{net}_{pj}} \cdot \frac{\partial \text{net}_{pj}}{\partial w_{ji}}. \tag{3.82}$$

We know that

$$\text{net}_{pj} = \sum_i w_{ji} o_{pi}. \tag{3.83}$$

Hence,

$$\frac{\partial \text{net}_{pj}}{\partial w_{ji}} = o_{pi}. \tag{3.84}$$

Now let us define

$$\delta_{pj} \triangleq -\frac{\partial E_p}{\partial \text{net}_{pj}}. \tag{3.85}$$

By comparing this with Eq. (3.76), we note that this is consistent with the definition of δ_{pj} used in the original delta rule, since $o_{pj} = net_{pj}$ for the case of linear units. Eq. (3.82) therefore becomes

$$-\frac{\partial E_p}{\partial w_{ji}} = \delta_{pj} o_{pi}. \tag{3.86}$$

So this states that, to implement gradient descent in E, we must make weight changes according to

$$\triangle_p w_{ji} = \eta \delta_{pj} o_{pi}. \tag{3.87}$$

The problem remains, how do we compute δ_{pj}? This can be done using Eq. (3.85) with the chain rule as follows:

$$\delta_{pj} = -\frac{\partial E_p}{\partial \text{net}_{pj}} = -\frac{\partial E_p}{\partial o_{pj}} \cdot \frac{\partial o_{pj}}{\partial \text{net}_{pj}}. \tag{3.88}$$

For the second term above, this simply equates to the derivative of the activation function:

$$\frac{\partial o_{pj}}{\partial \text{net}_{pj}} = \Phi'_j\left(\text{net}_{pj}\right). \tag{3.89}$$

However, for the first term, we must consider two separate cases.

1. When u_j is an output unit:

In this case

$$-\frac{\partial E_p}{\partial o_{pj}} = \left(t_{pj} - o_{pj}\right), \tag{3.90}$$

and hence

$$\delta_{pj} = \left(t_{pj} - o_{pj}\right)\Phi'_j\left(\text{net}_{pj}\right). \tag{3.91}$$

If we assume that the activation function is a logistic (sigmoid) function, as in Eq. (3.70), then for an output unit

$$\delta_{pj} = \left(t_{pj} - o_{pj}\right) o_{pj}\left(1 - o_{pj}\right). \tag{3.92}$$

2. When u_j is a hidden unit:

In this case no specific target exists but δ_{pj} can be calculated recursively in terms of the δ values of the units u_k that it connects to in the subsequent layer. These values are funneled back through the network, multiplied by the appropriate connecting weights. Thus instead of $\frac{\partial E_p}{\partial o_{pj}}$ we can substitute $\frac{\partial E_p}{\partial \text{net}_{pk}} \cdot \frac{\partial \text{net}_{pk}}{\partial o_{pj}}$, but we actually need to take account of the contribution of all K connected downstream nodes in the next layer; hence we use

$$\frac{\partial E_p}{\partial o_{pj}} \equiv \sum_k \frac{\partial E_p}{\partial \text{net}_{pk}} \cdot \frac{\partial \text{net}_{pk}}{\partial o_{pj}} \tag{3.93}$$

$$= \sum_k \frac{\partial E_p}{\partial \text{net}_{pk}} \cdot \frac{\partial\left(\sum_j w_{kj} o_{pj}\right)}{\partial o_{pj}} = \sum_k \frac{\partial E_p}{\partial \text{net}_{pk}} . w_{kj} = -\sum_k \delta_{pk} w_{kj}. \tag{3.94}$$

This makes sense as it says that the contribution of output o_j to the overall error for a given input pattern can be computed in terms of a weighted sum of the δ values of all the units in layer k that are connected to unit j. Hence we now have an expression for backpropagating errors:

$$\delta_{pj} = \Phi'_j\left(\text{net}_{pj}\right) . \sum_k \delta_{pk} w_{kj}. \tag{3.95}$$

Again, assuming that the activation function is a logistic (sigmoid) function, as in Eq. (3.70), then for a hidden unit

$$\delta_{pj} = o_{pj}\left(1 - o_{pj}\right) \sum_k \delta_{pk} w_{kj}, \tag{3.96}$$

and finally if the system has only one hidden layer (i.e., a three-layer system), then the weight update is given by

$$\triangle_p w_{ji} = \eta \delta_{pj} x_{pi}. \tag{3.97}$$

3.7.3 Deep neural networks

When popular, MLPs were generally fully connected and had relatively small numbers of hidden layers. This often led to a large total number of parameters to optimize, and made these networks prone to overfitting. DL has, over the past decade, emerged as a powerful development of the MLP, for finding patterns, analyzing information, and predicting future events. DNNs are capable of handling huge amounts of high-dimensional data and of discovering latent structures in unlabeled data. The word "deep" refers to the fact that the network comprises multiple hidden layers of neurons that have local connectivity with learnable weights and biases.[6] When the data being processed occupies multiple dimensions (images for example), convolutional neural networks (CNNs) are often employed. These are (loosely) a biologically inspired architecture, implement-limited connectivity and their results are tiled so that they overlap to obtain a better representation of the original inputs. The first CNN (the neocognitron) was designed by Fukushima in 1980 [21] as a tool for visual pattern recognition. This was a hierarchical architecture with multiple convolutional and pooling layers. In 1989, LeCun et al. applied the standard backpropagation algorithm [22] to a DNN aimed at handwritten ZIP code recognition. The real breakthrough however was driven by the availability of graphics processing units that could dramatically accelerate training. Since around 2012, CNNs have represented the state-of-the-art for complex problems such as image classification and recognition and are now emerging as a powerful force in compression.

As mentioned earlier, there is a huge literature on DL, and the topic is advancing rapidly. In this context we necessarily limit our coverage here to an overview of some of the important concepts. The reader is referred to the references in this chapter and in Chapter 13 for further details.

Convolutional neural networks (CNNs)

The basic CNN (or ConvNet) architecture as shown in Fig. 3.23 comprises a number of different modules that are interconnected according to the task at hand. The CNN can successfully capture spatial dependencies in an image through the application of localized and sparsely connected 2D convolution filters as described in Section 3.3.3. Generally, each layer in the CNN trains on features produced by the previous layer, those features becoming more complex and more semantically meaningful as the number of layers increases. For example, a CNN might learn to detect edges from raw pixels in the first layer, then use those edges to detect simple shapes in the next layer, and so on, building complexity through the feature hierarchy. This means that CNNs can exploit both low-level features and a higher-level understanding of what the data represent. The early layers in a CNN have been found to extract low-level features conceptually similar to visual basis functions found in the primary visual cortex [23].

[6] The number of layers in a deep network is unlimited but most current networks contain between 10 and 100 layers.

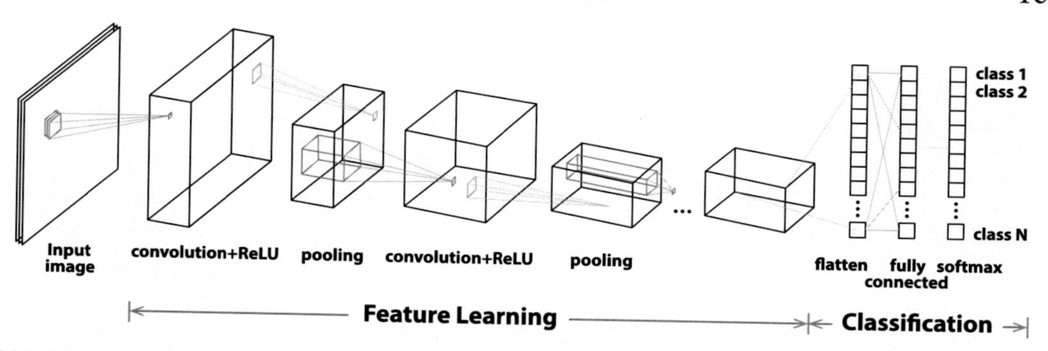

FIGURE 3.23 A basic CNN architecture (adapted from https://uk.mathworks.com/solutions/deep-learning/convolutional-neural-network.html).

The convolutional layers are inherently designed to take advantage of 2D structures, such as those found in images. They employ locally connected layers that apply convolution operations between a predefined-size kernel and an internal signal. As with the MLP, the weights of the filter are adjusted according to a loss function that assesses the mismatch (during training) between the network output and the ground truth values or labels. These errors are then backpropagated through multiple forward and backward iterations and the filter weights adjusted based on estimated gradients of the local error surface. This in turn drives what features are detected, associating them to the characteristics of the training data.

The CNN architecture in Fig. 3.23 shows the outputs from its convolution layers connected to a pooling layer, which combines the outputs of neuron clusters into a single neuron. This can be a simple maxpooling (taking the maximum value within a subregion) or applying averaging or L2-norm pooling. The pooling layer acts to reduce the spatial size of the representation, to reduce the number of network parameters, and hence to also control overfitting. Activation functions such as *tanh* (the hyperbolic tangent), *ReLU* (a rectified linear unit), or its variants are applied to introduce nonlinearity into the network [24]. The introduction of ReLU has been a major step forward in the training and performance of deep networks. ReLU is a piecewise linear function and hence preserves many of the properties that facilitate optimization using gradient descent and enable linear models to generalize well.

This structure is repeated with similar or different kernel sizes. The final layers effect the classification part of the network. These usually consist of fully connected layers and a softmax layer, which models the output class as a probability distribution – exponentially scaling the output between 0 and 1 (often called a normalized exponential function).

Some architectures omit pooling layers in order to create dense features providing an output with the same size as the input. Alternatively, the size of the feature maps can be enlarged to be the same as the input via deconvolutional layers, as shown in Fig. 3.24. This type of architecture is known generally as an autoencoder and has successfully been employed to problems such as image denoising [25] and superresolution [26]. Some architectures also add skip connections [27] so that the local and global features, as well as semantics, are connected and captured, providing improved pixel-wise accuracy. These techniques are widely used in object detection and tracking [28,29].

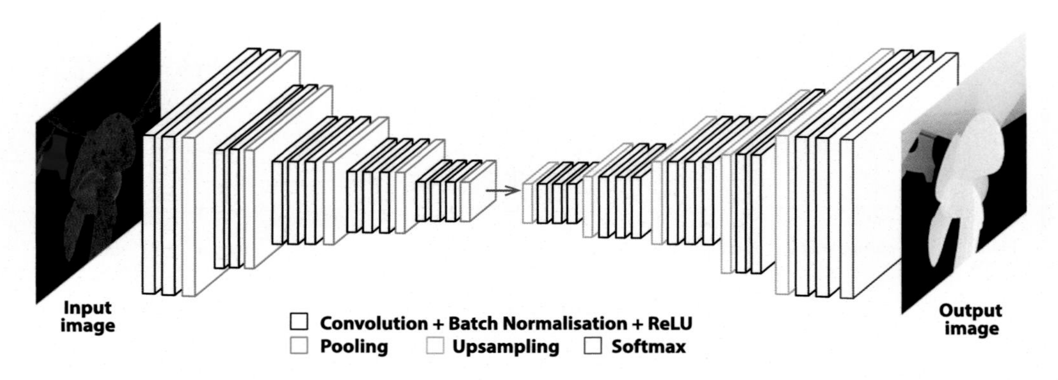

FIGURE 3.24 Basic autoencoder architecture.

Some architectures also introduce modified convolution operations for specific applications. For example, dilated convolution [30] enlarges the receptive field, to support feature extraction locally and globally. ResNet is an architecture developed for residual learning, comprising several residual blocks [31]. These residual blocks have two convolutional layers and a skip connection between the input and the output of the last convolution layer. This avoids the problem of vanishing gradients, enabling very deep CNN architectures. Residual learning has become an important part of the state-of-the-art in many applications, such as contrast enhancement, colorization, superresolution, and denoising.

Training of CNNs uses error backpropagation in the same way as described for the MLP above. However the situation is more complicated since the error backpropagation pass must also employ convolutions, but this time multiplying by the transpose of the convolution matrix. This process is scalable to networks of any depth. For further details the reader is referred to texts such as [15].

Generative adversarial networks (GANs)

An alternative architecture, referred to as a generative adversarial network (GAN), has been proposed by Goodfellow et al. [32]. GANs consist of two competing modules, where the first creates synthetic images (the generator) and the second (the discriminator) assesses whether the received image is real or created from the first module. The competitive nature of a GAN has been demonstrated to result in a reduction of deceptive results, and hence GAN technologies have found success across a number of applications, including those related to image and video compression (see Chapter 13).

The general GAN architecture is shown in Fig. 3.25. The generative network creates new candidates to increase the error rate of the discriminative network until it cannot tell whether these candidates are real or synthesized. The generator is typically a deconvolutional neural network, and the discriminator is a CNN. Recent successful applications of GANs include superresolution, contrast enhancement, and compression [33].

Variational autoencoders

Another form of deep generative model is the variational autoencoder (VAE). In this type of autoencoder, the encoding distribution is regularized to ensure the latent space has good properties to support the generative process. Then the decoder samples from this distribution to generate new data. VAEs tend to be more stable during training than GANs but do not always produce more realistic images. VAE performance for image generation is competitive with conventional GANs but with greater capacity to generate a diverse range of images. There have also been many attempts to merge GANs and VAEs so that the end-to-end network benefits from both good samples and good representation, for example using a VAE as the generator for a GAN [34]. However, the results of this have not yet demonstrated significant improvement in terms of overall performance, remaining an ongoing research topic.

A review of recent state-of-the-art GAN models and applications can be found in [35].

The need for data

An AI system effectively combines a computational architecture, a learning strategy, and a data environment in which it learns. Training databases are thus a critical component in optimizing the performance of ML processes and hence a significant proportion of the value of an ML system resides in them. A well-designed training database with appropriate size and coverage can thus significantly enhance model generalization and avoid problems of overfitting, ultimately enabling a DL system to deliver optimal performance.

Three types or partitions of dataset are normally defined:

1. *The training dataset*: This the dataset used for initial training of the network, where input–target pairs are presented, typically over many thousands of epochs until the network converges.
2. *The testing dataset*: The test dataset is used to provide an unbiased evaluation of model performance based on the final architecture and training data used. The testing dataset should be independent of the training dataset – i.e., its contents should not have been used for training.
3. *The validation dataset*: Often a third dataset will be employed independent of the two above. The validation dataset is used to provide an unbiased evaluation of a model fit after training dataset while tuning the model's hyperparameters or architecture.

Often a single dataset will be divided into training and validation partitions to support *cross-validation*. For example a dataset could be split into five equal partitions (fivefold cross-validation), where 80% of the data is used for training and 20% for validation, cycling through each partition as the validation set.

A more detailed review of datasets specifically designed for compression applications is included in Chapter 13.

3.8 Summary

This chapter has reviewed selected preliminary topics in signal processing, information theory, and ML that help to place the remainder of this book in context. We have not burdened the reader with series of proofs and theorems but have instead focused on the use of examples to aid understanding of some important topics. We examined some basic statistical measures and used these to characterize picture redundancy. Filters and transforms were introduced

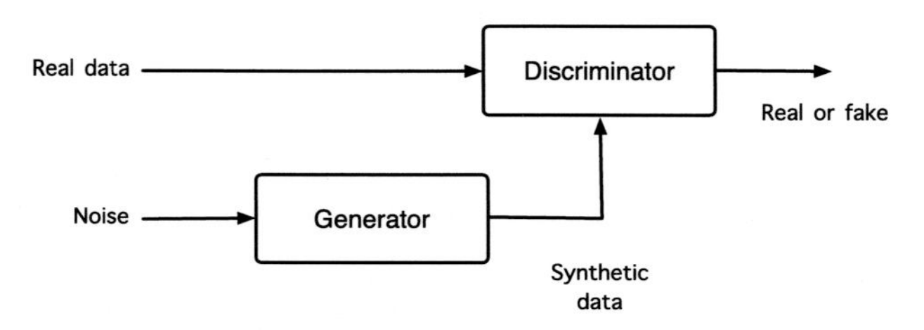

FIGURE 3.25 Basic GAN configuration.

as they form the basis of most compression methods, and these will be covered further in Chapters 5 and 6. We also revisit the area of quantization in the same chapters.

We spent some time here on prediction structures as these are used frequently in compression and can provide substantial coding gains over nonpredictive methods, both for spatial and temporal processing. Intra-prediction will be covered in Chapters 9 and 12 and motion prediction in Chapters 8 and 9. We reviewed the basics of information theory, explaining how its performance depends heavily on assumed source models and context. This forms the basis of symbol encoding as used in most compression standards. We specifically focus on entropy coding methods in Chapter 7 and follow this up in Chapter 12 in the context of standards. Finally, we have introduced the topic of ML and specifically DNNs. While these are not yet widely deployed in standardized image and video compression systems, they are attracting significant attention and there is clear evidence that they can offer significant performance gains and improved measures of perceptual image quality. However issues of computational complexity still remain which have limited uptake in current standards. The use of these data-driven optimization methods is further covered in Chapters 10 and 13.

References

[1] S. Mitra, Digital Signal Processing: A Computer Based Approach, 4e, McGraw Hill, 2011.
[2] J. Woods, Multidimensional Signal, Image and Video Processing and Coding, 2e, Academic Press, 2012.
[3] C. Shannon, Communication in the presence of noise, Proc. Inst. Radio Eng. 37 (1) (1949) 10–21.
[4] E. Dubois, The sampling and reconstruction of time-varying imagery with application in video systems, Proc. IEEE 73 (1985) 502–522.
[5] Y. Wang, J. Ostermann, Y-Q. Zhang, Video Processing and Communications, Prentice Hall, 2001.
[6] R. O'Callaghan, D. Bull, Combined morphological-spectral unsupervised image segmentation, IEEE Trans. Image Process. 14 (1) (2005) 49–62.
[7] P. Czerepiński, D. Bull, Enhanced interframe coding based on morphological segmentation, IEE Proc., Vis. Image Signal Process. 144 (4) (1997) 220–226.
[8] A. Oppenheim, A. Willsky, H. Nawab, Signals and Systems, 2e, Prentice Hall, 1999.
[9] J-R. Ohm, Multimedia Signal Coding and Transmission, Springer, 2015.
[10] S. Lloyd, Least squares quantization in PCM (1957), Reprint in IEEE Trans. Commun. 30 (1982) 129–137.
[11] Y. Linde, A. Buzo, R. Gray, An algorithm for vector quantizer design, IEEE Trans. Commun. 28 (1980) 84–95.
[12] M. Hayes, Statistical Digital Signal Processing and Modeling, Wiley, 1999.
[13] C. Shannon, A mathematical theory of communication, Bell Syst. Tech. J. 27 (3) (1948) 379–423.
[14] T. Cover, J. Thomas, Elements of Information Theory, Wiley, 2006.
[15] I. Goodfellow, Y. Bengio, A. Courville, Deep Learning, MIT Press, 2017.

[16] S. Russell, P. Norvig, Artificial Intelligence: A Modern Approach, 4e, Pearson, 2020.
[17] S. Ma, et al., Image and video compression with neural networks: a review, IEEE Trans. Circuits Syst. Video Technol. 30 (6) (2020) 1883–1898.
[18] N. Anantrasirichai, D. Bull, Artificial intelligence in the creative industries: a review, arXiv:2007.12391v2, 2020.
[19] W. McCulloch, W. Pitts, A logical calculus of the ideas immanent in nervous activity, Bull. Math. Biophys. 5 (1943) 115–133.
[20] D. Rumelhart, G. Hinton, R. Williams, Learning representations by backpropagating errors, Nature 323 (1986) 533–536.
[21] K. Fukushima, Neocognitron: a self-organizing neural network model for a mechanism of pattern recognition unaffected by shift in position, Biol. Cybern. 36 (1980).
[22] Y. LeCun, et al., Backpropagation applied to handwritten zip code recognition, Neural Comput. 1 (4) (1989) 541–551.
[23] M. Matsugu, et al., Subject independent facial expression recognition with robust face detection using a convolutional neural network, Neural Netw. 16 (5) (2003) 555–559.
[24] F. Agostinelli, et al., Learning activation functions to improve deep neural networks, in: Proc. International Conference on Learning Representations, 2015, pp. 1–9.
[25] K. Zhang, et al., Beyond a Gaussian denoiser: residual learning of deep CNN for image denoising, IEEE Trans. Image Process. 26 (7) (2017) 3142–3155.
[26] W. Shi, et al., Real-time single image and video super-resolution using an efficient sub-pixel convolutional neural network, IEEE Conf. on Computer Vision and Pattern Recognition (CVPR) (2016) 1874–1883.
[27] J. Long, E. Shelhamer, T. Darrell, Fully convolutional networks for semantic segmentation, in: IEEE Conf. on Computer Vision and Pattern Recognition (CVPR), 2015, pp. 3431–3440.
[28] N. Anantrasirichai, D. Bull, DefectNet: multi-class fault detection on highly imbalanced datasets, in: IEEE International Conf. on Image Processing (ICIP), 2019, pp. 2481–2485.
[29] A. Bochkovskiy, C. Wang, H. Liao, YOLOv4: optimal speed and accuracy of object detection, ArXiv, arXiv: 2004.10934 [abs], 2020.
[30] F. Yu, V. Koltun, Multi-scale context aggregation by dilated convolutions, in: International Conference on Learning Representations, 2016.
[31] K. He, et al., Deep residual learning for image recognition, in: IEEE Conference on Computer Vision and Pattern Recognition (CVPR), 2016, pp. 770–778.
[32] I. Goodfellow, et al., Generative adversarial nets, in: Z. Ghahramani, et al. (Eds.), Advances in Neural Information Processing Systems 27, Curran Associates, 2014, pp. 2672–2680.
[33] D. Ma, M. Afonso, F. Zhang, D. Bull, Perceptually-inspired super-resolution of compressed videos, in: Proc. SPIE 11137, Applications of Digital Image Processing XLII, 2019, pp. 310–318.
[34] C. Wan, et al., Crossing Nets: Combining GANs and VAEs with a shared latent space for hand pose estimation, in: IEEE Conf. on Computer Vision and Pattern Recognition (CVPR), 2017, pp. 1196–1205.
[35] D. Foster, Generative Deep Learning: Teaching Machines to Paint, Write, Compose, and Play, O'Reilly Media Inc., 2019.

CHAPTER

4
Digital picture formats and representations

Since the origin of digital video communications, there has been a need to understand the links between the human visual system (HVS) and picture representations. Building on Chapters 1, 2, and 3, this chapter describes the picture formats, processing techniques, and assessment methods that underpin the coding process; it introduces the processing methods and sampling structures that enable compression to be so effective.

Building on the mathematical preliminaries from Chapter 3, we first show how samples are represented in digital images and video signals, emphasizing the important aspects of color and spatio-temporal representations. We examine how samples are grouped to form pictures and how pictures are grouped to form sequences of pictures or videos.

The mapping of light levels to brightness as perceived by the HVS is considered in Section 4.4, showing how the signal from the camera must be transformed using a nonlinear mapping (gamma correction) to ensure best use of the bits available. Next, developing further the trichromacy theory of vision introduced in Chapter 2, we introduce the area of color spaces in Section 4.5; these are important since, with relatively little effort, we can create representations that reduce the bit rate for a digital video by 50%. Finally, we introduce the important topic of image and video quality assessment – both as a means of comparing the performance of different compression systems and as a means of controlling encoder modes during compression.

4.1 Pixels, blocks, and pictures

4.1.1 Pixels, samples, or pels

A still image or picture is a spatial distribution of sample values that is constant with respect to time. These are often referred to as pixels or pels.

Monochrome image

A monochrome image is a 2D matrix of luma samples, **S**, with spatial dimensions (often called resolution) $X \times Y$ (Fig. 4.1). If we zoom in on an actual image, as shown in Fig. 4.2, we see that the pixel boundaries become resolvable and we can observe pixelation effects. As we

Intelligent Image and Video Compression
https://doi.org/10.1016/B978-0-12-820353-8.00013-X

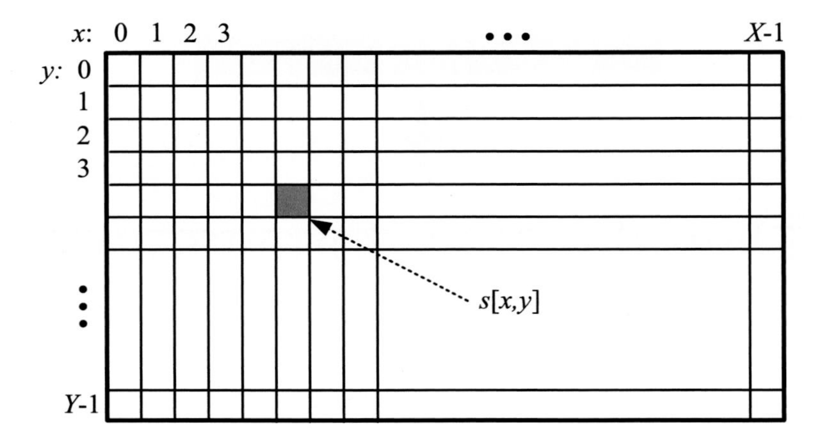

FIGURE 4.1 Image sample array.

zoom in still further, we lose the structure of the underlying image and the pixel structure dominates.

Fig. 4.3 presents this in a slightly different way, through the use of four different sample rates. If you move away from this figure you will see (at around 5 m) that the top two images start to look very similar. If you can get even further away (at around 10 m), all four images will take on a similar appearance. This emphasizes the fact that resolution is not about the number of horizontal and vertical samples, but about the angle that each pixel subtends at the viewer's retina. In other words, it depends on the number of pixels, the size of the display, and the distance of the observer.

Color image

In the case of a color picture, each element of **S** is itself a vector **s**, with three dimensions representing the intensities in a given color space (e.g., RGB, HSI, or YC_bC_r).

So we have in matrix form:

$$\mathbf{S} = \begin{bmatrix} \mathbf{s}[0,0] & \mathbf{s}[0,1] & \dots & \mathbf{s}[0,X-1] \\ \mathbf{s}[1,0] & \mathbf{s}[1,1] & \dots & \mathbf{s}[1,X-1] \\ \vdots & \vdots & & \vdots \\ \mathbf{s}[Y-1,0] & \mathbf{s}[Y-1,1] & \dots & \mathbf{s}[Y-1,X-1] \end{bmatrix}, \tag{4.1}$$

where (e.g., for an *RGB* signal)

$$\mathbf{s}[x,y] = \begin{bmatrix} s_R[x,y] & s_G[x,y] & s_B[x,y] \end{bmatrix}.$$

4.1.2 Moving pictures

A video signal is a spatial intensity pattern that changes with time. Pictures are captured and displayed at a constant frame rate, $1/T$ where T is the sampling period. In order to trade

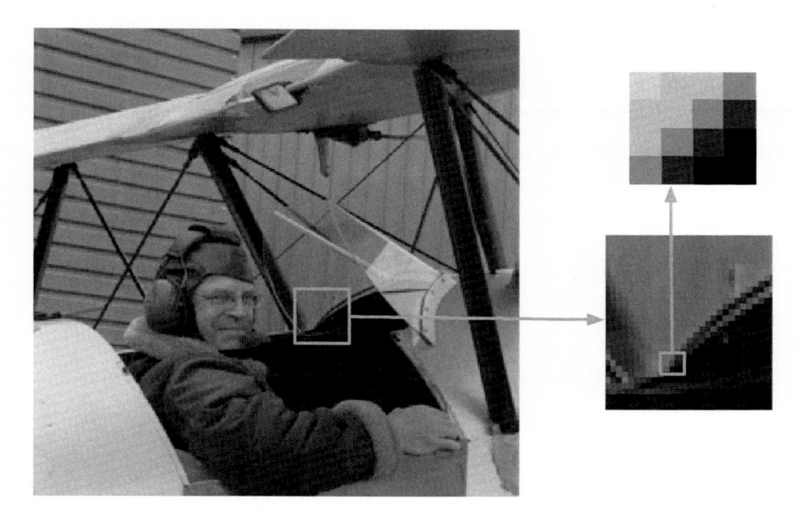

FIGURE 4.2 Image sampling.

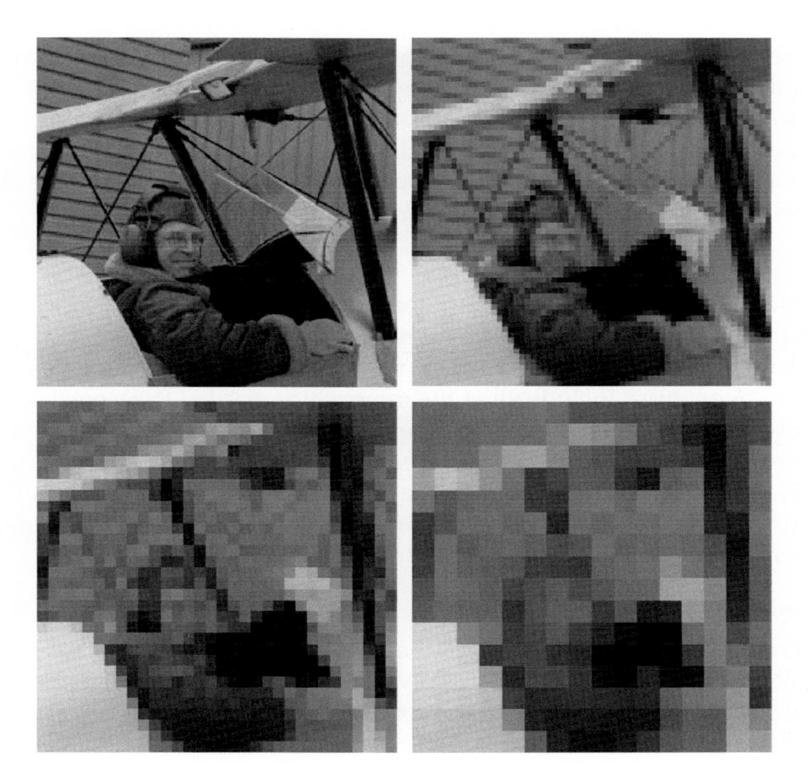

FIGURE 4.3 Pixelation at varying resolutions. Top left to bottom right: 256 × 256; 64 × 64; 32 × 32; 16 × 16.

off temporal artifacts (flicker, judder, etc.) against available bandwidth, a frame rate of 24, 25, or 30 Hz has conventionally been chosen. More recently, as screens have become larger and spatial resolutions have increased, higher frame rates have been introduced at 50 and 60 Hz. The UHDTV standard Rec.2020 specifies a maximum frame rate of 120 Hz. The reader is referred to Chapter 12 for a more in-depth discussion of frame rate requirements.

In terms of our discrete signal representation, we add an extra dimension (time or frame index), z, to the two spatial dimensions associated with a still image, thus

$$\mathbf{S} = \begin{bmatrix} \mathbf{s}[0,0,z] & \mathbf{s}[0,1,z] & \dots & \mathbf{s}[0,X-1,z] \\ \mathbf{s}[1,0,z] & \mathbf{s}[1,1,z] & \dots & \mathbf{s}[1,X-1,z] \\ \vdots & \vdots & & \vdots \\ \mathbf{s}[Y-1,0,z] & \mathbf{s}[Y-1,1,z] & \dots & \mathbf{s}[Y-1,X-1,z] \end{bmatrix}, \tag{4.2}$$

where (again for an RGB signal)

$$\mathbf{s}[x,y,z] = \begin{bmatrix} s_R[x,y,z] & s_G[x,y,z] & s_B[x,y,z] \end{bmatrix}.$$

4.1.3 Coding units and macroblocks

As we will see later, video compression algorithms rarely process information at the scale of a picture or a pixel. Instead, the basic coding unit is normally a square block of pixels. In standards up to and including H.264/AVC, this took the form of a 16×16 block, comprising luma and chroma information, called a macroblock.[1] In more recent standards, the size of this basic coding unit has been increased in line with the increase in picture resolutions. Let us examine this important coding structure in more detail.

Macroblocks

A typical macroblock structure is illustrated in Fig. 4.4. The macroblock shown corresponds to what is known as a 4:2:0 format (see Section 4.5) and comprises a 16×16 array of luma samples and two subsampled 8×8 arrays of chroma (color difference) samples. This structure, when coded, must include all of the information needed to reconstruct it – including transform coefficients, motion vectors, quantization information, and other information relating to how the block is partitioned for prediction purposes. In standards up to and including H.264/AVC, the base macroblock size is 16×16 pixels. In H.264/AVC itself, this block can be further partitioned for motion estimation (down to 4×4), whereas decorrelating transforms are normally applied at either 8×8 or 4×4 levels.

Coding tree units

The H.265/HEVC standard has extended the size of the macroblock (now referred to as a coding tree unit) to 64×64 samples and H.266/VVC has extended this still further to 128×128 samples in order to support higher spatial resolutions. Transform sizes have also increased, up to 32×32 (HEVC) and 64×64 (VVC). Both standards also provide much more flexibility

[1] We will examine the nature of color spaces and chroma subsampling in Section 4.5.

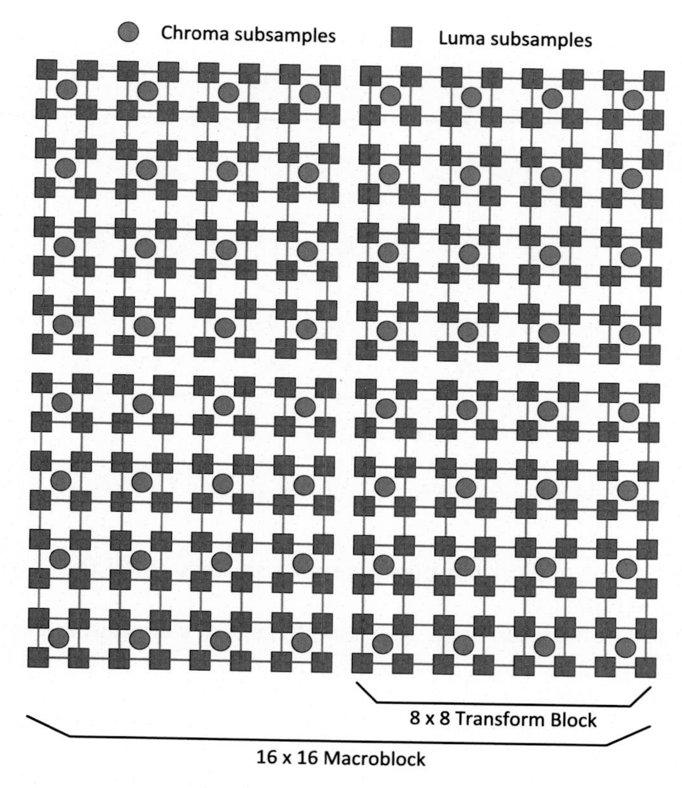

FIGURE 4.4 Typical macroblock structure.

in terms of block partitioning to support various new prediction modes. Further details on this are provided in Chapter 12.

4.1.4 Picture types and groups of pictures

Frame types

Most video coding formats support three types of picture (or frame[2]), classified in terms of their predictive properties. These are described below:

- **Intra-coded (I)-pictures:** I-pictures (or I-frames) are coded without reference to any other picture, but can be used as a reference for other pictures.
  - They provide anchor pictures and support random access.
  - They offer the highest resilience to errors as they do not propagate errors from previous pictures.

[2] These terms can generally be used interchangeably.

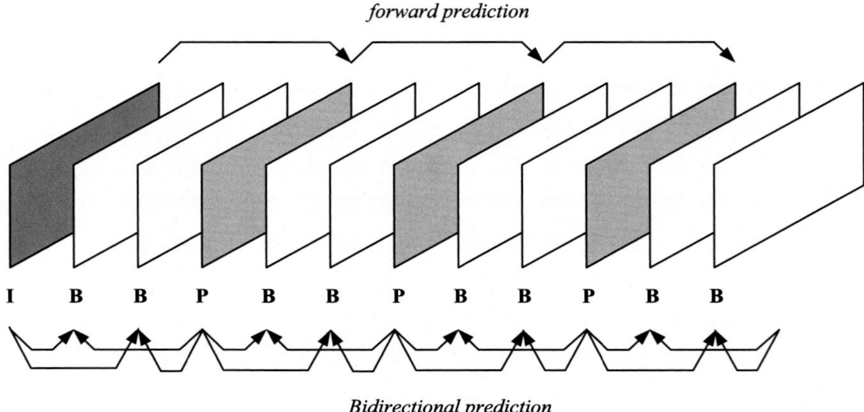

FIGURE 4.5 Typical group of pictures structure and prediction modes.

- Because they do not exploit temporal correlation, they have the lowest compression ratio of all frame types.
- **Predicted (P)-pictures:** P-pictures are inter-coded, i.e., they contain the compressed difference between current frame and a prediction of it based on one previous I- or P-frame.
  - Like I-frames, they can be used as a reference for other frames.
  - In order to decode a P-frame, the transformed residual must be accompanied by a set of motion vectors that provide information on the prediction reference location.
  - P-pictures are more efficient than I-pictures as they exploit both spatial and temporal correlations.
  - Unlike I-frames, they will propagate errors due to predictive coding.
- **Bipredicted (B)-pictures:** B-pictures comprise the compressed difference between the current frame and a prediction of it, based on up to two I-, P-, or B-frames.
  - In standards up until H.264/AVC, B-pictures were not used as a reference for predicting other pictures. However, H.264/AVC, HEVC, and VVC have generalized their definition and in these cases they can be used as a reference, although normally in a hierarchical fashion (see Chapter 12).
  - They require accompanying motion vectors.
  - They provide efficient coding due to exploitation of temporal and spatial redundancies and can be quantized more heavily if not used as a reference.
  - They will not propagate errors if not used as a reference for other pictures.

Groups of pictures (GOPs)

All video coding methods and standards impose some bitstream syntax in order for the bitstream to be decodable. Part of this relates to the order of encoding, decoding, and display of pictures and the way in which pictures of different types are combined. A typical group of pictures (GOP) structure that illustrates this (typical of standards such as MPEG-2) is shown in Fig. 4.5. Standard-specific variations on this structure are covered in Chapters 9 and 12.

Example 4.1. Intra- vs. interframe coding – the benefits of prediction

Consider a 512×512 @30 fps color video with 8-bit samples for each color channel. Compare the intra-compressed, inter-compressed, and uncompressed bit rates if the average numbers of bits per pixel after compression are as follows:

- Intraframe mode: 0.2 bpp (assume this is an average value that applies to all color channels).
- Interframe mode: 0.02 bpp (assume a GOP length of 100 pictures and that this value includes any overheads for motion vectors).

Solution.

a. Uncompressed:

The uncompressed bit rate can be calculated as

$$512 \times 512 \times 30 \times 24 \simeq 189 \text{ Mb/s}.$$

b. Compressed intra-mode:

In intra-mode, all frames are compressed independently to give a bit rate of

$$512 \times 512 \times 30 \times 0.2 \times 3 \simeq 4.7 \text{ Mb/s}.$$

c. Compressed inter-mode:

In inter-mode, 1 in every 100 frames are intraframes and the other 99 are interframes. Hence the bit rate is

$$512 \times 512 \times 30 \times \left(\frac{0.6 + 99 \times 0.06}{100} \right) \simeq 514 \text{ kb/s}.$$

This indicates that we can achieve significant savings (the figures given here are not unrealistic in practice) through the use of interframe coding to exploit temporal as well as spatial correlations.

4.2 Formats and aspect ratios

We examined the range of commonly used video formats in Chapter 1, in the context of their raw bit rates and the limitations of available transmission bandwidths. The video industry has come a long way since it started coding QCIF (176×144 luma samples) images for low-bit rate conferencing applications around 1989. We are now able, since the introduction of HEVC, to acquire, process, store, deliver, and display UHDTV formats, initially at a resolution of 3840×2160, but soon rising to 8k (7680×4320) for some applications. The relative sizes of these formats are illustrated in Fig. 4.6, where the size drawn is normalized to the viewing angle subtended for a single pixel at the retina. The top subfigure is particularly startling in showing the relative dimensions of QCIF and UHDTV formats.

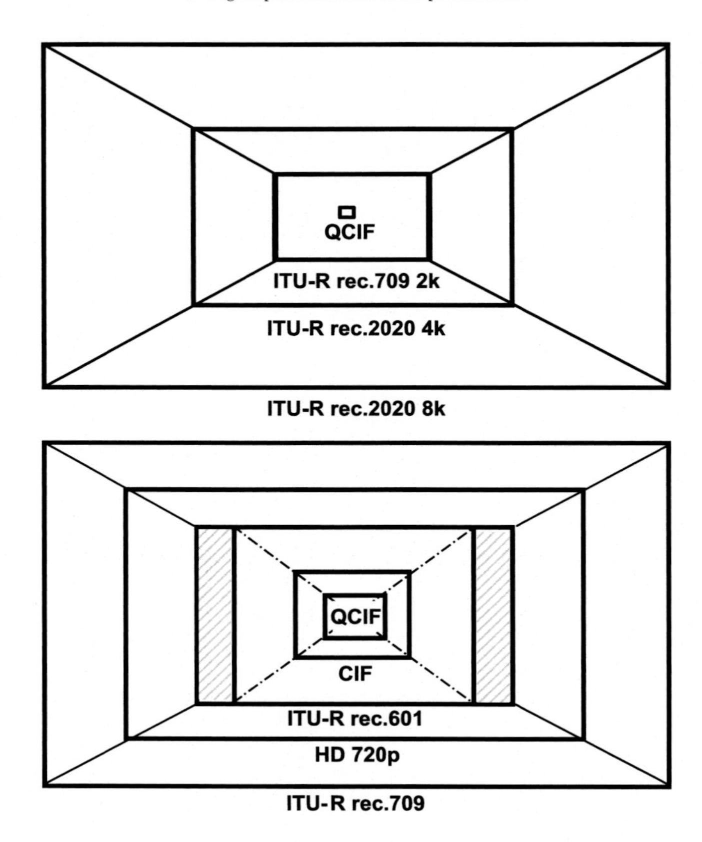

FIGURE 4.6 Aspect ratios of common formats, normalized according to resolution.

4.2.1 Aspect ratios

Table 4.1 illustrates some of the more common video format aspect ratios. The choice of aspect ratio has been the subject of much debate; choosing one closest to the ratio of viewing angles of the retina would seem the most obvious choice. These are approximately 95 degrees temporally and roughly 70 degrees above and below the horizontal meridian. It is not difficult to see, however, that this causes a problem – it argues that the aspect ratio should be infinite! So let us have another go – what about the foveal visual field? This approximately circular region of our retina covers 2–3 degrees of our visual field and visual acuity rapidly decreases away from the fovea, typically 20/20 in the fovea and 20/200 at a field angle of 20 degrees. So, because this is roughly circular, should we assume that a square aspect ratio is best (or even that screens should be circular)? The answer is clearly somewhere between 1:1 and infinity and, maybe we do not need rectangular screens! Some argue that the *golden ratio* of 1.61:1 is the most aesthetically pleasing, although the reason for this has not been clearly argued.

A practical solution, certainly for larger and closer screens, is to create the impression that the screen fills our peripheral vision, without completely doing so. This provides a better

TABLE 4.1 Aspect ratios of film and TV.

Format	Aspect ratio
SDTV	1.33 (4:3)
WSTV/HDTV/UHDTV	1.78 (16:9)
widescreen 35-mm film	1.85
70-mm film	2.10
Cinemascope anamorphic 35-mm film*	2.35
Cinemascope anamorphic 35-mm (1970+)	2.39

sense of immersion in the content while not requiring overly wide screens – hence the convergence on typical aspect ratios of 1.77 to 2.39. However none of this accounts for the changes in viewing pattern with vary large screens, where we exhibit a greater tendency to track objects with eye and head movements. This introduces different problems that we will touch on again in Chapter 13.

In practice the 16:9 format, selected for most consumer formats, is a compromise that optimizes cropping of both film and older 4:3 content. The 16:9 ratio (1.78) is very close to the geometric mean of 1.33 and 2.39, so all formats occupy the same image circle. At the end of the day, the best aspect ratio will depend a lot on composition and the relationships between the main subject, the edges of the frame, and the amount of uninteresting space. The art of the cinematographer is to compose the content in a manner that is compelling, regardless of frame aspect ratio.

Fig. 4.7 gives an example of some common aspect ratios superimposed on a scene. The preference, of course, is also dictated by the viewing conditions. Within the constraints of comfortable viewing, comparing 4:3 and 16:9 formats based on identical width, height, or area, most subjects prefer the format that provides the greatest viewing area.

For the case of a sensor or display with N_p pixels and an aspect ratio, r, it is straightforward to compute the horizontal and vertical pixel counts, X and Y, as follows:

$$X = \sqrt{rN_p},$$

$$Y = \sqrt{\frac{N_p}{r}}. \tag{4.3}$$

Field of view ratio

Let us consider the viewing angle ratio (or field of view ratio) rather than the aspect ratio. It can easily be shown that this is independent of screen size, s, but will vary nonlinearly with viewing distance, d. The commonly adopted TV and computer monitor format, 16:9, is used as an example in Fig. 4.8, but other aspect ratios exhibit similar characteristics. This figure plots the viewing angle ratio against the viewing distance in multiples of screen height H. A number of important things can be taken from this graph. Firstly, as d increases, the viewing angle ratio $\theta_w/\theta_h \to r$, where r is the aspect ratio of the screen. Secondly, as θ_w/θ_h decreases below $3H$ there is a rapid fall-off in this ratio down to unity at d=0. For a 16:9 aspect ratio, the viewing distance of $1.5H$ (specified for UHDTV) corresponds to a ratio of

FIGURE 4.7 Comparison of widescreen formats.

approximately 1.65, actually quite close to the golden ratio mentioned above! Example 4.2 examines this in more detail.

Example 4.2. Field of view ratio

Derive an expression for the field of view ratio for a screen of size s (diagonal dimension), with an aspect ratio r, viewed at a distance d.

Solution. Frequently the dimensions of a screen are given in terms of its diagonal dimension, s. It is a simple matter to calculate the screen height H and width W, using

$$H = \frac{s}{\sqrt{1+r^2}}, \quad W = \frac{s}{\sqrt{1+1/r^2}}. \tag{4.4}$$

The horizontal and vertical viewing angles that correspond to a screen of aspect ratio r ($W \times H$) viewed at a distance d are then

$$\theta_w = 2\arctan\left(\frac{W}{2d}\right), \tag{4.5}$$

$$\theta_h = 2\arctan\left(\frac{H}{2d}\right), \tag{4.6}$$

and their ratio, defined here as the field of view ratio, Ψ, is thus

$$\Psi = \frac{\arctan\left(\frac{W}{2d}\right)}{\arctan\left(\frac{H}{2d}\right)}. \tag{4.7}$$

This is plotted for a 16:9 aspect ratio screen for various viewing distances in Fig. 4.8.

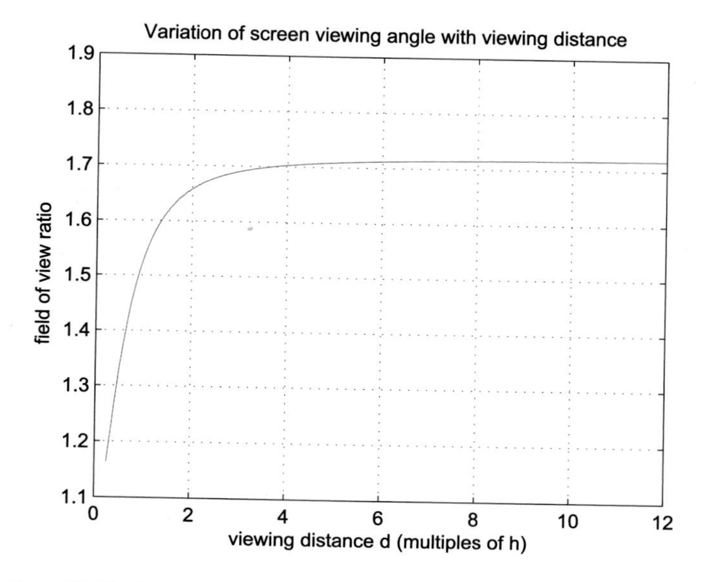

FIGURE 4.8 Variation of field of view with viewing distance (aspect ratio =16:9 here).

4.2.2 Displaying different formats

The existence of multiple formats creates significant extra work for content producers, especially film makers. It is quite common for films or older TV programs to be cropped or distorted in some way – which can be annoying for the consumer who feels that they are not getting the most out of their screen!

For example, displaying a 2.39:1 format on a 16:9 screen produces a *letter box* appearance. This can be avoided by distorting the original to fit or by zooming, but both have their obvious problems. In contrast, a 4:3 format displayed on a 16:9 screen produces a *pillar box* appearance. Again this can be stretched to produce short and fat subjects or zoomed to lose edge content.

Pan and scan and Active Format Description

Active Format Description (AFD) is a mechanism used in digital broadcasting to avoid the problems described above. This is a means of signaling the active content of a picture to enable intelligent processing of the picture at the receiver, in order to map the format to the display in the best way. Considering, for example, Fig. 4.7, if we wish to watch the 2.39:1 version on a 16:9 display, the blue box (representing the 16:9 picture) does not need to stay in the center of the 2.39:1 picture. It could instead move left or right to frame the action in the best way for the smaller screen.

This capability was introduced as a nonnormative feature in MPEG-2 on a GOP basis and is also supported by newer standards (e.g., through SEI data in H.264/AVC). AFD codes include information about the location of the active video and also about any protected area which must be displayed. The picture edges outside of this area can be discarded without major impact on the viewing experience.

4.3 Picture scanning

Interlaced vs. progressive scanning

Interlaced scanning is an attempt to trade off bandwidth, flicker, and resolution, by constructing each frame from two consecutive fields. Fields are sampled at different times such that consecutive lines belong to alternate fields; in the US a 60 Hz format is adopted, whereas in Europe a 50 Hz format is employed.

Early analog television systems needed to minimize bandwidth while maintaining flickerless viewing. Interlacing was devised as a compromise between a high temporal update rate (a field rate at 50 or 60 Hz), which reduces flicker for most content, and lower effective bandwidth (due to the 25 or 30 Hz frame rate). The persistence of the phosphor in early CRT displays, coupled with the persistence of human vision, meant that, provided the vertical dimension of a picture is scanned rapidly enough, the HVS does not notice the effect of alternating lines giving the experience of continuous motion. An illustration of the difference between interlaced and progressive scanning is given in Fig. 4.9. In modern digital cameras and displays the scanning, as shown in Fig. 4.9, is no longer performed by deflecting an electron beam. The row-wise and field-wise scanning pattern is however the same as indicated in the figure.

Notation

The notation we will adopt here to describe the picture format is that used by the European Broadcasting Union (EBU) and SMPTE:

$Y\pi/f$, where $Y =$ the number of vertical lines in the frame; $\pi =$ p (progressive) or i (interlaced); and $f =$ frame rate.

For example, 1080p/50 represents an HD progressive format with spatial resolution 1920×1080 and temporal sampling at 50 fps. Similarly 1080i/30 represents the same spatial resolution, but this time interlaced at 30 fps and 60 fields per second.

Problems caused by interlacing

The effects of interlacing become apparent in the presence of fast motion as illustrated by the frame shown in Fig. 4.10. The inset shows the *combing* artifacts that characterize this format, due to the temporal offset between the two constituent fields. Interlace also introduces a second problem of aliasing, commonly referred to as *twitter*. This occurs where information changes vertically at a rate similar to the field scanning rate. This becomes particularly prevalent when the vertical resolution of the content is similar to the resolution of the format. For example the clothes of a person wearing fine horizontal stripes will appear to twitter. Filtering is often applied to reduce this effect. An example of twitter can be observed at [2].

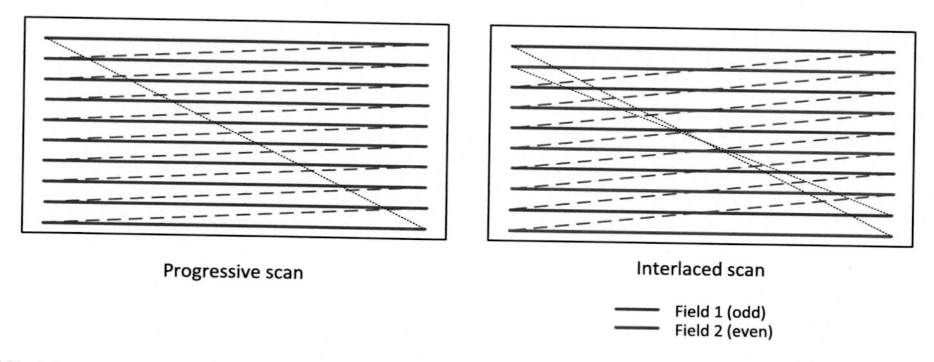

FIGURE 4.9 Interlaced vs. progressive frame scanning.

FIGURE 4.10 Example of effects of interlaced scanning with poor deinterlacing.

It is finally worth noting that the bandwidth advantages of interlacing associated with analog systems is reduced in the presence of digital video compression, especially if a frame with two fields is processed as one. MPEG-2 and subsequent standards have introduced an adaptive field-frame mode in an attempt to improve this situation.

The artifacts associated with interlacing can be minimized if the receiver and display employ good deinterlacing methods [7]. These employ local adaptive filtering to the two fields as part of the combining process.

4.3.1 Standards conversion

3:2 pull-down

The need to convert between film and TV formats has been a requirement since the beginnings of public television broadcasting. Film and digital cinema content is traditionally captured at 24 fps (using progressive scanning for digital formats). This needs to be converted

Algorithm 4.1 3:2 pulldown.

1. Play the film at 23.976 frames/s;
2. Split each cinema frame into video fields. At 23.976 frames/s, there are four frames of film for every five frames of 60-Hz video, i.e., 23.976/29.97 = 4/5;
3. Convert these four frames to five by alternately placing the first film frame across two fields, the next across three and so on.

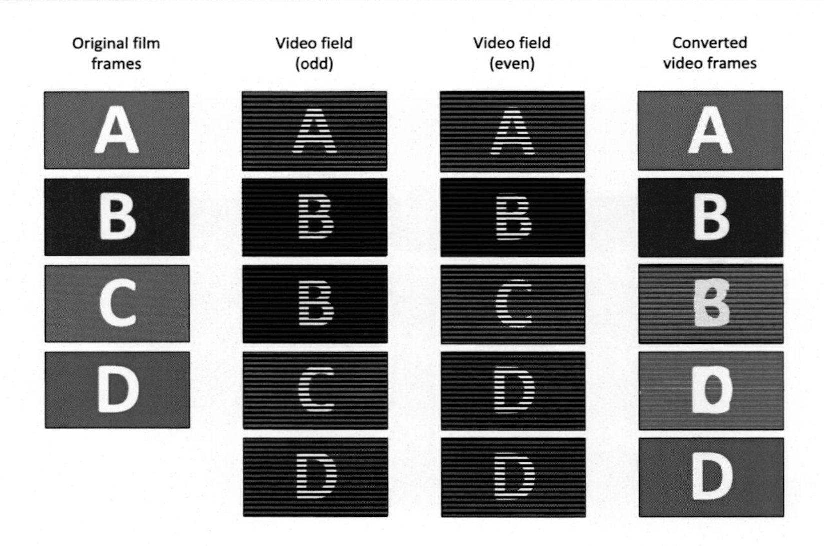

FIGURE 4.11 3:2 pull-down example.

to 29.97 fps (reduced from 30 fps in analog systems to avoid interference between the chroma subcarrier and the sound carrier) in the US based on their 60 Hz interlaced scanning and 25 fps in Europe based on 50 Hz. For the case of 25 fps it has been common practice to simply speed up the playback by a factor of 25/24. This produces some audiovisual changes but these are not generally noticeable. However, for motion to be accurately rendered at 29.97 fps, a telecine device must be employed to resample the 24-fps signal and a technique known as 3:2 pull-down is normally employed.

3:2 pull-down uses the approach described in Algorithm 4.1 and it is illustrated in Fig. 4.11.

Some streaming sites, such as Netflix, have strict requirements for their content and require an inverse telecine operation to detect and remove 3:2 pull-down from telecine video sources, thereby reconstructing the original 24-fps film format. This improves the quality for compatibility with noninterlaced displays and eliminates some of the redundant data prior to compression.

Motion-compensated standards conversion

More general standards or format conversion is provided by motion compensation methods such as those based on phase correlation (see Chapter 6). These are required to convert content between interlaced and progressive formats, between HD and SD and other formats, and between formats of different frame rates and color spaces. Phase correlation is a Fourier-based method that exploits duality between displacement in the time (or space) domain and phase shift in the frequency domain. It offers better immunity to noise and luminance changes than conventional block matching methods. Subpixel accuracy can easily be achieved, as is required for the generation of new frames in the target format.

4.4 Gamma correction

Gamma correction [7] is used to correct the differences between the way a camera captures content, the way a display displays content, and the way our visual system processes light. Our eyes do not respond to light in the same way that a camera captures it. In basic terms, if twice the number of photons hits a digital image sensor, then the output voltage will be twice as big. For the case of older CRT-based cameras this was not, however, the case and there was a highly nonlinear relationship between light intensity and output voltage. The HVS also has a very nonlinear response to luminance levels, being more sensitive to small changes in darker areas and much less sensitive in light areas (recall Weber's law from Chapter 2). Without correction, to avoid banding effects due to perceptible jumps between coding levels, around 11 bits would be needed. However, over most of the higher end of the scale, the coding would be highly redundant since coding levels would be imperceptible. With gamma precorrection we can get an acceptable representation with only 8 bits.

Photopic human vision characteristics can be approximated by a power law function. Correction of the signal output from the camera using this type of function ensures that not too many bits are allocated to brighter regions where humans are less discriminating and more to darker regions, where they are more so. Video cameras therefore normally perform a nonlinear mapping of illumination intensity to output value, providing a more uniform perceptual scale with finer increments for darker values. This is known as the gamma characteristic of the device as is described by the transfer function:

$$V = c_1 \Phi^{\gamma} + c_2, \tag{4.8}$$

where Φ is the luminous flux (normalized), V is the value (voltage) of the signal, c_1 is the camera sensitivity, and c_2 is an offset value. Fig. 4.12 shows transfer characteristic plots for three gamma values, $\gamma = \{0.45, 1, 2.2\}$. These values are typical of those used in modern TV systems. Fig. 4.13 shows some corrected examples of an image at various gamma levels.

It should be noted that modern formats such as Rec.709 use a piecewise transfer characteristic from the luminance to the output. This is linear at the lower region and then changes to the power law for the rest of the range. For Rec.709 this is

$$V = \begin{cases} 4.5\Phi, & \Phi < 0.018, \\ 1.099\Phi^{0.45} - 0.099, & \Phi \geq 0.018. \end{cases} \tag{4.9}$$

Although a topic of some confusion, it is becoming accepted to refer to the gamma-corrected luminance signal as *luma*.

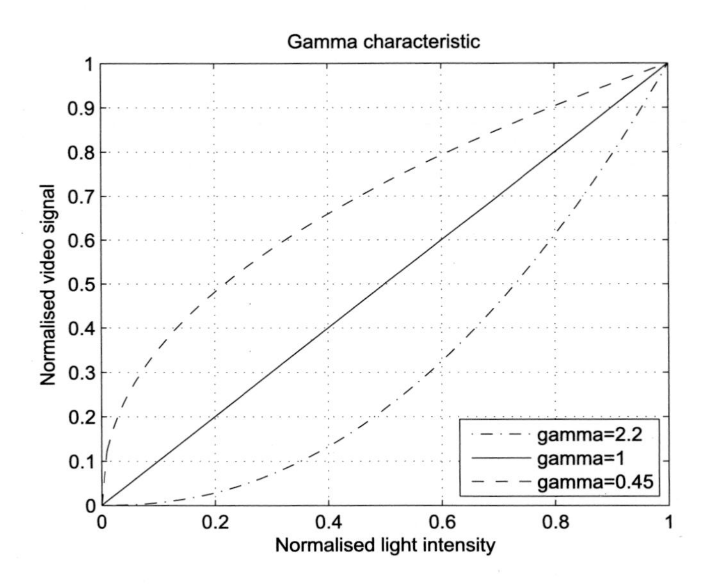

FIGURE 4.12 Gamma curves for $\gamma = \{0.45, 1, 2.2\}$.

4.5 Color spaces and color transformations

4.5.1 Color descriptions and the HVS

Trichromacy theory

As we have examined in Chapter 2, the cone cells in the retina are the main photoreceptors for normal photopic conditions and have sensitivities peaking around short (S: 420–440 nm), middle (M: 530–540 nm), and long (L: 560–580 nm) wavelengths (the low-light sensors [rods] peak around 490–495 nm). The sensation of color in humans is the result of electromagnetic radiation in the range 400–700 nm incident on the retina.

One of the challenges when color television was introduced in the 1960s was the trade-off between delivering realistic colors to the screen vs. conserving bandwidth. In simple terms, the transmission of color information (assuming the use of three color channels [RGB]) occupies three times the bandwidth of a monochrome transmission. The solution to this for analog TV was (inspired by the trichromatic theory of vision [see Chapter 2]) to encode the color information separately from the luma information, with significantly reduced resolution in order to conserve bandwidth. This had the advantage that the luma signal was backwards compatible with older monochrome TV sets while, for color sets, the higher resolution luma and the lower resolution color signals combined in the HVS to produce a perceptually high-resolution color image. As technology has progressed since the introduction of analog color television, there has been substantial effort invested in developing new color space representations appropriate for modern digital encodings, formats, and displays. Some of these are reviewed below. For further information, the reader is referred to Netravali and Haskell [1] and the very good introduction to color spaces presented by Poynton [3,7].

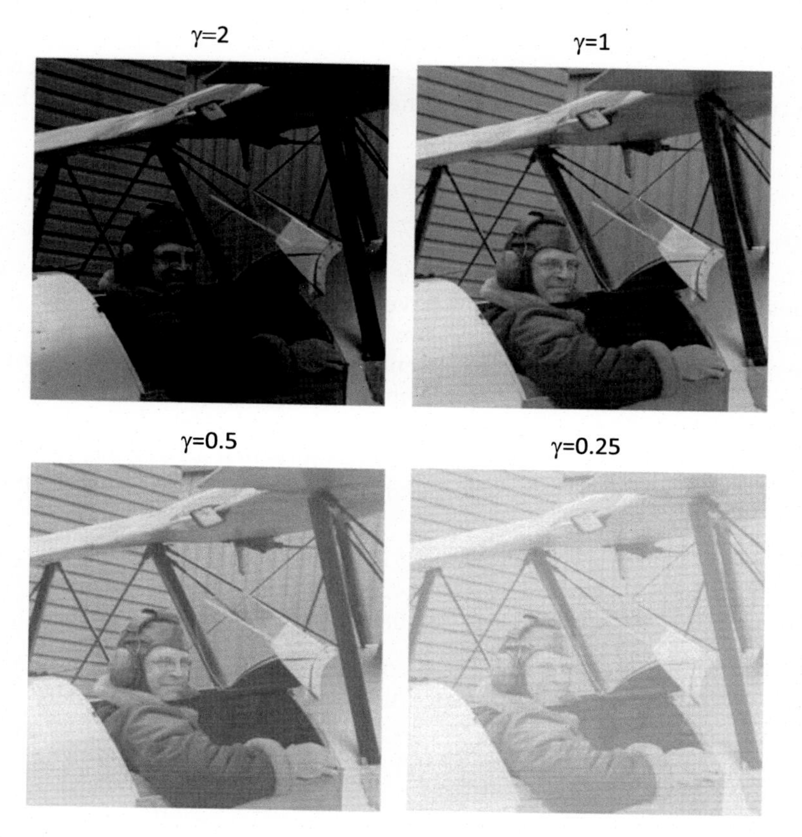

FIGURE 4.13 Examples of gamma correction.

Much of colorimetry is based on the tristimulus theory of color, i.e., any color can be formed from three primaries as long as they are orthogonal, but also on Grassman's laws, which reflect the linearity and additivity rules of color. These state that the color match between any two stimuli is constant when the intensities of the two stimuli are increased or decreased by the same factor. For example, if we have two sources, C_1 and C_2, then if

$$\begin{aligned} &C_1 = R_{C1}(R) + G_{C1}(G) + B_{C1}(B), \\ &C_2 = R_{C2}(R) + G_{C2}(G) + B_{C2}(B), \quad \text{and} \\ &C_3 = C_1 + C_2, \ \text{then} \end{aligned} \tag{4.10}$$

$$C_3 = [R_{C1} + R_{C2}](R) + [G_{C1} + G_{C2}](G) + [B_{C1} + B_{C2}](B). \tag{4.11}$$

Color spaces

We saw in Chapter 2 that the CIE has defined the coordinates of color spaces, based on a trichromatic representation, which describe the set of color sensations experienced by the human observer. The first of these, CIE 1931 (developed in 1931), was based on the concept of a standard observer, matching the observations from a large number of subjects (with normal vision) to controlled stimuli at a 2-degree field of view. These stimuli were based on the primaries

$$[\ R_0 \quad G_0 \quad B_0\ ] = [\ 700 \quad 546.1 \quad 435.8\ ]\ \text{nm}.$$

Color space transformations

According to the basic principles of colorimetry, various linear combinations of the RGB stimulus values can be derived that might have better properties than the original format. CIE introduced an alternative reference system (or color space), referred to as XYZ, which could provide a more flexible basis for representing tristimulus values. A color space maps a range of physically produced colors (e.g., from mixed lights) to an objective description of those color sensations based on a trichromatic additive color model, although not usually the LMS space. CIE XYZ can represent all the color sensations that an average person can experience and is a transformation of the $[R_0, G_0, B_0]$ space. The CIE have defined this transform as

$$\begin{bmatrix} X \\ Y \\ Z \end{bmatrix} = \begin{bmatrix} 2.365 & -0.515 & 0.005 \\ -0.897 & 1.426 & -0.014 \\ -0.468 & 0.089 & 1.009 \end{bmatrix} \begin{bmatrix} R_0 \\ G_0 \\ B_0 \end{bmatrix}. \tag{4.12}$$

CIE defined the XYZ space such that the Y component corresponds to the luminance signal, normalized to equal energy white. Also, unlike some RGB systems, where contributions have to be negative to match a color, the XYZ system is wholly positive. However, the disadvantage of this is that the XYZ primaries do not correspond to visible colors, i.e., they cannot be physically realized by an actual stimulus.

Chromaticity values can also be defined in the XYZ system as follows:

$$\begin{aligned} x &= \frac{X}{X+Y+Z}, \\ y &= \frac{Y}{X+Y+Z}, \\ z &= \frac{Z}{X+Y+Z}. \end{aligned} \tag{4.13}$$

Large x values correspond to red or orange hues, y values correspond to green, blue-green, or yellow-green, and large z corresponds to large blue, violet, or purple hues. Because $x+y+z=1$, only two of these values are actually required, so normally this is expressed as an Yxy system, where Y represents luminance and x and y independently represent chrominance.

Chromaticity diagrams

The chromaticity coordinates can be expressed on a 2D chromaticity diagram like that shown in Fig. 4.14. This diagram shows all the hues perceivable by the standard observer

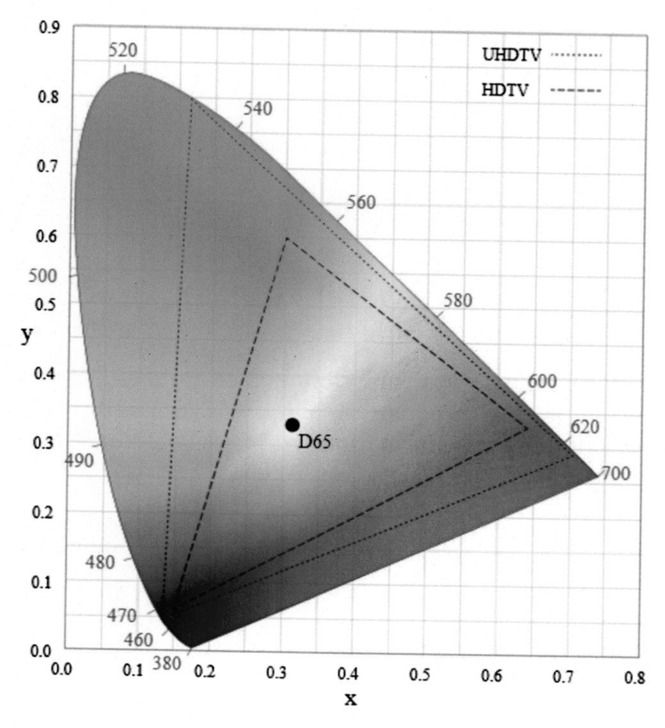

FIGURE 4.14 CIE1931 chromaticity diagram (reproduced with permission from Wikimedia Commons (http://commons.wikimedia.org/wiki/File:CIExy1931_Rec_2020_and_Rec_709.svg).

for various (x,y) pairs and also indicates the spectral wavelengths of the dominant single frequency colors. The D65 point shown on this diagram is the white point, which in this case corresponds to a color temperature of 6500K (representative of average daylight).

CIE XYZ or CIE 1931 is a standard reference which has been used as the basis for defining most other color spaces. This maps human color perception based on the two CIE parameters x and y. Although revised in 1960 and 1976, CIE 1931 is still the most commonly used reference.

If we consider again the CIE 1931 color space in Fig. 4.14, this also shows two triangles that correspond to the colors representable (the color gamut) by the Rec.709 (HDTV) and the Rec.2020 (UHDTV) color spaces. It is interesting to note that the latter can represent many colors that cannot be shown with Rec.709. The Rec.2020 primaries correspond to 630 nm (R), 532 nm (G), and 467 nm (B). The Rec.2020 space covers approximately 76% of the CIE 1931 space, whereas Rec.709 covers only 36%, thus offering better coverage of more saturated colors.

Color spaces for analog TV

In almost all cases, a format YC_1C_2 is used for analog and digital formats, where Y represents signal luminance. This is done for the following two reasons:

- compatibility with monochrome displays,
- possible bandwidth reduction on the chrominance signals with no loss of perceptual quality.

For example, in the case of US NTSC systems for color TV, the primaries are defined in terms of CIE xyz coordinates as follows:

$$\mathbf{m} = \begin{array}{cccc} & x & y & z \\ R: & 0.67 & 0.33 & 0.00 \\ G: & 0.21 & 0.71 & 0.08 \\ B: & 0.14 & 0.08 & 0.78 \end{array}. \tag{4.14}$$

The actual values used in the TV transmission format are those after gamma correction. Consequently, after various normalizations and rotations [1] the resulting color space, defined as YIQ, is given by the following transformation (where R', G', and B' represent the gamma-corrected camera output signals):

$$\begin{bmatrix} Y \\ I \\ Q \end{bmatrix} = \begin{bmatrix} 0.299 & 0.587 & 0.114 \\ 0.596 & -0.274 & -0.322 \\ 0.211 & -0.523 & 0.311 \end{bmatrix} \begin{bmatrix} R' \\ G' \\ B' \end{bmatrix}. \tag{4.15}$$

A similar mapping for PAL systems in Europe is given by

$$\begin{bmatrix} Y \\ U_t \\ V_t \end{bmatrix} = \begin{bmatrix} 0.299 & 0.587 & 0.114 \\ -0.147 & -0.289 & -0.436 \\ 0.615 & -0.515 & -0.100 \end{bmatrix} \begin{bmatrix} R' \\ G' \\ B' \end{bmatrix}. \tag{4.16}$$

Color spaces for digital formats

With the introduction of digital formats, starting with SDTV (Rec.601), there was a need to adjust the color space transformation to deal with these digital formats and finite word lengths. The components again are referenced against the gamma-corrected camera outputs and are quantized to 8 bits. The representation for the ITU-R Rec.601 system is known as YC_bC_r and is given by the following transformation from corrected 8-bit RGB signals in the range 0 to 255[3]:

$$\begin{bmatrix} Y \\ C_b \\ C_r \end{bmatrix} = \begin{bmatrix} 0.257 & 0.504 & 0.098 \\ -0.148 & -0.291 & 0.439 \\ 0.439 & -0.368 & -0.071 \end{bmatrix} \begin{bmatrix} R' \\ G' \\ B' \end{bmatrix} + \begin{bmatrix} 16 \\ 128 \\ 128 \end{bmatrix}. \tag{4.17}$$

In order to provide coding headroom, the full range of 256 values is not used. Instead ranges from 16–235 (Y) and 16–240 (C_b and C_r) are employed. It should be noted that some authors add a prime to the Y (luma) component above to differentiate it from the luminance signal, because it results after nonlinear encoding based on gamma-corrected RGB primaries. C_b and C_r are the blue-difference and red-difference chroma components.

[3] Studio $R'G'B'$ signals are in the range 16 to 235 and require a modified conversion matrix.

FIGURE 4.15 Color difference images using the YC_bC_r decomposition. Top left: Original. Right: Luma channel. Bottom left: C_b channel. Right: C_r channel.

Similarly, the ITU-R Rec.709 system for HDTV formats is given by

$$\begin{bmatrix} Y \\ C_b \\ C_r \end{bmatrix} = \begin{bmatrix} 0.183 & 0.614 & 0.062 \\ -0.101 & -0.338 & 0.439 \\ 0.439 & -0.399 & -0.040 \end{bmatrix} \begin{bmatrix} R' \\ G' \\ B' \end{bmatrix} + \begin{bmatrix} 16 \\ 128 \\ 128 \end{bmatrix}. \tag{4.18}$$

Consider as an example the image of Thala Beach shown in Fig. 4.15. This displays the Y, C_b, and C_r components. Several things can be observed from this – for example the white clouds are represented as middle values in both chroma components, the blue sky is strong in the blue channel, and the sand and browner aspects of the tree show strongly in the red channel. Furthermore, the two chroma components appear fuzzy to the human eye relative to the luma channel and this provides some evidence that we could downsample them without loss of overall perceived detail.

4.5.2 Subsampled color spaces

Most capture devices produce anRGB output, where each component requires high precision and high resolution to provide perceptually high-quality images. As discussed above, this format is rarely used for transmission as it requires three times the bandwidth of a monochrome channel. Hence, as we have seen, a color space transformation is applied to produce a luminance signal Y and two color difference signals, C_b and C_r. The motivations for this are the following.

1. **Chroma subsampling:** As discussed in Chapter 2, the HVS has a relatively poor response to color information. The effective resolution of our color vision is about half that of luminance information. Separation of luma and chroma components thus means that we can subsample the chroma by a factor of two without affecting the luma resolution or the perceived image quality. This provides an immediate compression benefit as we can reduce the effective bit rate by 50% even before we consider decorrelation processes.
2. **Luma processing:** As well as having compatibility with monochrome displays, the availability of an independent luma signal is useful during compression. For example, we normally perform motion estimation only on the luma component and use the resulting vectors to compensate both luma and chroma components. We also often base objective quality assessments only on the luma component.

Chroma subsampling

The human eye has a relatively poor response to chroma spatial detail. Chroma components are thus normally subsampled (usually by a factor of 2) to reduce bandwidth requirements. The most commonly used subsampling patterns are shown in Fig. 4.16. This shows the relationships between luma and chroma samples for a range of common subsampling formats. These are:

- **4:4:4** – This is the full resolution image with luma and chroma samples at all pixel locations. It is used when there is a requirement for the highest-quality studio work.
- **4:2:2** – This is the conventional format for studio production. Most professional cameras output video in a 4:2:2 format. It provides subsampling horizontally but not vertically. This is advantageous in the case of interlaced scanning as it eliminates problems with color degradation.
- **4:2:0** – This system is the most common format for broadcast delivery, internet streaming, and consumer devices. It subsamples the chroma signals by a factor of two in both horizontal and vertical directions, maximizing the exploitation of perceptual color redundancy.
- **4:1:1** – This system was used in DV format cameras. It subsamples by a factor of 4 horizontally but not vertically. Its quality is inferior to 4:2:0 formats, so it is rarely used now.
- **4:0:0** – This is not shown in the figure but represents the simple case of a luma-only or monochrome image.

The question that arises is, how are the chroma signals downsampled? The answer is by preprocessing with a linear-phase FIR filter designed to avoid aliasing. Similarly signals are upsampled for reconstruction using an appropriate linear-phase FIR interpolation filter. It should be noted that the subsampling filter will dictate where the chroma samples are sited relative to the luma samples.

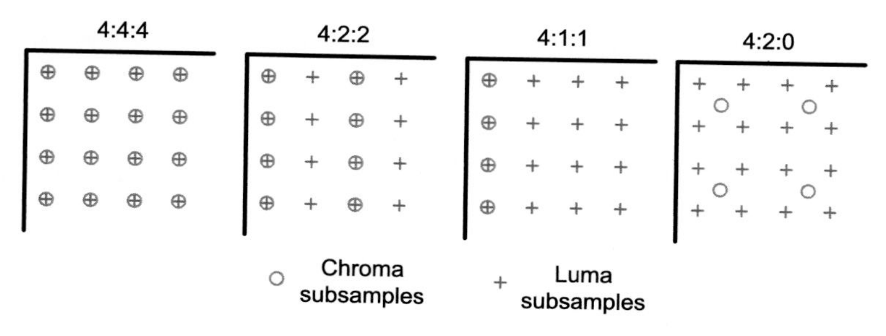

FIGURE 4.16 Common chroma subsampling formats.

Notation: 4:X:Y

This notation is confusing to most people and there have been several attempts to fit rules that explain its meaning, all of which fail under some circumstances. The reason for this is that historically, the 4:X:Y format was devised at the time of PAL and NTSC systems when subcarrier-locked sampling was being considered for component video. The first digit represented the luma horizontal sampling reference as multiple of $3\frac{3}{8}$ MHz. The second number X represented the horizontal sampling factor for chroma information relative to the first digit. The third number, Y, was similar to X, but in the vertical direction.

It is however obvious that the 4:2:0 and 4:1:1 formats do not comply with this definition. So other rules that do fit have been invented. Perhaps the best one is as follows (it is not very elegant but it works for most formats):

1. Assume a 4×2 block of pixels, with 4 luma samples in the top row.
2. X is the number of chroma samples in the top row.
3. Y is the number of additional chroma samples in the second row.

An issue of course is that with some 4:2:0 formats (as shown in Fig. 4.16), the chroma samples are not necessarily cosited with the luma samples).

Example 4.3. Color space transformations in ITU-R Rec.601

Consider the YC_bC_r format used with ITU-R Rec.601. Compute the component values for the following cases:

a. R′G′B′=[0 0 0].
b. R′G′B′=[220 0 0].
c. R′G′B′=[0 220 0].

Solution. Using Eq. (4.17), we can compute these values as follows:

a. YC_bC_r=[16 128 128].

A signal at the base black level in terms of luminance and at central and equal chroma levels.

b. YC_bC_r=[73 95 224].

A signal that is higher in the red color difference channel with mid levels of luminance and the blue channel.

c. YC_bC_r=[127 64 47].

Notice the difference between this and (b). The green channel is weighted more heavily as it contributes more to our sense of brightness.

Example 4.4. Chroma subsampling and compression

If a color movie with a duration of 120 minutes is represented using ITU-R.601 (720×576 @30 fps @8 bits, 4:2:0 format):

a. What hard disk capacity would be required to store the whole movie?
b. If the movie is MPEG-2 encoded at a compression ratio CR=50:1 and transmitted over a satellite link with 50% channel coding overhead, what is the total bit rate required for the video signal?

Solution. a. The bit rate is given by

$$720 \times 576 \times 30 \times 8 \times \frac{3}{2} = 149,299,200 \text{ b/s} = 149 \text{ Mb/s}.$$

For a 120-minute movie, the total storage space required is

$$120 \times 60 \times 149,299,200 = 1,074,954,240,000 \text{ bits} = 134 \text{ GB}.$$

b. With a compression ratio of 50:1, the bit rate would be approximately 3 Mbps. With 50% coding overhead, this represents a total bit rate of 4.5 Mbps.

4.5.3 Color sensing

High-end professional cameras in the past employed three CCD sensors, one for each color channel, illuminated via an optical splitter. However, with advances in CMOS technology, single-sensor solutions now dominate in both consumer and professional markets. A recent example is Samsung's 108-megapixel (Mp) ISOCELL Bright HMX, the industry's first mobile image sensor to exceed 100 million pixels. These sensors apply a color filter array (CFA – often referred to as a Bayer filter) to the image sensor elements in order to enable it to produce R, G, and B samples.

Bayer filtering

The Bayer CFA is a pattern of color filters [4] typically arranged as shown in Fig. 4.17. Half of the filter elements are green and the remainder are divided between blue and red. This

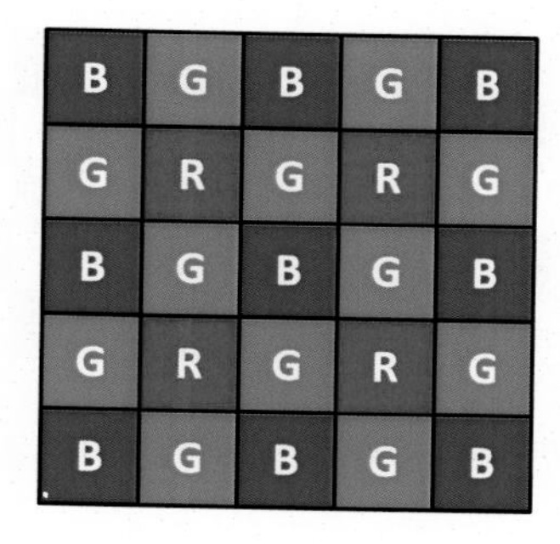

FIGURE 4.17 Bayer filter mosaic.

approximates human photopic vision, where the M and L cones combine to produce a bias in the green spectral region.

In general, the effective resolution of the Bayer CFA sensor is reduced compared to a three- or four-sensor array of the same resolution, and the degree of degradation is influenced by both the content being acquired and the demosaicing algorithm employed. As a rule of thumb, the effective sensor size in Mp is $1/\sqrt{2}$ times that of the actual sensor, so a 10-Mp array effectively becomes a 7-Mp array after demosaicing.

Bayer demosaicing

Consider the CFA decomposed into three color planes as shown in Fig. 4.18. For a simple bilinear interpolation, we process each pixel with the following convolution kernels:

$$K_R = K_B = \frac{1}{4}\begin{bmatrix} 1 & 2 & 1 \\ 2 & 4 & 2 \\ 1 & 2 & 1 \end{bmatrix}, \quad K_G = \frac{1}{4}\begin{bmatrix} 0 & 1 & 0 \\ 1 & 4 & 1 \\ 0 & 1 & 0 \end{bmatrix}, \tag{4.19}$$

where K_R, K_B, and K_G are used to generate the red, blue, and green values for each pixel location.

There are many classes of demosaicing algorithm and we will not consider them in detail here. These extend from relatively simple bilinear interpolators to bicubic or spline interpolation. These methods are however prone to artifacts, and more sophisticated methods, for example based on the exploitation of local pixel statistics or gradient estimation, can provide improvements. More advanced methods based on superresolution techniques and deep learning have been proposed that address these issues and eliminate aliasing [5,6].

4.6 Measuring and comparing picture quality

We introduce some basic topics in image and video quality assessment here. An excellent introduction to video quality metrics and subjective assessment is provided by Winkler in [8]. We also discuss these topics in much more detail in Chapter 10.

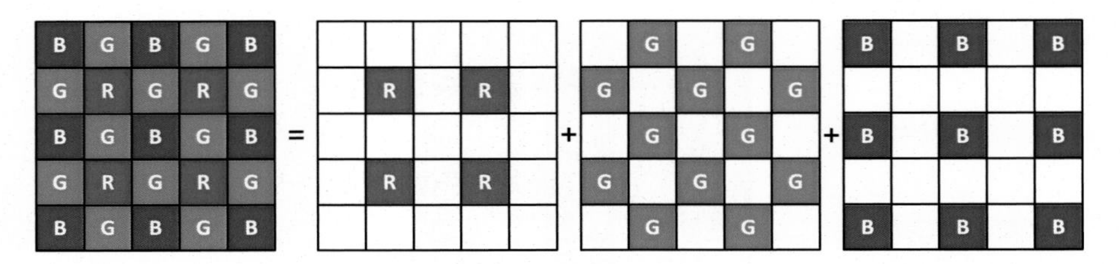

FIGURE 4.18 Bayer CFA decomposed into three color planes.

4.6.1 Compression ratio and bit rate

A very basic measure of compression performance is the compression ratio. This is simply defined as

$$C_1 = \frac{\text{no. of bits in original video}}{\text{no. of bits in compressed video}} = \frac{X \cdot Y \cdot B \cdot F}{\|\mathbf{c}\|} \quad \text{(unitless)}, \tag{4.20}$$

where B is the word length used in the original, F is the number of frames in the sequence, and $\mathbf{c}$ represents the compressed bitstream. This provides an absolute ratio of the compression performance, and only if used in the context of similar content, similar-sized images, and similar target qualities can it be used as a meaningful comparator. A commonly employed alternative used for still images is the normalized bit rate:

$$C_2 = \frac{\text{no. of bits in compressed video}}{\text{no. of pixels in original video}} = \frac{\|\mathbf{c}\|}{X \cdot Y \cdot F} \quad \text{(bits per pixel)}. \tag{4.21}$$

This will also vary according to content type and image resolution but also in terms of the original signal word length. It is just a rate indicator and provides no assessment of quality. A commonly used parameter for video is the actual bit rate:

$$C_3 = \frac{\text{no. of bits in compressed video}}{\text{no. of frames in original video}} \times \text{frame rate} = \frac{\|\mathbf{c}\| \cdot f}{F} \quad \text{(bits per second)}. \tag{4.22}$$

Again, this gives no information on the quality of the compressed signal, but it does provide us with an absolute value of how much channel capacity is taken up by the signal coded in a given manner.

In order for all of these measures to be useful they need to be combined with an associated measure of image or video quality. These are considered below.

4.6.2 Objective distortion and quality metrics

Mean squared error (MSE)

Consider an image $\mathbf{S} = s[x, y]$, $x = 0, \cdots, X-1$, $y = 0, \cdots, Y-1$ and its reconstructed version $\tilde{\mathbf{S}}$. If the area of the image (or equivalently for a monochrome image, the number of samples in it) is $A=XY$, then the MSE is given by

$$\mathrm{MSE} = \frac{1}{A}\left(\sum_{x=0}^{X-1}\sum_{y=0}^{Y-1}(s[x,y]-\tilde{s}[x,y])^2\right) = \frac{1}{A}\left(\sum_{x=0}^{X-1}\sum_{y=0}^{Y-1}(e[x,y])^2\right). \tag{4.23}$$

A signal to noise ratio (SNR) is also common in many communications applications:

$$\mathrm{SNR} = 10\cdot\log\left(\frac{E\left\{(s[x,y]-\mu_s)^2\right\}}{E\left\{(s[x,y]-\tilde{s}[x,y])^2\right\}}\right) = 20\cdot\log\left(\frac{\sigma_s}{\sigma_e}\right)\ \mathrm{dB}. \tag{4.24}$$

If we extend the MSE calculation to a video sequence with K frames, then the MSE is given by

$$\mathrm{MSE} = \frac{1}{KA}\left(\sum_{z=0}^{K-1}\sum_{x=0}^{X-1}\sum_{y=0}^{Y-1}(s[x,y,z]-\tilde{s}[x,y,z])^2\right). \tag{4.25}$$

Peak signal to noise ratio (PSNR)

MSE-based metrics are in common use as objective measures of distortion, due mainly to their ease of calculation. An image with a higher MSE will generally express more visible distortions than one with a low MSE. In image and video compression, it is however common practice to use peak SNR (PSNR) rather than MSE to characterize reconstructed image quality. The PSNR is of particular use if images are compared with different dynamic ranges and it is employed for a number of different reasons:

1. MSE values will take on different meanings if the word length of the signal samples changes.
2. Unlike many natural signals, the mean of an image or video frame is not normally zero and indeed will vary from frame to frame.
3. The PSNR normalizes MSE with respect to the peak signal value rather than signal variance, and in doing so enables direct comparison between the results from different codecs or systems.
4. The PSNR can never be less than zero.

The PSNR for an image **S** is given by

$$\mathrm{PSNR} = 10\cdot\log_{10}\left(\frac{As_{max}^2}{\sum_{x=0}^{X-1}\sum_{y=0}^{Y-1}(e[x,y])^2}\right), \tag{4.26}$$

where for a word length B, $s_{max} = 2^B - 1$. We can also include a binary mask, $b[x,y]$, so that the PSNR value for a specific arbitrary image region can be calculated. Thus,

$$\mathrm{PSNR} = 10 \cdot \log_{10}\left(\frac{s_{max}^2 \sum_{x=0}^{X-1}\sum_{y=0}^{Y-1} b[x,y]}{\sum_{x=0}^{X-1}\sum_{y=0}^{Y-1} (b[x,y] \cdot e[x,y])^2}\right). \tag{4.27}$$

Although there is no perceptual basis for the PSNR, it does fit reasonably well with subjective assessments, especially in cases where algorithms are compared that produce similar types of artifact. It remains the most commonly used objective distortion metric, because of its mathematical tractability and because of the lack of any widely accepted alternative. It is worth noting, however, that the PSNR can be deceptive, for example when:

1. There is a phase shift in the reconstructed signal. Even tiny phase shifts that the human observer would not notice will produce significant changes.
2. There is visual masking in the coding process that provides perceptually high quality by hiding distortions in regions where they are less noticeable.
3. Errors persist over time. A single small error in a single frame may not be noticeable, yet it could be annoying if it persists over many frames.
4. Comparisons are being made between different coding strategies (e.g., synthesis-based vs. block transform-based; see Chapters 10 and 13).

Consider the images in Fig. 4.19. All of these have the same PSNR value (16.5 dB), but most people would agree that they do not exhibit the same perceptual qualities. In particular, the bottom right image with a small spatial shift is practically indistinguishable from the original. In cases such as this, perception-based metrics can offer closer correlation with subjective opinions and these are considered in more detail in Chapter 10.

PSNR for color images and for video

For color images with three (R,G,B) values per pixel, the PSNR can be calculated, as above, separately for each color channel. Alternatively a value for the combined image can be calculated, except that the MSE is now based on the sum of squared value differences across R, G, and B images:

$$\mathrm{MSE} = \frac{1}{3A}\left(\sum_{RGB}\sum_{x=0}^{X-1}\sum_{y=0}^{Y-1}(s[x,y] - \tilde{s}[x,y])^2\right). \tag{4.28}$$

In practice, because the human eye is not equally sensitive to all channels, it is common to just compute the luma PSNR. This can easily be performed when using a format such as YC_bC_r, where the luma component represents a weighted average of the color channels. If one does want to explicitly incorporate color channels, for example in a subsampled 4:2:0 format, then computation of the MSE is generally performed as follows:

$$\mathrm{MSE}_{4:2:0} = \frac{1}{6}\left(4\mathrm{MSE}_Y + \mathrm{MSE}_{Cb} + \mathrm{MSE}_{Cr}\right). \tag{4.29}$$

Furthermore, it is important to note that the PSNR calculation for a video signal, based on averaging PSNR results for individual frames, is not the same as computing the av-

FIGURE 4.19 Quality comparisons for the same PSNR (16.5 dB). Top left to bottom right: original; AWGN (variance 0.24); grid lines; salt and pepper noise; spatial shift by 5 pixels vertically and horizontally.

erage MSE for all frames in the sequence and then computing the PSNR. The former is normally used for codec comparisons but can bias algorithms that produce SNR fluctuations, so normally the latter is used for comparison with perceptual metrics (see Chapter 10).

Typical values for the PSNR in lossy image and video compression are between 30 and 50 dB, where higher is better. Values over 40 dB are normally considered very good and those below 20 dB are normally unacceptable. When the two images are identical, the MSE will be zero, resulting in an infinite PSNR.

Further commentary on the PSNR and its variants are provided in Chapter 10.

Example 4.5. Calculating the PSNR

Consider a 3×3 image $\mathbf{S}$ of 8-bit values and its approximation after compression and reconstruction, $\tilde{\mathbf{S}}$. Calculate the MSE and the PSNR of the reconstructed block.

1	2	3
4	5	6
7	8	9

$\mathbf{S}$

1	2	2
4	4	8
7	8	8

$\tilde{\mathbf{S}}$

Solution. We can see that X=3, Y=3, s_{max}=255, and A=9. Using Eq. (4.23) the MSE is given by

$$\text{MSE} = \frac{7}{9},$$

and from Eq. (4.26), it can be seen that the PSNR for this block is given by

$$\text{PSNR} = 10 \cdot \log_{10}\left(\frac{9 \times 255^2}{7}\right) = 49.2 \text{ dB}.$$

Mean absolute difference (MAD) and sum of absolute differences (SAD)

MSE-based metrics, although relatively simple to implement, still require one multiplication and two additions per sample. In order to reduce complexity further, especially in some highly search-intensive operations such as motion estimation (see Chapters 8 and 9), other simpler metrics have been employed. One such metric is the mean absolute difference (MAD), which corresponds to the L_1 norm. The MAD between two images or regions, s_1 and s_2, is defined as

$$\text{MAD} = \frac{1}{A}\sum_{x=0}^{X-1}\sum_{y=0}^{Y-1}\left|s_1[x,y] - s_2[x,y]\right|, \tag{4.30}$$

where A is defined as previously. For most purposes the sum of absolute differences (SAD) metric offers the same functionality, but without the complexity of the additional division operation:

$$\text{SAD} = \sum_{x=0}^{X-1}\sum_{y=0}^{Y-1}\left|s_1[x,y] - s_2[x,y]\right|. \tag{4.31}$$

Sum of absolute transformed differences (SATD)

Calculating the absolute difference in a transform domain can improve the versatility of the cost function. Typically a Hadamard transform is applied prior to computing the SAD value. This metric not only serves to measure the overall difference between two images but also,

TABLE 4.2 ITU-R Rec. BT.500 subjective evaluation impairment scale.

Score	Impairment	Quality rating
5	Imperceptible	Excellent
4	Perceptible but not annoying	Good
3	Slightly annoying	Fair
2	Annoying	Poor
1	Very annoying	Bad

due to the transformation process, factors in an estimate of the coding cost of a residual signal. The SATD metric is thus a useful basis for rate-distortion optimization (RDO), particularly in the context of motion estimation residuals. It is given by

$$\text{SATD} = \sum_{n=1}^{N-1} \sum_{m=0}^{M-1} |r_H(n,m)|, \tag{4.32}$$

where $r_H(n,m)$ are the Hadamard transform coefficient values (see Chapter 5), given by

$$\mathbf{R}_{\mathrm{H}} = \mathbf{H} \cdot (\mathbf{S}_1 - \mathbf{S}_2) \cdot \mathbf{H}^{\mathrm{T}}. \tag{4.33}$$

4.6.3 Subjective assessment

Mathematical, distortion-based metrics such as the MSE and the PSNR can in some cases be poor indicators of subjective quality. Therefore, it has been necessary to develop perception-based criteria to achieve better (subjectively meaningful) quality assessments. One example of a perception-based quality assessment tool is the mean opinion score (MOS). With MOS measurements, a number of observers view an image and assess its quality on a five-point scale from bad to excellent. The subjective quality of the image (the MOS) is then characterized by performing statistical analysis on the ratings across a representative number of subjects.

Probably the most common testing procedure still is the double-stimulus continuous quality scale (DSCQS) approach, which is described in ITU-R Rec. BT.500 [9], although several multimedia quality evaluations will be based on single-stimulus tests. DSCQS testing employs a five-point impairment scale as shown in Table 4.2. ITU-R Rec. BT.500 sets out a wide range of testing criteria and methods, covering viewing conditions, choice of test material, characteristics of the observer ensemble, methods for conducting the test, and methods for analyzing the results. Further details are provided in Chapter 10.

The problem with subjective tests is that they are time consuming and can be expensive to run. Hence this provides additional impetus for the development of robust perceptual quality metrics that can be used, with confidence, in their place.

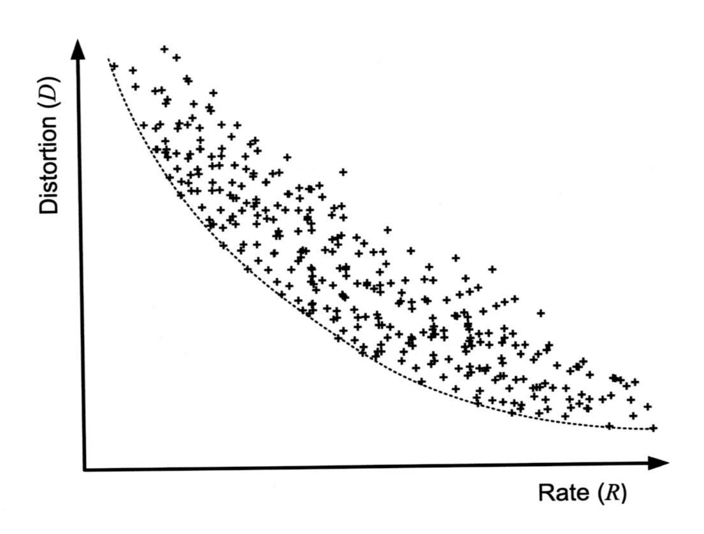

FIGURE 4.20 Rate-distortion plot for various coding parameter choices.

4.7 Rates and distortions

4.7.1 Rate-distortion characteristics

The bit rate for any compressed sequence will depend on:

- the encoding algorithm used (intra- vs. interframe, integer or subpixel motion estimation, coding modes or block sizes available);
- the content (high spatio-temporal activity will in general require more bits to code);
- the encoding parameters selected. At the coarsest level this includes things such as spatial resolution and frame rate. At a finer granularity, issues such as quantizer control, intra- vs. inter-modes, and block size choices will be key. The difference between an encoding based on good parameter choices vs. one based on poor choices is huge.

Following the argument in Section 4.6.1 we need a description of coder performance that captures how it trades rate against distortion. This is achieved using a graph that plots rate vs. distortion (or rate vs. quality) for a given codec and a given video file. This allows us to compare codec performances and to assess how parameter selections within a single codec can influence performance. However if we randomly select coding parameters, some of these combinations might produce good results and some will inevitably produce poor results. The plot of points might, for many different parameter choices, look something like that in Fig. 4.20. The aim is, of course, to ensure that the operating points produced by the codec in practice are those on the dotted curve – the Pareto curve that produces the lowest possible distortion for any given coding rate. To achieve this we must perform RDO or rate-quality optimization (RQO).

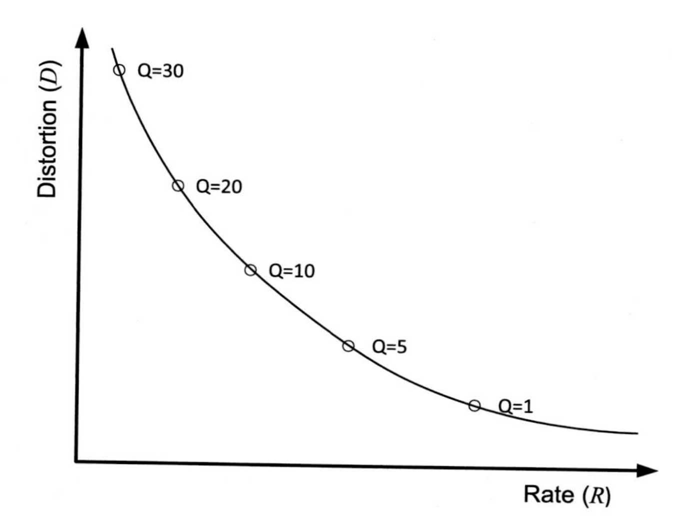

FIGURE 4.21 Rate-distortion plot showing quantizer controlled operating points.

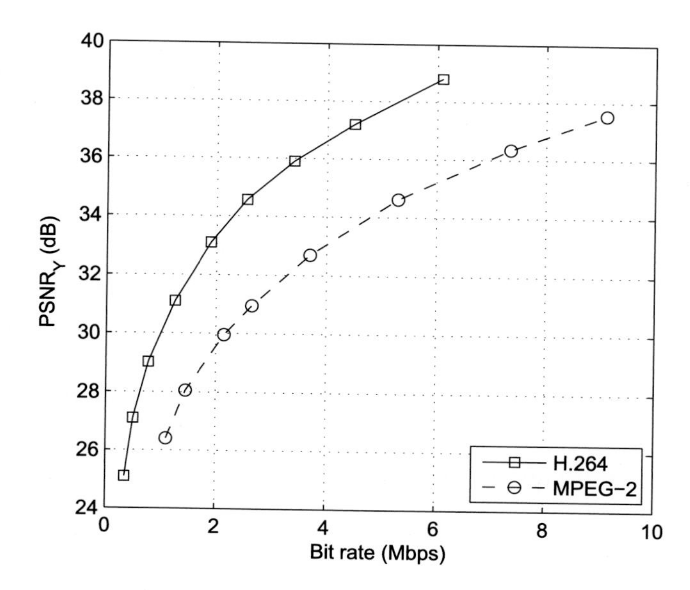

FIGURE 4.22 Example rate-quality comparison curves for typical standard-definition entertainment content.

Take for example a simple codec where the only control over RD performance is quantizer step size. In this case, adjusting the quantizer would produce different codec operating points as shown in Fig. 4.21.

An example rate-quality curve (based on the PSNR rather than distortion) is shown in Fig. 4.22. This compares the performance of two codecs (MPEG-2 and H.264/AVC).

4.7.2 Rate-distortion optimization

RDO aims to maximize image or video quality, subject to bit rate constraints. RDO requirements and methods are nonnormative in all video coding standards, but nonetheless are a key differentiator in terms of encoder performance. Many of the advanced features (e.g., block size selections) included in modern standards such as H.264/AVC and HEVC will not deliver savings without optimization included as part of the coder control process. The RDO must select the best modes and parameter sets for each region of the video. Often a Lagrangian optimization approach is adopted where the distortion measure, D, is based on sum of squared difference and the rate, R, includes all bits associated with the decision, including header, motion, side information, and transform data.

For all possible parameter vectors, $\mathbf{p}$, the aim is to solve the constrained problem

$$\min_{\mathbf{p}} D(\mathbf{p}) \text{ s.t. } R\left(\mathbf{p}\right) \leq R_T, \tag{4.34}$$

where R_T is the target bit rate, or to solve the unconstrained Lagrangian formulation

$$\mathbf{p}_{opt} = \arg\min_{\mathbf{p}} \left\{D\left(\mathbf{p}\right) + \lambda R\left(\mathbf{p}\right)\right\}, \tag{4.35}$$

where λ controls the rate-distortion trade-off. Further details on RDO methods can be found in Chapter 10.

4.7.3 Comparing video coding performance

To design and validate efficient compression algorithms, we must benchmark their performance against competing algorithms on representative datasets (discussed further in Chapter 10). In order to measure the coding gain of one codec relative to another over a range of bit rates, Gisle Bjøntegaard proposed a model to calculate the average PSNR and bit rate differences between two rate-distortion (R-D) curves [10]. This model can be used in two ways:

1. Bjøntegaard delta PSNR (BD-PSNR) – computes the average PSNR difference (dB) for the same bit rate range, and
2. Bjøntegaard delta rate (BD-rate) – computes the average bit rate difference (%) for the same PSNR range.

The method firstly fits a third order polynomial to the two RD curves being compared using four ln(Rate) vs. PSNR points (Fig. 4.23):

$$\text{PSNR} = f(\ln R) = a + b \cdot (\ln R) + c \cdot (\ln R)^2 + d.(\ln R)^3, \tag{4.36}$$

$$\ln R = g(\text{PSNR}) = a' + b'(\text{PSNR}) + c' \cdot (\text{PSNR})^2 + d' \cdot (\text{PSNR})^3. \tag{4.37}$$

An expression for the integral of each curve is then used to calculate BD-PSNR and BD-rate as follows:

$$\text{BD-PSNR} = \frac{\Delta S}{\Delta R}\,(\text{dB}), \tag{4.38}$$

$$\text{BD-rate} = \left(\mathrm{e}^{\Delta S'/\Delta\text{PSNR}} - 1\right) \times 100\,(\%). \tag{4.39}$$

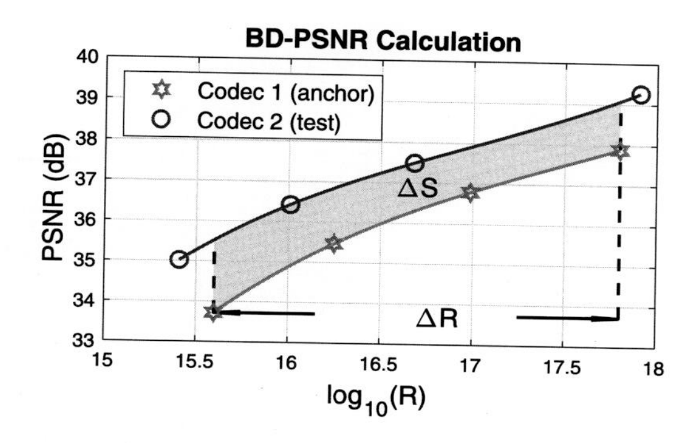

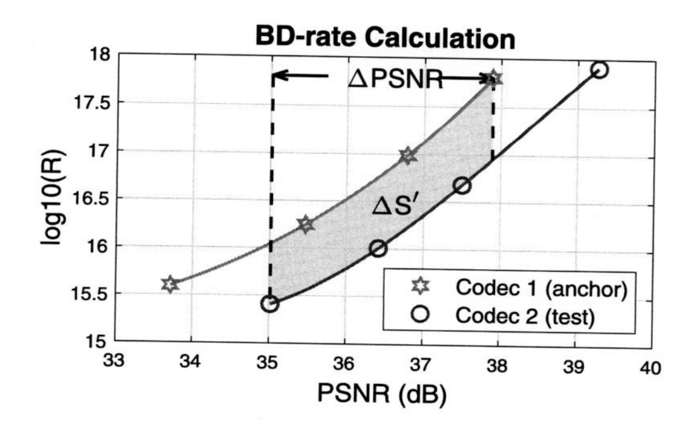

FIGURE 4.23 BD-PSNR and BD-rate calculations.

Example 4.6. Calculating BD-PSNR and BD-rate

Given the following four rate (Mbps)/PSNR (dB) points of the Campfire (2160p30, 10-bit) sequence encoded by H.254/JM 19.0 and VVC/VTM 3.0, calculate the BD-PSNR and BD-rate using H.264/JM as the anchor codec.

Codec	R1/Q1	R2/Q2	R3/Q3	R4/Q4
H.264	5.947/33.720	11.393/35.467	23.844/36.789	53.758/37.894
VVC	4.925/35.016	8.965/36.420	17.556/37.499	59.923/39.272

Solution. These curves are plotted in Fig. 4.23. Firstly we fit an RD curve for each codec using a third order polynomial function (computed using MATLAB®):

H.264:

$$\mathrm{PSNR} = f_1(\ln R) = -895.612 + 156.785(\ln R) - 8.853(\ln R)^2 + 0.168(\ln R)^3,$$

$$\ln R = g_1(\mathrm{PSNR}) = -49.274 + 6.801(\mathrm{PSNR}) - 0.241(\mathrm{PSNR})^2 + 0.003(\mathrm{PSNR})^3,$$

VVC:

$$\mathrm{PSNR} = f_2(\ln R) = -957.830 + 172.974(\ln R) - 10.091(\ln R)^2 + 0.198(\ln R)^3,$$

$$\ln R = g_2(\mathrm{PSNR}) = 710.547 - 55.603(\mathrm{PSNR}) + 1.467(\mathrm{PSNR})^2 - 0.013(\mathrm{PSNR})^3.$$

Calculate ΔS and ΔR for BD-PSNR:

$$\Delta S = \int_{15.598}^{17.800} (f_2(\ln R) - f_1(\ln R))d(\mathrm{lnR}) = 2.8448, \quad \Delta R = 2.202.$$

Then

$$\text{BD-PSNR} = \frac{\Delta S}{\Delta R} = 1.292\,(\text{dB}).$$

Calculate $\Delta S'$ and ΔPSNR for BD-rate:

$$\Delta S' = \int_{35.016}^{37.894} (g_2(\text{PSNR}) - g_1(\text{PSNR}))d(\text{PSNR}) = -2.151; \quad \Delta\text{PSNR} = 2.878.$$

Then

$$\text{BD-rate} = (e^{\frac{\Delta S'}{\Delta \text{PSNR}}} - 1) \times 100 = -52.643\%.$$

4.8 Summary

This chapter has introduced the digital representations, formats, processing techniques, and assessment methods that underpin the coding process. Techniques such as gamma correction and color space conversion have been shown to be essential in providing a digital description of a video signal that is best suited to further compression. Indeed, by exploiting the color processing and perception attributes of the HVS, we have shown that appropriate luma/chroma formats can be designed that reduce the bit rate for a digital video by 50%, even before we apply the conventional compression methods that are described in the following chapters. Finally, we have emphasized that assessment methods are an essential element in video codec design, both as a means of comparing the performance of different compression systems and as a basis for optimizing a codec's rate-distortion performance.

References

[1] A. Netravali, B. Haskell, Digital Pictures: Representation, Compression and Standards, 2e, Plenum Press, 1995.
[2] http://en.wikipedia.org/wiki/Interlaced_video, September 2013.
[3] C. Poynton, A guided tour of color space, available from http://www.poynton.com/PDFs/Guided_tour.pdf, August 2013.
[4] B. Bayer, Color imaging array, US Patent 3,971,065, 1976.
[5] X. Li, B. Gunturk, L. Zhang, Image demosaicing: a systematic survey, in: Proc. SPIE 6822, Visual Communications and Image Processing, 2008.
[6] N. Syu, Y. Chen, Y. Chuang, Learning deep convolutional networks for demosaicing, arXiv preprint, arXiv: 1802.03769, 2018.
[7] C. Poynton, Digital Video and HD, 2e, Morgan Kaufmann, 2012.
[8] S. Winkler, Digital Video Quality, John Wiley, 2005.
[9] Recommendation ITU-R BT.500-13, Methodology for the subjective assessment of the quality of television pictures, ITU-R, 2012.
[10] G. Bjøntegaard, Calculation of average PSNR differences between RD-curves, in: VCEG-M33, VCEG Meeting of ITU-T SG16, 2001.

CHAPTER

5

Transforms for image and video coding

As we saw in Chapter 3, most natural scenes captured using regular sampling will exhibit high levels of interpixel correlation. This means that the actual entropy of the pixels in a region of an image is likely to be significantly lower than that given by the first order entropy or, in other words, that the spatial representation is redundant. If our aim is to compress the image, it is therefore likely that an alternative representation may be available that is more amenable to reducing redundancy. The aim of image transformation is to create such a representation by decorrelating the pixels in an image or image region. This is achieved by mapping from the image domain to an alternative domain where the pixel energy is redistributed and concentrated into a small number of coefficients. This process is sometimes referred to as energy compaction.

As we will see, this redistribution of energy does not result in data compression as the transform itself is a lossless operation. What we can achieve, however, is a concentration of energy into a small number of high-valued coefficients. These larger coefficients are likely to have a higher psychovisual significance than their low-valued counterparts and can be coded accordingly.

This chapter firstly introduces the principles and properties of decorrelating transforms and explains their relationship to principal component analysis (PCA) and eigenanalysis. We introduce the optimal Karhunen–Loeve transform (KLT) in Section 5.4.2 and, after discussing its limitations, focus the remainder of the chapter on deriving and characterizing the discrete cosine transform (DCT). The DCT is introduced in Section 5.5 and its extension to 2D in Section 5.5.3. Quantization of DCT coefficients is presented in Section 5.6 and performance comparisons are provided in Section 5.7. We conclude with a discussion of DCT implementation and complexity in Section 5.8 and with a brief overview of JPEG in Section 5.9.

5.1 The principles of decorrelating transforms

5.1.1 The basic building blocks

Transformation presents a convenient basis for compression and this comes about through three mechanisms:

1. It provides data decorrelation and creates a frequency-related distribution of energy allowing low energy coefficients to be discarded.

Intelligent Image and Video Compression
https://doi.org/10.1016/B978-0-12-820353-8.00014-1

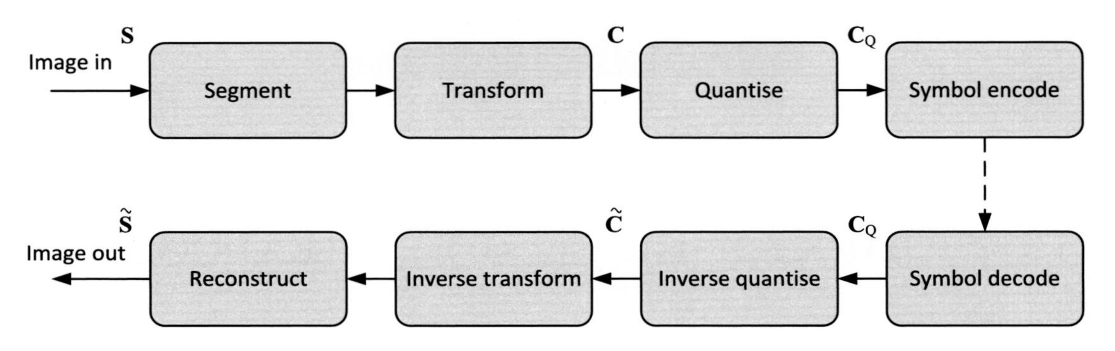

FIGURE 5.1 Typical transform-based image compression architecture.

2. Retained coefficients can be quantized, using a scalar quantizer, according to their perceptual importance.
3. The sparse matrix of all remaining quantized coefficients exhibits symbol redundancy, which can be exploited using variable-length coding.

The components in a typical transform-based image coding system are shown in Fig. 5.1. As we will explain later, for the purposes of transform coding an input image is normally segmented into small $N \times N$ blocks, where the value of N is chosen to provide a compromise between complexity and decorrelation performance. The components shown in this figure perform the following functions:

1. **Segmentation**: This divides the image into $N \times N$ blocks. This is a reversible one-to-one mapping.
2. **Transformation** (map or decorrelate): This transforms the raw input data into a representation more amenable to compression. Again this is normally a reversible one-to-one mapping.
3. **Quantization**: This reduces the dynamic range of the transformed output, according to a fidelity and/or bit rate criterion, to reduce psychovisual redundancy. For correlated spatial data, the resulting block of coefficients will be sparse. This is a many-to-one mapping and is not reversible; hence once quantized, the original signal cannot be perfectly reconstructed. This is the basis of lossy compression.
4. **Symbol encoding** (codeword assigner): The sparsity of the quantized coefficient matrix is exploited (typically by run-length coding) to produce a string of symbols. The symbol encoder assigns a codeword (a binary string) to each symbol. The code is designed to reduce coding redundancy and it normally uses variable-length codewords. This operation is reversible.

5.1.2 Principal components and axis rotation

The purpose of transformation in the context of data compression is energy compaction. This can be achieved through identification of trends in the data and then modifying the principal axes to optimally decorrelate it. It is best explained through a simple example. Consider

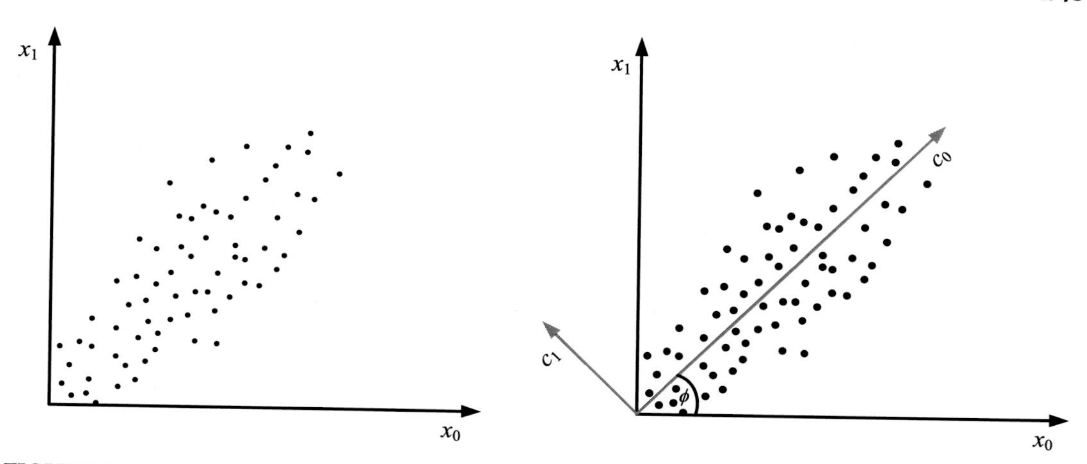

FIGURE 5.2 Plot of correlated adjacent pixel data (left) and decorrelation through rotation of principal axes (right).

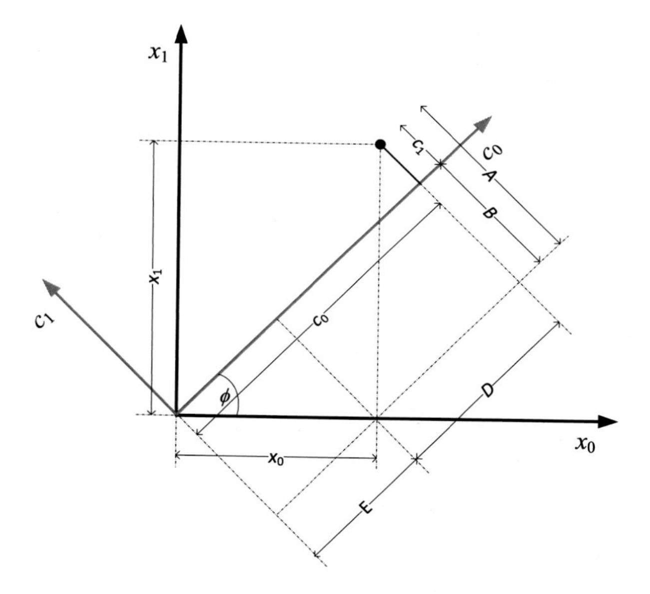

FIGURE 5.3 Relationship between original and transformed data.

the correlation between any two adjacent pixels in a natural image and let us represent the amplitudes of these pixels by $\{x_0, x_1\}$. A scatter plot of such adjacent pixels from a typical correlated image might look like that shown in Fig. 5.2.

In Fig. 5.2 (left) we observe a strong correlation characterized by the relationship $\hat{x}_1 = x_0$, as would be expected from a natural image. We can exploit this correlation by introducing a new set of axes, c_0 and c_1, as shown in Fig. 5.2, which are oriented such that c_0 aligns with the principal axes of the data. Using the annotations in Fig. 5.3 we can deduce that

$$A = x_1 \cos\phi,$$

$$B = x_0 \sin\phi,$$

$$D = x_1 \sin\phi,$$

$$E = x_0 \cos\phi.$$

Furthermore,

$$c_0 = D + E = x_0 \cos\phi + x_1 \sin\phi,$$

$$c_1 = A - B = -x_0 \sin\phi + x_1 \cos\phi.$$

Rearranging gives

$$\begin{bmatrix} c_0 \\ c_1 \end{bmatrix} = \begin{bmatrix} \cos\phi & \sin\phi \\ -\sin\phi & \cos\phi \end{bmatrix} \begin{bmatrix} x_0 \\ x_1 \end{bmatrix}, \quad (5.1)$$

or in matrix vector form

$$\mathbf{c} = \mathbf{A}\mathbf{x}.$$

For correlated pixels at $\phi = 45$ degrees,

$$\mathbf{A} = \frac{1}{\sqrt{2}} \begin{bmatrix} 1 & 1 \\ -1 & 1 \end{bmatrix}. \quad (5.2)$$

As we will see in Section 5.3.2, this bears a striking resemblance to the two-point discrete Walsh–Hadamard transform (DWHT). We will also see in Section 5.4.2 that this process is equivalent to extracting the eigenvectors for this dataset, i.e., $(1/\sqrt{2}, 1/\sqrt{2})$ $(-1/\sqrt{2}, 1/\sqrt{2})$.

From Eq. (5.2), we can see that the row vectors for this simple transform are $\frac{1}{\sqrt{2}}\begin{bmatrix} 1 & 1 \end{bmatrix}$ and $\frac{1}{\sqrt{2}}\begin{bmatrix} -1 & 1 \end{bmatrix}$. These are effectively sum and difference operations, respectively, and can be interpreted as basic low- and high-pass filters. For example, in the high-pass case,

$$H(z) = \frac{1}{\sqrt{2}}\left(1 - z^{-1}\right), \quad (5.3)$$

which has a pole at z=0 and a zero at z=1. This alternative view of the transform will stand us in good stead when we consider wavelets and subband filters in Chapter 6.

Let us reflect on what we have achieved so far. Firstly, we note that the operation of rotating the axes effectively decorrelates the data, with many of the c_1 values being very small. In practice, we may set small values to zero, leaving only a sparse set of coefficients which embody and preserve the main attributes of the original data. These can then be quantized to reduce their dynamic range and coded for transmission or storage. They can then be decoded, rescaled, and inverse transformed to yield an approximation to the original data which is optimal in the mean squared error (MSE) sense. This is the essence of transform-based data compression as depicted in Fig. 5.1.

5.2 Unitary transforms

5.2.1 Basis functions and linear combinations

We know from our introduction to Fourier analysis in Chapter 3 that we can approximate a signal using a linear combination of basis functions, such as harmonically related sines and cosines or complex exponentials. Let us assume we have N such basis functions and indicate the contribution of each basis function $\mathbf{b}_k$ that is needed to approximate the block of data by a value $c(k)$, where k represents the index of the basis function. Considering a 1D $N \times 1$ block of data $\mathbf{x}$, we can write this relationship in matrix vector form as follows:

$$\begin{bmatrix} x[0] \\ x[1] \\ \vdots \\ x[N-1] \end{bmatrix} = \begin{bmatrix} b_{0,0} & b_{1,0} & \cdots & b_{N-1,0} \\ b_{0,1} & b_{1,1} & \cdots & b_{N-1,1} \\ \vdots & \vdots & \ddots & \vdots \\ b_{0,N-1} & b_{1,N-1} & \cdots & b_{N-1,N-1} \end{bmatrix} \begin{bmatrix} c(0) \\ c(1) \\ \vdots \\ c(N-1) \end{bmatrix}. \tag{5.4}$$

However, if we are given an appropriate set of basis functions and some input data, then what we usually want to compute are the coefficient values or weights $c(k)$ as it is these that we will be processing, storing, and transmitting in the context of image compression. So let us rewrite Eq. (5.4) as follows:

$$\begin{bmatrix} c(0) \\ c(1) \\ \vdots \\ c(N-1) \end{bmatrix} = \begin{bmatrix} a_{0,0} & a_{0,1} & \cdots & a_{0,N-1} \\ a_{1,0} & a_{1,1} & \cdots & a_{1,N-1} \\ \vdots & \vdots & \ddots & \vdots \\ a_{N-1,0} & a_{N-1,1} & \cdots & a_{N-1,N-1} \end{bmatrix} \begin{bmatrix} x[0] \\ x[1] \\ \vdots \\ x[N-1] \end{bmatrix},$$

$$c(k) = \sum_{i=0}^{N-1} x[i]\, a_{k,i}, \quad k = 0 \ldots N-1,$$

or

$$\mathbf{c} = \mathbf{A}\mathbf{x}. \tag{5.5}$$

We have already defined the original signal in terms of our coefficients as

$$\mathbf{x} = \mathbf{B}\mathbf{c}.$$

But from Eq. (5.5) we can also see that the original signal can be reconstructed using the inverse transform:

$$\mathbf{x} = \mathbf{A}^{-1}\mathbf{c}. \tag{5.6}$$

Hence $\mathbf{B}=\mathbf{A}^{-1}$. $\mathbf{A}$ is an $N \times N$ matrix referred to as the transform matrix and $\mathbf{c}$ is an $N \times 1$ vector of transform coefficients. The columns of the matrix $\mathbf{A}^{-1}$ are the basis functions of the transform. It can thus be observed that the signal $\mathbf{x}$ is represented by a linear combination of weighted basis functions, where the weights are given by the coefficients in the column vector $\mathbf{c}$.

5.2.2 Orthogonality and normalization

Orthogonality ensures that the basis functions of a transform are independent and hence that they provide decorrelation of the input vector. Basis functions $\mathbf{a}_i$ and $\mathbf{a}_j$ are defined as orthogonal if their inner product is zero. That is,

$$\mathbf{a}_i^{\mathrm{H}}\mathbf{a}_j = 0;\ \forall i, j\ (i \neq j). \tag{5.7}$$

Orthonormality imposes further constraints on the norm of the basis vectors in that they must have unit length, i.e., $\|\mathbf{a}_i\| = 1$. Orthonormality is useful when we wish to use the same operations for forward and inverse transformation as it preserves energy between domains, i.e.,

$$\sum_{k=0}^{N-1} c^2(k) = \sum_{i=0}^{N-1} x^2[i]. \tag{5.8}$$

Furthermore, as shown above, it ensures that the forward and inverse basis function matrices are related through transposition:

$$\mathbf{a}_i^{\mathrm{H}}\mathbf{a}_j = \begin{cases} 1, & i = j, \\ 0, & i \neq j. \end{cases} \tag{5.9}$$

If the transform is orthonormal (or unitary), then it has a unique and computable inverse given by

$$\mathbf{A}^{-1} = \mathbf{A}^{\mathrm{H}}, \tag{5.10}$$

and hence $\mathbf{A}^{\mathrm{H}}\mathbf{A} = \mathbf{I}$. Clearly if the $\mathbf{A}$ matrix is real-valued, then

$$\mathbf{A}^{-1} = \mathbf{A}^{\mathrm{T}} \tag{5.11}$$

and the inverse transform then becomes

$$\mathbf{x} = \mathbf{A}^{\mathrm{T}}\mathbf{c}. \tag{5.12}$$

In this special case, the rows of the transform matrix are the basis functions. Unitary transforms have some useful properties:

1. The autocorrelation (or covariance for nonzero mean) matrices in the signal and transform domains are related by

$$\mathbf{R}_{\mathrm{c}} = \mathbf{A}\mathbf{R}_{\mathrm{x}}\mathbf{A}^{\mathrm{H}}. \tag{5.13}$$

2. The total energy of the original and transformed vectors are equal (see Section 5.4.1).
3. If we select a subset of k transform coefficients, then the approximation error for a particular vector is minimized if we choose the k largest coefficients.

5.2.3 Extension to 2D

In the case of a 2D signal, let us assume that the transform is applied to an $N \times N$ block of real data $\mathbf{X}$. Assuming a separable transform, as discussed in Chapter 3 and Section 5.5.3,

$$\mathbf{C} = \mathbf{A}\mathbf{X}\mathbf{A}^{\mathrm{T}} \tag{5.14}$$

and

$$\mathbf{X} = \mathbf{A}^{\mathrm{T}}\mathbf{C}\mathbf{A}. \tag{5.15}$$

Note here that Eq. (5.14) requires two matrix multiplications of size $N \times N$ instead of (in the nonseparable case) one multiplication with a matrix of size $N^2 \times N^2$. As we will see later, the existence and exploitation of separability is important in reducing the complexity of forward and inverse transforms.

It follows that the basis functions of a 2D transform are themselves 2D functions, and the 2D transform can be interpreted as expansions in terms of matrices that are obtained from the outer product of the individual 1D basis functions, $\mathbf{a}_j$. So if the 2D basis functions are denoted as $\boldsymbol{\alpha}_{i,j}$ and the 1D basis functions are real-valued column vectors, then the 2D functions are formed from the outer product of $\mathbf{a}_i$ and $\mathbf{a}_j$:

$$\boldsymbol{\alpha}_{i,j} = \mathbf{a}_i \mathbf{a}_j^{\mathrm{T}}, \tag{5.16}$$

$$\boldsymbol{\alpha}_{i,j} = \begin{bmatrix} a_{i,0}a_{j,0} & a_{i,0}a_{j,1} & \cdots & a_{i,0}a_{j,N-1} \\ a_{i,1}a_{j,0} & a_{i,1}a_{j,1} & \cdots & a_{i,1}a_{j,N-1} \\ \vdots & \vdots & \ddots & \cdots \\ a_{i,N-1}a_{j,0} & a_{i,N-1}a_{j,1} & \cdots & a_{i,N-1}a_{j,N-1} \end{bmatrix}. \tag{5.17}$$

Example 5.1. Orthonormality of basis functions

Consider the simple transform matrix from Section 5.1.2:

$$\mathbf{A} = \frac{1}{\sqrt{2}}\begin{bmatrix} 1 & 1 \\ -1 & 1 \end{bmatrix}.$$

Show that this is a unitary transform.

Solution. The basis functions are orthogonal because

$$\frac{1}{2}\begin{bmatrix} 1 & 1 \end{bmatrix}\begin{bmatrix} -1 \\ 1 \end{bmatrix} = 0,$$

and all basis functions have unit length since

$$\|\mathbf{a}_i\| = \frac{1}{\sqrt{2}}\sqrt{(1)^2 + (1)^2} = 1.$$

5.3 Basic transforms

5.3.1 The Haar transform

Consider the simple Haar transform:

$$\mathbf{H}_4 = \frac{1}{2}\begin{bmatrix} 1 & 1 & 1 & 1 \\ 1 & 1 & -1 & -1 \\ \sqrt{2} & -\sqrt{2} & 0 & 0 \\ 0 & 0 & \sqrt{2} & -\sqrt{2} \end{bmatrix}. \tag{5.18}$$

The first basis function creates a running sum of the input data, the second creates a difference between the first two and the second two data samples, the third creates a difference between the first two data points, and similarly the basis function in the bottom row does the same for the last two data points.

Example 5.2. Energy compaction and the Haar transform

Consider the input vector

$$\mathbf{x} = \begin{bmatrix} 1.0 & 0.5 & -0.5 & -1.0 \end{bmatrix}^T.$$

Show that the application of the Haar transform provides energy compaction.

Solution. A reduced number of nonzero coefficients results from the Haar transform, together with a modest amount of energy compaction; thus

$$\mathbf{c} = \frac{1}{2}\begin{bmatrix} 0 & 3 & \frac{\sqrt{2}}{2} & \frac{\sqrt{2}}{2} \end{bmatrix}^T.$$

As we will see later, other transforms exist that can provide much greater energy compaction and decorrelation.

5.3.2 The Walsh–Hadamard transform

The Walsh and Walsh–Hadamard transforms are simple but effective ways of compressing data. They have the significant advantage that the basic transform requires no multiplications, only sums and differences. While their coding gain is lower than that of transforms

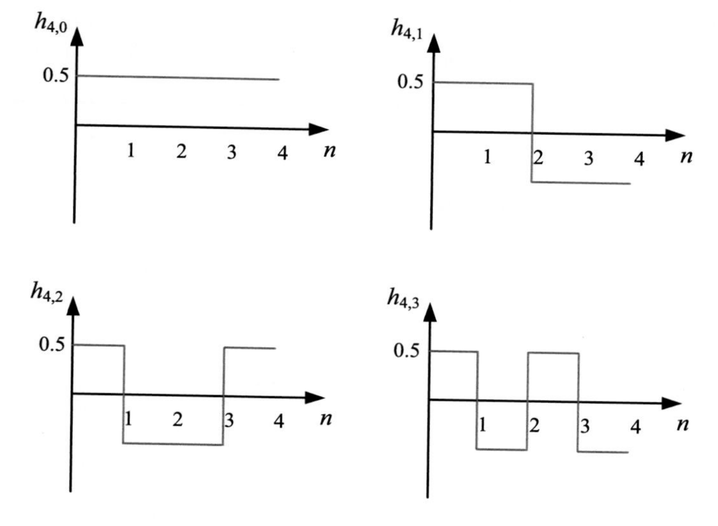

FIGURE 5.4 DWHT basis functions for N=4.

such as the DCT, which we will consider shortly, they do find application in modern video compression standards, where they are used in intraframe coding and as a metric for RDO.

The DWHT is obtained from a simple rearrangement of the discrete Hadamard matrix. The Hadamard matrix is an $N \times N$ matrix with the property $\mathbf{H}\mathbf{H}^{\mathrm{T}} = N\mathbf{I}$. Higher order matrices can be found by iteratively applying the following operation:

$$\mathbf{H}_{2N} = \begin{bmatrix} \mathbf{H}_N & \mathbf{H}_N \\ \mathbf{H}_N & -\mathbf{H}_N \end{bmatrix}. \tag{5.19}$$

For example,

$$\mathbf{H}_1 = 1, \ \mathbf{H}_2 = \begin{bmatrix} 1 & 1 \\ 1 & -1 \end{bmatrix}, \ \mathbf{H}_4 = \begin{bmatrix} 1 & 1 & 1 & 1 \\ 1 & -1 & 1 & -1 \\ 1 & 1 & -1 & -1 \\ 1 & -1 & -1 & 1 \end{bmatrix}. \tag{5.20}$$

The DWHT is obtained from the corresponding Hadamard matrix by normalization and rearranging the rows in sequence order (i.e., in terms of the number of sign changes). Therefore the four-point DWHT is given by

$$\mathbf{H}_4 = \frac{1}{2}\begin{bmatrix} 1 & 1 & 1 & 1 \\ 1 & 1 & -1 & -1 \\ 1 & -1 & -1 & 1 \\ 1 & -1 & 1 & -1 \end{bmatrix}. \tag{5.21}$$

The basis functions for the 1D DWHT are depicted in Fig. 5.4.

In the case of the 2D transform, the basis functions are formed from the outer product of the individual 1D basis functions, $\mathbf{h}_j$. The basis functions for the 2D 4 × 4 DWHT are shown in Fig. 5.5.

Example 5.3. 1D DWHT

Compute the 1D DWHT for the data vector $\mathbf{x} = \begin{bmatrix} 5 & 6 & 4 & 8 \end{bmatrix}^T$,

$$\begin{bmatrix} c(0) \\ c(1) \\ c(2) \\ c(3) \end{bmatrix} = \frac{1}{2} \begin{bmatrix} 1 & 1 & 1 & 1 \\ 1 & 1 & -1 & -1 \\ 1 & -1 & -1 & 1 \\ 1 & -1 & 1 & -1 \end{bmatrix} \begin{bmatrix} 5 \\ 6 \\ 4 \\ 8 \end{bmatrix}.$$

Solution. We have

$$\mathbf{c} = \begin{bmatrix} 11.5 & -0.5 & 1.5 & -2.5 \end{bmatrix}^T.$$

Example 5.4. 2D DWHT

Compute the 2D DWHT of the following input image block **S**:

$$\mathbf{S} = \begin{bmatrix} 5 & 6 & 8 & 10 \\ 6 & 6 & 5 & 7 \\ 4 & 5 & 3 & 6 \\ 8 & 7 & 5 & 5 \end{bmatrix}.$$

Solution. Since the 2D DWHT is separable, we can use Eq. (5.14). Letting $\mathbf{C} = \mathbf{C}'\mathbf{H}^T$, where $\mathbf{C}' = \mathbf{HS}$,

$$\mathbf{C}' = \frac{1}{2} \begin{bmatrix} 1 & 1 & 1 & 1 \\ 1 & 1 & -1 & -1 \\ 1 & -1 & -1 & 1 \\ 1 & -1 & 1 & -1 \end{bmatrix} \begin{bmatrix} 5 & 6 & 8 & 10 \\ 6 & 6 & 5 & 7 \\ 4 & 5 & 3 & 6 \\ 8 & 7 & 5 & 5 \end{bmatrix} = \begin{bmatrix} 11.5 & 12 & 10.5 & 14 \\ -0.5 & 0 & 2.5 & 3 \\ 1.5 & 1 & 2.5 & 1 \\ -2.5 & -1 & 0.5 & 2 \end{bmatrix}.$$

Then

$$\mathbf{C} = \begin{bmatrix} 24 & -0.5 & 1.5 & 2 \\ 2.5 & 3 & 0 & -0.5 \\ 3 & -0.5 & -0.5 & 1 \\ -0.5 & -3 & 0 & -1.5 \end{bmatrix}.$$

The energy compaction and decorrelation properties of the DWHT can clearly be seen in this simple example. Recall that, in the case of the original image block, the energy was distributed fairly uniformly across the block. After transformation, the data has been decorrelated horizontally and vertically and one dominant coefficient (top left) now contains some 93% of the energy.

FIGURE 5.5 Basis functions for the 2D DWHT.

Example 5.5. 2D DWHT compression

Taking the result of Example 5.4, set all small-valued coefficients (when $|c_{i,j}| \leq 2$) to zero and perform inverse transformation. What is the PSNR of the reconstruction if the original signal was represented with a 6-bit word length?

Solution. Now

$$\tilde{\mathbf{C}} = \begin{bmatrix} 24 & 0 & 0 & 0 \\ 2.5 & 3 & 0 & 0 \\ 3 & 0 & 0 & 0 \\ 0 & -3 & 0 & 0 \end{bmatrix}$$

and

$$\tilde{\mathbf{S}} = \mathbf{H}^{\mathrm{T}}\tilde{\mathbf{C}}\mathbf{H},$$

so

$$\tilde{\mathbf{S}} = \begin{bmatrix} 7 & 7 & 7 & 7 \\ 7 & 7 & 4 & 4 \\ 3 & 3 & 6 & 6 \\ 6 & 6 & 6 & 6 \end{bmatrix}.$$

Assuming a 6-bit word length, MSE=3, and the PSNR of the reconstruction is

$$\text{PSNR} = 10\log_{10}\left[\frac{16(2^6-1)^2}{\sum_{\forall(i,j)}(s[i,j]-\tilde{s}[i,j])^2}\right] = 31.2 \text{ dB}.$$

5.3.3 So why not use the discrete Fourier transform?

As a starting point for selecting a transform that decorrelates data we might consider the familiar discrete Fourier transform (DFT). After all, we know that the DFT is very good at representing sinusoidal data with very few coefficients. However, things are not quite that simple.

With the DFT, a windowed finite-length data sequence is naturally extended by periodic extension prior to transformation. Applying the discrete-time Fourier series (DTFS) to this new periodic function produces the DFT coefficients. The first problem with this is that it produces a set of complex coefficients. Although this is quite useful for capturing the magnitude and phase of the underlying signal in the frequency domain, it is not particularly useful for compression since (for the case of a real input signal) we are doubling the amount of information!

Putting the complex nature of the output to one side, periodic extension can be acceptable for long stationary sequences or for shorter sequences where the extension process does not introduce discontinuities. However, for shorter sequences like those typical in compression, discontinuities are common and can have a major impact on performance as they produce ringing or spectral leakage in the frequency domain. This is exactly what we do not want in a compression system as it introduces more energy in the coefficient space, requiring more bits to code.

The effect of the rectangular window on the sampling and transformation processes is illustrated in Fig. 5.6. Recall that multiplication of the input signal by the window function in the time (or spatial) domain is equivalent to the convolution of their spectra in the frequency domain, i.e.,

$$w[n]x[n] \Longleftrightarrow W(\Omega) * X(\Omega). \tag{5.22}$$

The top two subfigures show a length N window function, alongside its spectrum, while the lower two subfigures show a sampled and windowed input signal alongside its spectrum. Since the DFT is effectively the DTFS of the periodic signal obtained by concatenating an infinite number of windowed signal samples, discontinuities exist at the window boundaries as shown in Fig. 5.6. These cause spectral leakage or ripples in the frequency domain. The effects of windowing on the resulting DFT coefficients can be clearly seen.

Ringing is normally addressed in spectral analysis applications by applying a non-rectangular window function (such as a Hamming window) to the input data prior to extension. Such windows exhibit reduced frequency domain sidelobes compared with the conventional rectangular window, but also have a wider central lobe. Hence they reduce spectral ringing but smear sharp frequency domain edges, distorting the characteristics of the underlying signal. Again, this is exactly what we do not want in a compression system!

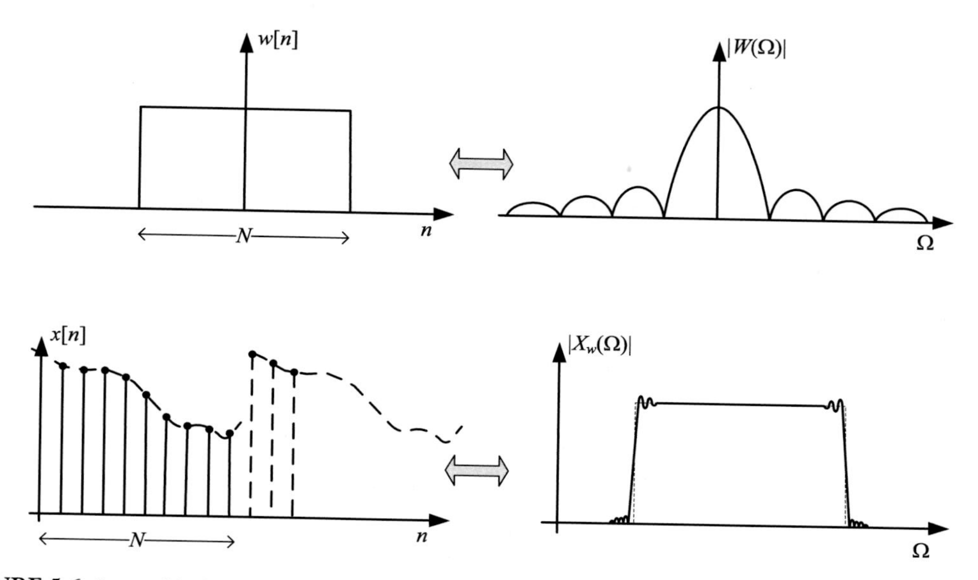

FIGURE 5.6 Spectral leakage due to windowing and periodic extension with the DFT: the rectangular window function and its magnitude spectrum (top) and the input signal showing periodic extension and its spectrum (bottom).

5.3.4 Desirable properties of an image transform

So what properties should we seek in a transform intended for image or video compression? We have already seen the benefits of orthonormality, the problems of ringing due to windowing, and the processing disadvantages of complex coefficient values. The primary desirable characteristics are thus:

1. **Good energy compaction**: For correlated sources in order to support quantization and matrix sparsity.
2. **Orthonormality**: To support energy balancing and efficient implementation and transposition.
3. **Basis functions independent of the signal**: To avoid the need to recompute and transmit basis functions to the decoder.
4. **Basis functions related to spatial frequency**: To support frequency-dependent perceptual quantization.
5. **Visually pleasing basis functions**: To avoid distracting artifacts caused by compression or transmission loss.
6. **Minimal spectral leakage**: To avoid oscillations at signal transitions due to windowing.
7. **Separability**: To reduce complexity for fast 2D computations.
8. **Real-valued basis functions and coefficients**: To ease arithmetic complexity and support simple matrix transposition.

5.4 Optimal transforms

5.4.1 Discarding coefficients

Let us assume that we want to define a transform based on a set of basis functions that pack the most energy into the fewest transform coefficients. If we then approximate our signal by a limited number (M) of coefficients, setting the remaining coefficients to constant value, k_i, firstly we ask, what should the value of k_i be? Following the approach of Clarke [1],

$$\tilde{\mathbf{x}}_M = \sum_{i=0}^{M-1} c(i)\,\mathbf{a}_i + \sum_{i=M}^{N-1} k_i\,\mathbf{a}_i,$$

and the error which results from this can be computed as follows:

$$\mathbf{e}_M = \mathbf{x}_M - \tilde{\mathbf{x}}_M = \sum_{i=M}^{N-1} (c(i) - k_i)\mathbf{a}_i.$$

We now wish to compute the energy in this residual signal and attempt to minimize it:

$$E\left\{e_M^2\right\} = E\left\{(c(M) - k_M)^2\mathbf{a}_M^{\mathrm{T}}\mathbf{a}_M + (c(M) - k_M)(c(M+1) - k_{M+1})\mathbf{a}_M^{\mathrm{T}}\mathbf{a}_{M+1} + \cdots\right.$$
$$\left.\cdots + (c(N-1) - k_{N-1})(c(N-1) - k_{N-1})\mathbf{a}_{N-1}^{\mathrm{T}}\mathbf{a}_{N-1}\right\}.$$

Assuming that the transform is orthonormal (and noting that the basis functions are column vectors here),

$$\mathbf{a}_i^{\mathrm{T}}\mathbf{a}_j = \begin{cases} 1, & i = j, \\ 0, & i \neq j. \end{cases}$$

Hence the equation for error energy reduces to

$$E\left\{e_M^2\right\} = E\left\{\sum_{i=M}^{N-1} (c(i) - k_i)^2\right\}. \tag{5.23}$$

If we minimize this by partial differentiation with respect to k_i and set the result equal to 0, we can determine the optimal value of k_i. We have

$$\frac{\partial}{\partial k_i} E\left\{e_M^2\right\} = \frac{\partial}{\partial k_i} E\left\{\sum_{i=M}^{N-1} (c(i) - k_i)^2\right\} = E\left\{-2(c(i) - k_i)\right\} = 0 \text{ at minimum.}$$

Hence at the minimum

$$k_i = E\{c(i)\} = \mathbf{a}_i^{\mathrm{T}} E\{\mathbf{x}\}.$$

If we assume a zero-mean sequence, the solution becomes

$$k_i = 0 \; (i = M, \cdots, N-1)$$

and hence our best approximation is

$$\tilde{\mathbf{x}}_M = \sum_{i=0}^{M-1} c(i)\,\mathbf{a}_i. \tag{5.24}$$

Now we can compute the basis functions which minimize the error energy in Eq. (5.23). We have

$$E\left\{e_M^2\right\} = E\left\{\sum_{i=M}^{N-1} (c(i))^2\right\}. \tag{5.25}$$

5.4.2 The Karhunen–Loeve transform (KLT)

Also known as the Hotelling or eigenvector transform, the basis vectors of the KLT transformation matrix, **A**, are obtained from the eigenvectors of the signal autocorrelation matrix (note the similarity to principal component analysis). Assuming the signal is derived from a random process, the resulting transform coefficients will be uncorrelated. The KLT provides the best energy compaction of any transform and yields optimal performance in the statistical (MSE) sense for a Gaussian source, by minimizing the geometric mean of the variance of the transform coefficients. For a signal $\mathbf{x}$ (biased to remove the mean for convenience), the basis functions $\mathbf{a}_k$ can be shown to be eigenvectors of the signal autocorrelation matrix; thus,

$$\mathbf{R}_x \mathbf{a}_k = \lambda_k \mathbf{a}_k, \tag{5.26}$$

where λ_k are the corresponding eigenvalues. Consider a signal $\mathbf{x} = \{x[0], x[1], \cdots, x[N-1]\}$. Assuming that the signal is a stationary random zero-mean sequence, the autocorrelation matrix for this block of pixels is given by

$$\mathbf{R}_x = \begin{bmatrix} r_{xx}(0) & r_{xx}(1) & \cdots & r_{xx}(N-1) \\ r_{xx}(1) & r_{xx}(0) & \cdots & r_{xx}(N-2) \\ \vdots & \vdots & \ddots & \vdots \\ r_{xx}(N-1) & r_{xx}(N-2) & \cdots & r_{xx}(0) \end{bmatrix}, \tag{5.27}$$

where

$$r_{xx}(k) = E\{x[n+k]x[n]\} = r_{xx}(-k).$$

As stated above, the KLT is the unitary transform that uses the eigenvectors from Eq. (5.26) as its basis vectors. Replacing the basis function matrix **A** in Eq. (5.13) with the matrix of eigenvectors, Ø, we obtain

$$\mathbf{R}_c = Ø^{\mathrm{H}} \mathbf{R}_x Ø \tag{5.28}$$

$$= \begin{bmatrix} \boldsymbol{\phi}_1^{\mathrm{H}} \\ \boldsymbol{\phi}_2^{\mathrm{H}} \\ \vdots \\ \boldsymbol{\phi}_N^{\mathrm{H}} \end{bmatrix} \mathbf{R}_{\mathrm{x}} \begin{bmatrix} \boldsymbol{\phi}_1 & \boldsymbol{\phi}_2 & \cdots & \boldsymbol{\phi}_N \end{bmatrix},$$

but from Eq. (5.26) this can be rewritten as

$$\mathbf{R}_{\mathrm{c}} = \begin{bmatrix} \boldsymbol{\phi}_1^{\mathrm{H}} \\ \boldsymbol{\phi}_2^{\mathrm{H}} \\ \vdots \\ \boldsymbol{\phi}_N^{\mathrm{H}} \end{bmatrix} \begin{bmatrix} \lambda_1\boldsymbol{\phi}_1 & \lambda_2\boldsymbol{\phi}_2 & \cdots & \lambda_N\boldsymbol{\phi}_N \end{bmatrix} = \begin{bmatrix} \lambda_1 & 0 & \cdots & 0 \\ 0 & \lambda_2 & 0 & 0 \\ \vdots & \vdots & \vdots & \vdots \\ 0 & 0 & 0 & \lambda_N \end{bmatrix}. \tag{5.29}$$

Hence we can see that the autocorrelation matrix is diagonalized by the KLT. It can be shown, for any autocorrelation matrix with diagonal entries, $\sigma_{c,k}^2$, that the following inequality holds:

$$\det[\mathbf{R}_c] \leq \prod_{\forall k} \sigma_{c,k}^2. \tag{5.30}$$

But from Eq. (5.13) we can see that

$$\det[\mathbf{R}_c] = \det[\mathbf{R}_x].$$

Therefore, for any unitary transform the following inequality holds:

$$\prod_{\forall k} \sigma_{c,k}^2 \geq \det[\mathbf{R}_x]. \tag{5.31}$$

But, for the case of the KLT we can see that

$$\prod_{\forall k} \sigma_{c,k}^2 = \det[\mathbf{R}_x] = \det[\mathbf{R}_c]. \tag{5.32}$$

Hence the KLT minimizes the geometric mean and, as a consequence (see Eq. (5.48)), maximizes the coding gain for a Gaussian source. It also provides the minimum approximation error, according to Eq. (5.25), if a subset of coefficients are use to represent the signal.

5.4.3 The KLT in practice

The main drawback of the KLT is the dependence of the transform on the signal and thus the need to compute and transmit the transformation matrix to the decoder prior to signal reconstruction. This can be a significant overhead in the case of small block sizes and signal nonstationarity. Furthermore, for correlated data (such as an image), other transforms such as the *discrete cosine transform* (DCT) are a close approximation to the KLT. For these reasons, the DCT is the transform of choice for most image and video coding applications.

For further details on the KLT, the reader is referred to the excellent text of Clarke [1].

Example 5.6. 1D KLT

Compute the 1D KLT of vector **s** (with zero mean):

$$\mathbf{s} = \begin{bmatrix} -5 & 0 & 1 & 4 \end{bmatrix}^T.$$

Solution. We first calculate the autocorrelation matrix of the vector **s** using Eq. (5.27):

$$\mathbf{R}_x = \begin{bmatrix} 10.5 & 1 & -1.25 & -5 \\ 1 & 10.5 & 1 & -1.25 \\ -1.25 & 1 & 10.5 & 1 \\ -5 & -1.25 & 1 & 10.5 \end{bmatrix}.$$

The eigenvector matrix, $\boldsymbol{\phi}$, and its transpose can then be obtained through singular value decomposition:

$$\boldsymbol{\phi} = \begin{bmatrix} -0.67 & -0.22 & 0.22 & 0.67 \\ 0.03 & -0.71 & -0.71 & 0.03 \\ 0.22 & -0.67 & 0.67 & -0.22 \\ 0.71 & 0.03 & 0.03 & 0.71 \end{bmatrix}.$$

So the KLT coefficient vector $\mathbf{s}_c$ can be calculated as

$$\begin{aligned} \mathbf{S}_c &= \boldsymbol{\phi}\mathbf{s} \\ &= \begin{bmatrix} -0.67 & -0.22 & 0.22 & 0.67 \\ 0.03 & -0.71 & -0.71 & 0.03 \\ 0.22 & -0.67 & 0.67 & -0.22 \\ 0.71 & 0.03 & 0.03 & 0.71 \end{bmatrix} \begin{bmatrix} -5 \\ 0 \\ 1 \\ 4 \end{bmatrix} = \begin{bmatrix} 6.25 \\ -0.74 \\ -1.31 \\ -0.68 \end{bmatrix}. \end{aligned}$$

5.5 Discrete cosine transform (DCT)

5.5.1 Derivation of the DCT

The DCT was first introduced by Ahmed et al. in 1974 [2] and is the most widely used unitary transform for image and video coding applications. Like the DFT, the DCT provides information about a signal in the frequency domain. However, unlike the DFT, the DCT of a real-valued signal is itself real-valued and importantly it also does not introduce artifacts due to periodic extension of the input data. The DCT asymptotically approximates the KLT for Markov-1 signals with high correlation.

With the DFT, a finite-length data sequence is naturally extended by periodic extension. Discontinuities in the time (or spatial) domain therefore produce ringing or spectral leakage in the frequency domain. This can be avoided if the data sequence is symmetrically (rather than periodically) extended prior to application of the DFT. This produces an even sequence which has the added benefit of yielding real-valued coefficients. The DCT is not as useful as the DFT for frequency domain signal analysis due to its deficiencies when representing pure sinusoidal waveforms. However, in its primary role of signal compression, it performs exceptionally well.

As we will see, the DCT has good energy compaction properties and its performance approaches that of the KLT for correlated image data. Furthermore, unlike the KLT, its basis functions are independent of the signal. The 1D DCT,[1] in its most popular form, is given by

$$c(k) = \sqrt{\frac{2}{N}}\varepsilon_k \sum_{n=0}^{N-1} x[n] \cos\left(\frac{\pi k}{N}\left(n+\frac{1}{2}\right)\right), \tag{5.33}$$

and in 2D by

$$c(k,l) = 2\frac{\varepsilon_k \epsilon_l}{\sqrt{NM}} \sum_{m=0}^{M-1}\sum_{n=0}^{N-1} x[m,n] \cos\left(\frac{\pi k}{N}\left(m+\frac{1}{2}\right)\right) \cos\left(\frac{\pi l}{N}\left(n+\frac{1}{2}\right)\right), \tag{5.34}$$

$$\varepsilon_k = \begin{cases} \frac{1}{\sqrt{2}}, & k=0, \\ 1, & \text{otherwise.} \end{cases}$$

DCT derivation

Consider the 1D signal shown in Fig. 5.7. The lower subfigure illustrates how a 1D signal can be symmetrically extended (or mirrored) to form a sequence of length $2N$. This has the clear advantage that no discontinuities are introduced and so, when we apply the DFT to create a frequency domain representation, ringing artifacts are absent.

All N elements of the original signal, $x[n]$, are duplicated to give a new sequence $x_1[n]$:

$$x_1[n] = \begin{cases} x[n], & 0 \le n \le N-1, \\ x[2N-1-n], & N \le n \le 2N-1. \end{cases}$$

The DFT of $x_1[n]$ is then given by

$$\begin{aligned} \mathcal{X}_1(k) &= \sum_{n=0}^{N-1} x[n] W_{2N}^{nk} + \sum_{n=N}^{2N-1} x[2N-1-n] W_{2N}^{nk} \\ &= \sum_{n=0}^{N-1} x[n]\left(W_{2N}^{nk} + W_{2N}^{-(n+1)k}\right), \quad 0 \le k \le 2N-1. \end{aligned}$$

[1] This is formally the DCT-II transform, but is commonly referred to simply as the DCT (see Section 5.5.4).

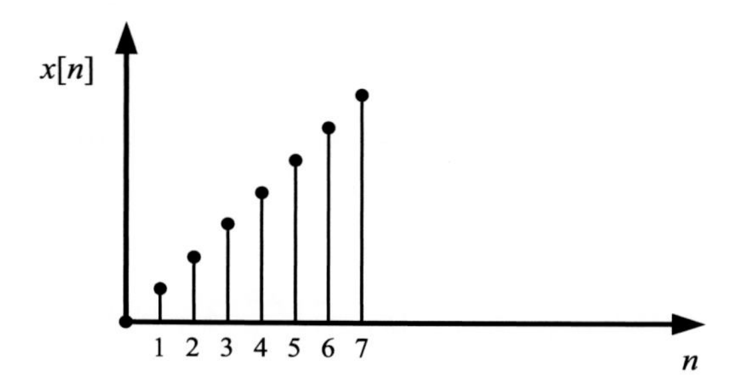

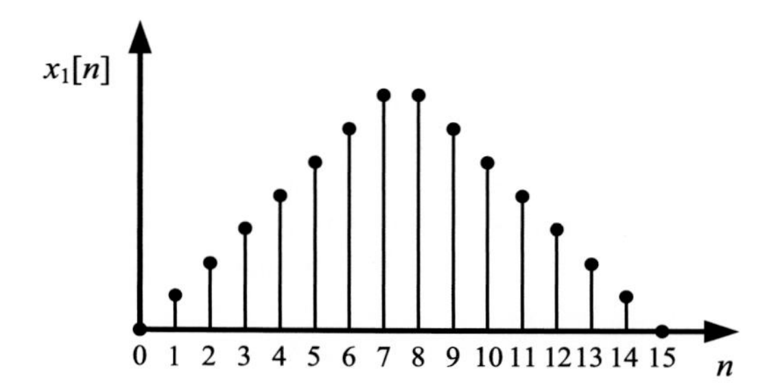

FIGURE 5.7 Symmetrical signal extension for the DCT.

Multiplying the numerator and denominator by $W_{2N}^{0.5k}$,

$$\begin{aligned}\mathcal{X}_1(k) &= W_{2N}^{-k/2} \sum_{n=0}^{N-1} x[n] \left(W_{2N}^{(n+0.5)k} + W_{2N}^{-(n+0.5)k} \right) \\ &= 2W_{2N}^{-k/2} \sum_{n=0}^{N-1} x[n] \cos\left(\frac{\pi(n+0.5)k}{N} \right).\end{aligned}$$

This is starting to look attractive as a basis for compression, apart from the rotation factor which makes the result complex and nonsymmetric. Since this term contributes nothing in terms of compression performance, the DCT is normally defined without it as

$$c(k) = W_{2N}^{k/2} \mathcal{X}_1(k)$$

and then normalized to give

$$c(k) = \sqrt{\frac{2}{N}} \varepsilon_k \sum_{n=0}^{N-1} x[n] \cos\left(\frac{\pi k}{N} (n+0.5) \right), \tag{5.35}$$

where

$$\varepsilon_k = \begin{cases} \frac{1}{\sqrt{2}}, & k = 0, \\ 1, & \text{otherwise.} \end{cases}$$

Similarly, the inverse DCT is defined by

$$x[n] = \sqrt{\frac{2}{N}} \sum_{k=0}^{N-1} \varepsilon_k c(k) \cos\left(\frac{\pi n}{N}(k + 0.5)\right). \tag{5.36}$$

5.5.2 DCT basis functions

The basis functions are the column vectors of the inverse transform matrix. It is these that are weighted by the transform coefficients and linearly combined to form the signal approximation. However, in the case of a real-valued orthonormal transform, we have already seen that these are identical to the row vectors in the forward transform matrix.

Using Eq. (5.35) we can see that the DCT basis functions are given by

$$a(k, n) = \sqrt{\frac{2}{N}} \varepsilon_k \cos\left(\frac{\pi k}{N}(n + 0.5)\right), \; 0 \le k, \; n \le N - 1. \tag{5.37}$$

Example 5.7. Four-point DCT basis functions

Compute the basis functions and transform matrix for the four-point DCT.

Solution. Applying Eq. (5.37) we have

$$\mathbf{A} = \sqrt{\frac{2}{N}} \varepsilon_k \begin{bmatrix} \cos(0) & \cos(0) & \cos(0) & \cos(0) \\ \cos(\frac{\pi}{8}) & \cos(\frac{3\pi}{8}) & \cos(\frac{5\pi}{8}) & \cos(\frac{7\pi}{8}) \\ \cos(\frac{\pi}{4}) & \cos(\frac{3\pi}{4}) & \cos(\frac{5\pi}{4}) & \cos(\frac{7\pi}{4}) \\ \cos(\frac{3\pi}{8}) & \cos(\frac{9\pi}{8}) & \cos(\frac{15\pi}{8}) & \cos(\frac{21\pi}{8}) \end{bmatrix}$$

$$= \sqrt{\frac{1}{2}} \varepsilon_k \begin{bmatrix} \cos(0) & \cos(0) & \cos(0) & \cos(0) \\ \cos(\frac{\pi}{8}) & \cos(\frac{3\pi}{8}) & -\cos(\frac{3\pi}{8}) & -\cos(\frac{\pi}{8}) \\ \cos(\frac{\pi}{4}) & -\cos(\frac{\pi}{4}) & -\cos(\frac{\pi}{4}) & \cos(\frac{\pi}{4}) \\ \cos(\frac{3\pi}{8}) & -\cos(\frac{\pi}{8}) & \cos(\frac{\pi}{8}) & -\cos(\frac{3\pi}{8}) \end{bmatrix}. \tag{5.38}$$

Returning to the four-point DCT, let us convince ourselves that this transform is orthonormal and confirm the value of ε_k. Consider, for example, the first basis function. We can rewrite this as the vector $\mathbf{a}_0$, as follows:

$$\mathbf{a}_0 = k_0 \begin{bmatrix} 1 & 1 & 1 & 1 \end{bmatrix}.$$

Now we know that for orthonormality the basis functions must have unity norm; thus,

$$\|\mathbf{a}_0\| = k_0 \sqrt{1^2 + 1^2 + 1^2 + 1^2} = 2k_0 = 1.$$

Hence,

$$k_0 = \frac{1}{2}.$$

Similarly, for the second basis function,

$$\|\mathbf{a}_1\| = k_1\sqrt{\cos^2(\frac{\pi}{8}) + \cos^2(\frac{3\pi}{8}) + \cos^2(\frac{3\pi}{8}) + \cos^2(\frac{\pi}{8})} = \sqrt{2}k_1 = 1.$$

Hence,

$$k_1 = \frac{1}{\sqrt{2}}.$$

And it can also be shown that

$$k_2 = k_3 = \frac{1}{\sqrt{2}}.$$

So in general we have, as expected,

$$k_i = \frac{1}{\sqrt{2}}\varepsilon_i, \ \varepsilon_i = \begin{cases} 1/\sqrt{2}, & 1 = 0, \\ 1, & i \neq 0. \end{cases}$$

Since we now know that the DCT is orthonormal, we can simplify computation of the inverse transform since it is straightforward to show from Eq. (5.11) that

$$\mathbf{A}^{-1} = \mathbf{A}^{\mathrm{T}} = \sqrt{\frac{1}{2}}\begin{bmatrix} \frac{1}{\sqrt{2}} & \cos(\frac{\pi}{8}) & \cos(\frac{\pi}{4}) & \cos(\frac{3\pi}{8}) \\ \frac{1}{\sqrt{2}} & \cos(\frac{3\pi}{8}) & -\cos(\frac{\pi}{4}) & -\cos(\frac{\pi}{8}) \\ \frac{1}{\sqrt{2}} & -\cos(\frac{3\pi}{8}) & -\cos(\frac{\pi}{4}) & \cos(\frac{\pi}{8}) \\ \frac{1}{\sqrt{2}} & -\cos(\frac{\pi}{8}) & \cos(\frac{\pi}{4}) & -\cos(\frac{3\pi}{8}) \end{bmatrix}.$$

Example 5.8. Eight-point DCT basis functions

We can easily compute the basis functions for other-length transforms. For example, compute the eight-point DCT basis functions.

Solution. The eight-point DCT basis functions can be computed using Eq. (5.37). These are shown in Fig. 5.8.

5.5.3 Extension to 2D: Separability

The 2D DCT basis functions are, as described earlier, obtained from the outer product of the 1D basis functions. These are shown for the case of the 8×8 DCT in Fig. 5.9.

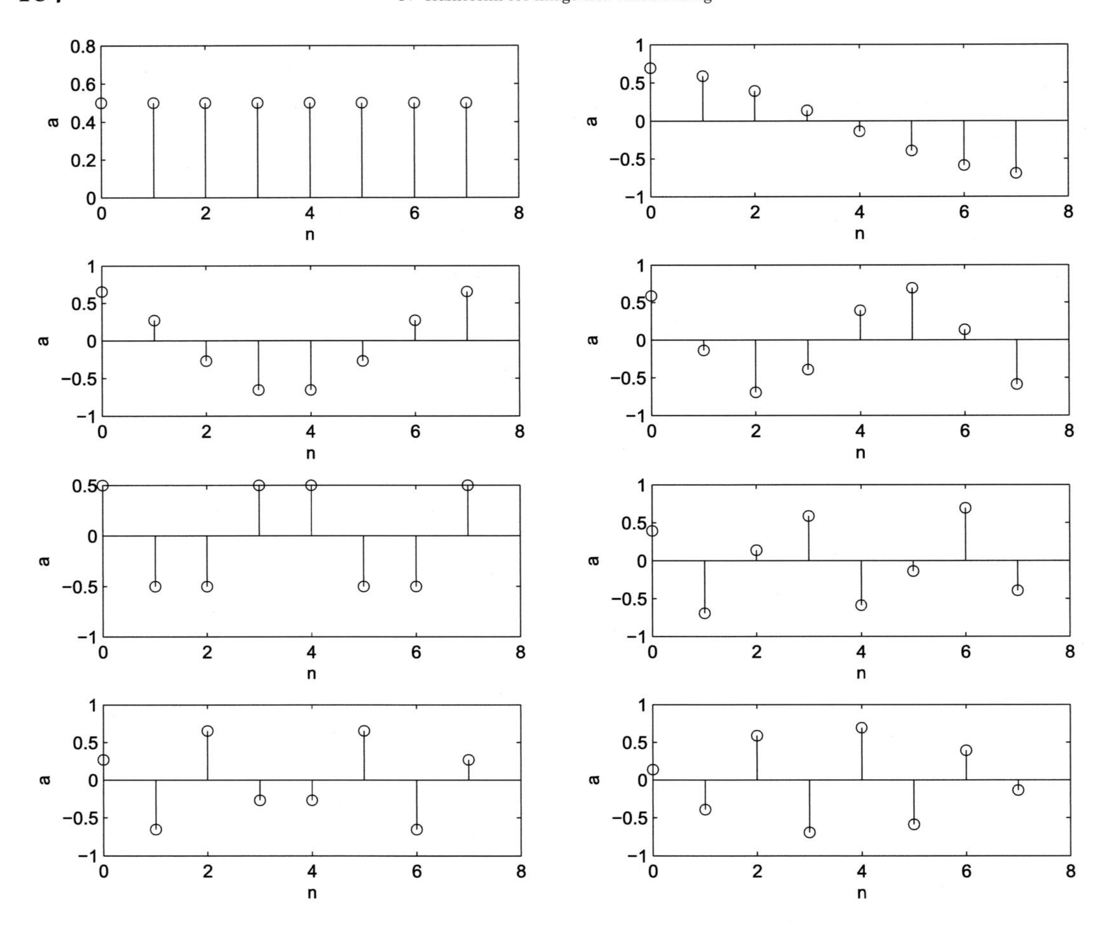

FIGURE 5.8 DCT basis functions for N=8.

Let us now consider the issue of separability in the context of the 2D DCT. Using the 8×8 transform as an example,

$$c_2(m,n) = \frac{\varepsilon_m \varepsilon_n}{4} \sum_{i=0}^{7} \sum_{j=0}^{7} s[i,j] \cos\left(\frac{(2i+1)m\pi}{16}\right) \cos\left(\frac{(2j+1)n\pi}{16}\right). \tag{5.39}$$

Now the 1D DCT is given by

$$c_1(m) = \frac{\varepsilon_m}{2} \sum_{i=0}^{7} s[i,j] \cos\left(\frac{(2i+1)\pi m}{16}\right).$$

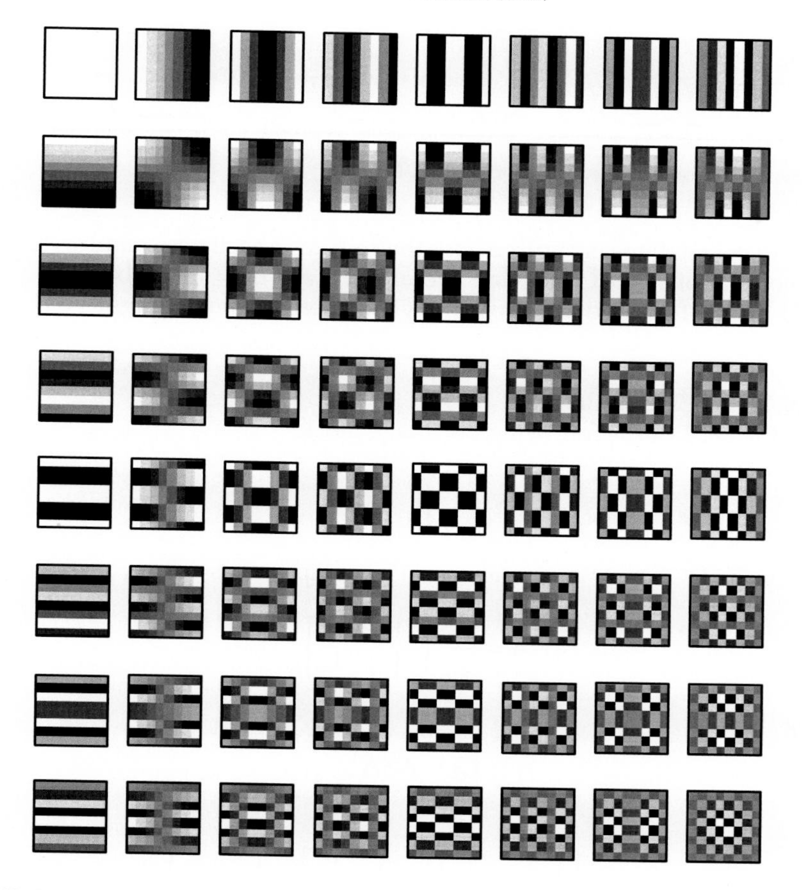

FIGURE 5.9 2D DCT basis functions for N=8.

Eq. (5.39) can thus be rearranged as follows:

$$c_2(m,n) = \frac{\varepsilon_n}{2}\sum_{j=0}^{7} c_1(m)\cos\left(\frac{(2j+1)\pi n}{16}\right).$$

In other words, the 2D DCT can be formed in two passes, firstly in the m direction and then in the n direction; thus,

$$c_2(m,n) = \mathrm{DCT}_n^{1\mathrm{D}}(\mathrm{DCT}_m^{1\mathrm{D}}). \tag{5.40}$$

Separable computation of the 2D DCT (or indeed any separable transform) can be visualized as shown in Fig. 5.10.

Separability has two main advantages:

FIGURE 5.10 Separable computation of the forward DCT.

1. The number of calculations is reduced – for the 8×8 case, each pass requires 8×64 multiply accumulate operations (MACs), giving a total of $2 \times 8 \times 64 = 1024$ MACs.
2. The 1D DCT can be further decomposed to yield even greater savings (see Section 5.8.2).

Example 5.9. 2D DCT calculation

Calculate the DCT of the following 2D image block:

$$\mathbf{S} = \begin{bmatrix} 8 & 9 \\ 10 & 11 \end{bmatrix}.$$

Solution. The 2D DCT transform matrix can be computed as

$$\mathbf{A} = \frac{1}{\sqrt{2}} \begin{bmatrix} 1 & 1 \\ 1 & -1 \end{bmatrix}.$$

Recalling that

$$\mathbf{C}_2 = \mathbf{A}\mathbf{S}\mathbf{A}^{\mathrm{T}},$$

we have

$$\begin{aligned} \mathbf{C}_2 &= \frac{1}{2} \begin{bmatrix} 1 & 1 \\ 1 & -1 \end{bmatrix} \begin{bmatrix} 8 & 9 \\ 10 & 11 \end{bmatrix} \begin{bmatrix} 1 & 1 \\ 1 & -1 \end{bmatrix} \\ &= \begin{bmatrix} 19 & -1 \\ -2 & 0 \end{bmatrix}. \end{aligned}$$

Now if we perform the inverse DCT on this result, we should recover the original signal:

$$\tilde{\mathbf{S}} = \frac{1}{2} \begin{bmatrix} 1 & 1 \\ 1 & -1 \end{bmatrix} \begin{bmatrix} 19 & -1 \\ -2 & 0 \end{bmatrix} \begin{bmatrix} 1 & 1 \\ 1 & -1 \end{bmatrix} = \begin{bmatrix} 8 & 9 \\ 10 & 11 \end{bmatrix} = \mathbf{S}.$$

5.5.4 Variants on sinusoidal transforms

As described by Britanak [4], the DCT is not just a single transform, but part of a family of related orthonormal sinusoidal transforms with attractive properties for image and video compression. This family includes eight DCT transforms, DCT-I to DCT-VIII, and eight discrete sine transforms (DSTs), DST-I to DST-VIII. These include both the forward and inverse transforms. In image and video compression, the commonly used transforms are DCT-II and DST-VII. The forward transform derived earlier in this section is formally known as DCT-II (DCT-III is its inverse), but is commonly referred to simply as the DCT. This pair is generally used for intra-coding and for residual coding after prediction. The DST-VII pair (generally referred to as the DST) has gained popularity in the current HEVC and VVC video coding standards, particularly for residual coding after intra-prediction, where 7–10% savings have been reported using a hybrid DCT/DST approach [5].

The DST was first introduced by Ahmed et al. [2] in their seminal 1974 paper. Its derivation is similar to the DCT but instead based on asymmetric (odd) extension; hence it is equivalent to the imaginary parts of a DFT with approximately twice the length. Like the DCT, it has real-valued basis functions. The DST-VII basis functions are defined by

$$a(k,n) = \frac{2}{\sqrt{2N+1}} \sin\left(\frac{\pi(2n+1)(k+1)}{2N+1}\right), \quad 0 \le k,\ n \le N-1, \tag{5.41}$$

and hence the forward transform is given by

$$c(k) = \frac{2}{\sqrt{2N+1}} \sum_{n=0}^{N-1} x[n] \sin\left(\frac{\pi(2n+1)(k+1)}{2N+1}\right), \quad 0 \le k \le N-1. \tag{5.42}$$

5.6 Quantization of DCT coefficients

5.6.1 The basics of quantization

As we have discussed previously, the process of energy compaction in itself does not provide any data compression. For example, if we have an 8×8 block of spatial domain pixel values with a dynamic range of 8 bits and we transform these to an 8×8 block of transform domain coefficients, also requiring 8 bits each, then we have gained little. What we do achieve, however, is a concentration of energy into a small number of high-valued coefficients. These larger coefficients are likely to have a higher psychovisual significance than their low-valued counterparts and can be coded accordingly.

Quantization is an important step in lossy compression; however, it is irreversible and causes signal degradation. As we saw in Chapter 3, the quantizer comprises a set of decision levels and a set of reconstruction levels. One of the challenges is to perform quantization in such a way as to minimize its psychovisual impact.

The primary advantages of transform coding is that it provides both energy compaction and decorrelation. This means that the transform coefficients can be approximated as originating from a memoryless source (e.g., i.i.d. Laplacian) for which simple quantizers exist. Intraframe transform coefficients are therefore normally quantized using a uniform quantizer, with the coefficients preweighted to reflect the frequency dependent sensitivity of the human visual system (HVS). A general expression which captures this is given in the following equation, where Q is the quantizer step size, k is a constant, and $\mathbf{W}$ is a coefficient-dependent weighting matrix obtained from psychovisual experiments:

$$c_Q(i,j) = \left\lfloor \frac{kc(i,j)}{Qw_{i,j}} \right\rceil. \tag{5.43}$$

After transmission or storage, we must rescale the quantized transform coefficients prior to inverse transformation; thus,

$$\tilde{c}(i,j) = \frac{c_Q(i,j)Qw_{i,j}}{k}. \tag{5.44}$$

First let us consider a simple example (Example 5.10), before progressing to the more realistic 8×8 case (Example 5.11).

Example 5.10. Recalculating Example 5.9 with quantization

Quantize the DCT result from Example 5.8 using the following quantization matrix (assume k=1 and Q=1):

$$\mathbf{W} = \begin{bmatrix} 4 & 8 \\ 8 & 8 \end{bmatrix}.$$

Solution. Now

$$\mathbf{C}_Q = \lfloor \mathbf{C}_2./\mathbf{W} \rceil.$$

So

$$\mathbf{C}_Q = \begin{bmatrix} 5 & 0 \\ 0 & 0 \end{bmatrix}.$$

Now compute the inverse 2D DCT after rescaling:

$$\tilde{\mathbf{S}} = \frac{1}{2}\begin{bmatrix} 1 & 1 \\ 1 & -1 \end{bmatrix}\begin{bmatrix} 20 & 0 \\ 0 & 0 \end{bmatrix}\begin{bmatrix} 1 & 1 \\ 1 & -1 \end{bmatrix} = \begin{bmatrix} 10 & 10 \\ 10 & 10 \end{bmatrix} \neq \mathbf{S}.$$

This leaves an error signal of

$$\mathbf{E} = \begin{bmatrix} 2 & 1 \\ 0 & 1 \end{bmatrix}.$$

Assuming 6-bit signals, what is the PSNR?

Example 5.11. 8×8 **2D DCT calculation with quantization**
Consider the following 8×8 image block, $\mathbf{S}$:

$$\mathbf{S} = \begin{bmatrix} 168 & 163 & 161 & 150 & 154 & 168 & 164 & 154 \\ 171 & 154 & 161 & 150 & 157 & 171 & 150 & 164 \\ 171 & 168 & 147 & 164 & 164 & 161 & 143 & 154 \\ 164 & 171 & 154 & 161 & 157 & 157 & 147 & 132 \\ 161 & 161 & 157 & 154 & 143 & 161 & 154 & 132 \\ 164 & 161 & 161 & 154 & 150 & 157 & 154 & 140 \\ 161 & 168 & 157 & 154 & 161 & 140 & 140 & 132 \\ 154 & 161 & 157 & 150 & 140 & 132 & 136 & 128 \end{bmatrix}.$$

First of all, calculate the 8×8 block of transform coefficients, $\mathbf{C}$, and then quantize the resulting DCT coefficients using the quantization matrix, $\mathbf{W}$ (this is actually the default JPEG quantization matrix). Assuming k=1 and Q=1,

$$\mathbf{W} = \begin{bmatrix} 16 & 11 & 10 & 16 & 24 & 40 & 51 & 61 \\ 12 & 12 & 14 & 19 & 26 & 58 & 60 & 55 \\ 14 & 13 & 16 & 24 & 40 & 57 & 69 & 56 \\ 14 & 17 & 22 & 29 & 51 & 87 & 80 & 62 \\ 18 & 22 & 37 & 56 & 68 & 109 & 103 & 77 \\ 24 & 35 & 55 & 64 & 81 & 104 & 113 & 92 \\ 49 & 64 & 78 & 87 & 103 & 121 & 120 & 101 \\ 72 & 92 & 95 & 98 & 112 & 100 & 103 & 99 \end{bmatrix}.$$

Finally compute the rescaled DCT coefficient matrix and perform the inverse DCT to obtain the reconstructed image block, $\tilde{\mathbf{S}}$.

Solution.

$$\mathbf{C} = \begin{bmatrix} 1239 & 50 & -3 & 20 & -10 & -1 & 0 & -6 \\ 35 & -24 & 11 & 13 & 4 & -3 & 14 & -6 \\ -6 & -3 & 8 & -9 & 2 & -3 & 5 & 10 \\ 9 & -10 & 4 & 4 & -15 & 10 & 5 & 6 \\ -12 & 5 & -1 & -2 & -15 & 9 & -6 & -2 \\ 5 & 10 & -7 & 3 & 4 & -7 & -14 & 2 \\ 2 & -2 & 3 & -1 & 1 & 3 & -3 & -4 \\ -1 & 1 & 0 & 2 & 3 & -2 & -4 & -2 \end{bmatrix}.$$

Quantize the DCT coefficients using the quantization matrix, **W**. Assuming k=1 and Q=1, we have

$$\mathbf{C_Q} = \begin{bmatrix} 77 & 5 & 0 & 1 & 0 & 0 & 0 & 0 \\ 3 & -2 & 1 & 1 & 0 & 0 & 0 & 0 \\ 0 & 0 & 1 & 0 & 0 & 0 & 0 & 0 \\ 1 & -1 & 0 & 0 & 0 & 0 & 0 & 0 \\ -1 & 0 & 0 & 0 & 0 & 0 & 0 & 0 \\ 0 & 0 & 0 & 0 & 0 & 0 & 0 & 0 \\ 0 & 0 & 0 & 0 & 0 & 0 & 0 & 0 \\ 0 & 0 & 0 & 0 & 0 & 0 & 0 & 0 \end{bmatrix}.$$

Rescaling the DCT coefficient matrix,

$$\tilde{\mathbf{C}} = \begin{bmatrix} 1232 & 55 & 0 & 16 & 0 & 0 & 0 & 0 \\ 36 & -24 & 14 & 19 & 0 & 0 & 0 & 0 \\ 0 & 0 & 16 & 0 & 0 & 0 & 0 & 0 \\ 14 & -17 & 0 & 0 & 0 & 0 & 0 & 0 \\ -18 & 0 & 0 & 0 & 0 & 0 & 0 & 0 \\ 0 & 0 & 0 & 0 & 0 & 0 & 0 & 0 \\ 0 & 0 & 0 & 0 & 0 & 0 & 0 & 0 \\ 0 & 0 & 0 & 0 & 0 & 0 & 0 & 0 \end{bmatrix},$$

and inverse transformation gives the reconstructed image block, $\tilde{\mathbf{S}}$:

$$\tilde{\mathbf{S}} = \begin{bmatrix} 173 & 162 & 150 & 149 & 158 & 164 & 164 & 160 \\ 176 & 166 & 156 & 154 & 160 & 163 & 160 & 154 \\ 173 & 165 & 158 & 156 & 160 & 157 & 150 & 143 \\ 163 & 159 & 155 & 154 & 154 & 150 & 142 & 135 \\ 158 & 157 & 156 & 157 & 155 & 151 & 143 & 138 \\ 161 & 161 & 161 & 160 & 157 & 152 & 146 & 143 \\ 163 & 163 & 161 & 156 & 149 & 143 & 139 & 137 \\ 161 & 160 & 156 & 148 & 138 & 131 & 127 & 126 \end{bmatrix}.$$

As we can see, the resulting matrix after quantization is close in value to the original signal. The key thing to notice is that the matrix $\tilde{\mathbf{C}}$ is now sparse in nonzero values and, as expected (because the input data is highly correlated), the energy in the DCT coefficients is compacted in the top left corner of the matrix.

As an additional exercise, calculate the PSNR for this approximation – assuming a signal word length of 8 bits.

An example of the effects of coefficient quantization on reconstruction quality for a range of block types is shown in Fig. 5.11. It can be observed, as expected, that more textured blocks require larger numbers of coefficients in order to create a good approximation to the original content. The best reconstruction is achieved with fewer coefficients for the case of untextured blocks, as shown for the left-hand block in the figure.

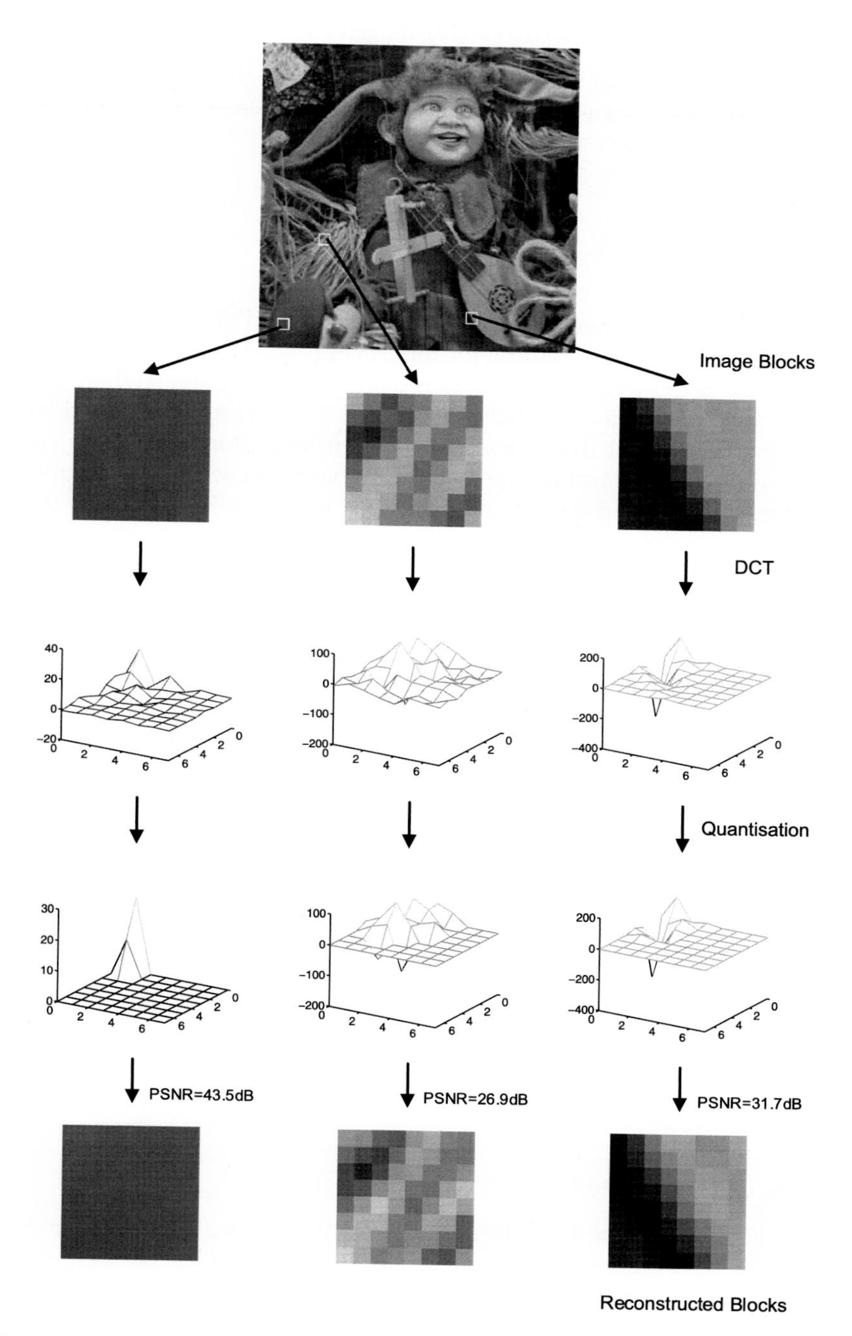

FIGURE 5.11 Effects of coefficient quantization on various types of data block.

5.6.2 Perceptually optimized quantization matrices

We can see from Eq. (5.43) that quantizer performance is highly dependent on the design of the weighting matrix **W**. As an example, for the case of H.265/HEVC, the parameters in this matrix are based on a HVS contrast sensitivity function (CSF) [13], in which each weight is derived from

$$\frac{1}{w_{i,j}} = \begin{cases} 2.2(0.192 + 0.114 f(i,j)) exp(-(0.114 f'(i,j))^{1.1}, & \text{if } f'(i,j) > f_{\max}, \\ 1.0, & \text{otherwise.} \end{cases} \tag{5.45}$$

This model was originally proposed by Mannos and Sakrison [14] using a 2D HVS-CSF (see Chapter 2 for HVS and CSF) with parameter modifications. Here $f(i,j)$ denotes the radial spatial frequency (cycles per degree), and f_{max} is 8 (cycles per degree) – where the sensitivity function achieves its peak value. The radial frequencies $f(i,j)$ are calculated based on the discrete vertical and horizontal frequencies in the transform domain, which are related to viewing distance, display size, and image resolution. Based on sensible assumptions for the values of these parameters, the default 8×8 weighting matrix in HEVC is given by

$$\left[\frac{1}{w_{i,j}}\right] = \begin{bmatrix} 1.0000 & 1.0000 & 1.0000 & 1.0000 & 0.9599 & 0.8746 & 0.7684 & 0.6571 \\ 1.0000 & 1.0000 & 1.0000 & 1.0000 & 0.9283 & 0.8404 & 0.7371 & 0.6306 \\ 1.0000 & 1.0000 & 0.9571 & 0.8898 & 0.8192 & 0.7371 & 0.6471 & 0.5558 \\ 1.0000 & 1.0000 & 0.8898 & 0.7617 & 0.6669 & 0.5912 & 0.5196 & 0.4495 \\ 0.9599 & 0.9283 & 0.8192 & 0.6669 & 0.5419 & 0.4564 & 0.3930 & 0.3393 \\ 0.8746 & 0.8404 & 0.7371 & 0.5912 & 0.4564 & 0.3598 & 0.2948 & 0.2480 \\ 0.7684 & 0.7371 & 0.6471 & 0.5196 & 0.3930 & 0.2948 & 0.2278 & 0.1828 \\ 0.6571 & 0.6306 & 0.5558 & 0.4495 & 0.3393 & 0.2480 & 0.1828 & 0.1391 \end{bmatrix}. \tag{5.46}$$

According to Eq. (5.43), the quantization matrix can be determined when the quantization step size Q is provided. The relationship between the quantization parameter (QP) and quantization step Q is given by

$$Q(\text{QP}) = 2^{\frac{\text{QP}-4}{6}}. \tag{5.47}$$

More details on HEVC quantization matrix design can be found in [15,16].

5.7 Performance comparisons

5.7.1 DCT vs. DFT revisited

Earlier we explained why the DFT does not perform well as a basis for compression. We are now in a position to compare the performance of the DCT and the DFT. Fig. 5.12 shows an input signal $x[n]$ which is compressed with both an eight-point DFT and an eight-point DCT. In both cases compression is achieved by truncating the transform coefficient vector to leave only the first four coefficients, the remainder being set to zero. It can be clearly seen that the energy compaction of the DCT is superior to the DFT, as is the reconstructed signal quality

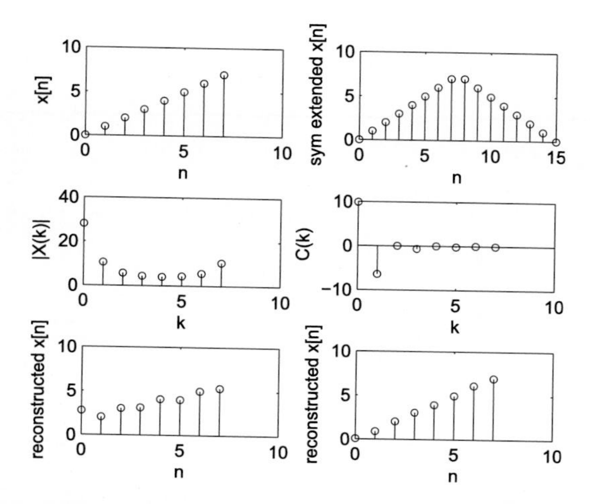

FIGURE 5.12 Comparison of DFT and DCT for compression. Top: Input signal $x[n]$ and the symmetrically extended sequence $x_1[n]$. Middle: Eight-point DFT $\{x[n]\}$ (magnitude only) and the eight-point DCT $\{x[n]\}$. Bottom: IDFT $\{X(k)\}$ (magnitude only) and IDCT $\{C(k)\}$.

for a given compression strategy. Recall also that the DFT coefficients are complex and thus require approximately double the bandwidth of the DCT coefficients.

5.7.2 Comparison of transforms

Fig. 5.13 shows the effects of transformation on a natural image (256 × 256) using the DFT, DWHT, and DCT. As expected, the DFT magnitude spectrum exhibits multiple energy concentrations, whereas the performance of the DCT provides significant energy compaction without any of the disadvantages of the DFT. The DWHT also provides good performance on this image despite its very simple structure.

Computation of the KLT for a 256 × 256 image is impractical, so Fig. 5.14 compares its performance with the DCT based on the use of 8 × 8 transform blocks. This figure shows two cases – in the first case, only 9 of the 64 coefficients in each block are retained for reconstruction, and in the second case, 25 are used. As can be seen, the DCT comes close to the KLT in terms of quality, but importantly without the need for the transmission of basis functions. This provides additional justification of why the DCT remains so popular in image and video compression applications.

5.7.3 Rate-distortion performance of the DCT

The benefit of block transform-based coding with the DCT is that, for correlated data, the transform compacts most of the energy in the original signal into a small number of significant coefficients. The coding gain for the block transform is defined in terms of the ratio of

FIGURE 5.13 Comparison of DFT, DWHT, and DCT transformations of a natural image.

the arithmetic and geometric means of the transformed block variances as follows:

$$G_{TC} = 10\log_{10}\left[\frac{\frac{1}{N}\sum_{i=0}^{N-1}\sigma_i^2}{\left(\prod_{i=0}^{N-1}\sigma_i^2\right)^{1/N}}\right]. \tag{5.48}$$

Fig. 5.15 shows an example of the performance of DCT compression (8 × 8 blocks) for a 256 × 256 resolution image at 0.3 bpp, whereas Fig. 5.16 shows a comparison between the same 256 × 256 image and an equivalent 512×512 image. It can clearly be seen that, for the same compression ratio, the larger image is much less distorted. There are two fairly obvious reasons for this: (i) the 512 × 512 image, despite being compressed at the same ratio, has more bits and (ii) the 512 × 512 image will exhibit higher correlation over an 8 × 8 block.

Fig. 5.17 shows an example of the rate-distortion performance of the DCT coder after entropy coding (see Chapter 7) based on the puppet image.

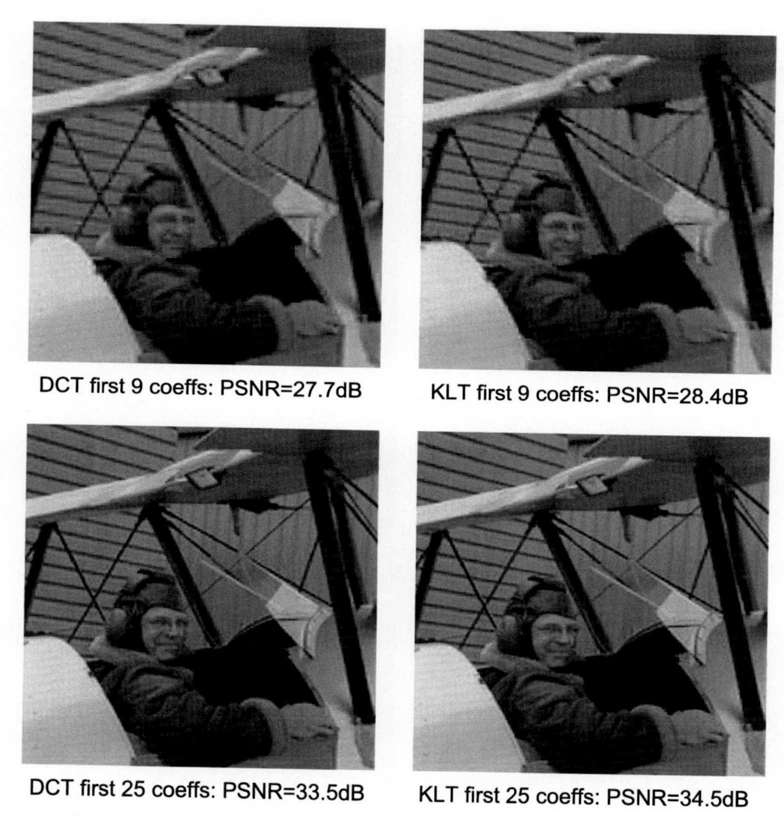

FIGURE 5.14 Comparison of KLT and DCT performance.

5.8 DCT implementation

5.8.1 Choice of transform block size

What is the optimal block size for transform coding? This question is best answered by considering the data being processed and the complexity of the transform itself. We normally seek the best compromise between decorrelation performance and complexity. As we will see below, the typical size historically has been 8×8 pixels, although sizes down to 4×4 are used in H.264/AVC and transform block sizes up to 32×32 are available in H.265/HEVC and 64×64 in VVC; these enable recent standards to better cope with UHD image resolutions. Variable block sizes are now becoming commonplace and the size is selected as part of the rate-distortion optimization (RDO) process (see Chapter 10).

It is also worth noting that the choice of transform can be data-dependent. For example, the *discrete Sine transform* (DST) can outperform the DCT for cases of lower correlation or for data with certain symmetry properties [5].

FIGURE 5.15 DCT performance. Left: Puppet original 256 × 256. Right: Puppet compressed at 0.5 bpp.

FIGURE 5.16 DCT performance. Left: Puppet 256 × 256, 0.5 bpp; Right: Puppet 512 × 512, 0.5 bpp.

DCT complexity

Using the standard 2D DCT, the computation of each coefficient in an $N \times N$ block requires N^2 multiplications and N^2 additions. This is indicated by a complexity of order $O(N^4)$. This can be reduced to $2N$ multiplies per coefficient when separability is exploited (i.e., $O(N^3)$). Hence for the case of 4 × 4 blocks we require 8 multiplies per coefficient, for an 8 × 8 block

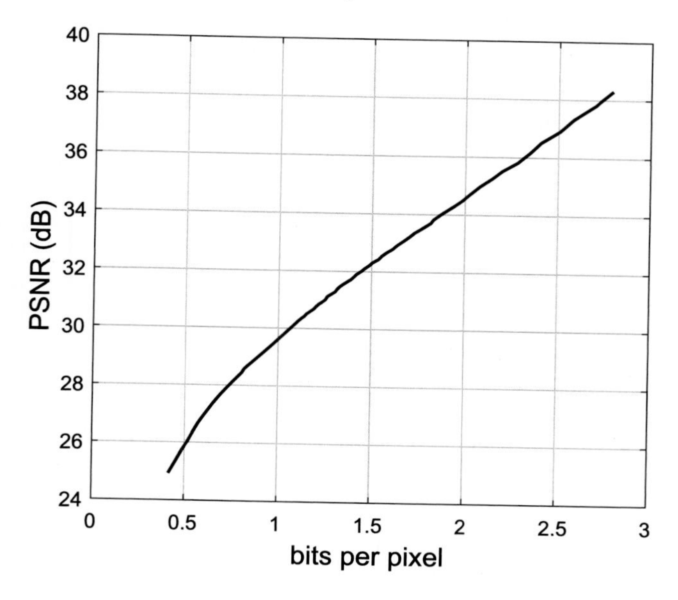

FIGURE 5.17 DCT rate-distortion performance for 256 × 256 puppet image using Huffman encoding.

we require 16 multiplies per coefficient, and for a 16 × 16 block we require 32 multiplies per coefficient.

Taking account of both complexity and autocorrelation surface characteristics, a choice between 4 × 4 and 32 × 32 is normally used in practice.

5.8.2 DCT complexity reduction

Due to its ubiquity, many fast algorithms exist for computing the DCT. Early fast algorithms operated indirectly, exploiting the fast Fourier transform, with $O(N \log_2 N)$ complexity. It can be shown [1,2] that the 1D DCT can be expressed as

$$c(k) = \frac{2}{N}\mathbb{R}\left[e^{\frac{-jk\pi}{2N}} \sum_{n=0}^{2N-1} x[n]\, W^{kn} \right]. \tag{5.49}$$

This type of approach is not so popular now due to the requirement for complex multiplications and additions. Other approaches take into account the unitary property of the DCT, which allows it to be factorized into products of relatively sparse matrices [6]. Recursive algorithms have also been proposed [3,7]. The complexity of these approaches is typically $O(N^2 \log_2 N)$. Later algorithms took direct advantage of the properties of the transform matrix, through factorization or matrix partitioning or by exploiting its symmetries in a divide and conquer methodology. One such approach is described below.

McGovern algorithm

In McGovern's algorithm [9], the matrix rows are decomposed, exploiting odd and even symmetries using a divide and conquer approach. The aim of this method is to reduce the number of multiplications per coefficient. We have

$$\begin{bmatrix} c(0) \\ c(1) \\ c(2) \\ c(3) \\ c(4) \\ c(5) \\ c(6) \\ c(7) \end{bmatrix} = \begin{bmatrix} a_4 & a_4 & a_4 & a_4 & a_4 & a_4 & a_4 & a_4 \\ a_1 & a_3 & a_5 & a_7 & -a_7 & -a_5 & -a_3 & -a_1 \\ a_2 & a_6 & -a_6 & -a_2 & -a_2 & -a_6 & a_6 & a_2 \\ a_3 & -a_7 & -a_1 & -a_5 & a_5 & a_1 & a_7 & -a_3 \\ a_4 & -a_4 & -a_4 & a_4 & a_4 & -a_4 & -a_4 & a_4 \\ a_5 & -a_1 & a_7 & a_3 & -a_3 & -a_7 & a_1 & -a_5 \\ a_6 & -a_2 & a_2 & -a_6 & -a_6 & a_2 & -a_2 & a_6 \\ a_7 & -a_5 & a_3 & -a_1 & a_1 & -a_3 & a_5 & -a_7 \end{bmatrix} \begin{bmatrix} x[0] \\ x[1] \\ x[2] \\ x[3] \\ x[4] \\ x[5] \\ x[6] \\ x[7] \end{bmatrix}, \quad (5.50)$$

where

$$a_i = \cos\left(i\frac{\pi}{16}\right),$$

$$\begin{bmatrix} c(0) \\ c(2) \\ c(4) \\ c(6) \end{bmatrix} = \begin{bmatrix} a_4 & a_4 & a_4 & a_4 \\ a_2 & a_6 & -a_6 & -a_2 \\ a_4 & -a_4 & -a_4 & a_4 \\ a_6 & -a_2 & a_2 & -a_6 \end{bmatrix} \begin{bmatrix} x[0]+x[7] \\ x[1]+x[6] \\ x[2]+x[5] \\ x[3]+x[4] \end{bmatrix}, \quad (5.51)$$

$$\begin{bmatrix} c(1) \\ c(3) \\ c(5) \\ c(7) \end{bmatrix} = \begin{bmatrix} a_1 & a_3 & a_5 & a_7 \\ a_3 & -a_7 & -a_1 & -a_5 \\ a_5 & -a_1 & a_7 & a_3 \\ a_7 & -a_5 & a_3 & -a_1 \end{bmatrix} \begin{bmatrix} x[0]-x[7] \\ x[1]-x[6] \\ x[2]-x[5] \\ x[3]-x[4] \end{bmatrix}. \quad (5.52)$$

The decomposition in Eq. (5.51) reduces the number of multiplications required from 64 ($=N^2$) in Eq. (5.50) to only 32 ($=2(N/2)^2$). Note that the matrix required to compute the even components of **x** is the same as that required to compute the four-point DCT. The resultant matrices can be further decomposed into a sparse form as follows:

$$\begin{bmatrix} c(0) \\ c(4) \\ c(2) \\ c(6) \\ c(7) \\ c(5) \\ c(1) \\ c(3) \end{bmatrix} = \begin{bmatrix} a_4 & 0 & 0 & 0 & 0 & 0 & 0 & 0 \\ 0 & a_4 & 0 & 0 & 0 & 0 & 0 & 0 \\ 0 & 0 & a_6 & a_2 & 0 & 0 & 0 & 0 \\ 0 & 0 & -a_2 & a_6 & 0 & 0 & 0 & 0 \\ 0 & 0 & 0 & 0 & a_7 & a_5 & a_1 & a_3 \\ 0 & 0 & 0 & 0 & a_5 & -a_1 & a_3 & a_7 \\ 0 & 0 & 0 & 0 & a_1 & a_3 & -a_7 & a_5 \\ 0 & 0 & 0 & 0 & a_3 & a_7 & -a_5 & -a_1 \end{bmatrix} \begin{bmatrix} g_0 \\ g_1 \\ g_2 \\ g_3 \\ g_4 \\ g_5 \\ g_6 \\ g_7 \end{bmatrix}, \quad (5.53)$$

where

$$
\begin{aligned}
g_0 &= x[0] + x[1] + x[2] + x[3] + x[4] + x[5] + x[6] + x[7], \\
g_1 &= x[0] + x[7] + x[3] + x[4] - x[1] - x[6] - x[2] - x[5], \\
g_2 &= x[1] + x[6] - x[2] - x[5], \\
g_3 &= x[0] + x[7] - x[3] - x[4], \\
g_4 &= x[0] - x[7], \\
g_5 &= x[6] - x[1], \\
g_6 &= x[3] - x[4], \\
g_7 &= x[2] - x[5].
\end{aligned}
$$

This reduces the number of multiplications required from 64 to 22. This number can be further reduced to 14 using *standard rotator products,*

$$
\begin{bmatrix} x & w \\ w & -y \end{bmatrix} \begin{bmatrix} r_0 \\ r_1 \end{bmatrix} = \begin{bmatrix} 1 & 1 & 0 \\ 0 & 1 & -1 \end{bmatrix} \begin{bmatrix} (x-w)\,r_0 \\ w\,(r_0 + r_1) \\ (w+y)\,r_1 \end{bmatrix},
$$

and still further reduced to 12 multiplications if we divide through by a_4. A bit-serial architecture for this algorithm is shown in Fig. 5.18 [9], where the gray boxes represent synchronizing delay operators.

Although the DST family does not share the same symmetry properties as the DCT, factorizations can still be employed to reduce complexity, as demonstrated in reference [8].

5.8.3 Field vs. frame encoding for interlaced sequences

In the case of interlaced image sequences, if motion occurs between the field scans, large-valued spurious DCT coefficients can occur at high spatial frequencies. These will be coarsely quantized during compression, reducing the quality of the encoded sequence. Standards since MPEG-2 have supported the option of performing a DCT either on each field independently or on the combined frame. In the case of field pictures, all the blocks in every macroblock come from one field, whereas with an (interlaced) frame picture, frame or field DCT coding decisions can be made adaptively on a macroblock-by-macroblock basis.

When an interlaced macroblock from an interlaced frame picture is frame DCT coded, each of its four blocks has pixels from both fields. However, if the interlaced macroblock from an interlaced frame picture is field coded, each block consists of pixels from only one of the two fields.

Martin and Bull [10] produced an efficient algorithm using the primitive operator methodology of Bull and Horrocks [11]. This supports coding either one 8×8 DCT or two independent 4×4 DCT blocks. The result was compared with alternative implementations using an area-time metric and significant complexity savings were observed.

5.8.4 Integer transforms

While the DCT is in principle lossless, it can suffer from numerical mismatches between the encoder and decoder, leading to drift. This can be overcome through the use of integer

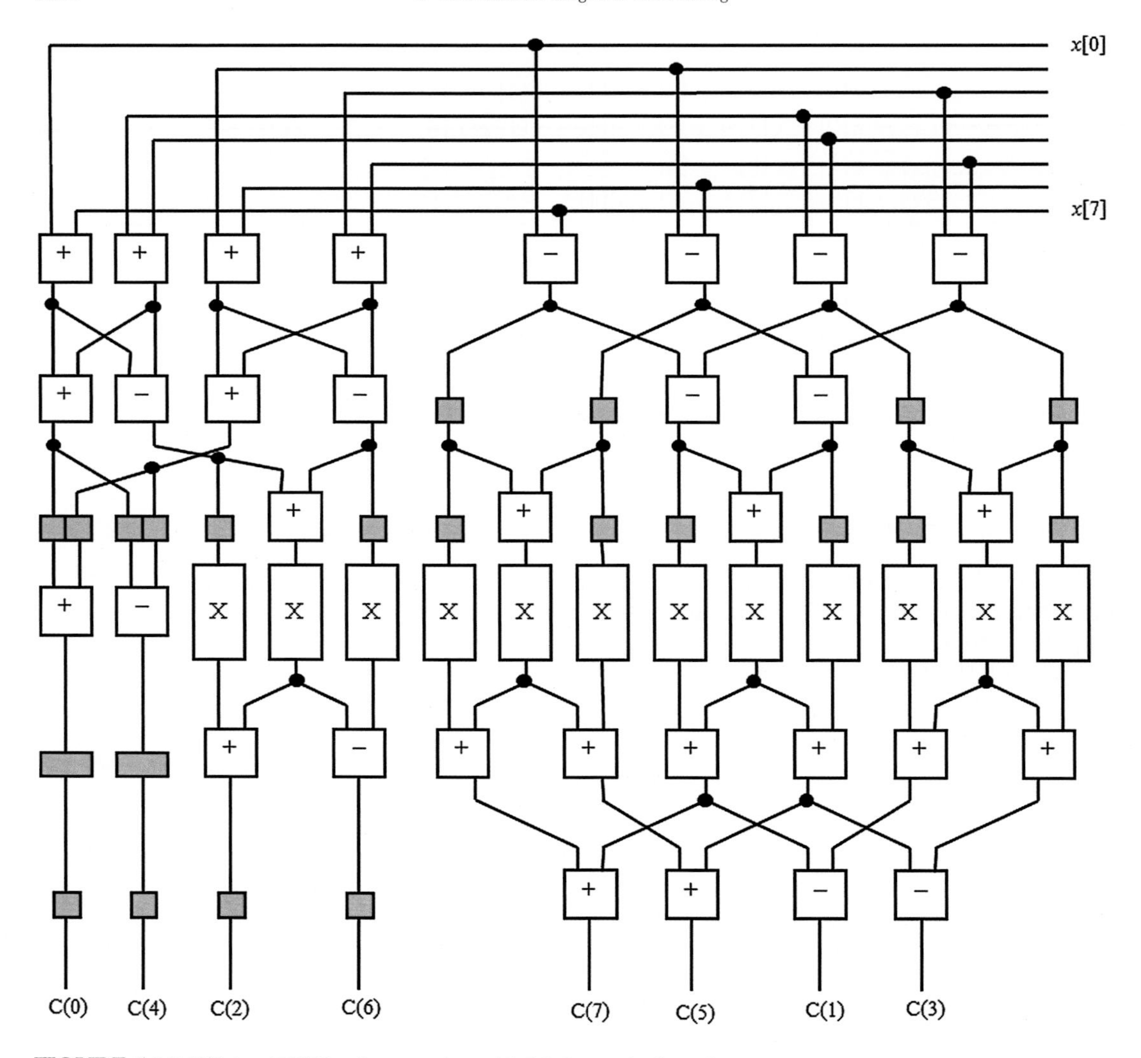

FIGURE 5.18 Efficient DCT implementation with McGovern's algorithm.

transforms where integer arithmetic is employed to ensure drift-free realizations. This approach is employed in the basic H.264/AVC 4×4 integer transform, which is itself derived from the 4×4 DCT. This transform is given in Eq. (5.54) and is discussed further in Chapter 9. We have

$$\mathbf{A} = \begin{bmatrix} 1 & 1 & 1 & 1 \\ 2 & 1 & -1 & -2 \\ 1 & -1 & -1 & 1 \\ 1 & -2 & 2 & -1 \end{bmatrix}. \tag{5.54}$$

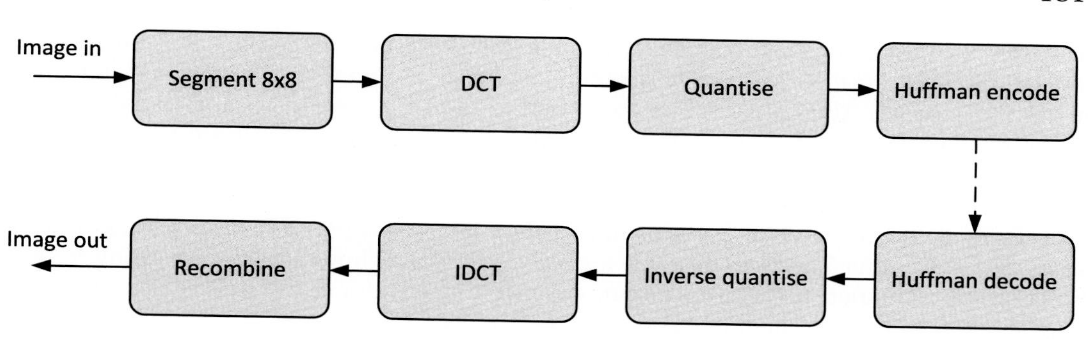

FIGURE 5.19 JPEG baseline system diagram.

5.8.5 DCT DEMO

An informative DCT demonstration based on MATLAB® software offers readers of this book the opportunity to visualize the effect of selecting and deselecting DCT coefficients, showing their influence on the quantized DCT matrix and on the reconstructed image. Readers can download this as part of an integrated demonstration package from https://fan-aaron-zhang.github.io/Intelligent-Image-and-Video-Compression.htm.

5.9 JPEG

The JPEG standard, created originally in 1992 [12], is still ubiquitous. Despite the emergence of JPEG2000 as a serious contender with superior performance in 2000 and various intra-coding modes offered by video standards such as H.265/HEVC, JPEG has maintained its position as the dominant codec for internet and consumer use. JPEG has four primary modes:

- baseline (sequential DCT plus Huffman coding),
- sequential lossless (predictive coding plus Huffman or arithmetic coding),
- progressive DCT (bit plane progressive),
- hierarchical.

Baseline JPEG is based on a simple block-based DCT with scalar quantization and Huffman coding. It is applicable to gray-scale and color images with up to 8 bits per sample and is described in Fig. 5.19. We will examine the complete JPEG coding process in more detail in Chapter 7, after describing entropy coding.

5.10 Summary

In this chapter we have introduced the reader to transform coding, explaining its basic concepts and mathematical underpinnings. We have shown that the optimal transform, while providing the best energy compaction for a given dataset, suffers when the data statistics are nonstationary, because of the need to continually update the transform matrix and transmit this to the decoder. Alternative transforms, such as the DCT, provide near-optimal performance for correlated images with fixed (signal-independent) bases and are the preferred choice for practical image and video compression. We have derived the DCT and demonstrated its excellent performance and its low-complexity implementations.

In the next chapter we will explore another data transformation based on wavelet filterbanks and in Chapter 7 we will describe lossless coding and show how decorrelating transforms can be combined with entropy (symbol) encoding to achieve impressive compression ratios while retaining excellent perceptual image quality.

References

[1] R. Clarke, Transform Coding of Images, Academic Press, 1985.

[2] N. Ahmed, T. Natarajn, K. Rao, Discrete cosine transform, IEEE Trans. Comput. 23 (1974) 90–93.

[3] B. Lee, A new algorithm to compute the discrete cosine transform, IEEE Trans. Acoust. Speech Signal Process. 32 (1984) 1243–1245.

[4] V. Britanak, et al., Discrete Cosine and Sine Transforms: General Properties, Fast Algorithms and Integer Approximations, Academic Press, 2006.

[5] J. Han, A. Saxena, K. Rose, Towards jointly optimal spatial prediction and adaptive transform in video/image coding, in: Proc. IEEE Intl. Conf. on Acoustics, Speech and Signal Processing, 2010.

[6] W. Chen, C. Smith, S. Fralick, A fast computational algorithm for discrete cosine transform, IEEE Trans. Commun. 25 (9) (1977) 1004–1009.

[7] P. Lee, F.-Y. Huang, Restructured recursive DCT and DST algorithms, IEEE Trans. Signal Process. 42 (7) (1994) 1600–1609.

[8] R. Chivukula, Y. Reznik, Fast computing of discrete cosine and sine transforms of types VI and VII, Proc. SPIE Applications of Digital Image Processing, vol. XXXIV, 2011, p. 813505.

[9] F. McGovern, R. Woods, M. Yan, Novel VLSI implementation of (8×8) point 2-d DCT, Electron. Lett. 30 (8) (1994) 624–626.

[10] F. Martin, D. Bull, An improved architecture for the adaptive discrete cosine transform, in: IEEE Intl. Symp. on Circuits and Systems, 1996, pp. 742–745.

[11] D. Bull, D. Horrocks, Primitive operator digital filters, IEE Proc., Circuits Devices Syst. 138 (3) (1991) 401–412.

[12] ISO/IEC International Standard 10918-1, Information technology – digital and coding of continuous-tone still images – requirements and guidelines, 1992.

[13] C. Wang, S. Lee, L. Chang, Designing JPEG quantization tables based on human visual system, Signal Process. Image Commun. 16 (5) (2001) 501–506.

[14] J. Mannos, F. Sakrison, The effects of a visual fidelity criterion of the encoding of images, IEEE Trans. Inf. Theory 20 (4) (1974) 525–536.

[15] M. Wien, High Efficiency Video Coding, Springer, 2015.

[16] M. Budagavi, et al., Core transform design in the high efficiency video coding (HEVC) standard, IEEE J. Sel. Top. Signal Process. 7 (6) (2013) 1029–1041.

CHAPTER

6

Filter-banks and wavelet compression

Wavelets are mathematical functions that can be used as an alternative to transforms such as the discrete cosine transform (DCT) to split a signal into different frequency bands (subbands) prior to processing and quantization. Wavelets are usually implemented as banks of filters that enable each subband to be analyzed at a resolution matched to its scale. This, as we will see later, is normally achieved by combining filtering operations with subsampling, which also has the benefit of maintaining a constant overall sample rate at all stages in the decomposition. This approach has significant advantages over Fourier-based methods, especially when signals are not harmonically related, are of short duration with discontinuities, or are nonstationary. It also has benefits in terms of subband coefficient quantization since the artifacts produced tend to be perceptually less annoying than the blocking artifacts associated with the DCT.

This chapter firstly examines the basic two-channel filter-bank structure and then investigates the filter properties that enable perfect reconstruction in the absence of quantization. We then go on to consider the more general case of multirate filtering, showing how, through the appropriate combination of filters and up- and downsampling operations, we can achieve critical sampling. We next examine specific cases of useful subband and wavelet filters and then go on to extend the basic architecture to the case of a multistage (multiscale) decomposition and to two dimensions as required for image compression applications. Finally, we examine bit allocation strategies and scalability in the context of the JPEG2000 still image coding standard.

6.1 Introduction to multiscale processing

As the name suggests, wavelets are small waves with compact support. They can be designed to have specific and useful properties and, in the same way as Fourier or DCT bases, can be linearly combined to synthesize a signal. The difference with wavelets is that the basis functions at different scales are related through translation and dilation of the basic building block functions. Wavelets can thus be localized in both spatial (or temporal) and frequency domains and, as such, offer important multiresolution properties. They provide much more flexibility in tiling the time-frequency or spatial frequency plane than conventional transforms, offering a means of analyzing signals at different scales and with different resolutions. For example, the tiles can vary in size to better match the characteristics of the human vi-

Intelligent Image and Video Compression
https://doi.org/10.1016/B978-0-12-820353-8.00015-3

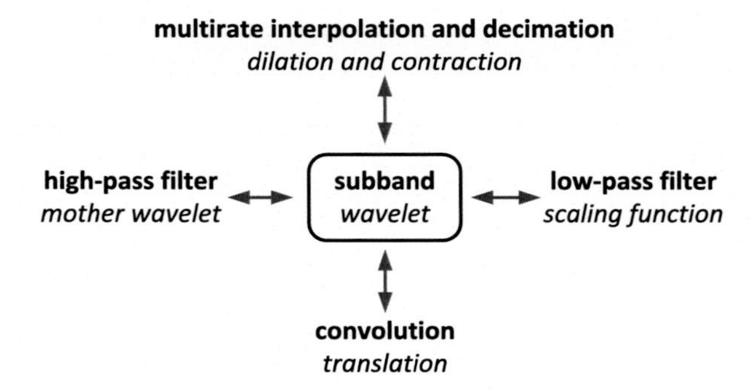

FIGURE 6.1 Equivalences between subband filtering and wavelet nomenclature.

sual system (HVS) for image coding (or the auditory system in the case of speech coding), thus creating a basis for improved quantization strategies and hence improved compression performance.

Historically, development of the theory of wavelets (although they were not originally called wavelets) preceded that of filter-banks due to the early work of Haar over 100 years ago [1]. The area was then dormant for many years, until the emergence of subband coding methods for speech, audio, and image processing in the 1980s. One of the major challenges identified at that time was the design of appropriate channel splitting filters, and the earliest contributions in this respect were by Crosier et al., who introduced the concept of the quadrature mirror filter (QMF) [2]. This then led to the development of perfect reconstruction filters with significant contributions from Smith and Barnwell [3,4], Vetterli [5], Woods and O'Neill [6], and Vaidyanathan [7].

Work on wavelets developed in parallel, perhaps starting with the contribution in 1984 from Grossman and Morlet [8], who were the first to use the term "wavelet." This initiated a major body of work on the topic, with major contributions from Daubechies [9] and several others. Wavelet theories and their associated design methods were developed into a comprehensive framework by Mallat [10,11] in 1989.

Although we will primarily adopt a subband filtering rather than a wavelet-oriented nomenclature here, it is worth comparing the terms commonly used in each domain. This is illustrated in Fig. 6.1. Many excellent texts exist on the topics of wavelets and subband filtering [12–14] and the reader is referred to these for further information.

6.1.1 The short-time Fourier transform and the Gabor transform

One of the main problems with conventional transforms (such as the Fourier transform) is that they are constructed around a set of infinite-length basis functions such as sinusoids or complex exponentials. This means that any temporally (or spatially) localized changes or features are spread over the whole frequency domain. This is a particular problem for nonstationary signals such as those typical of images and videos.

One approach to overcome this has been to divide the input signal into shorter time duration sections. While this solves the issue of nonstationarity, it introduces additional problems due to boundary effects and edge distortions. As we saw in Chapter 5, these can distort the underlying spectrum and, in the context of compression, make perfect reconstruction of the original signal impossible. To reduce the influence of these boundary effects it has been conventional in spectrum analysis to window each block of data. The window function reduces to zero at the edges of the block, hence eliminating temporal discontinuities and any associated spectral ringing effects. If we consider the case of the short-time Fourier transform (STFT), this is defined in the following equation, where if the window function $w(t)$ is a Gaussian function, then this is referred to as the Gabor transform:

$$X(\omega, \tau) = \int_{-\infty}^{\infty} x(t)w(t-\tau)e^{-j\omega t}dt. \tag{6.1}$$

A primary issue with the STFT is that the window size is fixed, so any variation in the temporal extent of any nonstationarities cannot be adequately represented. Also, no good basis functions exist for the STFT. Another problem is that the window function distorts the underlying signal, which means that perfect reconstruction is not possible unless overlapping windows are used.

Clearly, because of the uncertainty principle, we cannot have both arbitrarily high temporal resolution and frequency resolution. We can however trade off temporal and frequency resolution by altering the window duration used and this is considered next.

6.1.2 What is a wavelet?

Ideally we would like a set of basis functions that can isolate localized high-frequency components in a signal while, at the same time, capturing lower-frequency components of longer duration. Consider the example in Fig. 6.2, where two signals – a sinusoid and an impulse – are analyzed in the frequency and temporal domains. The time domain analysis clearly picks out the impulse, whereas the frequency domain analysis localizes the sine wave. In contrast, frequency domain analysis cannot localize the impulse and time domain analysis cannot localize the sine wave. If we were able to tile the time-frequency plane more flexibly, as shown in the bottom right subfigure, then it would be possible to pick out both short-duration high-frequency signal components and longer-duration low-frequency components.

In practice, we can do this by combining short high-frequency basis functions with longer lower-frequency ones. Wavelets achieve this in a very elegant way, through the dilation and translation of a single prototype function. Consider Fig. 6.3, where a basic function of duration T is dilated by a factor of 2 and translated by an amount T_0. Clearly the relationship between $x_1(t)$ and $x_2(t)$ in the figure is

$$x_2(t) = x_1\left(\frac{t - T_0}{2}\right).$$

We can imagine a linear combination of weighted versions of these translated and dilated functions being used to approximate a signal of interest. However, unlike conventional basis functions constructed from cosine terms or complex exponentials, these exhibit compact support as well as oscillatory behavior. They thus offer the potential to capture the local features as well as the large-scale characteristics of a signal.

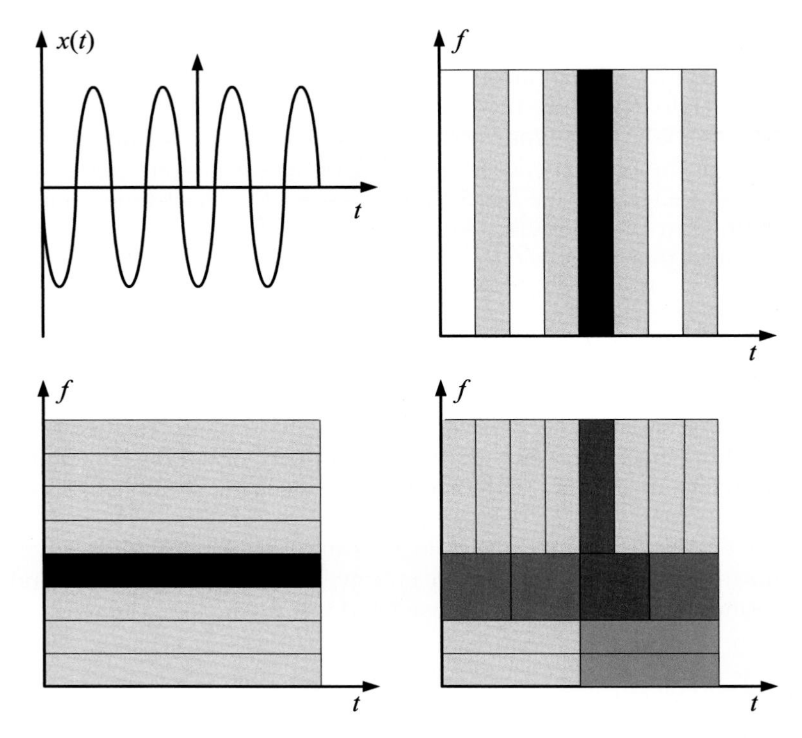

FIGURE 6.2 Comparison of time domain (top right), frequency domain (bottom left), and time-frequency analysis (bottom right) applied to the combination of a sinusoidal signal and an impulse function (top left) (example adapted from Vetterli and Kovacevic [12]).

The continuous wavelet transform (CWT)

We can define a single general prototype function (Eq. (6.2)) that can be stretched and shifted to form new functions which can be linearly combined to synthesize a signal of interest. This is the basic wavelet transform as given in Eq. (6.3):

$$h_{a,b}(t) = \frac{1}{\sqrt{a}} h\left(\frac{t-b}{a}\right), \tag{6.2}$$

$$X(a,b) = \frac{1}{\sqrt{a}} \int_{-\infty}^{\infty} h\left(\frac{t-b}{a}\right) x(t)\, dt. \tag{6.3}$$

Here a is the scale parameter, b is the translation parameter, and $h(t)$ is called the mother wavelet (i.e., the prototype function from which all the other bases are formed).

This ability to deal with local variations as well as big image features in the same framework is clearly attractive for image compression. However, in order to apply it, we need to be able to evaluate it in discrete form. The continuous-time wavelet transform is highly redundant, so the scale space can be sampled dyadically using a wavelet series approach. This

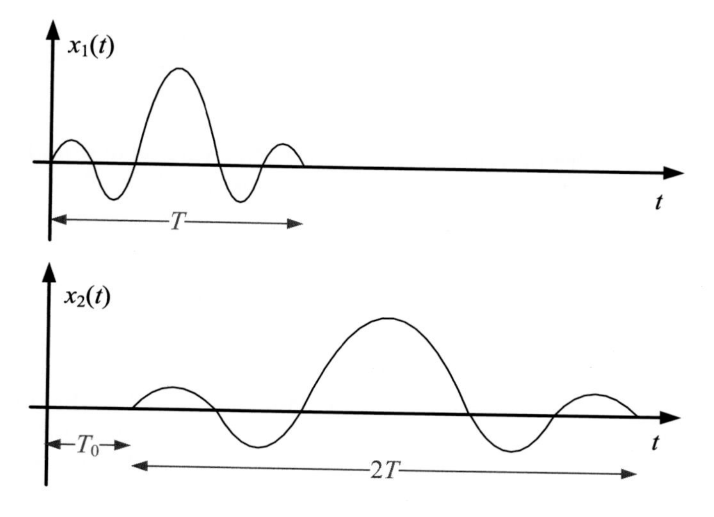

FIGURE 6.3 Wavelet translation and dilation.

however is not a true discrete transform, but fortunately one does exist and we will consider that next.

The discrete wavelet transform (DWT)

Although the continuous wavelet transform (CWT) can be computed discretely as described above, this is still only a sampled version of the CWT and not a proper discrete transform. The discrete wavelet transform (DWT) is instead based on a set of discrete functions or filters that are convolved with the input signal to effect translation and that are effectively dilated by subsampling the signal as it passes through each scale. It is normal to use octave band decompositions with dyadic sampling. This means that the wavelet filters form a set of band-pass responses that provide constant Q filtering. As the width of the octave bands is reduced, more localized high-frequency details are captured by the representation. To achieve this, we constrain the scale and translation terms to $a = 2^j$ and $b = k2^j$, respectively, in Eq. (6.3), where j and k are integers.

Wavelet decomposition is usually realized in the form of a filter-bank, as shown (for the case of a simple two-band split) in Fig. 6.4. The input signal is spectrally decomposed into distinct bands in an analysis section which uses carefully designed filters in conjunction with downsampling[1] to split the signal without increasing the effective sample rate. After analysis, the information in each band can be processed (e.g., quantized) independently according to the characteristics of the source and/or the application. The signal is then passed though a matched synthesis section that combines filtering with upsampling to reconstruct the original signal or (in the case of lossy compression) an approximation to it.

The wavelet transform and its filter-bank realization do not simply divide a signal into multiple frequency bands, but it does this iteratively, normally only iterating the low-pass

[1] Downsampling is sometimes referred to as decimation.

output at each scale. It thus produces a series of band-pass outputs which are, in fact, the wavelet transform coefficients. The role of the wavelet is played by the high-pass filter and the cascade of low-pass filters followed by a high-pass filter. This will be discussed in more detail later in Section 6.4.3 and is illustrated in Fig. 6.19.

As indicated above, this framework has been formalized by Mallat and others, and the reader is referred to references [10] and [12] for further details.

6.1.3 Wavelet and filter-bank properties

As we have discussed above, the DWT is normally implemented as a filter-bank and, in this respect, it is identical to the conventional subband filtering structure. Wavelet filter-banks can be considered as a subclass of subband filter-banks where the filters used have some specific properties associated with regularity. This means that as the filters are iterated, their characteristic shape converges to a regular function, i.e., one that is continuous or continuously differentiable. A necessary condition in this respect is that the filter should have sufficient zeros at the half sampling frequency (i.e., at $z = -1$). This ensures that the spectral images introduced during each filtering stage are attenuated. An excellent review of the requirements for convergence is provided by Rioul and Vetterli in [15].

Before we look in detail at the construction of wavelet filter-banks, let us summarize some of their advantages. Wavelet coding offers a number of potential benefits over other transforms. These include:

- **Constant relative bandwidth processing**: As illustrated in Fig. 6.2, this facilitates more flexible and perceptually accurate tiling of the spatial frequency (or time-frequency) plane.
- **Improved perceptual quality**: As discussed previously, the main disadvantage of transform coding is the subjective degradation at lower bit rates where the viewer is able to perceive the outline of the transform blocks. Subband approaches are however (generally) free of block distortions. By decomposing the signal into a number of individual (possibly overlapping) frequency bands, each band may be quantized differently according to subjective HVS or energy distribution criteria.
- **Scalability**: As well as improved perceptual performance, wavelet methods also provide a basis for scalability,[2] both for images – using for example set partitioning in hierarchical trees (SPIHT) [16] or embedded block coding with optimized truncation (EBCOT) [17] – and for video – using 3D wavelets and motion-compensated temporal filtering [18].
- **Error resilience**: A single error during transmission can be modeled as an impulse that excites the synthesis filters to produce a weighted impulse response, which is superimposed on the correct response. With smooth short filters, this produces a localized ringing effect which is perceptually preferable to the blocking errors common with the DCT. Of course error propagation due to prediction loops and variable-length coding is still possible. Techniques such as wavelet-based pyramid vector quantization [19] overcome this rather successfully by using fixed-length codes (see Chapter 11).

[2] Scalability refers to the property where, after encoding, a compressed bitstream can be manipulated to reduce its bit rate without decoding. This is useful in order to adapt to prevailing channel conditions without the need for explicit transcoding.

Although not selected for incorporation in modern video coding standards, subband (wavelet) decomposition techniques can offer improved performance and flexibility (with similar implementation complexity) compared to their block transform counterparts, such as the DCT. This is particularly true for image coding, especially at lower bit rates. These benefits were recognized through the adoption of wavelet coding methods in the still image coding standard JPEG2000 [20,21].

6.2 Perfect reconstruction filter-banks

6.2.1 Filter and decomposition requirements

A simple two-channel filter-bank is shown in Fig. 6.4. Here, as we will shortly see, $H_0(z)$ and $H_1(z)$ are the low- and high-pass analysis filters and $G_0(z)$ and $G_1(z)$ are the corresponding synthesis filters. The circles in Fig. 6.4 represent the downsampling (decimation) and upsampling (interpolation) operators. These are nonlinear operators that:

1. ensure critical sampling, i.e., the effective sampling rate throughout the filter-bank is kept constant; this is important in compression applications as otherwise there is an expansion in the amount of data being processed;
2. are responsible for introducing aliasing and imaging; this, as we will see later, must be managed through the careful design of the filters used.

In designing filters for filter-bank use, we must therefore consider two important issues:

1. The filters used in practice do not have perfect transition characteristics and so there will be spectral cross-talk across the bands.
2. The sample rate changing operations will introduce aliasing and spectral imaging that must be eliminated during reconstruction if perfect reconstruction (a property essential for lossless compression) is to be achieved.

6.2.2 The 1D filter-bank structure

The fundamental building block of subband or wavelet filtering is the two-channel filter-bank of Fig. 6.4. We will examine this structure and its properties in detail later. In the meantime, let us consider how it comes about using a very simple approach to filtering and sampling.

Intuitive development of the two-channel filter-bank

The basic idea of the two-band split is to divide a signal into high-pass and low-pass components such that they can be analyzed or quantized independently in a manner related to signal scale. Consider a simple signal that comprises a steadily changing function with some superimposed random fluctuations, for example

$$x[n] = \{2, 4, 8, 8, 10, 14, 12, 6, 8, 4\}.$$

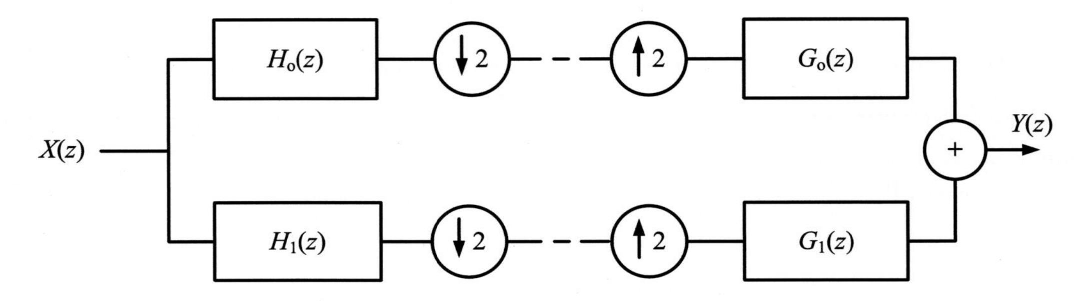

FIGURE 6.4 Two-channel perfect reconstruction filter-bank.

We could estimate the underlying function through low-pass filtering and isolate the noise through high-pass filtering. Perhaps the simplest low-pass filter is the 2-tap moving average as given by Eq. (6.4). The residual signal, $r[n]$, remaining after applying Eq. (6.4), is the high-pass component and can be characterized using Eq. (6.5):

$$l[n] = \frac{x[n] + x[n-1]}{2}, \tag{6.4}$$

$$r[n] = x[n] - l[n] = \frac{x[n] - x[n-1]}{2}. \tag{6.5}$$

Let us examine the frequency domain characteristics of these two simple filters in more detail, starting with the moving average or low-pass filter. From Eq. (6.4) we can see that

$$L(z) = 0.5\left(X(z) + X(z)z^{-1}\right),$$

and thus

$$H_l(z) = 0.5\left(\frac{z+1}{z}\right). \tag{6.6}$$

This, as expected, provides a simple low-pass filter response due to its pole at the origin in the z plane and its zero at $\Omega = \pi$. Similarly, the difference operation, or high-pass filter, is given by Eq. (6.7), which also has a pole at the origin and a zero at $\Omega = 0$:

$$R(z) = 0.5\left(X(z) - X(z)z^{-1}\right),$$

$$H_r(z) = 0.5\left(1 - z^{-1}\right) = 0.5\left(\frac{z-1}{z}\right). \tag{6.7}$$

The frequency responses of this complementary pair are shown in Fig. 6.5. These responses together form an all-pass function – which is not entirely surprising since their time domain equivalents are also complementary, adding to exactly reconstruct the original signal.

So we now begin to see a relationship between perfect reconstruction and a pair of complementary filters that cross at $\Omega = \pi/2$. It can be seen that the filter responses are symmetrical,

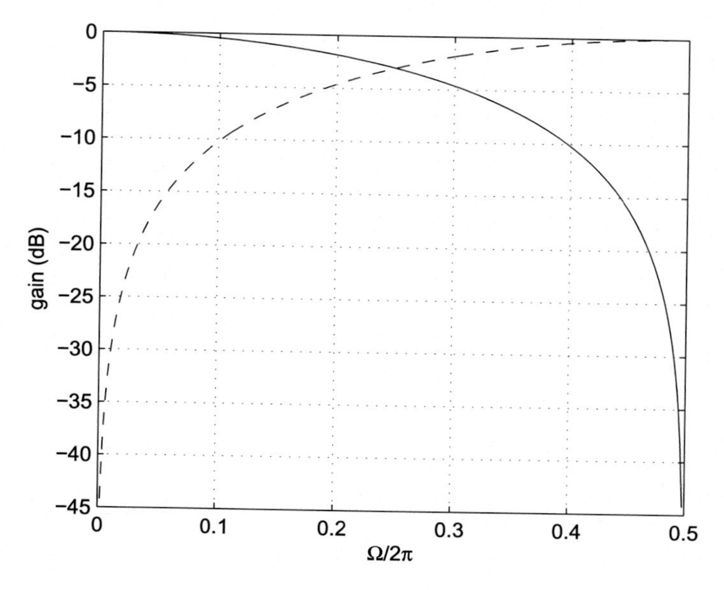

FIGURE 6.5 Complementary half-band filters from Eqs. (6.4) (solid) and (6.5) (dotted).

crossing with −3 dB attenuation at $\Omega = \pi/2$. Filters with this characteristic are often referred to as half-band filters or, more specifically, quadrature mirror filters (QMFs). We will explore this relationship further when we examine filter design and performance later.

We can apply these simple filters to our example data sequence, $x[n]$, to produce the following outputs (assuming that $x[-1] = 0$):

$$l[n] = \{1, 3, 6, 8, 9, 12, 13, 9, 7, 6\},$$

$$r[n] = \{1, 1, 2, 0, 1, 2, -1, -3, 1, -2\}.$$

The signals $l[n]$ and $r[n]$ together completely characterize the signal $x[n]$ and could be used as a representation for $x[n]$ that emphasizes its low-pass and high-pass (trend and difference) components. However, one obvious problem arises: $l[n]$ and $r[n]$ together contain twice as many samples as $x[n]$ alone. Doubling the size of the input signal is clearly not a great starting point for compression!

So what can we do about this? Firstly, there is obviously redundancy in the $\{l[n], r[n]\}$ combination – this must be the case as we have doubled the size without introducing any new information. We can observe this redundancy by splitting $l[n]$ and $r[n]$ into even and odd sequences, i.e., $\{l[2n-1], l[2n]\}$ and $\{r[2n-1], r[2n]\}$. Considering for example only the even sequences, we can reconstruct the original sequence $x[n]$ as follows:

$$l[2n] = \frac{x[2n] + x[2n-1]}{2},$$

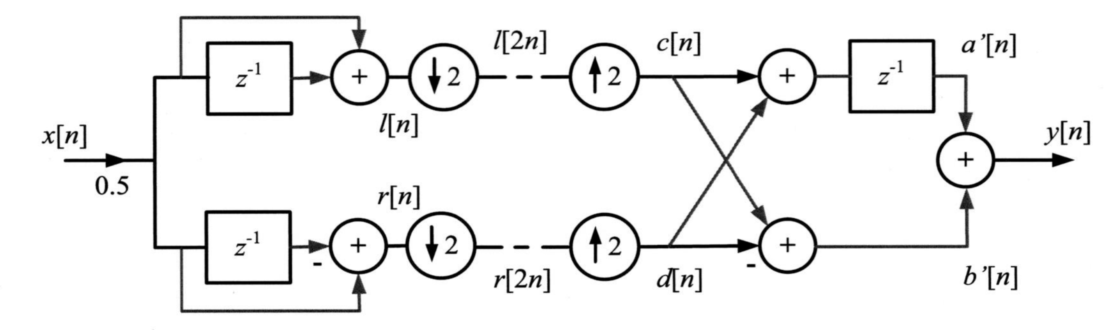

FIGURE 6.6 Basic system diagram for the two-channel filter-bank.

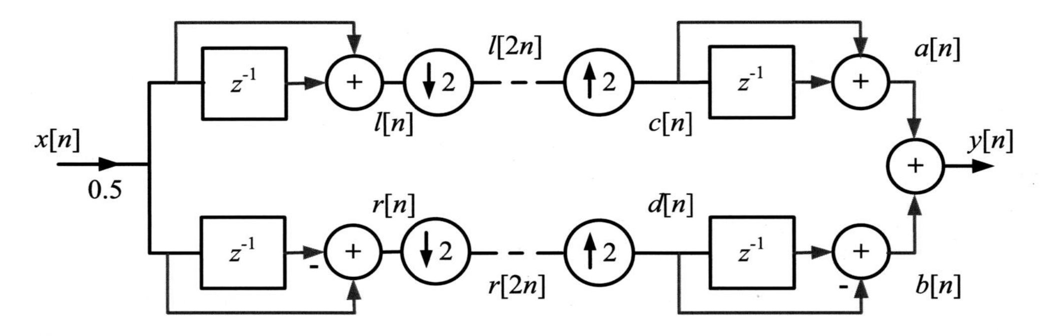

FIGURE 6.7 The two-channel filter-bank.

$$r\,[2n] = \frac{x\,[2n] - x\,[2n-1]}{2}.$$

Thus,

$$x\,[2n] = l\,[2n] + r\,[2n], \tag{6.8}$$

$$x\,[2n-1] = l\,[2n] - r\,[2n]. \tag{6.9}$$

Hence, we can perfectly reconstruct the original sequence $x\,[n]$, using only half of the data produced by the filters, by combining the downsampled sequence $x\,[2n-1]$ with the one-sample delayed version, $x\,[2n]$. The system diagram for this process is given in Fig. 6.6. This shows, on the left of the diagram, the analysis stage which produces the filtered outputs from Eq. (6.4) and Eq. (6.5) and downsamples these by a factor of two in order to maintain a constant overall bit rate. On the right side of the diagram is the synthesis section. This recombines the filtered and downsampled samples through upsampling and reconstruction based on Eq. (6.8) and Eq. (6.9).

A refined system diagram, obtained through reconfiguration of the synthesis section to yield a more recognizable filter-bank structure, is given in Fig. 6.7. Let us now validate the operation of this system through an example (Example 6.1).

Example 6.1. Operation of the simple two-channel perfect reconstruction filter-bank

Verify the operation of the filter-bank in Fig. 6.7 using the example data sequence $x[n] = \{2, 4, 8, 8, 10, 14, 12, 6, 8, 4\}$.

Solution. The operation of the system for this data is captured in the table below. This illustrates that the filter-bank does indeed provide perfect reconstruction, with the output delayed by one sample with respect to the input.

n	0	1	2	3	4	5	6	7	8	9
$x[n]$	2	4	8	8	10	14	12	6	8	4
$l[n]$	1	3	6	8	9	12	13	9	7	6
$r[n]$	1	1	2	0	1	2	−1	−3	1	−2
$l[2n]$	1	6	9	13	7	0	0	0	0	0
$r[2n]$	1	2	1	−1	1	0	0	0	0	0
$c[n]$	1	0	6	0	9	0	13	0	7	0
$d[n]$	1	0	2	0	1	0	−1	0	1	0
$a[n]$	1	1	6	6	9	9	13	13	7	7
$b[n]$	−1	1	−2	2	−1	1	1	−1	−1	1
$y[n]$	0	2	4	8	8	10	14	12	6	8

6.3 Multirate filtering

Subband decomposition techniques are based on multirate processing. Multirate theory enables a signal to be represented in the form of multiple band-pass components whilst maintaining a constant overall bit rate. Consider, for example, the two-band split. Since the convolution of a signal with a filter impulse response does not affect bit rate,[3] for an input bit rate of k bps, the overall output bit rate without subsampling will be $2k$ bps. Multirate theory however enables us to subsample the filter outputs in order to maintain a constant bit rate while preserving overall signal content (this is known as critical decimation or critical sampling).

The two primary operations used in multirate systems are interpolation and decimation. Decimation is normally employed in the encoder (or analysis stage) to reduce the sampling rate by a factor of M at each node relative to the input rate, whereas interpolation is normally used in the decoder (or synthesis stage) to increase it during reconstruction. These operations facilitate the removal of redundant information and hence help to manage the processing requirements.

6.3.1 Upsampling

Increasing the sample rate of a signal is often referred to as interpolation and is shown diagrammatically in Fig. 6.8. The operation of the upsampler is characterized according to its

[3] Assuming an infinite-length sequence.

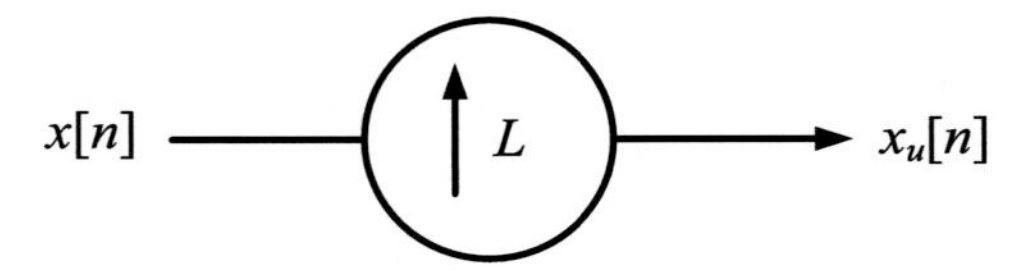

FIGURE 6.8 Upsampling by a factor L.

upsampling ratio, L, and is defined as follows:

$$x_u[n] = \begin{cases} x[\frac{n}{L}], & n = mL, \\ 0, & \text{otherwise.} \end{cases} \tag{6.10}$$

So, for example with L=2, we insert $L - 1 = 1$ zeros between each of the original samples. Thus if $x[n] = \{1, 2, 3, 4, 5, 6\}$, then $x_u[n] = \{1, 0, 2, 0, 3, 0, 4, 0, 5, 0, 6\}$. The z transform of the upsampled signal is then given by

$$X_u(z) = \sum_{n=-\infty}^{\infty} x_u[n] z^{-n} = \sum_{n=-\infty}^{\infty} x\left[\frac{n}{L}\right] z^{-n}, \quad {}^{n}/_{L} \text{ integer.} \tag{6.11}$$

Substituting $m = \frac{n}{L}$ in Eq. (6.11),

$$\sum_{m=-\infty}^{\infty} x[m] z^{-mL} = X\left(z^L\right). \tag{6.12}$$

Let us consider what the implications of this are in the frequency domain. In the case of L=2, the mapping $X(z^2)$ provides a factor 2 "compression" of the spectrum in the frequency domain. This can be verified by evaluating Eq. (6.12) on the unit circle in the z plane, i.e.,

$$X\left(z^2\right)\big|_{z=e^{j\Omega}} = X\left(e^{j2\Omega}\right),$$

and plotting the result, as illustrated in Fig. 6.9. This figure shows that an additional spectral image is introduced at $\Omega = \pi$ as a result of upsampling. This must be removed using a subsequent interpolation filter, as shown in Fig. 6.10. An example signal showing the spectral effect of upsampling is shown in Fig. 6.11.

6.3.2 Downsampling

Decreasing the sample rate of a signal is often referred to as decimation (although strictly this should refer only to downsampling by a factor of 10). The downsampling operation is shown diagrammatically in Fig. 6.12. The operation of the downsampler is characterized according to its downsampling ratio, M, and is defined as follows:

$$x_d[n] = x[nM]. \tag{6.13}$$

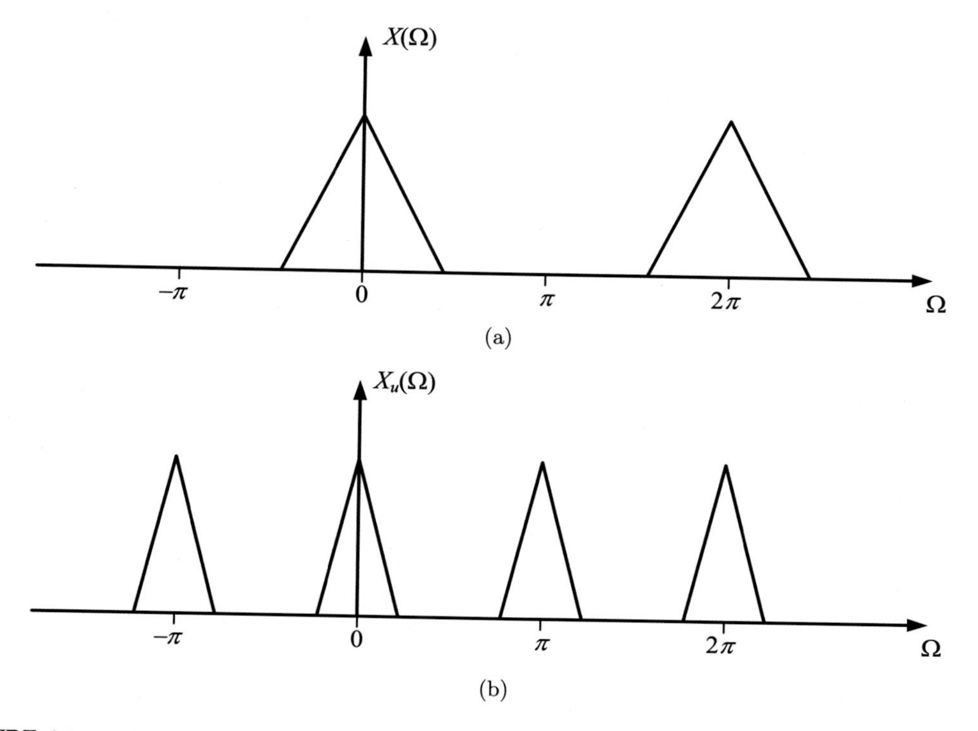

FIGURE 6.9 Spectral effect of upsampling. (a) Input. (b) Output.

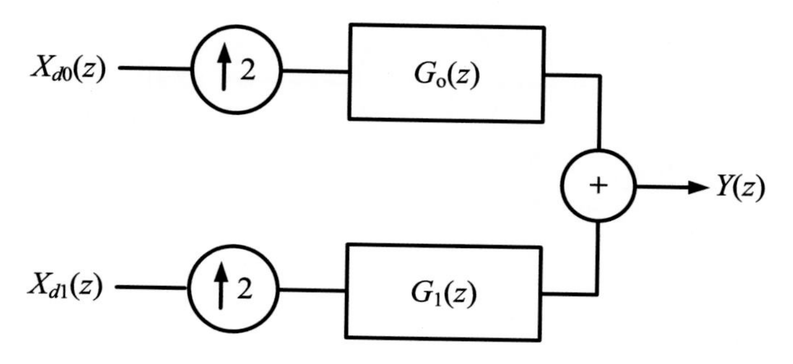

FIGURE 6.10 Two-channel synthesis filter-bank.

So, for example with M=2, we delete every Mth sample in the original sequence. Thus if $x[n] = \{1, 2, 3, 4, 5, 6\}$, then $x_d[n] = \{1, 3, 5\}$. The z transform of the downsampled signal is then given by

$$X_d(z) = \sum_{n=-\infty}^{\infty} x_d[n] z^{-n} = \sum_{n=-\infty}^{\infty} x[nM] z^{-n}. \tag{6.14}$$

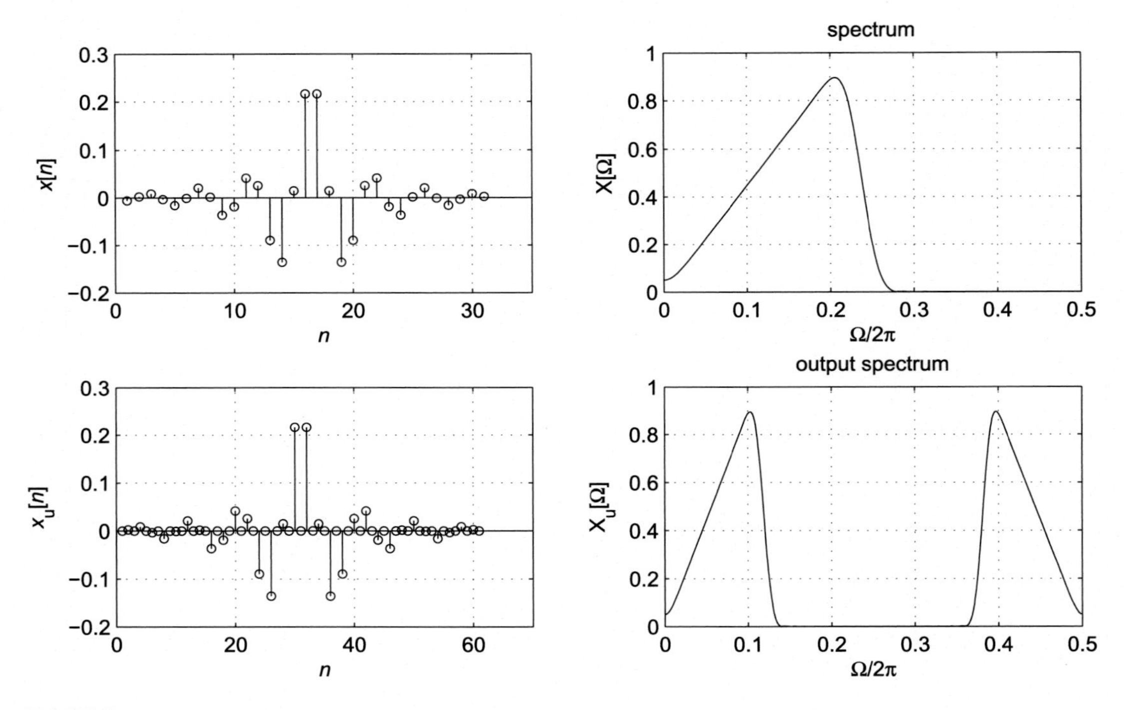

FIGURE 6.11 Spectral effects of upsampling where $L=2$.

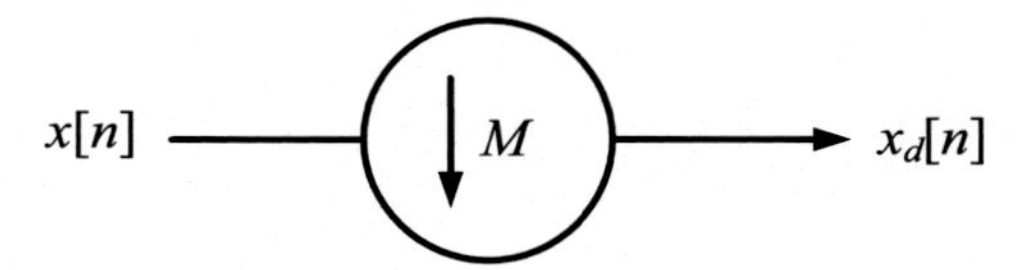

FIGURE 6.12 Downsampling by a factor of M.

Substituting $m=nM$ in Eq. (6.14),

$$X_d(z) = \sum_{m=-\infty}^{\infty} x'[m] z^{-m/M} = X'\left(z^{1/M}\right). \tag{6.15}$$

Note that $x'[n]$ is not the same as $x[n]$. They can however be related by $x'[n] = c[n]x[n]$, where

$$c[n] = \begin{cases} 1, & n = 0, \pm M, \pm 2M \ldots, \\ 0, & \text{otherwise,} \end{cases}$$

where $c[n]$ can be represented, using the DFT, as

$$c[n] = \frac{1}{M}\sum_{k=0}^{M-1} W_M^{nk}. \tag{6.16}$$

So, for example, in the case of M=2, we have

$$c[n] = \frac{1}{2}\left(1 + (-1)^n\right).$$

Now the z transform of $x'[n]$ is given by

$$X'(z) = \frac{1}{M}\sum_{n=-\infty}^{\infty}\left(\sum_{k=0}^{M-1} W_M^{nk}\right) x[n] z^{-n} = \frac{1}{M}\sum_{k=0}^{M-1}\left(\sum_{n=-\infty}^{\infty} x[n] W_M^{nk} z^{-n}\right) \tag{6.17}$$

$$= \frac{1}{M}\sum_{k=0}^{M-1} X\left(zW_M^{-k}\right). \tag{6.18}$$

Eq. (6.18) can now be substituted into Eq. (6.15) to give

$$X_d(z) = \frac{1}{M}\sum_{k=0}^{M-1} X\left(z^{1/M} W_M^{-k}\right). \tag{6.19}$$

Consider the case where M=2:

$$X_d(z) = \frac{1}{2}\left(X\left(z^{1/2}\right) + X\left(-z^{1/2}\right)\right). \tag{6.20}$$

Evaluating this on the unit circle in the z plane gives the frequency response; thus,

$$X_d\left(e^{j\Omega}\right) = \frac{1}{2}\left(X\left(e^{j\Omega/2}\right) + X\left(-e^{j\Omega/2}\right)\right), \tag{6.21}$$

or equivalently,

$$X_d\left(e^{j\Omega}\right) = \frac{1}{2}\left(X\left(e^{j\Omega/2}\right) + X\left(e^{j(\Omega/2+\pi)}\right)\right). \tag{6.22}$$

A plot of Eq. (6.22) is shown in Fig. 6.13. This shows that the original spectrum is expanded by a factor of two in the frequency domain, but also that an additional spectral image is introduced at $\Omega = 2\pi$ as a result of the second term in Eq. (6.22). It can be seen that, in order to avoid aliasing, which would occur if the original spectrum extended beyond $\Omega = \pi/2$, the input signal must be antialias filtered prior to downsampling. This is achieved using the analysis filter in Fig. 6.14. An example of downsampling with an actual signal is shown in Fig. 6.15.

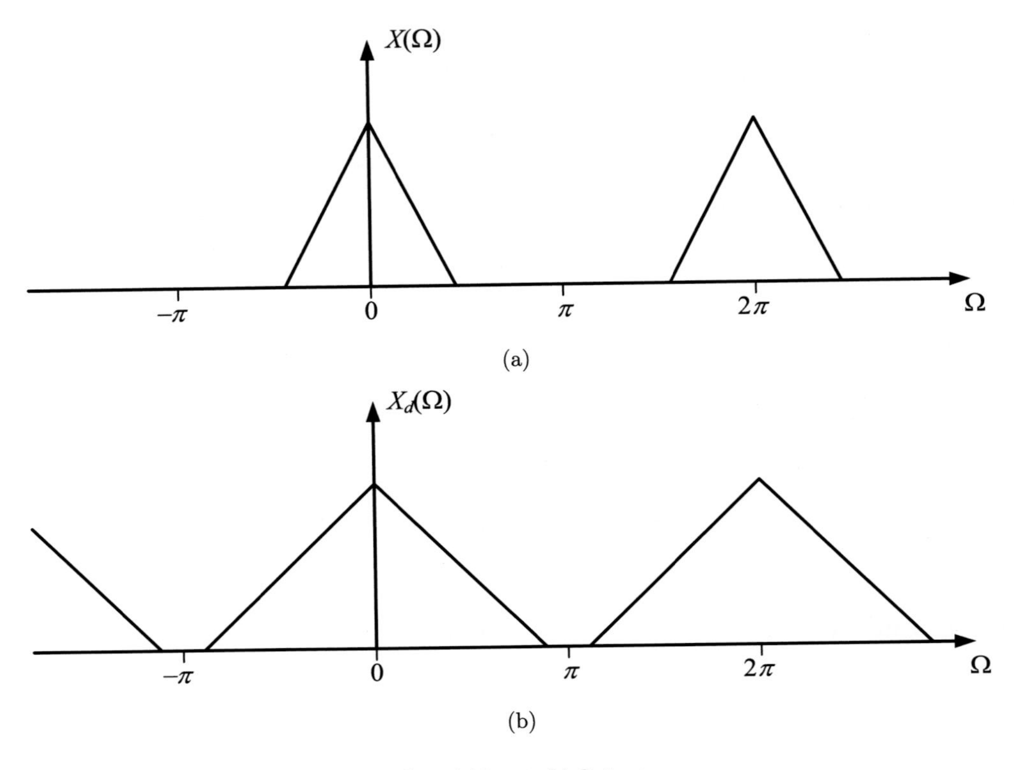

FIGURE 6.13 Spectral effects of downsampling. (a) Input. (b) Output.

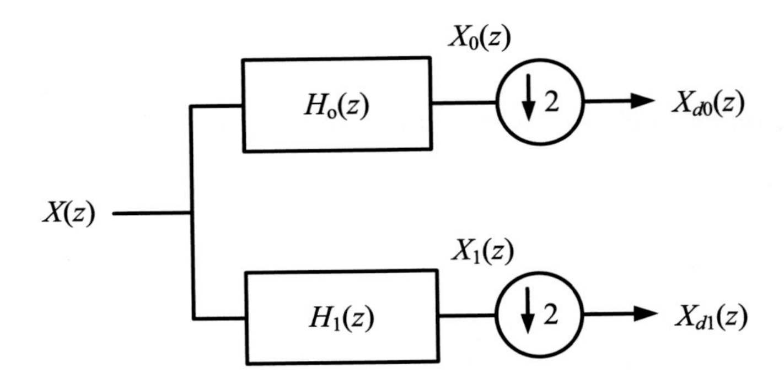

FIGURE 6.14 Two-channel analysis filter-bank.

6.3.3 System transfer function

We are now in a position to derive a transfer function for the two-channel filter-bank that was introduced in Fig. 6.4. This combines the analysis stage of Fig. 6.14 and the synthesis

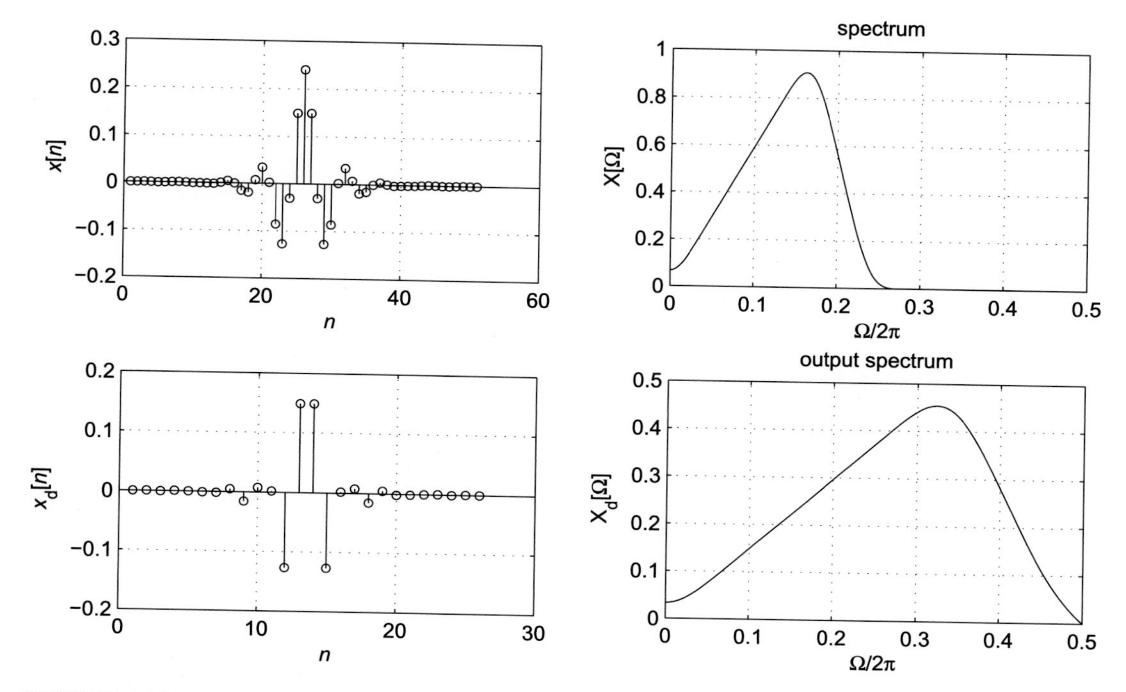

FIGURE 6.15 Spectral effects of downsampling with M=2.

stage of Fig. 6.10. Considering the top path in the analysis stage only at first, we have

$$X_0(z) = X(z)\,H_0(z) \tag{6.23}$$

and

$$X_{d0}(z) = \frac{1}{2}\left(X_0\left(z^{1/2}\right) + X_0\left(-z^{1/2}\right)\right). \tag{6.24}$$

Substituting Eq. (6.23) in Eq. (6.24) gives

$$X_{d0}(z) = \frac{1}{2}\left(X\left(z^{1/2}\right)H_0\left(z^{1/2}\right) + X\left(-z^{1/2}\right)H_0\left(-z^{1/2}\right)\right). \tag{6.25}$$

Now considering the synthesis stage,

$$X_{ud0}(z) = X_{d0}\left(z^2\right),$$

$$X_{ud0}(z) = \frac{1}{2}\left(X(z)\,H_0(z) + X(-z)\,H_0(-z)\right).$$

We define the filtered output on the upper path as $Y_0(z)$:

$$Y_0(z) = X_{ud0}(z)\,G_0(z), \tag{6.26}$$

$$Y_0(z) = \frac{1}{2}\left(X(z)\,H_0(z)\,G_0(z) + X(-z)\,H_0(-z)\,G_0(z)\right). \tag{6.27}$$

Similarly, for the filtered output on the lower path, we have

$$Y_1(z) = \frac{1}{2}\left(X(z)\,H_1(z)\,G_1(z) + X(-z)\,H_1(-z)\,G_1(z)\right). \tag{6.28}$$

Finally, combining Eq. (6.27) and Eq. (6.28) at the output of the filter-bank gives

$$Y(z) = \frac{1}{2}X(z)\left[H_0(z)\,G_0(z) + H_1(z)\,G_1(z)\right] + \frac{1}{2}X(-z)\left[H_0(-z)\,G_0(z) + H_1(-z)\,G_1(z)\right]. \tag{6.29}$$

So we now have the ability to analyze the end-to-end operation of the two-channel filter-bank. Let us consider this further in Example 6.2.

Example 6.2. A simple filter-bank transfer function

Consider the two-band split in Fig. 6.4, where the filters are given as follows:

$$H_0(z) = 1 + z^{-1}, \qquad H_1(z) = 1 - z^{-1},$$
$$G_0(z) = 1 + z^{-1}, \qquad G_1(z) = -1 + z^{-1}.$$

These filters are identical to those defined in our earlier example, except that the multiplicative factor of 0.5 in the analysis stage has been omitted for numerical convenience. Using Eq. (6.29), compute the output of the filter-bank for an input $x[n] = \{1, 2, 3, 4\}$.

Solution. The effects of analysis filtering, downsampling, upsampling, and synthesis filtering can be seen in the diagram below, yielding an output that is the same as the input, but scaled by a factor of two and delayed by one sample.

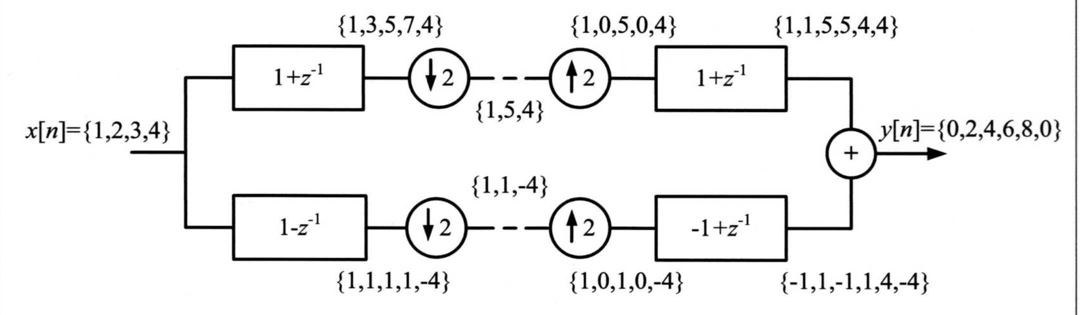

Rather than compute the filter-bank response in the time domain, as shown above, we can instead verify its operation in the z domain using Eq. (6.29). Thus we have

$$Y(z) = \frac{1}{2}\left(1 + 2z^{-1} + 3z^{-2} + 4z^{-3}\right)\left[\left(1 + 2z^{-1} + z^{-2}\right) + \left(-1 + 2z^{-1} - z^{-2}\right)\right] + \cdots$$

$$\cdots + \frac{1}{2}\left(1 - 2z^{-1} + 3z^{-2} - 4z^{-3}\right)\left[\left(1 - z^{-2}\right) + \left(-1 + z^{-2}\right)\right]$$
$$= \frac{1}{2}\left(1 + 2z^{-1} + 3z^{-2} + 4z^{-3}\right)\left[\left(4z^{-1}\right)\right] + \frac{1}{2}\left(1 - 2z^{-1} + 3z^{-2} - 4z^{-3}\right)[(0)]$$
$$= \left(2z^{-1} + 4z^{-2} + 6z^{-3} + 8z^{-4}\right).$$

Note that the leading and trailing zeros shown in the filter-bank output are redundant due to the expansive nature of the convolution operation.

6.3.4 Perfect reconstruction

Eq. (6.29) can be rewritten as

$$Y(z) = F_0(z)\, X(z) + F_1(z)\, X(-z), \tag{6.30}$$

where

$$F_0(z) = \frac{1}{2}\left[H_0(z)\, G_0(z) + H_1(z)\, G_1(z)\right]$$

and

$$F_1(z) = \frac{1}{2}\left[H_0(-z)\, G_0(z) + H_1(-z)\, G_1(z)\right],$$

where $F_1(z)$ represents the alias components produced by the overlapping filter frequency responses. If $F_1(z) = 0$, then the filter bank is alias-free. Moreover, $F_0(z)$ represents the quality of the reconstruction. If $\left|F_0\left(e^{j\Omega}\right)\right| = 1$ for all frequencies, then the filter-bank is free of amplitude distortion. If $F_0\left(e^{j\Omega}\right)$ is linear-phase, then the filter bank is free of phase distortion. In general we say that if $F_1(z) = 0$ and $F_0(z) = cz^{-k}$ (where c and k are constants), then the filter-bank provides perfect reconstruction.

6.3.5 Spectral effects of the two-channel decomposition

A perhaps more intuitive visualization of the operation of the two-channel perfect reconstruction filter-bank is provided by Fig. 6.16. This shows the spectral characteristics of a signal, $x[n]$, which is bandlimited to the Nyquist frequency, as it passes through the filter-bank. For simplicity in the diagram, we assume that the filters have brickwall responses with no leakage between bands. However, as we have already seen, with appropriate filter design, this is not a requirement for perfect reconstruction.

Working across the system from left to right we can see firstly that $H_0(z)$ and $H_1(z)$ select the low-pass and high-pass bands, respectively. In order to preserve critical sampling, these signals are then downsampled by a factor of two. The results of this operation, $C(\Omega)$ and $D(\Omega)$, present spectra that are expanded by a factor of two and that also include aliases, as expected. The signals would then typically be quantized for transmission, but we ignore this step at the moment and move straight to the decoder.

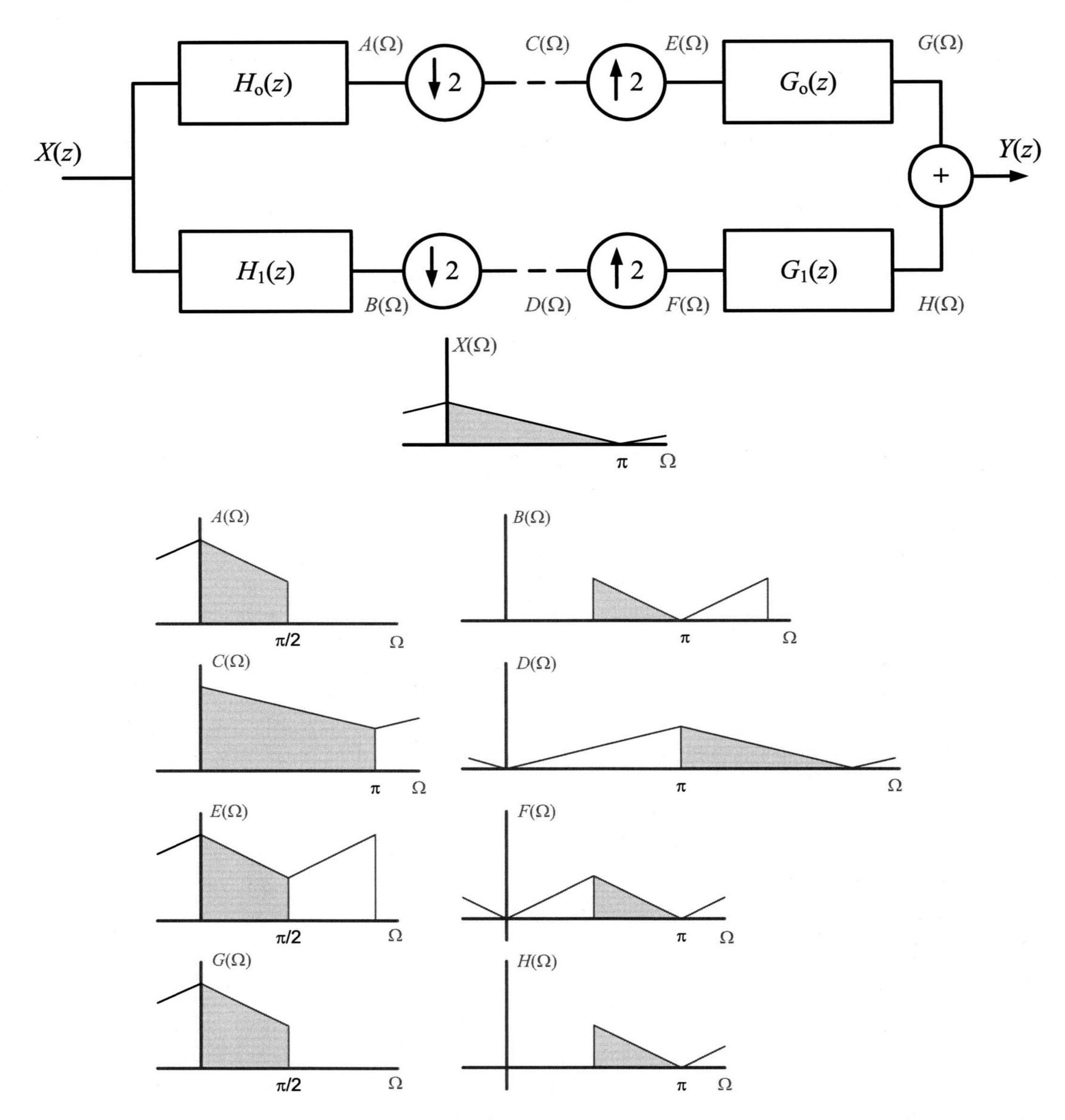

FIGURE 6.16 Spectral relationships within the two-channel perfect reconstruction filter-bank. The spectral characteristics at signal points *A–H* are shown.

At the decoder, the signals are upsampled in preparation for recombination. The effect of this is to "compress" the spectra by a factor of two along the frequency axis as shown by $E(\Omega)$ and $F(\Omega)$. Finally these signals are passed through the interpolation filters $G_0(z)$ and $G_1(z)$, which remove the aliases by low-pass and high-pass filtering, respectively, to yield

$G(\Omega)$ and $H(\Omega)$. As can be observed, $G(\Omega)$ and $H(\Omega)$ sum at the output to provide a perfect reconstruction of the input signal.

6.4 Useful filters and filter-banks

For compression applications it is desirable for the filters to exhibit certain characteristics. These are:

1. **Perfect reconstruction**: To enable distortion-free reconstruction of the original signal at the output of the filter-bank, in the absence of quantization.
2. **Compact support**: To ensure that distortions due to quantization and transmission loss are localized.
3. **Smoothness**: To better match the characteristics of natural images.
4. **Regularity**: To ensure convergence and maintain smoothness when iterated.
5. **Orthogonality and orthonormality**: To enable matching of analysis and synthesis sections and to remove transform redundancy.[4]

Some commonly used filters are examined below.

6.4.1 Quadrature mirror filters

QMFs are an important class of alias-cancelation filters that were used in early subband processing applications [2].

Aliasing elimination

QMFs start by eliminating the aliasing term in Eq. (6.30) by using the following relationships:

$$G_0(z) = H_1(-z), \quad G_1(z) = -H_0(-z).$$

QMFs are based on a pair of analysis filters that are symmetrical about $\Omega = \pi/2$. This can be achieved if $H_0(z) = H(z)$ and $H_1(z) = H(-z)$. We can therefore base our QMF design on the following relationships:

$$H_0(z) = H(z), \quad H_1(z) = H(-z),$$

$$G_0(z) = H(z), \quad G_1(z) = -H(-z).$$

[4] The imposition of orthogonality conditions places severe constraints on the filters (or basis functions) used in the filter-bank. An alternative condition, known as *biorthogonality*, can however be satisfied more generally, where the analysis filters are independent but do not form an orthogonal pair. This means that the inverse or synthesis transform is no longer given by the transpose of the analysis transform. In the case of biorthogonality, a dual basis can be found where projection on to an orthogonal set is possible.

If we substitute these relationships into Eq. (6.29), we can see that the aliasing term is zero; thus,

$$F_1(z) = \frac{1}{2}\left[H(-z)H(z) - H(z)H(-z)\right] = 0. \tag{6.31}$$

Amplitude distortion

The distortion term for the QMF filter-bank is given by

$$F_0(z) = \frac{1}{2}\left[H(z)H(z) - H(-z)H(-z)\right]. \tag{6.32}$$

Furthermore, if the prototype filter is linear-phase, $H(z) = A(z)z^{-(N-1)/2}$, where $A\left(e^{j\Omega}\right)$ is a real zero-phase frequency response, then

$$F_0(z) = \frac{1}{2}\left[A^2(z) + A^2(-z)\right]z^{-(N-1)}.$$

The QMF filter-bank will therefore exhibit amplitude distortion unless $\left|F_0\left(e^{j\Omega}\right)\right|$ is constant for all Ω. As we will see in Example 6.3, QMFs can be designed that exhibit perfect reconstruction.

Example 6.3. Simple QMFs

Consider the following example prototype filter:

$$H(z) = \frac{1}{\sqrt{2}}\left(1 + z^{-1}\right).$$

Generate the corresponding set of QMFs and show that this filter-bank exhibits perfect reconstruction.

Solution. This prototype yields the following set of filters, which hopefully the reader will recognize as the half-band filters from our previous example:

$$H_0(z) = \frac{1}{\sqrt{2}}\left(1 + z^{-1}\right), \quad H_1(z) = \frac{1}{\sqrt{2}}\left(1 - z^{-1}\right),$$
$$G_0(z) = \frac{1}{\sqrt{2}}\left(1 + z^{-1}\right), \quad G_1(z) = \frac{1}{\sqrt{2}}\left(-1 + z^{-1}\right).$$

We can see from Eq. (6.31) that this filter-bank is alias-free. We can further show, from Eq. (6.32), that the system provides a constant delay with unity magnitude, since

$$F_0(z) = \frac{1}{4}\left[\left(1 + z^{-1}\right)\left(1 + z^{-1}\right) - \left(1 - z^{-1}\right)\left(1 - z^{-1}\right)\right] = z^{-1}.$$

In fact these are the only causal finite impulse response QMF analysis filters yielding exact perfect reconstruction.

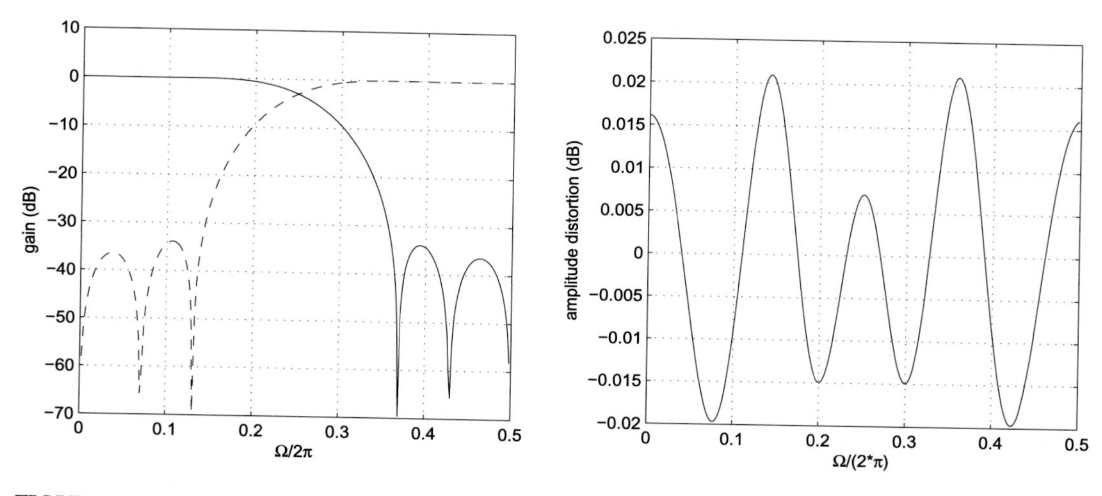

FIGURE 6.17 QMF pair frequency response.

Practical QMFs

In practice, approximate QMFs can be usefully designed where the characteristic does not exactly provide perfect reconstruction. In such cases, the amplitude distortion can be minimized using an optimization procedure to iteratively adjust the filter coefficients. We ideally want the magnitude response of the filter characteristic to have a sharp transition so that subbands are spectrally distinct while providing an overall system response that is approximately all-pass. This means that the filters have to be designed so that each branch delivers an attenuation of 6 dB (0.5) at the point where the low- and high-pass responses intersect – at one-quarter of the sampling frequency. Each of the four filters must therefore have an attenuation of 3 dB at half the Nyquist frequency. Such filters have been designed by Johnson [22] and others. Johnson's method minimizes the residual amplitude distortion using an objective measure, based upon stop-band attenuation and the requirement to maintain the power symmetry property of the filters. Low-amplitude distortions typically less than 0.02 dB result. Example Johnson filters are

$$h_0[0] = -0.006443977 = h_0[11], \quad h_0[1] = 0.02745539 = h_0[10],$$
$$h_0[2] = -0.00758164 = h_0[9], \quad h_0[3] = -0.0013825 = h_0[8],$$
$$h_0[4] = 0.09808522 = h_0[7], \quad h_0[5] = 0.4807962 = h_0[6].$$

The mirror responses of these filters are shown in Fig. 6.17. It can be seen that the overall distortion characteristic, although small, is not zero. Hence these filters do not provide perfect reconstruction. In practice, for compression applications where significant quantization error is introduced prior to transmission of the subband coefficients, this small distortion may not be a major issue. However, we will see shortly how useful filters can be designed that do offer perfect reconstruction.

6.4.2 Wavelet filters

The wavelet filter-bank can form a perfect reconstruction system that is completely defined by the low-pass filters. The high-pass filters are obtained by shifting and modulating the low-pass filters; thus,

$$H_1(z) = z^{-l} G_0(-z) \text{ and } G_1(z) = z^l H_0(-z), \tag{6.33}$$

and provided that l is odd, we have

$$F_1(z) = \frac{1}{2}\left[H_0(-z)\, G_0(z) + (-z)^{-l}\, G_0(z)\, z^l H_0(-z)\right] = 0,$$

i.e., the system is alias-free.

We can now address the distortion term $F_0(z)$. If $F_0(z) = \frac{1}{2}\left[H_0(z)\, G_0(z) + H_1(z)\, G_1(z)\right] = cz^{-k}$, then the filter-bank will exhibit perfect reconstruction. If we let $P(z) = H_0(z)\, G_0(z)$ and we assume zero-phase response filters, then we can rewrite $F_0(z)$ in the form (absorbing the $\frac{1}{2}$ term into the constant c)

$$P(z) + P(-z) = c. \tag{6.34}$$

Since all terms in odd powers of z will cancel, for Eq. (6.34) to hold true, with the exception of the z^0 term, all the terms in even powers of z must sum to zero.

To design filters we can now factorize the polynomial $P(z)$ into the two low-pass filters, $H_0(z)$ and $G_0(z)$, resulting in a completely defined filter-bank. The question is, how do we factorize it, and how do we ensure that we have good filter characteristics? We have already discussed the desirable property of regularity and the associated requirement to have multiple roots at $z = -1$. This gives us a good starting point. So if we have

$$P(z) = \left(1 + 2z^{-1} + z^{-2}\right),$$

then this can be factorized into our familiar simple perfect reconstruction filter-bank where both low-pass filters are identical; thus,

$$P(z) = \left(1 + z^{-1}\right)\left(1 + z^{-1}\right).$$

There are several filter combinations which (in the absence of quantization) offer perfect reconstruction. A performance comparison of these is provided by Villasenor [23] and a good overview of design methods is provided by Kingsbury in [24]. We will consider two example sets of filters here – both of which have been selected for use in JPEG2000. These are the LeGall 5/3 wavelet and the Daubechies 9/7 wavelet. They are both compact, have symmetric responses (this helps to avoid boundary artifacts), and are highly regular (they have a maximum number of vanishing moments).

LeGall 5/3 filters

One example of a simple, yet effective, set of wavelet filters is the biorthogonal LeGall (5/3) pair [25]. These can be designed using the above approach and are defined in terms of their low-pass filter responses as follows[5]:

$$H_0(z) = \frac{1}{8}z\left(1+z^{-1}\right)^2\left(-z-z^{-1}+4\right),$$

$$G_0(z) = \frac{1}{2}z\left(1+z^{-1}\right)^2. \tag{6.35}$$

The filter coefficients are thus

$$h_0[n] = \{-1, 2, 6, 2, -1\}/8, \quad g_0[n] = \{1, 2, 1\}/2,$$
$$h_1[n] = \{1, -2, 1\}/2, \quad g_1[n] = \{1, 2, -6, 2, 1\}/8.$$

Recall from Eq. (6.29) that

$$\begin{aligned} Y(z) &= \frac{1}{2}X(z)\left[G_0(z)H_0(z) + G_1(z)H_1(z)\right] + \frac{1}{2}X(-z)\left[G_0(z)H_0(-z) + G_1(z)H_1(-z)\right] \\ &= F_0(z)X(z) + F_1(z)X(-z). \end{aligned} \tag{6.36}$$

For perfect reconstruction we require $F_0(z) = constant$ and $F_1(z) = 0$. So for the distortion term (assuming causal filters) we have

$$G_0(z)H_0(z) = \left(-1+0z^{-1}+9z^{-2}+16z^{-3}+9z^{-4}+0z^{-5}-z^{-6}\right)/16,$$

$$G_1(z)H_1(z) = \left(+1+0z^{-1}-9z^{-2}+16z^{-3}-9z^{-4}+0z^{-5}+z^{-6}\right)/16.$$

Thus the distortion is constant for all frequencies:

$$F_0(z) = z^{-3}.$$

In the case of the alias term we have

$$G_0(z)H_0(-z) = \left(-1-4z^{-1}+z^{-2}+8z^{-3}+z^{-4}-4z^{-5}-z^{-6}\right)/16,$$

$$G_1(z)H_1(-z) = \left(1+4z^{-1}-z^{-2}-8z^{-3}-z^{-4}+4z^{-5}+z^{-6}\right)/16.$$

Thus the alias term is given by

$$F_1(z) = 0.$$

Hence the filter-bank is also alias-free and satisfies the conditions required for perfect reconstruction. The frequency responses of the four LeGall (5/3) biorthogonal filters are shown in Fig. 6.18.

The LeGall wavelet has rational coefficients and is recommended for reversible or lossless compression in JPEG2000.

[5] These LeGall filters are defined as noncausal. They are sometimes defined where both are premultiplied by a factor 1/4 rather than 1/8 and 1/2.

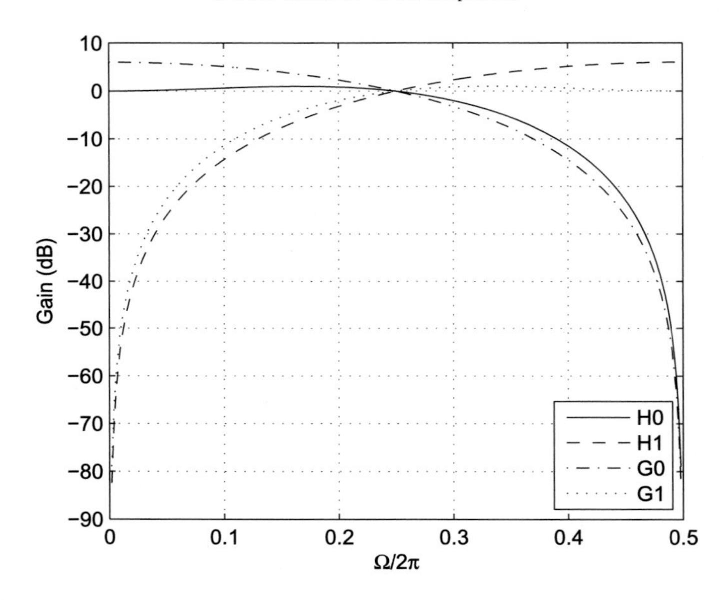

FIGURE 6.18 LeGall filter frequency responses.

Daubechies 9/7 filters

The Daubechies 9/7 filters [26] are recommended for lossy compression in JPEG2000. They are defined as

$$H_0(z) = \frac{2^{-5}}{\frac{64}{5\rho} - 6 + \rho} z^2 \left(1 + z^{-1}\right)^4 \left(z^2 + z^{-2} - (8 - \rho)\left(z + z^{-1}\right) + \frac{128}{5\rho} + 2\right), \tag{6.37}$$

$$G_0(z) = \frac{2^{-3}}{\rho - 2} z^2 \left(1 + z^{-1}\right)^4 \left(-z - z^{-1} + \rho\right),$$

where $\rho \approx 3.3695$. In comparisons of short-length filters by Villasenor [23], the Daubechies 9/7 filters were ranked as the best performing for lossy image compression.

6.4.3 Multistage (multiscale) decompositions

One important class of subband decomposition, commonly (although, as described previously, not always accurately) referred to as a DWT, is achieved by cascading a series of two-band systems, recursively applied to the low band. Using this method, the high-frequency bands have good spatial resolution and poorer frequency resolution, while the low-frequency bands have good frequency localization but with inferior spatial localization.

A 1D three-stage decomposition is shown in Fig. 6.19 and the associated time frequency tilings at all stages of decomposition are given in Fig. 6.20. This figure shows an input of eight consecutive time domain samples, subsequently filtered and downsampled to yield four samples in each of two bands (b). The lower-frequency band is then further filtered and

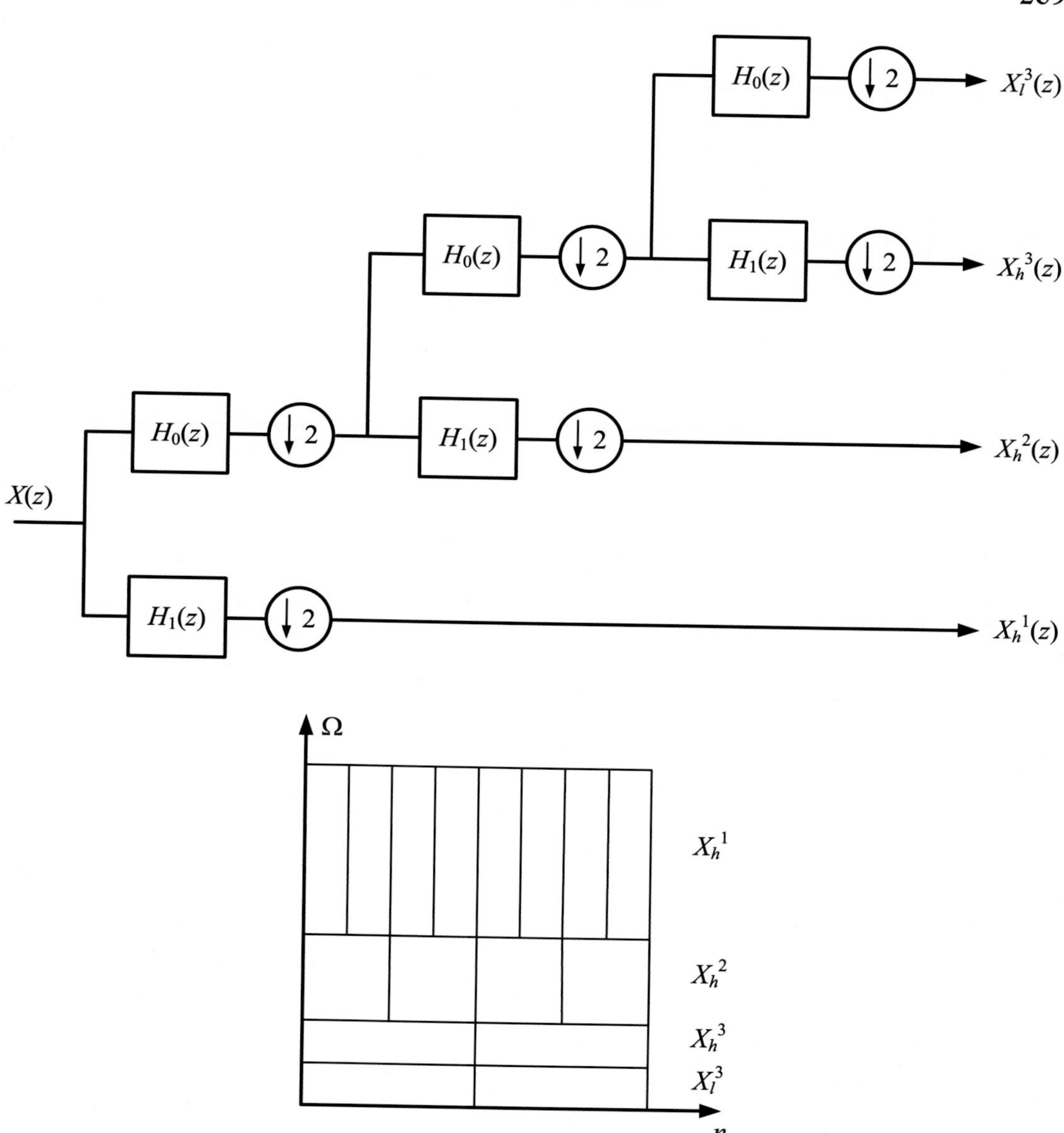

FIGURE 6.19 1D three-stage wavelet decomposition: Top: Block diagram. Bottom: Associated time-frequency tiling.

downsampled to yield two more bands, each with two samples (c). Finally in (d), the low-pass band is once again filtered and downsampled to provide the tiling shown. This clearly shows the property of critical sampling as the number of samples (eight in this case) is identical at each decomposition stage.

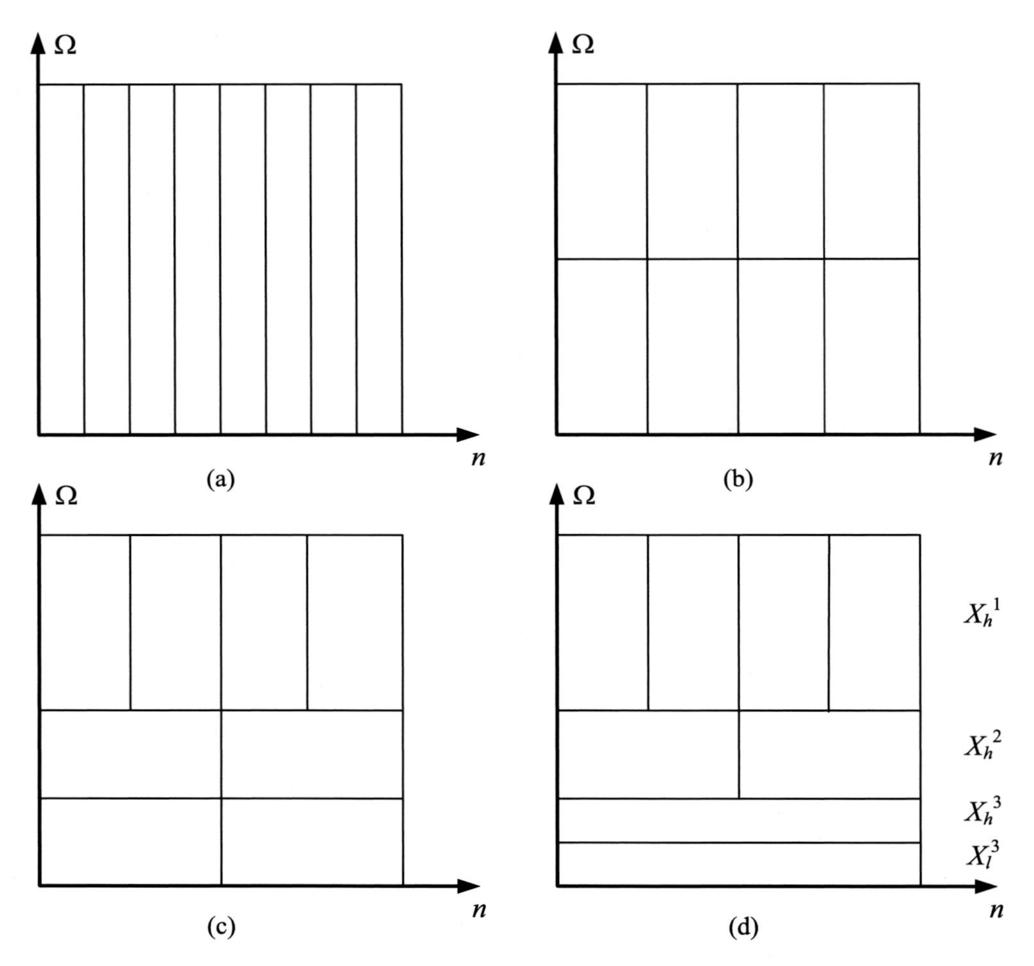

FIGURE 6.20 Time-frequency tiling for progressive stages of decomposition in a 1D wavelet filter-bank. (a) Input of eight samples. (b) After the first stage. (c) After the second stage. (d) After the third stage.

In general, for an input sequence of length N, the relationship between the bandwidth and the number of samples in each subband is given in Table 6.1. This is shown for a five-stage decomposition, although the pattern for more or fewer stages should be obvious.

An alternative view of multistage decomposition

We can see from Fig. 6.19 that each output from the analysis section is a series of filtering and downsampling operations. This can be restructured as shown in Fig. 6.21. This is straightforward to show, based on the two alternative configurations at the bottom of Fig. 6.21. For example, in the case of the downsampler followed by a filter $H(z)$ we have

$$Y(z) = \frac{1}{2}H(z)\left(X\left(z^{\frac{1}{2}}\right) + X\left(-z^{\frac{1}{2}}\right)\right),$$

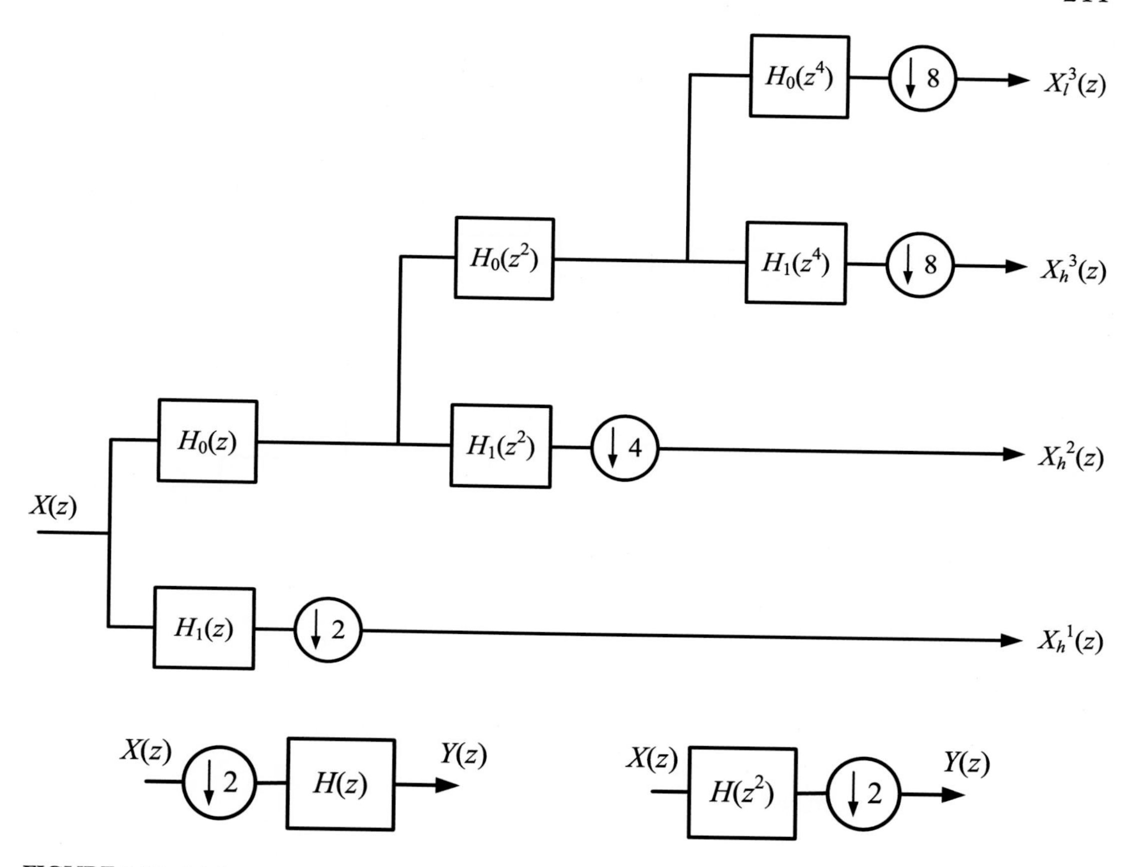

FIGURE 6.21 1D three-stage wavelet decomposition – modified downsampling structure.

and in the case of the modified filter followed by the downsampler, we have

$$Y(z) = \frac{1}{2}\left(X\left(z^{\frac{1}{2}}\right)H(z) + X\left(-z^{\frac{1}{2}}\right)H(z)\right).$$

A similar relationship exists for an upsampler–filter combination.

6.4.4 Separability and extension to 2D

A two-level subband decomposition applied in two dimensions (separably) is shown in Fig. 6.22 and the corresponding tiling of the spatial frequency plane is shown in Fig. 6.23. The upper left subband in Fig. 6.23 contains the lowest-frequency information and this would be recognizable as a small subsampled version of the original image. Intuitively, therefore, this subband is important from an HVS point of view and must be coded accurately (usually using differential pulse code modulation). The other subbands correspond to mid- and high-

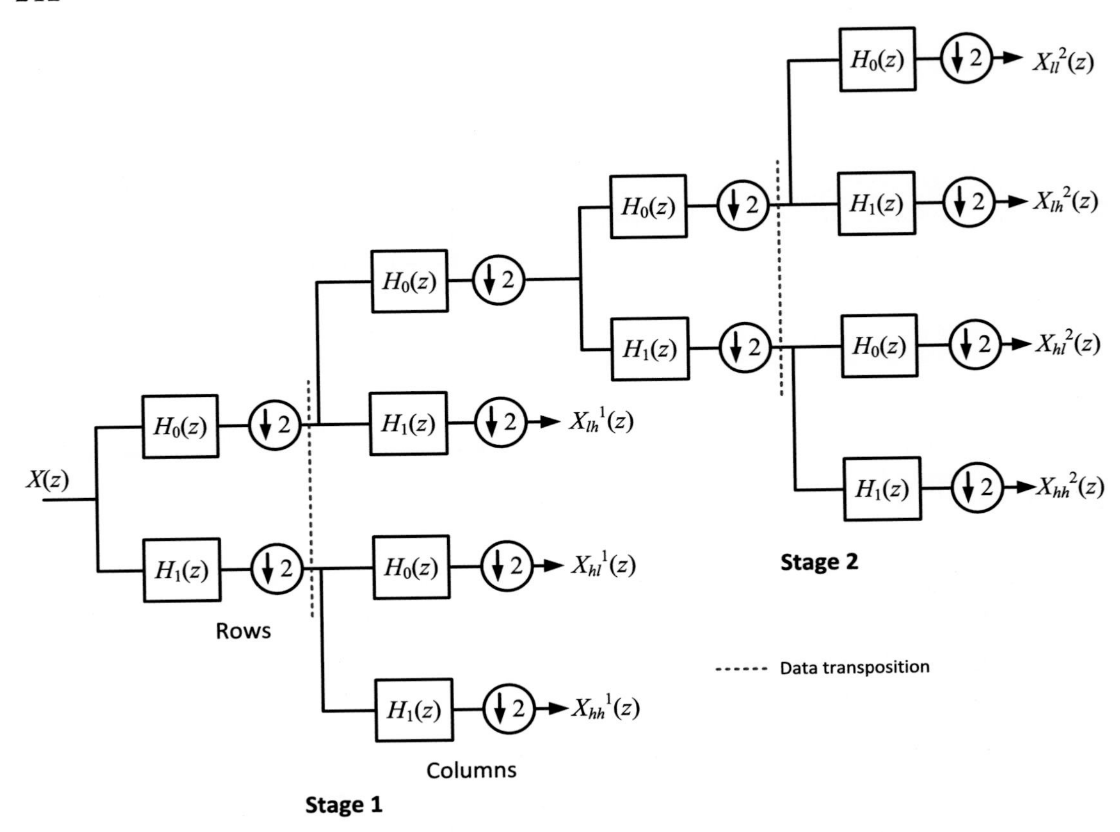

FIGURE 6.22 2D two-channel two-stage wavelet decomposition. Data transposition is required between row and column processing.

TABLE 6.1 Relationship between subband frequency range and number of samples for each subband in a five-stage decomposition.

Subband	Bandwidth	No. of samples
X_h^1	$\pi/2 \cdots \pi$	$N/2$
X_h^2	$\pi/4 \cdots \pi/2$	$N/4$
X_h^3	$\pi/8 \cdots \pi/4$	$N/8$
X_h^4	$\pi/16 \cdots \pi/8$	$N/16$
X_h^5	$\pi/32 \cdots \pi/16$	$N/32$
X_l^5	$0 \cdots \pi/32$	$N/32$

range spatial frequencies and usually contain relatively little energy. Thus the encoder can allocate fewer (or no) bits to coding this information. We will look at bit allocation strategies and entropy coding of these bands a little later.

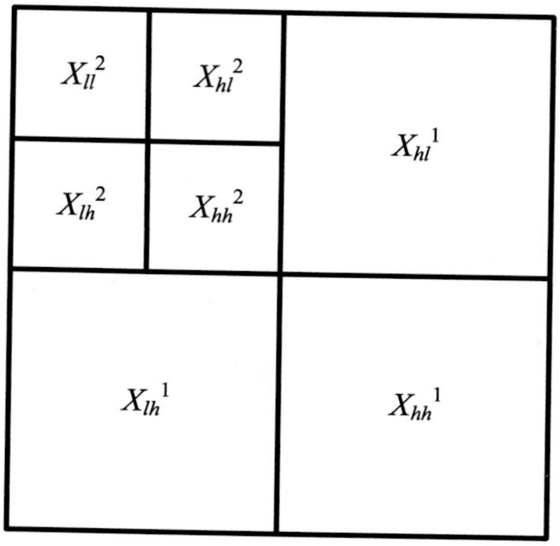

X_{ll}^2 X_{hl}^2 X_{hl}^1

X_{lh}^2 X_{hh}^2

X_{lh}^1 X_{hh}^1

FIGURE 6.23 2D wavelet – frequency plane tiling.

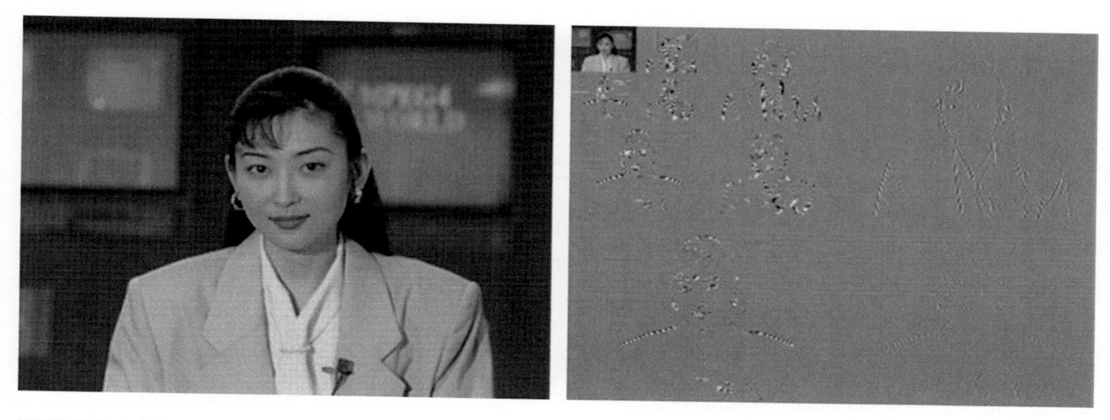

FIGURE 6.24 Frame from the Akiyo sequence (left) wavelet transformed (right) using a three-stage decomposition with LeGall 5/3 filters.

An example showing a three-level 2D wavelet decomposition of one frame in the Akiyo sequence is shown in Fig. 6.24. This decomposition employs the LeGall 5/3 filters described earlier.

6.4.5 Finite-length sequences, edge artifacts, and boundary extension

When we discussed critical sampling previously, we assumed the context of infinitely long sequences. It is clear in the case of finite-length sequences, such as those typical in image coding, that the number of subband samples will be greater than the number of input samples. If we maintain strict critical sampling, then distortion would be introduced by the filter-bank, due to discontinuities at the signal boundaries, and perfect reconstruction would not

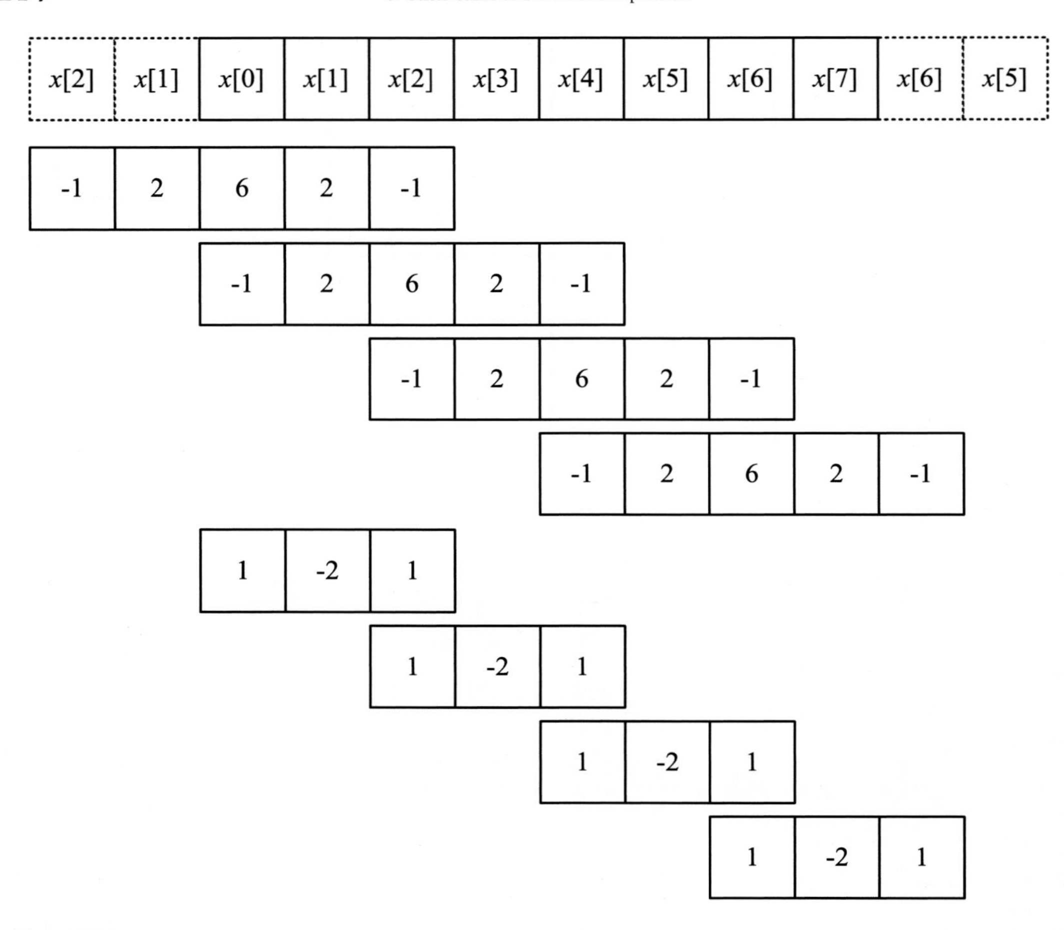

FIGURE 6.25 Illustration of boundary extension.

be achieved. This can be reduced by extending the boundary samples either periodically or symmetrically or by zero padding. Symmetrical extension is normally preferred as this does not create discontinuities and has the effect of introducing lower boundary distortions.

We can easily maintain critical sampling with boundary extension and an example for an eight-sample input with 5/3 filters is shown in Fig. 6.25. This illustrates the effective convolution shifts for each filter taking account of subsampling.

Example 6.4. Boundary effects and critical sampling

Consider a 1D two-channel single-stage filter-bank based on the LeGall 5/3 filters. Compute the signal values at all intermediate nodes for the case of an input signal $x[n] = \{1, 2, 3, 4, 5, 6, 7, 8\}$.

Solution. We can compute the convolution of the filters with the input sequence at all intermediate nodes, as defined in the following figure.

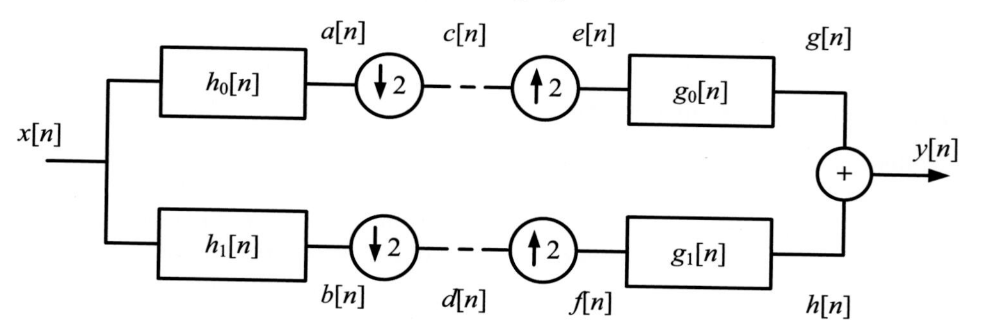

Let us consider first the case where critical sampling is not strictly observed. The signal values at each intermediate node are given in the following table.

$x[n]$	1	2	3	4	5	6	7	8							
$a[n]$	−1	0	7	16	24	32	40	48	65	56	9	−8			
$b[n]$	1	0	0	0	0	0	0	0	−9	8					
$c[n]$	−1	7	24	40	65	9									
$d[n]$	1	0	0	0	−9										
$e[n]$	−1	0	7	0	24	0	40	0	65	0	9				
$f[n]$	1	0	0	0	0	0	0	0	−9						
$g[n]$	−1	−2	6	14	31	48	64	80	105	130	74	18	9	0	0
$h[n]$	1	2	−6	2	1	0	0	0	−9	−18	54	−18	−9		
$y[n]$	**0**	**0**	**0**	**16**	**32**	**48**	**64**	**80**	**96**	**112**	**128**	**0**	**0**	**0**	**0**

Note that the filter gain terms have been ignored here for convenience. However it is clear that the answer is correct when we divide the output by 16. The problem with this is that we have 11 samples at the output of the analysis stage (c and d) rather than 8. If we maintain strict critical sampling, then distortion would be introduced by the filter-bank due to discontinuities at the signal boundaries and perfect reconstruction would not be achieved.

Let us now consider what happens if we keep only 8 samples for processing by the synthesis section. Let us discard the first two samples of the a and b signals as these are associated with the filter transients. Then we have the following.

$c[n]$	7	24	40	65							
$d[n]$	0	0	0	−9							
$e[n]$	7	0	24	0	40	0	65				
$f[n]$	0	0	0	0	0	0	−9				
$g[n]$	7	14	31	48	64	80	105	130	65	0	0
$h[n]$	0	0	0	0	0	0	−9	−18	54	−18	−9
$y[n]$	**7**	**14**	**31**	**48**	**64**	**80**	**96**	**112**	**119**	**−18**	**−9**

We can now see that the filter-bank output is corrupted at its boundaries. We can improve this by boundary extension using, for example, symmetric extension.

6.4.6 Wavelet compression performance

An excellent comparison between the performances of various types of wavelet filter is provided by Villasenor in [23]. Unser and Blu [27] also provide a rigorous comparison between the LeGall 5/3 and Daubechies 9/7 filters in the context of JPEG2000. Both papers show that the 9/7 filters provide excellent performance with compact support for lossy coding. Considering their size, the 5/3 filters also perform exceptionally well in an image coding context, with the added advantage that lossless reconstruction can always be guaranteed because of their use of integer coefficients.

A comparison between wavelet (JPEG2000) and DCT (JPEG) compression methods (based on the peak signal to noise ratio [PSNR]) is provided for the puppet image in Fig. 6.30 and Fig. 6.31. Although this shows clear benefits of wavelet methods over their DCT counterparts, we must be careful of reading too much into this graph because of the significant differences in the coding methods used and in the artifacts produced by them. In subjective testing, however, it has been widely reported that wavelet coding (for example in JPEG2000) significantly outperforms DCT coding (as in JPEG), especially at lower bit rates.

6.5 Coefficient quantization and bit allocation

6.5.1 Bit allocation and zonal coding

In order to quantize subband coefficients we can, in a similar manner to our approach with the DCT, apply different quantizer weights to different subbands according to their perceptual importance. This is shown in Fig. 6.26, where different weights, W_0 to W_6, are used to differentiate both subband orientation and scale.

Similar to the way that we applied zig-zag scanning to a DCT block in Chapter 5, we can also scan subband coefficients from high to low subbands and in a manner that is matched to the orientation of energy in the subband. This is shown in Fig. 6.27, where, for each of the three orientations, a data structure is used that supports efficient quantization. For example, subbands 2, 5, and 8 are likely to be highly correlated since subband 2 is a coarse approximation of subband 5 and subband 5 is a coarse approximation of subband 8. The same is true for subbands 4, 7, and 10 and subbands 3, 6, and 9. We can observe that the scanning patterns have been devised to exploit these relationships.

In the case of subband 1 (DC), coefficients will typically be encoded differentially and then {size, value} symbols entropy coded. For the AC coefficients, after scanning, a {run-length/size, value} format, similar to that used with JPEG, would normally be employed to yield symbols for entropy coding.

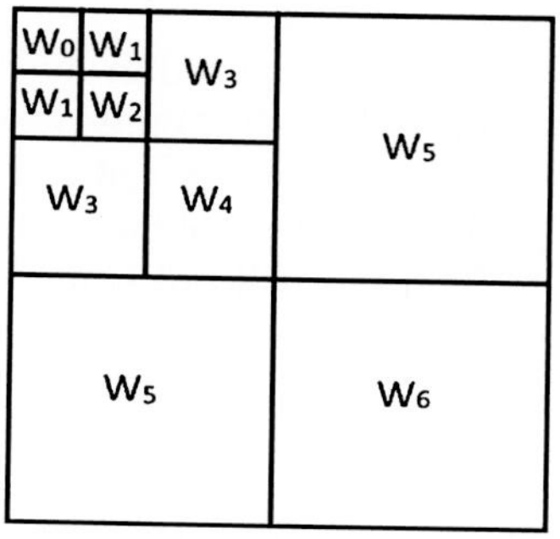

FIGURE 6.26 Subband weighting.

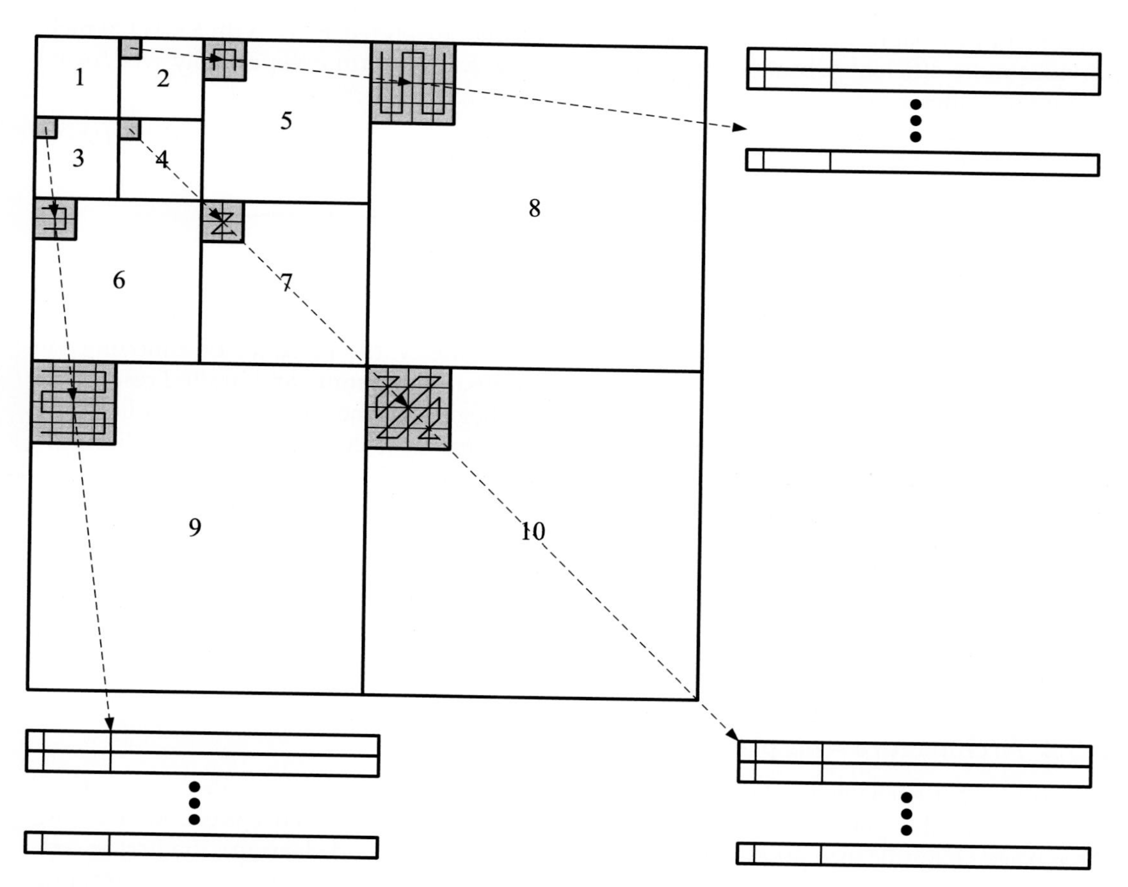

FIGURE 6.27 Scanning of wavelet subbands.

6.5.2 Hierarchical coding

Alternative scanning methods have been proposed in the context of wavelet compression that provide the basis for scalable encoding using embedded bitstreams. In particular the reader is referred to the work of Shapiro [28], who introduced the concept of scalable coding using embedded zerotree coding (EZW), and that of Said and Pearlman [16], who extended this idea into an approach called SPIHT. We will not consider these methods in detail, but provide a brief introduction below.

Embedded coding methods generally follow the approach illustrated in Fig. 6.28, where multiscale dependencies (self-similarities) are exploited across the wavelet subbands when coding the coefficients. The tree structure in the lower subfigure captures these dependencies. In EZW, Shapiro showed how these dependencies could be exploited using a succession of coding passes through the tree, in each pass coding significant coefficients that exceed a threshold. As the threshold decreases, more bits are included and a natural bit plane data structure results. In EZW, arithmetic coding is essential in order to compress the data resulting from the significance passes. Said and Pearlman extended EZW with SPIHT which provides more efficient subset partitioning and which can perform exceptionally well, even without arithmetic coding.

6.6 JPEG2000

6.6.1 Overview

JPEG has been a big success, with around 80% of all images still stored in this format. However, in the late 1990s, JPEG's limited coding efficiency, the presence of annoying visual blocking artifacts at high compression ratios, limited color gamut, and limited resolution were all reasons why work on an improved standard began. Furthermore, many applications were emerging that required functionalities not supported by JPEG, for example spatial scalability, SNR scalability, and region of interest (ROI) coding.

The aim of JPEG2000 was to achieve a 30% bit rate saving for the same quality compared to JPEG, supporting 2 to 16 million colors and both lossless and lossy compression within the same architecture at bit depths greater than 8 bits. Additional features to support ROI coding, error resilience, and data security were also included and a high emphasis was placed on scalability and progressive transmission. JPEG2000 Part 1 became an International Standard (ISO/IEC 15444-1) in December 2000 [21,29,31].

6.6.2 Architecture – bit planes and scalable coding

In JPEG2000, each subband is split into a number of code blocks. The encoder uses EBCOT [17,21] to code the quantized coefficients. Each bit plane of each code block is encoded during three passes (significance propagation, magnitude refinement, and cleanup), the first encoding bits (and signs) of insignificant coefficients with significant neighbors, the second refining bits of significant coefficients, and finally coding coefficients without significant neighbors. In the JPEG2000 lossless mode, all bit planes must be encoded, whereas in lossy mode, some of

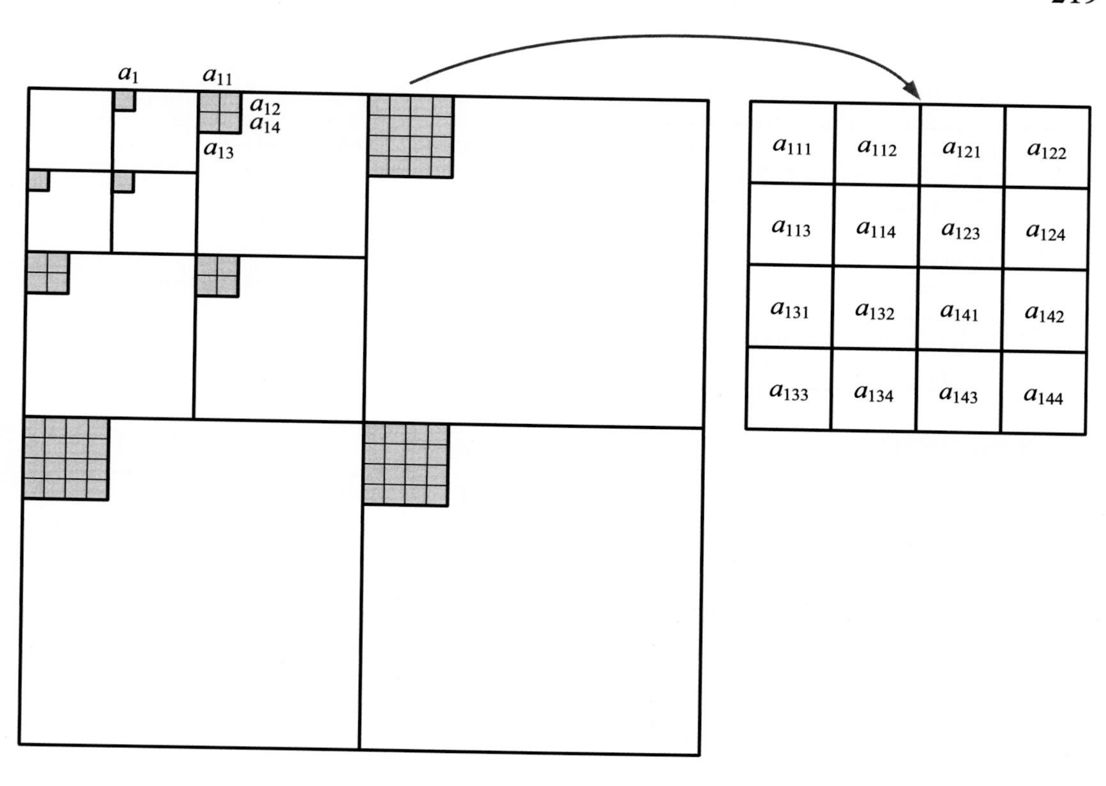

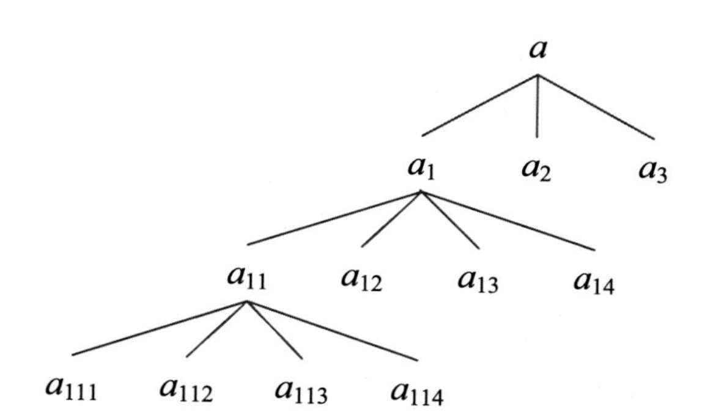

FIGURE 6.28 Tree structures for embedded wavelet coding.

the bit planes can be dropped. The bits selected during EBCOT are passed to a binary MQ encoder (a context-driven binary arithmetic coder), where the context of each coefficient is related to the state of its nine neighbors in the code block.

An overview of the main features of JPEG2000 is given in Table 6.2 and an architecture diagram is given in Fig. 6.29.

TABLE 6.2 Technical overview of JPEG2000.

Property	Attribute
Wavelet transform	Dyadic decomposition
Filters	5/3 integer filter (enables strictly reversible compression) [25]
	9/7 floating point filter [26]
Block structure	Each subband is divided into code-blocks
Quantization	Scalar quantization
Entropy coding	Block-based (EBCOT) and context-based arithmetic coding
Regions of interest	Maxshift with multiple regions supported – ROI shape not encoded
Error resilience	JPWL for wireless transmission
Security tools	JPSEC
Streaming tools	JPIP

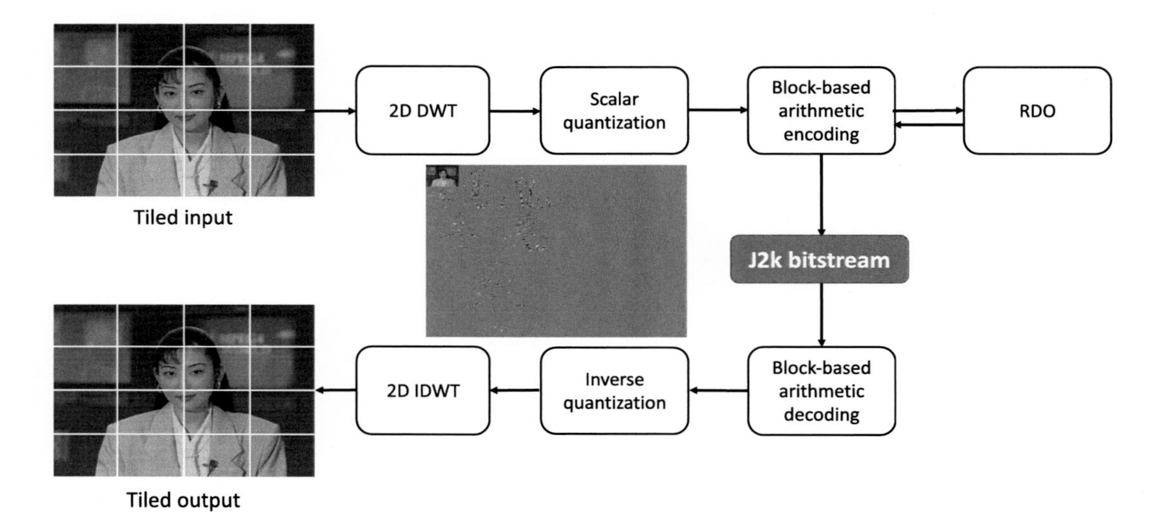

FIGURE 6.29 JPEG2000 architecture.

6.6.3 Coding performance

Examples of JPEG2000 coding are provided in Fig. 6.30. This shows the puppet test image (512×512), encoded and decoded at compression ratios of 64:1 and 128:1, compared to JPEG. It can be seen that the JPEG2000 codec significantly outperforms JPEG at high compression ratios. Ebrahimi et al. [30] undertook a comparison between JPEG and JPEG2000 and demonstrated, using the PSNR, that JPEG2000 is indeed superior at lower bit rates, but surprisingly that JPEG can actually offer small improvements at high bit rates. This is shown in Fig. 6.31.

In terms of compression performance, JPEG2000 achieves similar results to the intra-mode of H.264/AVC and is generally inferior to HEVC intra (HEIC). The latter two methods benefit from variable block sizes and sophisticated intra-prediction techniques. This perhaps also provides an explanation of why wavelet methods have not been widely adopted for video

FIGURE 6.30 JPEG2000 encoding of the *puppet* test image (512×512). Top left to bottom right: JPEG at 64:1 (PSNR=26.8 dB); JPEG2000 at 64:1 (PSNR=28.1 dB); JPEG at 128:1 (PSNR=21.5 dB); JPEG2000 at 128:1 (PSNR=25.5 dB).

coding, where advanced prediction modes and adaptive coding structures, used in combination with block-based transforms, can often provide equivalent or superior results.

6.6.4 Region of interest coding

ROI processing in JPEG2000 enables the coding of important portions of an image at higher qualities than the surrounding content (or background). Before entropy coding, the bit planes of the coefficients belonging to the ROI mask are shifted up. This scaling corresponds to a local increase of dynamic range, which is then prioritized in the rate allocation process. At the decoder, the quantized wavelet coefficients within each ROI are scaled back to their original values. This process is illustrated in Fig. 6.32.

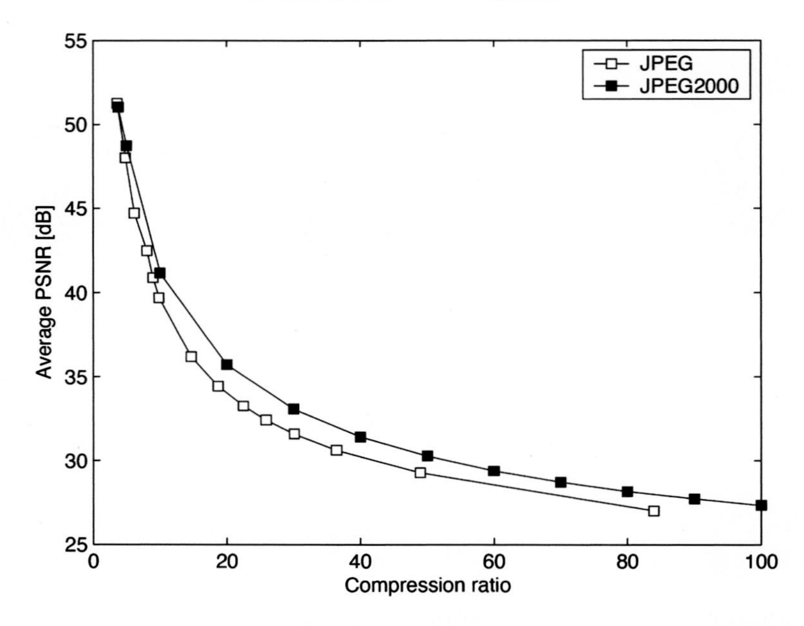

FIGURE 6.31 Performance comparison: JPEG vs. JPEG2000 across a range of test images (reproduced from [30]).

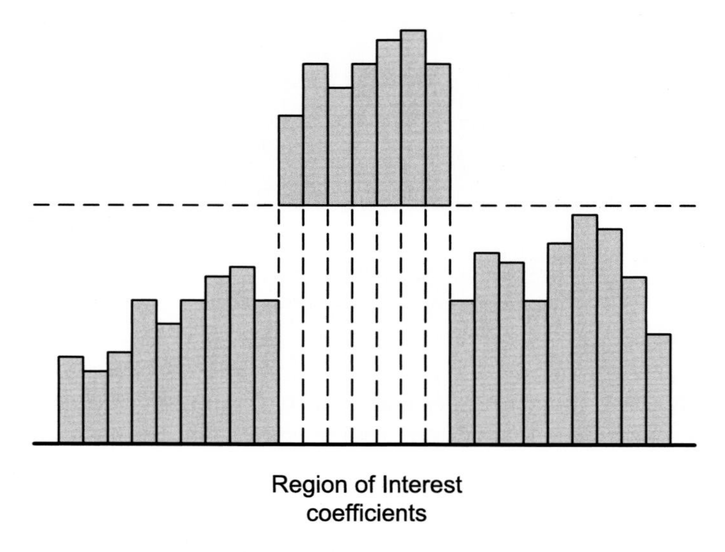

FIGURE 6.32 JPEG2000 ROI coefficient coding using Maxshift.

6.6.5 Benefits and status

Although the core image coding part of JPEG2000 is intended by ITU-T to be royalty-free, it is not patent-free. It has been reported that one of the reasons why the standard has not gained the success it perhaps deserved, is because of fears of patent infringement. However, as the standard has existed for some 20 years, all patents will now have lapsed. JPEG2000 has however gained favor in many military and surveillance applications, where its improved performance at lower bit rates and its ability to offer scalable delivery have proved attractive.

6.7 Summary

This chapter has explored an approach to image decorrelation and source coding based on multiscale analysis and synthesis. Through the combination of appropriate filters with up- and downsampling, we have seen how critical constant sample rates can be maintained throughout the filter-bank. We have also demonstrated how practical filters can be designed that, in the absence of quantization, achieve perfect reconstruction. We finally considered specific cases of wavelet filters, particularly in the context of the JPEG2000 still image coding standard.

When used in a lossy image compression system, wavelet-based coding generally produces less annoying visual artifacts than block-based transforms and it can also cope better with localized nonstationarities.

References

[1] A. Haar, Zur theorie der orthogonalen funktionensysteme, Math. Anal. 69 (1910) 331–371.
[2] A. Crosier, D. Esteban, C. Galand, Perfect channel splitting by use of interpolation/decimation techniques, in: Proc IEEE Conf. on Information Science and Systems, 1976.
[3] M. Smith, T. Barnwell, A procedure for designing exact reconstruction filterbanks for tree-structured sub-band coders, in: Proc. IEEE Intl. Conf. on Acoustics, Speech and Signal Processing, 1984.
[4] M. Smith, T. Barnwell, Exact reconstruction techniques for tree structured subband coders, IEEE Trans. Acoust. Speech Signal Process. 34 (3) (1986) 434–441.
[5] M. Vetterli, Filter banks allowing perfect reconstruction, Signal Process. 10 (3) (1986) 219–244.
[6] J. Woods, S. O'Neill, Sub-band coding of images, IEEE Trans. Acoust. Speech Signal Process. 34 (1986) 1278–1288.
[7] P. Vaidyanathan, Theory and design of M-channel maximally decimated quadrature mirror filters with arbitrary M, having the perfect reconstruction property, IEEE Trans. Acoust. Speech Signal Process. 35 (4) (1987) 476–492.
[8] A. Grossman, J. Morlet, Decomposition of Hardy functions into square-integrable wavelets of constant shape, SIAM J. Math. Anal. 15 (4) (1984) 723–736.
[9] I. Daubechies, Orthonormal bases of compactly supported wavelets, Commun. Pure Appl. Math. 41 (1988) 909–996.
[10] S. Mallat, A theory for multiresolution signal decomposition: the wavelet representation, IEEE Trans. Pattern Anal. Mach. Intell. 11 (7) (1989) 674–693.
[11] S. Mallat, Multifrequency channel decomposition of images and wavelet models, IEEE Trans. Acoust. Speech Signal Process. 37 (12) (1989) 2091–2110.
[12] M. Vetterli, J. Kovacevic, Wavelets and Subband Coding, Prentice Hall, 1995 (reissued 2007).
[13] R. Rao, A. Bopardikar, Wavelet Transforms – Introduction to Theory and Applications, Addison Wesley, 1998.
[14] G. Strang, T. Nguyen, Wavelets and Filter Banks, Wellesley-Cambridge Press, 1996.
[15] O. Rioul, M. Vetterli, Wavelets and signal processing, IEEE Signal Process. Mag. 8 (4) (1991) 14–38.
[16] A. Said, W. Pearlman, A new, fast, and efficient image codec based on set partitioning in hierarchical trees, IEEE Trans. Circuits Syst. Video Technol. 6 (3) (1996) 243–250.

[17] D. Taubman, High performance scalable image compression with EBCOT, IEEE Trans. Image Process. 9 (7) (2000) 1151–1170.
[18] D. Taubman, A. Secker, Highly scalable video compression with scalable motion coding, IEEE Trans. Image Process. 13 (8) (2004) 1029–1041.
[19] S. Bokhari, A. Nix, D. Bull, Rate-distortion-optimized video transmission using pyramid vector quantization, IEEE Trans. Image Process. 21 (8) (2012) 3560–3572.
[20] P. Schelkens, A. Skodras, T. Ebrahimi, The JPEG 2000 Suite, Wiley, Series: Wiley-IS&T Series in Imaging Science and Technology, 2009.
[21] D. Taubman, M. Marcellin, JPEG 2000: Image Compression Fundamentals, Standards and Practice, Kluwer International Series in Engineering and Computer Science, 2001.
[22] J. Johnson, A filter family designed for use in quadrature mirror filter banks, in: Proc. IEEE Intl. Conf. on Acoustics, Speech and Signal Processing, 1980, pp. 291–294.
[23] J. Villasenor, B. Belzer, J. Liao, Wavelet filter evaluation for image compression, IEEE Trans. Image Process. 4 (8) (1995) 1053–1060.
[24] N. Kingsbury, Good filters / wavelets, Connexions 8 (June 2005), http://cnx.org/content/m11139/2.3/, 2005.
[25] D. LeGall, A. Tabatabai, Sub-band coding of digital images using symmetric short kernel filters and arithmetic coding techniques, in: Proc. IEEE Intl. Conf. on Acoustics, Speech and Signal Processing, 1988, pp. 761–765.
[26] M. Antonini, M. Barlaud, P. Mathieu, I. Daubechies, Image coding using wavelet transform, IEEE Trans. Image Process. 1 (2) (1992) 205–220.
[27] M. Unser, T. Blu, Mathematical properties of the JPEG2000 wavelet filters, IEEE Trans. Image Process. 12 (9) (2003) 1080–1090.
[28] J. Shapiro, Embedded image coding using zerotrees of wavelet coefficients, IEEE Trans. Signal Process. 41 (12) (1993) 3445–3462.
[29] JPEG 2000 image coding system — Part 1: Core coding system, ISO/IEC 15444-1:2019 Information technology, 2019.
[30] F. Ebrahimi, M. Chamik, S. Winkler, JPEG vs. JPEG2000: an objective comparison of image encoding quality, in: Proc. SPIE Conf. Applications of Digital Signal Processing, vol. 5558, 2004.
[31] http://www.itu.int/rec/T-REC-T.800/en.

CHAPTER

7

Lossless compression methods

This chapter describes methods for coding images without loss; this means that, after compression, the input signal can be exactly reconstructed. These have two primary uses. Firstly, they are useful when it is a requirement of the application that loss of any kind is unacceptable. Examples include some medical applications, where compression artifacts could be considered to impact upon diagnoses, or perhaps in legal cases where documents are used as evidence. Secondly, and more commonly, they are used as a component in a lossy compression technique, in combination with transformation and quantization, to provide entropy-efficient representations of the resulting coefficients.

We first review the motivations and requirements for lossless compression and then focus our attention on the two primary methods, Huffman coding and arithmetic coding. Huffman coding is used widely, for example in most JPEG codecs, and we show it can be combined with the discrete cosine transform (DCT) to yield a compressed bitstream with minimum average codeword length. We explore the properties and limitations of Huffman coding in Section 7.3 and then introduce an alternative method – arithmetic coding – in Section 7.6. This elegant and flexible, yet simple, method supports encoding strings of symbols rather than single symbols as in Huffman coding. Arithmetic coding has now been universally adopted as the entropy coder of choice in all modern compression standards. Finally, Section 7.7 presents some performance comparisons between Huffman and arithmetic coding.

7.1 Motivation for lossless image compression

For certain image compression applications, it is important that the decoded signal is an exact replica of the original. Such methods are known as "lossless" since there is no loss of information between the encoder and the decoder. They do not exploit psychovisual redundancy, but they do exploit statistical redundancy to achieve a bit rate close to the entropy of the source.

The arguments for using lossless compression can be summarized as follows:

- With some applications there can be uncertainty as to what image features are important and how these would be affected by lossy image coding. For example in critical medical applications, errors in a diagnosis or interpretation could be legally attributed to artifacts introduced by the image compression process.

Intelligent Image and Video Compression
https://doi.org/10.1016/B978-0-12-820353-8.00016-5

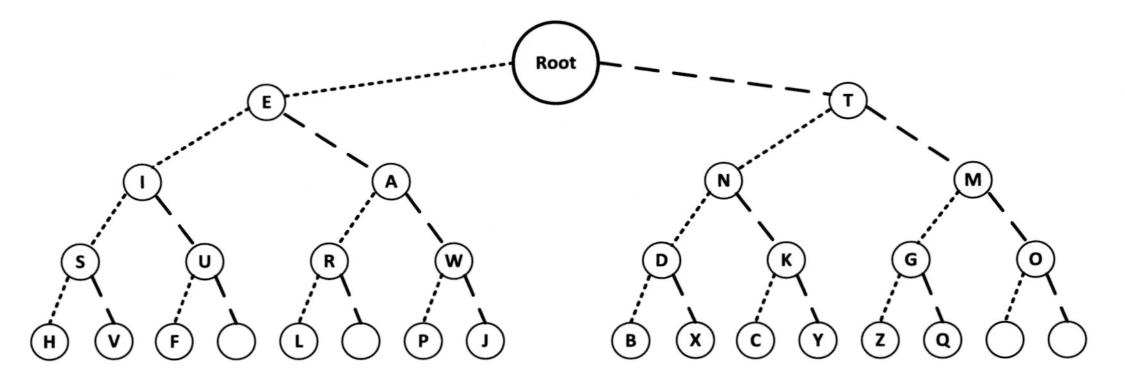

FIGURE 7.1 Symbol-code mappings for the 26 letters from the English alphabet in Morse code.

- In some applications it is essential that data is preserved intact – for example text, or numerical information in a computer file. In such cases, if the data is corrupted, it may lose value or malfunction.
- Lossless compression methods can provide an efficient means of representing the sparse matrices that result from quantization of transform coefficients in a lossy compression system. Sparsity is introduced through the use of decorrelation and quantization to exploit visual redundancy and lossless methods can further exploit the statistics of symbols, to achieve a bit rate close to the source entropy.

7.1.1 Applications

A good example of early lossless compression is the Morse code. In order to reduce the amount of data sent over the telegraph system, Samuel Morse, in the 19th century, developed a technique where shorter codewords were used to represent more common letters and numbers. His binary (digraph) system was based on dots and dashes with a space to delimit letters and an even longer space to indicate boundaries between words. For example, "a" and "e" are common letters while "q" and "j" occur less frequently; hence Morse used the mappings

$$e \mapsto ., \quad a \mapsto .-, \quad q \mapsto --.-, \quad \text{and} \quad j \mapsto .--- \quad .$$

In practice, Morse signaling was effected by switching on and off a tone from an oscillator, with a longer tone representing a dash and a shorter duration for a dot. The full structure of Morse code for the letters of the English alphabet is given in Fig. 7.1. Not shown are the other symbols represented such as numbers and special characters, including common pairs of letters such as "CH." Note the tree-structured representation used in the figure with left and right branches labeled as dots and dashes, respectively. We will see a similar structure later when we examine Huffman codes. It should also be noted that the Morse mapping demands a space between codewords – because otherwise the reader could not differentiate between two "E"s and an "I." We will see later how Huffman encoding cleverly overcomes this limitation of Morse's variable-length coding.

Example 7.1. Morse code

What is the Morse code for "MORSE CODE"?

Solution. It can be seen from Fig. 7.1 that the phrase MORSE CODE is represented by

– –   – – –   • – •   • • •   •   – • – •   – – –   – • •   •.

If we compare this with, let us say, a 5-bit fixed-length coding (actually there are 66 symbols in the Morse alphabet) which would require 45 bits (ignoring word delimiters), the Morse representation requires only 23 bits (ignoring letter and word delimiters), a saving of approximately 50%.

Unknown to many, this encoding system is still widely used today – every radio navigation beacon used in aviation has a Morse identifier, as does every licensed airfield. Any pilot who selects such a beacon will always identify it prior to using it for navigation. This is to ensure that it is operational and that they have selected the correct beacon frequency!

7.1.2 Approaches

Several methods exist that can exploit data statistics to provide lossless compression. These include:

- **Huffman coding** [1]: This is a method for coding individual symbols at a rate close to the first order entropy, often used in conjunction with other techniques in a lossy codec.
- **Arithmetic coding** [2–4]: This is a more sophisticated method which is capable of achieving fractional bit rates for symbols, thereby providing greater compression efficiency for more common symbols.
- **Predictive coding**: This exploits data correlations, as opposed to symbol redundancies, and can be used without quantization to provide lossless image compression. As we saw in Chapter 3 and will examine further in this chapter, predictive coding is often used as a preprocessor for entropy coding.
- **Dictionary-based methods**: Algorithms such as Lempel–Ziv (LZ) and Lempel–Ziv–Welch (LZW) [5,6] are well suited to applications where the source data statistics are unknown. They are frequently used in text and file compression.

Frequently, the above methods are used in combination. For example, DC DCT coefficients are often encoded using a combination of differential pulse code modulation (DPCM) and either Huffman or arithmetic coding. Furthermore, as we will see in Chapter 8, motion vectors are similarly encoded using a form of predictive coding to condition the data prior to entropy coding.

Lossless encoding alone cannot normally provide the compression ratios required for most modern storage or transmission applications – we saw in Chapter 1 that requirements for compression ratios of 100:1 or even 200:1 are not uncommon. In contrast, lossless compression methods, if used in isolation, are typically restricted to ratios of around 4:1.[1]

[1] This is highly data- and application-dependent.

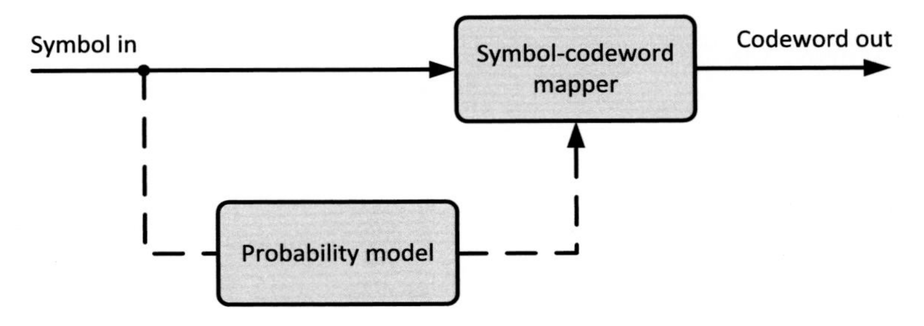

FIGURE 7.2 Generic model of lossless coding.

7.1.3 Dictionary methods

Dictionary methods such as LZW (including LZ77 and LZ78) are, in general, less suitable for image coding applications where the data statistics are known or can be estimated in advance. They do however find widespread use in more general data compression applications for text and file compression, in methods such as ZIP, UNIX compress, and Graphics Interchange Format (GIF). While these methods do not perform particularly well for natural images, methods such as Portable Network Graphics (PNG) improve upon them by preprocessing the image to exploit pixel correlations. PNG outperforms GIF and, for some images, can compete with arithmetic coding. We will not consider these techniques further here. The interested reader is referred to Sayood's book [7] for an excellent introduction.

7.2 Symbol encoding

7.2.1 A generic model for lossless compression

A generic model for lossless encoding is shown in Fig. 7.2. This comprises a means of mapping the input symbols from the source into codewords. In order to provide compression, the symbol to codeword mapper must be conditioned by a representative probability model. A probability model can be derived directly from the input data being processed, from a priori assumptions, or from a combination of both. It is important to note that compression is not guaranteed – if the probability model is inaccurate, then no compression or even data expansion may result. As discussed in Chapter 3, the ideal symbol-to-code mapper will produce a codeword requiring $\log_2(1/P_i)$ bits.

7.2.2 Entropy, efficiency, and redundancy

As described in Chapter 3, H is known as the source entropy, and represents the minimum number of bits per symbol required to code the signal. For a given encoder which codes a

TABLE 7.1 Alternative codes for a given probability distribution.

Letter	Probability	C_1	C_2	C_3
a_1	0.5	0	0	0
a_2	0.25	0	1	10
a_3	0.125	1	00	110
a_4	0.125	10	11	111
	Av. length	1.125	1.25	1.75

signal at an average rate (i.e., an average codeword length) $\bar{l}$, the coding efficiency can be defined as

$$E = \frac{H}{\bar{l}}. \tag{7.1}$$

Thus, if we code a source in the most efficient manner and the code is unambiguous (uniquely decodable), then the rate of the code will equal its entropy.

A coding scheme can also be characterized by its redundancy, R:

$$R = \frac{\bar{l} - H}{H} \cdot 100\%. \tag{7.2}$$

7.2.3 Prefix codes and unique decodability

We noted earlier that Morse code needs spaces to delineate between its variable-length codewords. In modern digital compression systems we cannot afford to introduce a special codeword for "space," as it creates overhead. The way we overcome this is to use what are referred to as prefix codes. These have the property of ensuring that no codeword is the prefix of any other, and hence (in the absence of errors) that each codeword is uniquely decodable. Consider the alternative codes for the four-letter alphabet $\{a_1, a_2, a_3, a_4\}$ in Table 7.1.

Let us examine the three codes in this table. We can see that, in the case of $\mathbf{C}_1$, there is ambiguity as $a_2a_3 = a_1a_3$. Similarly for $\mathbf{C}_2$, $a_1a_1 = a_3$. In fact, only $\mathbf{C}_3$ is uniquely decodable and this is because it is a prefix code, i.e., no codeword is formed by a combination of other codewords. Because of this, it is self-synchronizing – a very useful property. The question is, how do we design self-synchronizing codes for larger alphabets? This is addressed in the next section.

Example 7.2. Coding redundancy

Consider a set of symbols with associated probabilities and codewords as given in the table below. Calculate the first order entropy of the alphabet and the redundancy of the encoding.

Symbol	Probability	Codeword
s_0	0.06	0110
s_1	0.23	10
s_2	0.30	00
s_3	0.15	010
s_4	0.08	111
s_5	0.06	0111
s_6	0.06	1100
s_7	0.06	1101

Solution. 1. Average codeword length

The table is repeated below, including the average length of each codeword. The average length $\bar{l}$ in this case is 2.71 bits/symbol.

Symbol	Prob.	Code	Av. length
s_0	0.06	0110	0.24
s_1	0.23	10	0.46
s_2	0.30	00	0.60
s_3	0.15	010	0.45
s_4	0.08	111	0.24
s_5	0.06	0111	0.24
s_6	0.06	1100	0.24
s_7	0.06	1101	0.24
Overall average length			**2.71**

2. First order entropy

The first order entropy for this alphabet is given by

$$\begin{aligned} H &= -\sum P \log_2 P \\ &= -(0.06 \times \log_2 0.06 + 0.023 \times \log_2 0.23 + 0.3 \times \log_2 0.3 + \ldots + 0.06 \times \log_2 0.06) \\ &= 2.6849 \text{ bits/symbol.} \end{aligned}$$

3. Redundancy

The redundancy of this encoding is

$$R = \frac{\bar{l} - H}{H} \times 100 = \frac{2.71 - 2.6849}{2.6849} \approx 1\%.$$

7.3 Huffman coding

7.3.1 The basic algorithm

Huffman coding represents each symbol s_i with a variable-length binary codeword c_i. The length of c_i is determined by rounding H_i up to the nearest integer. Huffman codes are prefix

Algorithm 7.1 Huffman tree formation.

1. Create ranked list of symbols $s_0 \cdots s_N$ in decreasing probability order;
2. REPEAT
   3. Combine pair of symbols with lowest probabilities, i.e., those at bottom of list (indicated as s_i and s_j here);
   4. Update probability of new node: $P(s_i') = P(s_i) + P(s_j)$;
   5. Sort intermediate nodes: order s_i' in the ranked list;
6. UNTIL a single root node is formed with P=1;
7. Label the upper path of each branch node with a binary 0 and the lower path with a 1;
8. Scan tree from root to leaves to extract Huffman codes for each symbol;
9. END.

codes, i.e., no valid codeword is the prefix of any other. The decoder can thus operate without reference to previous or subsequent data and does not require explicit synchronization signals.

Huffman codes are generated using a tree-structured approach as will be described below. The decoder decodes codewords using a decoding tree of a similar form – this means that the exact tree structure must be transmitted prior to decoding or must be known a priori. For each symbol, the decoder starts at the tree root and uses each bit to select branches until a leaf node is reached which defines the transmitted symbol.

Huffman coding provides an efficient method for source coding provided that the symbol probabilities are all reasonably small. For coding of subband data (e.g., within JPEG) typical efficiencies between 90% and 95% are achievable. If any one of the symbol probabilities is high, then H will be small, and the efficiency of the codec is likely to be lower. Huffman coding requires that both the encoder and decoder have the same table of codewords. In some systems (e.g., JPEG) this table is transmitted prior to the data (see Fig. 7.7).

The procedure for creating a basic Huffman tree is as described by Algorithm 7.1. It should be clear that this procedure ensures that the result is a prefix code because, unlike Morse Code, each symbol is represented by a unique path through the graph from root to leaf.

Example 7.3. Basic Huffman encoding

Consider a source which generates five symbols $\{s_1, s_2, s_3, s_4, s_5\}$, where $P(s_1) = 0.2$, $P(s_2) = 0.4$, $P(s_3) = 0.2$, $P(s_4) = 0.15$, $P(s_5) = 0.05$. Compute the Huffman tree for this alphabet.

Solution. The solution is described in stages below. In the first stage the symbols are ordered in terms of their probability of occurrence and the bottom two symbols on the list are combined to form a new node s_4' (0.2), where the value in parentheses is the sum of the probabilities of the two constituent nodes. We then ensure that the newly formed list of nodes is also in descending probability order as shown below.

$s_2(0.4)$ — $s_2(0.4)$

$s_1(0.2)$ — $s_1(0.2)$

$s_3(0.2)$ — $s_3(0.2)$

$s_4(0.15)$ — $s'_4(0.2)$

$s_5(0.05)$

In stage 2, we repeat this process on the newly formed list of nodes, combining the bottom two nodes to form a new node s'_3 (0.4). This new node is placed in the output list in a position to ensure descending probability values. This is shown below.

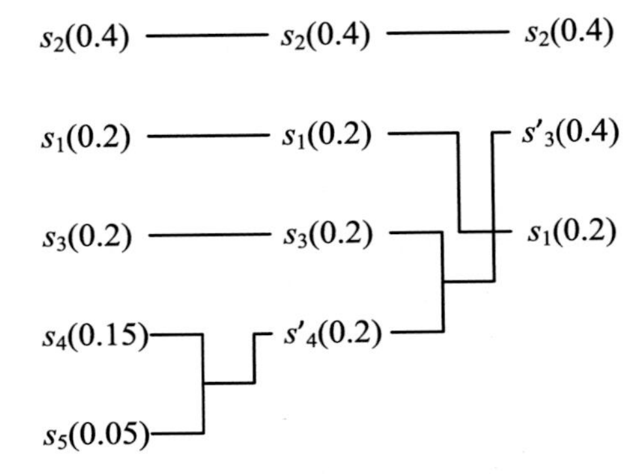

Finally in the third stage we form s''_3 (0.6) and combine this with the only other remaining node s_2 (0.4) to form the single root node with, as expected, a combined probability of 1.

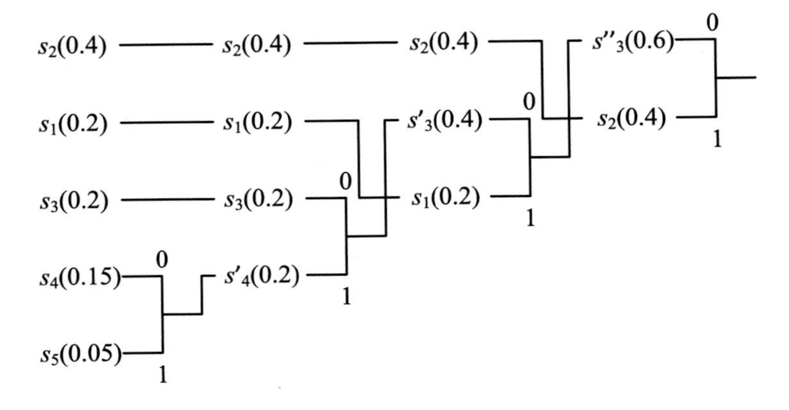

By tracing each path through the graph, we can see that the Huffman codewords for this alphabet are

$$s_1 = 01,\ s_2 = 1,\ s_3 = 000,\ s_4 = 0010,\ s_5 = 0011.$$

This gives an average codeword length of 2.2 bits/symbol.

7.3.2 Minimum variance Huffman coding

Minimum variance Huffman coding minimizes the length of the longest codeword, thus minimizing dynamic variations in transmitted bit rate. The procedure for achieving this is almost identical to the basic approach, except that when forming a new ranked list of nodes, the newly formed node is inserted in the list as high up as possible, without destroying the ranking order. This ensures that its reuse will be delayed as long as possible without destroying the ranking, thus minimizing the longest path through the graph. The revised algorithm is given in Algorithm 7.2.

Example 7.4. Minimum variance Huffman encoding

Consider the same problem as in Example 7.3, where a source generates five symbols $\{s_1, s_2, s_3, s_4, s_5\}$ with $P(s_1) = 0.2,\ P(s_2) = 0.4,\ P(s_3) = 0.2,\ P(s_4) = 0.15,\ P(s_5) = 0.05$. Compute the minimum variance set of Huffman codes for this alphabet.

Solution. In the first stage the symbols are ordered in terms of their probability of occurrence and the bottom two symbols on the list are combined to form a new node $s'_4\,(0.2)$, where the value in parentheses is the sum of the probabilities of the two constituent nodes. We ensure that the newly formed node is inserted in the ordered list as high up as possible as shown below.

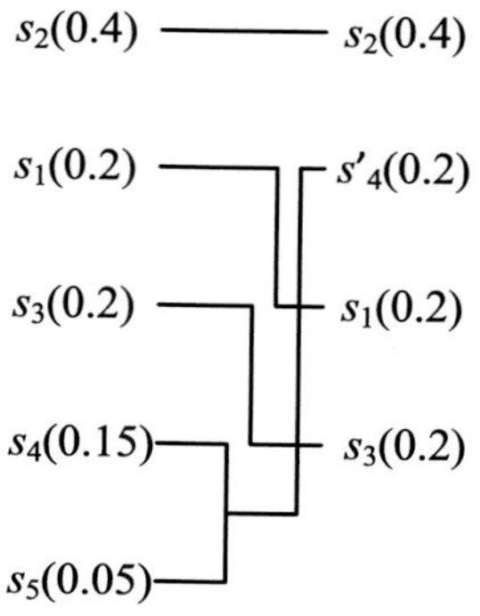

$s_2(0.4)$ — $s_2(0.4)$

$s_1(0.2)$ — $s'_4(0.2)$

$s_3(0.2)$ — $s_1(0.2)$

$s_4(0.15)$ — $s_3(0.2)$

$s_5(0.05)$

This process continues in a similar manner in stage 2.

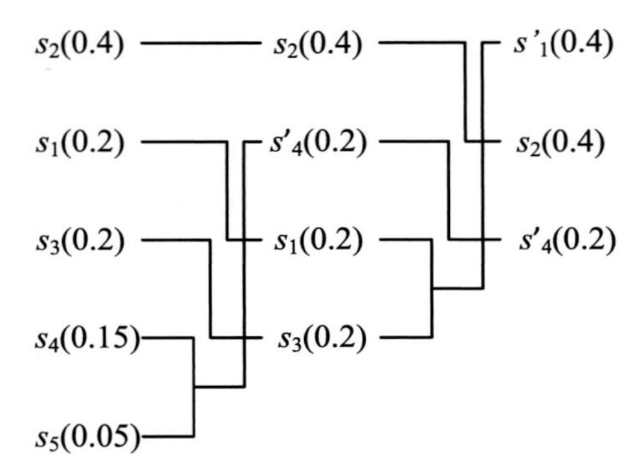

The complete tree is formed as shown below.

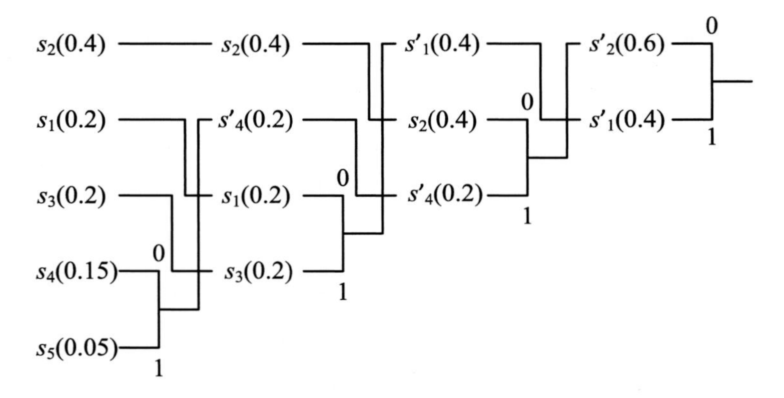

By tracing each path through the graph, we can see that the Huffman codewords for this alphabet are

$$s_1 = 10,\ s_2 = 00,\ s_3 = 11,\ s_4 = 010,\ s_5 = 011.$$

This again gives an average codeword length of 2.2 bits/symbol. However, this time the minimum variance approach has reduced the maximum codeword length to 3 bits rather than the 4 bits in the previous case.

7.3.3 Huffman decoding

Huffman decoding is a simple process which takes the input bitstream and, starting at the root node, uses each bit in sequence to switch between the upper and lower path in the graph. This continues until a leaf node is reached when the corresponding symbol is decoded. This is described for a bitstream of K bits representing a sequence of encoded symbols, by Algorithm 7.3. For ease of understanding, this algorithm references the Huffman tree as the data structure during decoding. It should however be noted that this will normally be implemented using a state machine.

Algorithm 7.2 Minimum variance Huffman tree formation.

1. Create ranked list of symbols $s_0 \cdots s_N$ in decreasing probability order;
2. REPEAT
   3. Combine pair of symbols with lowest probabilities, i.e., those at the bottom of the list (indicated as s_i and s_j here);
   4. Update probability of new node: $P(s_i') = P(s_i) + P(s_j)$;
   5. Sort intermediate nodes: insert s_i' as high up as possible in the ranked list;
6. UNTIL a single root node is formed with $p = 1$;
7. Label the upper path of each branch node with a binary 0 and the lower path with a 1;
8. Scan tree from root to leaves to extract Huffman codes for each symbol;
9. END.

Algorithm 7.3 Huffman decoding process.

1. INPUT bitstream representing a sequence of symbols: $b[0], b[1] \cdots b[K\text{-}1]$; INPUT Huffman tree structure;
2. $i = 0$;
3. REPEAT
   4. REPEAT
      5. Start at root node;
      6. IF $b[i] = 0$ THEN select upper branch, ELSE select lower branch;
      7. $i = i + 1$;
   8. UNTIL leaf node reached;
   9. OUTPUT *symbol*;
10. UNTIL $i = K$;
11. END.

7.3.4 Modified Huffman coding

In practical situations, the alphabet for Huffman coding can be large. In such cases, construction of the codes can be laborious and the result is often not efficient, especially for biased probability distributions. For example, consider an 8-bit source with the following probabilities:

value (x)	P(x)	I(x)	code	bits
0	0.20	2.232	10	2
+1	0.10	3.322	110	3
−1	0.10	3.322	1110	4
+2	0.05	4.322	11110	5
−2	0.05	4.322	11111	5
All other values	0.50	1	0+value	9

This implies that we could generate short and efficient codewords for the first five symbols and assign an extra 1-bit flag as an escape code to indicate any symbol that is not one of the first five. These other values can then be fixed-length coded with (in this case) an 8-bit value. The average code length is then 6.1 bits.

An alternative approach, sometimes referred to as *extended Huffman coding*, can also be used to reduce the coding redundancy and code the alphabet at a rate closer to its entropy. Rather than coding each symbol independently, this method groups symbols together prior to code formation. As we saw in Chapter 3, this can provide significant advantages. However, there are more efficient methods for doing this, such as *arithmetic coding*; we will look at these in Section 7.6.

7.3.5 Properties of Huffman codes

The Kraft inequality

An alphabet with N symbols $S = \{s_i\}$, if encoded using a binary prefix code and ranked according to their probability, P_i, will have $l_1 \leq l_2 \leq l_3 \leq \cdots \leq l_N$. The Kraft inequality states that these lengths must satisfy the following expression:

$$\sum_{i=1}^{N} 2^{-l_i} \leq 1. \tag{7.3}$$

This follows from the information content of each symbol being

$$l_i = -\log_2 P_i = \log_2 {}^{1}/_{P_i}.$$

In the case of a dyadic source, where all codeword lengths are integers, we have

$$\bar{l} = \sum l_i P_i = H(S).$$

Length of Huffman codes

The average length of a codeword in a Huffman encoded alphabet will depend primarily on the distribution of probabilities in the alphabet and on the number of symbols in the alphabet. We have seen above that, for integer codeword lengths, the lower bound on the average codeword length is the source entropy, H. However, in cases where the codeword length is noninteger, for practical purposes they must be rounded up to the next largest integer value. Hence,

$$\bar{l} = \sum \lceil \log_2 {}^{1}/_{P_i} \rceil P_i < \sum \left(\log_2 ({}^{1}/_{P_i}) + 1 \right) P_i < H(S) + 1.$$

The Huffman code for an alphabet, S, therefore has an average length $\bar{l}$ bounded by

$$H(S) \leq \bar{l} \leq H(S) + 1. \tag{7.4}$$

Optimality of Huffman codes

In order for the code to be optimal it must satisfy the following conditions [7,8]:

1. For any two symbols s_i and s_j in an alphabet S, if $P(s_i) > P(s_j)$, then $l_i \leq l_j$, where l represents the length of the codeword.
2. The two symbols with the lowest probabilities have equal and maximum lengths.
3. In a tree corresponding to the optimal code, each intermediate node must have two branches.

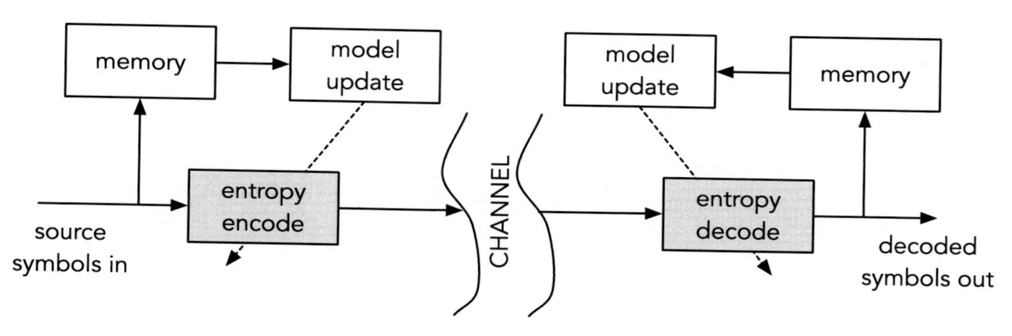

FIGURE 7.3 Adaptive entropy coding.

4. If we collapse several leaf nodes connected to a single intermediate node into a single composite leaf node, then the tree will remain optimal.

It can clearly be seen that Huffman codes satisfy these conditions and represent an optimal representation for coding of individual symbols. We will however see later that greater efficiencies can be obtained by encoding sequences of symbols rather than individual symbols.

7.3.6 Adaptive methods for Huffman encoding

There are two basic limitations of Huffman coding:

1. It encodes one symbol at a time.
2. It depends on a precomputed encoding and decoding structure.

As we will see later, both of these limitations are addressed by *arithmetic coding*. However, *adaptive Huffman coding* methods have also been devised which enable dynamic updating of the Huffman tree according to prevailing source statistics. Methods include those by Gallagher [9] and more recently by Vitter [10]. The essential element, as with any adaptive system, is that the encoder and decoder models initialize and update sympathetically, using identical methods. The basis for this approach is shown in Fig. 7.3.

In a simple case, both encoder and decoder would initialize their trees with a single node and both would know the symbol set in use. The update process assigns a weight W to each leaf node which equals the number of times that symbol has been encountered. Symbols are initially transmitted using a predetermined set of basic codes and, as frequencies of occurrence change, the tree is updated according to an agreed procedure.

An example of this procedure is illustrated in Algorithm 7.4. We first compute the number of nodes, M, required to represent all N symbols (i.e., $2N-1$). We label each node with a number $1 \cdots M$ and append the current W value according to the frequency of occurrence of the corresponding symbol (intermediate nodes are assigned values equal to the sum of their siblings). We define an escape (ESC) node initially as the root node with label N and weight $W=0$. When the first symbol s_i is encountered, it is encoded using the preagreed initial encoding for the alphabet. The ESC node then creates a new root node (label N) with two siblings: that for s_i labeled $N-1$ and a new ESC node labeled $N-2$. The new codeword for s_i is now 1 and that for ESC is 0. So when a new

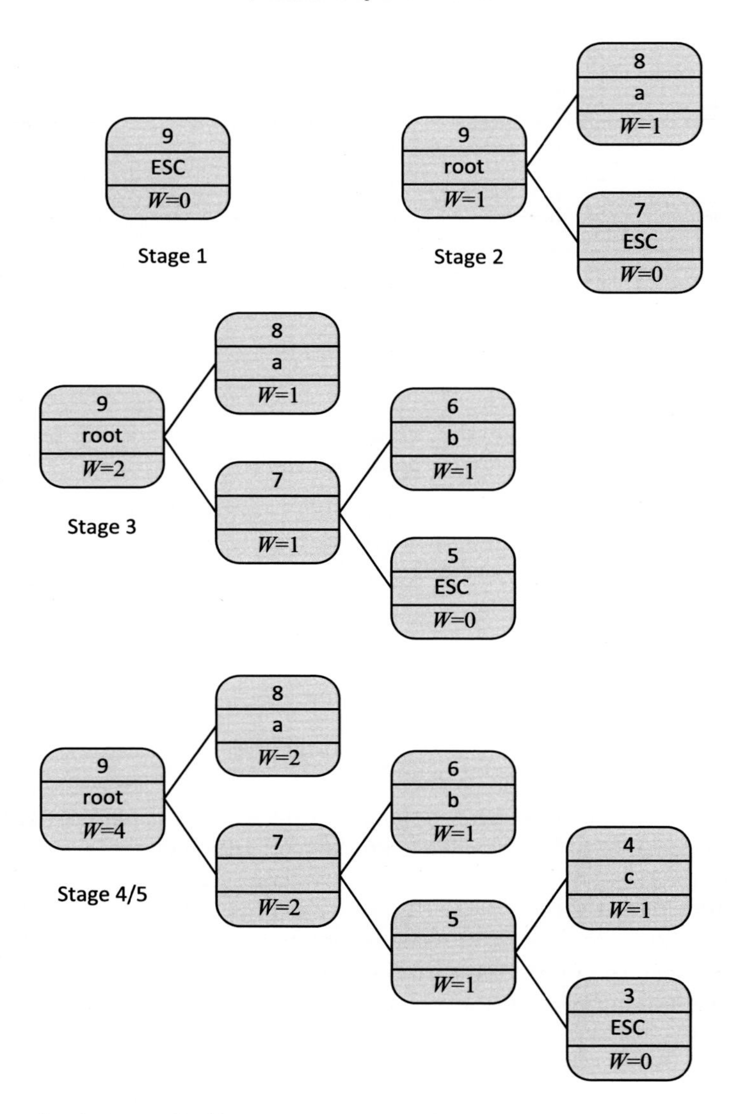

FIGURE 7.4 Example of adaptive Huffman encoding tree evolution.

symbol is encountered, it must be encoded as ESC (0 currently), followed by the preagreed code. This process continues, spawning new nodes as new symbols are encountered.

As the frequency count of a node increments, the tree will need to be restructured to account for this. This is achieved by swapping a node with weight W with the highest numbered node with weight $W-1$ and then rearranging nodes to provide a minimum variance structure.

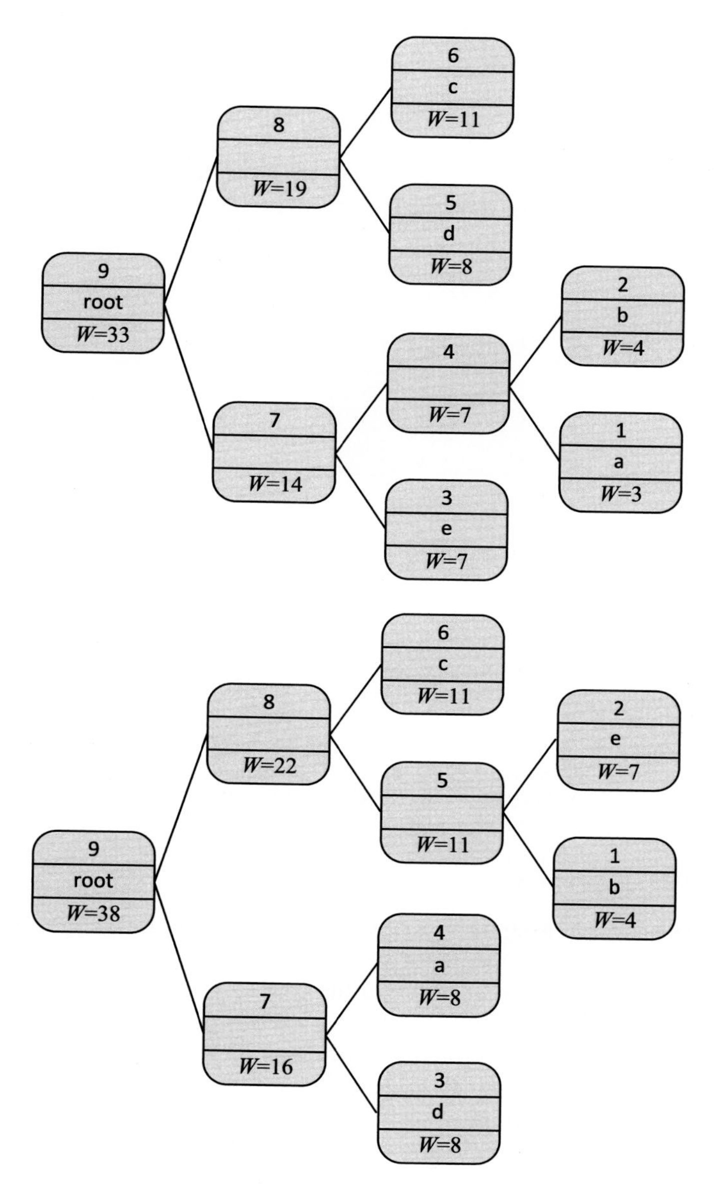

FIGURE 7.5 Example of adaptive Huffman encodingshowing node swapping (bottom) after weight updating symbol *a* by five counts.

This process is capable of continually adapting to dynamic variations in symbol statistics, providing optimal codewords.

Fig. 7.4 provides a simple example of the process of tree evolution for an alphabet of five symbols $\{a,b,c,d,e\}$, with the transmitted sequence $\{a,b,c,a\}$. Fig. 7.5 illustrates the case of an

Algorithm 7.4 Adaptive Huffman coding: tree updating.

1. INPUT symbol s_i;
2. If symbol s_i is new then create new node with two siblings: one for s_i and a new ESC node;
3. Update weight for node s_i: $W=W+1$;
4. Swap node with weight W with the highest numbered node with weight $W-1$ and then rearrange nodes to provide a minimum variance structure.

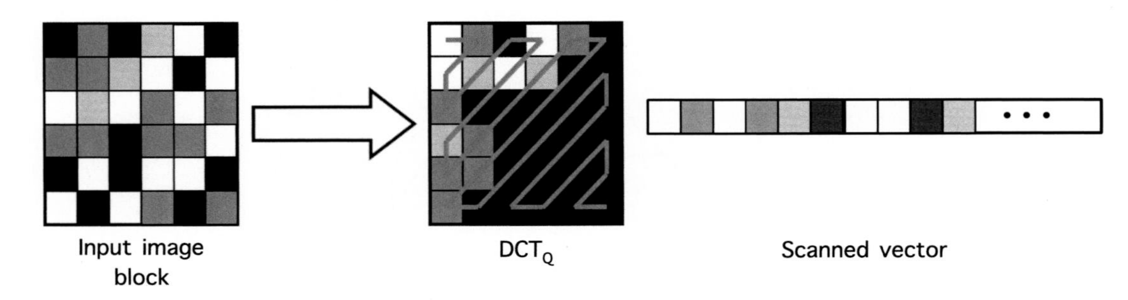

FIGURE 7.6 Zig-zag scanning prior to variable-length coding.

established tree where node swapping is used to update the tree after the weight for symbol a is incremented by five counts.

7.4 Symbol formation and encoding

7.4.1 Dealing with sparse matrices

Many data sources contain long strings of one particular symbol. For example, the quantized subband data generated by a transform or wavelet-based compression system is typically sparse. We have seen in Chapter 5 that, after applying a forward transform and quantization, the resulting matrix contains a relatively small proportion of nonzero entries with most of its energy compacted toward the lower frequencies (i.e., the top left corner of the matrix). In such cases, *run-length coding* (RLC) can be used to efficiently represent long strings of identical values by grouping them into a single symbol which codes the value and the number of repetitions. This is a simple and effective method of reducing redundancies in a sequence. A slight variant on this for sparse matrices is to code runs of zeros in combination with the value of the next nonzero entry.

In order to perform run-length encoding we need to convert the 2D coefficient matrix into a 1D vector and furthermore we want to do this in such a way that maximizes the runs of zeros. Consider for example the 6×6 block of data and its transform coefficients in Fig. 7.6. If we scan the matrix using a zig-zag pattern, as shown in the figure, then this is more energy-efficient than scanning by rows or columns.

Example 7.5. Run-length encoding of sparse energy-compact matrices

Consider the 4×4 quantized DCT matrix below. Perform zig-zag scanning to produce a compact 1D vector and run-length code this using $\{run/value\}$ symbols. The matrix is

$$\mathbf{C}_Q = \begin{bmatrix} 8 & 5 & 2 & 0 \\ 3 & 0 & 0 & 0 \\ 0 & 0 & 0 & 0 \\ 1 & 0 & 0 & 0 \end{bmatrix}.$$

Solution. Perform zig-zag scanning to produce a 1D vector:

$$\mathbf{c}_Q = \begin{bmatrix} 8 & 5 & 3 & 0 & 0 & 2 & 0 & 0 & 0 & 1 & 0 & 0 & 0 & 0 & 0 & 0 \end{bmatrix}.$$

Next represent this vector using $\{run/value\}$ symbols:

$$\mathbf{s}_Q = \begin{bmatrix} (0/8) & (0/5) & (0/3) & (2/2) & (3/1) & \text{EOB} \end{bmatrix}.$$

Note that, to avoid encoding the last run of zeros, it is normal to replace these with an end of block (EOB) symbol.

7.4.2 Symbol encoding in JPEG

JPEG remains the universally adopted still image coding standard. It was first introduced in 1992 and has changed little since then [11]. The operation of the baseline JPEG encoder is shown in Fig. 7.7. First of all the input image, usually in 4:2:0 YC_bC_r format, is segmented into nonoverlapping 8×8 blocks and these are decorrelated using an 8×8 DCT and quantized as discussed in Chapter 5. The entropy coding of these quantized coefficients is normally based on Huffman coding[2] and proceeds as follows:

- **DC coefficients**: These are DPCM modulated with respect to the corresponding value from the previous block. The *size* category for the prediction residual (as defined in Table 7.4) is then Huffman encoded using a set of tables specifically designed for DC coefficients. The amplitude of the prediction residual is then appended to this codeword in 1s complement form.
- **AC coefficients**: These are first of all zig-zag scanned as described above and then run-length encoded. The run-length symbols used for each nonzero AC coefficient are $(run/size)$, where *run* is the number of zeros preceding the next nonzero coefficient and*size* relates to the value of that coefficient, as defined in Table 7.4. These symbols are then entropy encoded using a further set of Huffman tables designed for the AC coefficients and the coefficient value is again appended to the codeword in 1s complement form.

These processes are described in more detail in Algorithm 7.5 and Algorithm 7.6. The approach is justified by the graph in Fig. 7.8, which shows the distribution of size values for various run-lengths for a JPEG encoded Lena image. The figure shows statistics for 0.5 bpp but

[2] It should be noted that the JPEG standard also supports arithmetic coding but this is rarely used in practice.

Algorithm 7.5 JPEG DC coefficient symbol encoding.

1. Form DPCM residual for the DC coefficient in the current block DC_i, predicted relative to the DC coefficient from the previously encoded block, DC_{i-1}: $DC_{i(pred)} = DC_i - DC_{i-1}$;
2. Produce $symbol_{DC} = (size)$ for the DC coefficient prediction;
3. Huffman encode $symbol_{DC}$ using the table for DC coefficient differences;
4. OUTPUT bits to JPEG file or bitstream;
5. OUTPUT *amplitude* value of the coefficient difference in 1s complement format to the file or bitstream;
6. END.

Algorithm 7.6 JPEG AC coefficient symbol encoding.

1. Form a vector of zig-zag scanned AC coefficient values from the transformed data block as described in Fig. 7.6;
2. $i = 0$;
3. REPEAT
    4. $i = i + 1$;
    5. Produce $symbol_{AC,i} = (run/size)$ for the next nonzero AC coefficient in the scan;
    6. Huffman encode $symbol_{AC,i}$ using the table for AC coefficients;
    7. OUTPUT codeword bits to the JPEG file or bitstream;
    8. OUTPUT *amplitude* value of the nonzero coefficient in 1s complement form to the JPEG file or bitstream;
9. UNTIL $symbol_{AC,i-1} = (0/0)$;
10. END.

similar characteristics are obtained for other compression ratios. It can be seen that probabilities fall off rapidly as *run-length* increases and as *size* decreases. Hence longer codewords can be justified in such cases, with shorter codewords reserved for lower values of *run* and *size*.

JPEG actually restricts the number of bits allocated to run and size to 4 bits each, giving a range of $0 \cdots 15$ for both. If a run is greater than 15 zeros (which is unlikely), then it is divided into multiples of 15 plus the residual length. This is effective in constraining the size of the entropy code tables. Hence in JPEG, the *run/size* symbol is actually represented by *RRRRSSSS*. Thus the only *run/size* symbols with a size of 0 are EOB (0/0) and ZRL (15/0).

A subset of the default JPEG Huffman codes for DC and AC luminance coefficients is provided in Tables 7.2 and 7.3. A full set can be obtained from reference [11].

Example 7.6. JPEG coefficient encoding

Consider an AC luminance coefficient value of −3 preceded by five zeros. What is the JPEG symbol for this run and coefficient value? Using the default JPEG Huffman tables, compute the Huffman code for this coefficient and the resulting output bitstream.

Solution. From Table 7.4, it can be seen that the *size*(−3)=2. The symbol is thus s=(5/2).

From the default JPEG Huffman table for luminance AC coefficients, the Huffman code for (5/2) is 11111110111. The value of the nonzero coefficient is −3, which in 1s complement format is 00.

The output bitstream produced for this sequence is thus

1111111011100.

TABLE 7.2 Table for luminance DC coefficient differences.

Category	Code length	Codeword
0	2	00
1	3	010
2	3	011
3	3	100
4	3	101
5	3	110
6	4	1110
7	5	11110
8	6	111110
9	7	1111110
10	8	11111110
11	9	111111110

TABLE 7.3 Table for luminance AC coefficients, showing *run/size* (R/S) values, Huffman codes, and their lengths (L).

R/S	L	Codeword	R/S	L	Codeword
0/0	4	1010 (EOB)	2/1	5	11100
0/1	2	00	2/2	8	11111001
0/2	2	01	2/3	10	1111110111
0/3	3	100	2/4	12	111111110100
0/4	4	1011	2/5	16	1111111110001001
0/5	5	11010	2/6	16	1111111110001010
0/6	7	1111000	2/7	16	1111111110001011
0/7	8	11111000	2/8	16	1111111110001100
0/8	10	1111110110	2/9	16	1111111110001101
0/9	16	1111111110000010	2/A	16	1111111110001110
0/A	16	1111111110000011			
1/1	4	1100	3/1	6	111010
1/2	5	11011	3/2	9	111110111
1/3	7	1111001	3/3	12	111111110101
1/4	9	111110110	3/4	16	1111111110001111
1/5	11	11111110110	3/5	16	1111111110010000
1/6	16	1111111110000100	3/6	16	1111111110010001
1/7	16	1111111110000101	3/7	16	1111111110010010
1/8	16	1111111110000110	3/8	16	1111111110010011
1/9	16	1111111110000111	3/9	16	1111111110010100
1/A	16	1111111110001000	3/A	16	1111111110010101

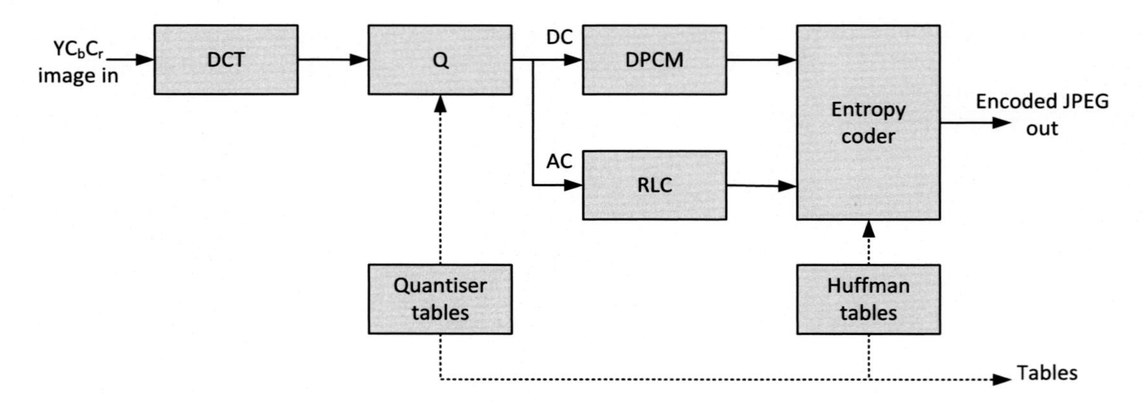

FIGURE 7.7 JPEG baseline image encoder architecture.

TABLE 7.4 JPEG coefficient size categories.

Category	Amplitude range
0	0
1	−1,+1
2	−3,−2,+2,+3
3	−7...−4,+4...+7
4	−15...−8,+8...+15
5	−31...−16,+16...+31
6	−63...−32,+32...+63
etc.	etc.

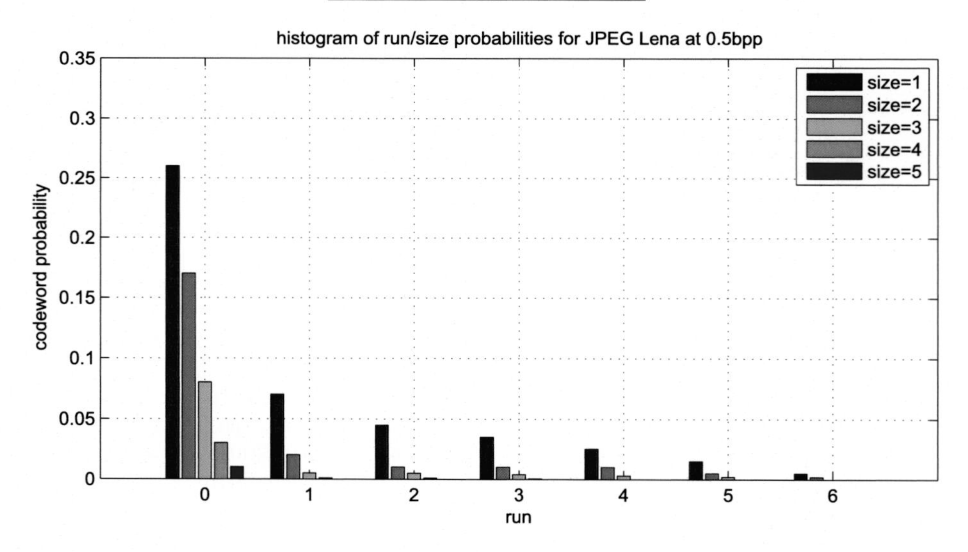

FIGURE 7.8 Distribution of *run*/*size* values in a typical JPEG encoding.

Example 7.7. JPEG matrix encoding

Consider the following quantized luminance DCT matrix. Assuming that the value of the DC coefficient in the preceding block is 40, what is the JPEG symbol sequence for this matrix? Using the default JPEG Huffman tables, compute its Huffman codes and the resulting output bitstream. Assuming the original pixels are 8-bit values, what is the compression ratio for this encoding?

$$\mathbf{C}_Q = \begin{bmatrix} 42 & 16 & 0 & 0 & 0 & 0 & 0 & 0 \\ -21 & -15 & 0 & 0 & 0 & 0 & 0 & 0 \\ 10 & 3 & -3 & 0 & 0 & 0 & 0 & 0 \\ -2 & 2 & 2 & 0 & 0 & 0 & 0 & 0 \\ 0 & -1 & 0 & 0 & 0 & 0 & 0 & 0 \\ 0 & 0 & 0 & 0 & 0 & 0 & 0 & 0 \\ 0 & 0 & 0 & 0 & 0 & 0 & 0 & 0 \\ 0 & 0 & 0 & 0 & 0 & 0 & 0 & 0 \end{bmatrix}.$$

Solution. 1. DC coefficient:

The prediction residual for the current DC coefficient is $\mathrm{DC}_{i(\mathrm{pred})} = \mathrm{DC}_i - \mathrm{DC}_{i-1} = 2$. Thus *size*=2 and *amplitude*=2.

The Huffman codeword from Table 7.2 is 011. Hence the bitstream for the DC coefficient in this case is 01110.

2. AC coefficients:

The zig-zag scanned vector for the matrix above is

$$\begin{matrix} \mathbf{c}_Q = [& 16 & -21 & 10 & -15 & 0 & 0 & 0 & 3 & -2 & 0 & \cdots \\ \cdots & 2 & -3 & 0 & 0 & 0 & 0 & 0 & 2 & -1 & 0 & \cdots]. \end{matrix}$$

For each of these nonzero coefficients, we can now compute the (*run*/*size*) symbol, its Huffman code (from Table 7.3), and the corresponding 1s complement amplitude representation. These are summarized in the table below.

3. Output bitstream:

The output bitstream for this matrix is the concatenation of DC and AC codes as follows:

(01110110101000011010010101011101010110000 11111...

...0111110101110111001001111111011110000 1010).

4. Compression ratio:

Ignoring any header or side information, the total number of bits for this block is $5 + 82 = 87$ bits. In contrast, the original matrix has 64 entries, each of 8 bits. Hence it requires 512 bits, giving a compression ratio of 5.9:1 or alternatively 1.36 bpp.

Coeff	run/size	Huffman code	Amplitude	#bits
16	0/5	11010	10000	10
−21	0/5	11010	01010	10
10	0/4	1011	1010	8
−15	0/4	1011	0000	8
3	3/2	111110111	11	11
−2	0/2	01	01	4
2	1/2	11011	10	7
−3	0/2	01	00	4
2	5/2	11111110111	10	13
−1	0/1	00	0	3
EOB	0/0	1010	-	4
Total AC bits				**82**

7.4.3 JPEG performance examples

Some examples that demonstrate the performance of JPEG encoding are shown in Figs. 7.9 and 7.10 for a highly textured image (*maple*) and a more structured image (*puppet*). Both original images have a word length of 8 bits. As can be observed, the results begin to be psychovisually acceptable beyond 0.25 bpp and become good at 1 bpp. Subjectively at 0.22 bpp the result for *puppet* is more acceptable than that for *maple*, due to the lower contribution from high-frequency textures that are more heavily quantized during compression.

7.4.4 Lossless JPEG mode

A simple lossless mode is defined in the JPEG standard, which combines predictive coding with Huffman or arithmetic coding. The prediction framework is shown in Fig. 7.11 and the choices of predictor are shown in Table 7.5. Predictor modes are selected from the table and signaled as side information in the header – and are normally kept constant for the whole scan. After prediction, residuals are Huffman or arithmetically encoded. In practice the reconstructed pixel values, rather than the actual pixel values, are used in the prediction to eliminate numerical mismatches between the encoder and decoder.

TABLE 7.5 Prediction modes for lossless JPEG.

Mode	Predictor
0	No prediction
1	$\hat{X} = A$
2	$\hat{X} = B$
3	$\hat{X} = C$
4	$\hat{X} = A + B - C$
5	$\hat{X} = A + (B - C)/2$
6	$\hat{X} = B + (A - C)/2$
7	$\hat{X} = (A + B)/2$

FIGURE 7.9 JPEG results for various compression ratios for the 512 × 512 *maple* image. Top left: Original. Top right: 0.16 bpp. Middle left: 0.22 bpp. Middle right: 0.5 bpp. Bottom left: 1 bpp. Bottom right: 2 bpp.

7.4.5 Context-adaptive variable-length coding (CAVLC) in H.264/AVC

In its baseline and extended profiles, H.264/AVC uses two methods of entropy coding. The first, used for all syntax elements (SEs) apart from transform coefficients, is based on a single codeword table for all SEs. The mapping of the SE to this table is adapted according

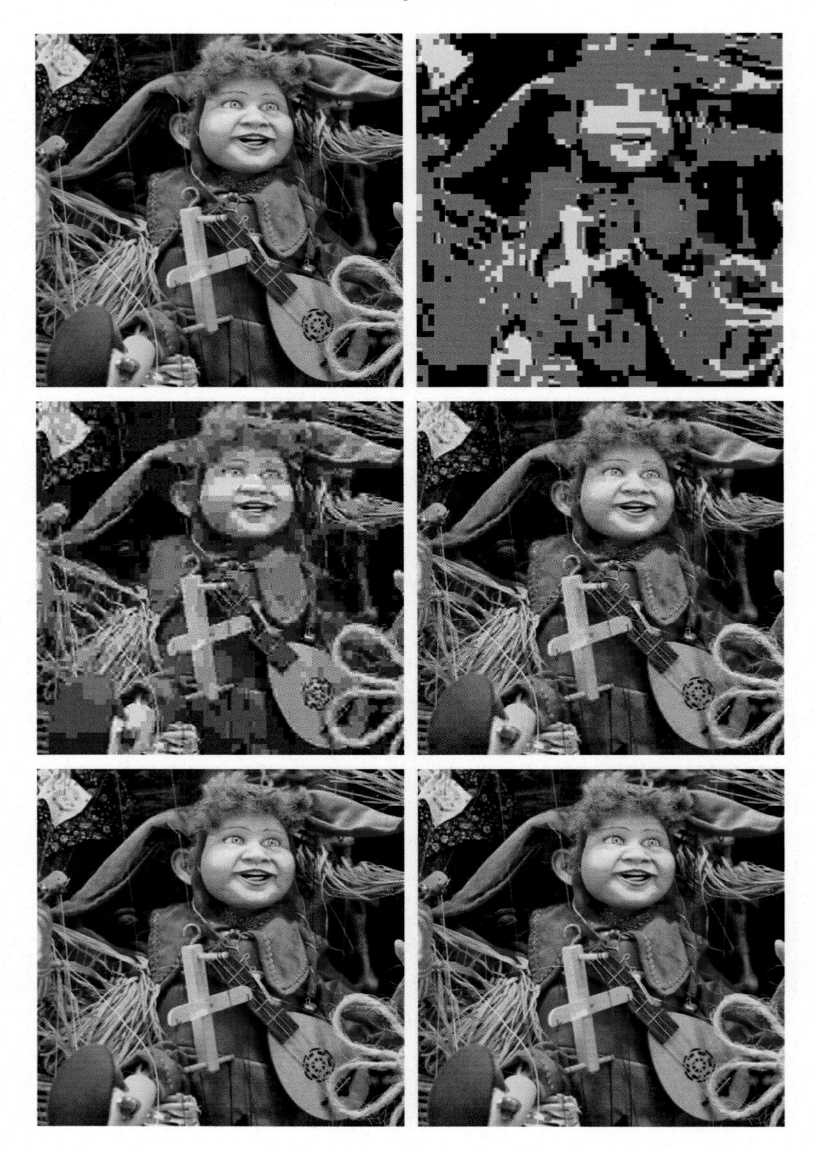

FIGURE 7.10 JPEG results for various compression ratios for the 512 × 512 *puppet* image. Top left: Original. Top right: 0.16 bpp. Middle left: 0.22 bpp. Middle right: 0.5 bpp. Bottom left: 1 bpp. Bottom right: 2 bpp.

to the data statistics. An Exp-Golomb code (see Section 7.5) is used which is effective while offering a simple decoding process.

In the case of transform coefficients, context-adaptive variable-length coding (CAVLC) is used. This provides better performance and supports a number of SE-dependent tables that can be switched between, according to the prevailing data statistics. Although CAVLC is not

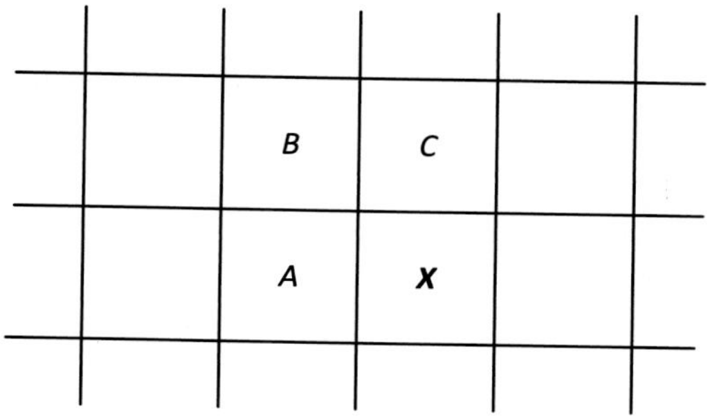

B C
A X

FIGURE 7.11 Prediction samples for lossless JPEG.

based on Huffman coding, it is worthy of a brief mention here since it shows the power of context switching. Good examples of CAVLC encoding are given in [12].

CAVLC exploits the sparsity of the transform matrix as we have seen previously, but also exploits other characteristics of the scanned coefficient vector, most notably: (i) that the scan often ends with a series of ± 1 values and (ii) that the numbers of coefficients in neighboring blocks are correlated. The method uses different VLC tables according to the local context.

CAVLC codes the number of nonzero coefficients and the size and position of these coefficients separately. Based on the characteristics of the data, a number of different coding modes can be invoked. For example, CAVLC can encode:

- The number of nonzero coefficients and the *Trailing 1s* (T1s): T1s are the number of coefficients with a value of ± 1 at the end of the scan (scans often end with runs of ± 1). Four VLC tables are used for this, based on the number of coefficients in neighboring blocks.
- The values of nonzero coefficients: T1s can be represented as signs; other coefficients are coded in reverse scan order because variability is lower at the end of the scan. Six Exp-Golomb code tables are used for adaptation according to the magnitudes of recently coded coefficients.
- TotalZeroes and RunBefore: These specify the number of zeros between the start of the scan and its last nonzero coefficient. RunBefore indicates how these zeros are distributed.

A good example of CAVLC coding is provided by Richardson in [13].

7.5 Golomb coding

The Golomb family includes Unary, Golomb, Golomb–Rice, and exponential Golomb codes [14,15]. These are variable-length codes which are generated via a fixed mapping between an index and a codeword. In the same way as with Huffman codes, shorter codewords can be assigned to symbols with higher probabilities and vice versa. Golomb codes have the advantage that they can be simply encoded and decoded algorithmically without the need for look-up tables. They also possess the property that, for any geometric distribution, it is

TABLE 7.6 Golomb codes for index values $i = 0 \cdots 9$.

i	Unary	Golomb–Rice (m=4)	Exp-Golomb
0	1	1|00	1|
1	01	1|01	01|0
2	001	1|10	01|1
3	0001	1|11	001|00
4	00001	01|00	001|01
5	000001	01|01	001|10
6	0000001	01|10	001|11
8	000000001	01|11	0001|000
9	0000000001	001|00	0001|001
⋮	⋮	⋮	⋮

possible to find an encoding that is an optimal prefix code. Adaptive methods have thus been developed and exploited in standards such as the lossless form of JPEG – JPEG-LS.

7.5.1 Unary codes

The simplest type of unary code encodes a nonnegative integer, i, by i 0s followed by a single 1 (or the other way around). The first few unary codes are shown in Table 7.6, where it is assumed that the index values correspond to the alphabet symbols ranked in terms of decreasing probability. This encoding is generally not very efficient except when the probabilities are successive powers of 2.

7.5.2 Golomb and Golomb–Rice codes

Golomb codes divide all index values i into equal-sized groups of size m. The codeword is then constructed from a unary code that characterizes each group, followed by a fixed-length code ν_i that specifies the remainder of the index that has been encoded. The Golomb code residual can be simply generated using the following equation, where ν_i represents the remainder for index i:

$$\nu_i = i - \left\lfloor \frac{i}{m} \right\rfloor m. \tag{7.5}$$

The Golomb–Rice code is a special case where $m = 2^k$ and is shown in Table 7.6. It can be observed that Golomb–Rice codes grow more slowly than unary codes and are therefore generally more efficient.

7.5.3 Exponential Golomb codes

Exponential Golomb codes were proposed by Teuhola in 1978 [16], and whereas Golomb–Rice codes divide the alphabet into equal sized groups, Exp-Golomb codes divide it into exponentially increasing group sizes. The comparative performance of Golomb–Rice and

Exp-Golomb codes will depend on the size of the alphabet and the choice of m, but for larger alphabets, Exp-Golomb coding has significant advantages.

The Exp-Golomb code for a symbol index i comprises a unary code (ζ_i zeros, followed by a single 1), concatenated with a fixed-length code, ν_i. It can be simply generated as follows:

$$\begin{cases} \nu_i = 1 + i - 2^{\zeta_i}, \\ \zeta_i = \lfloor \log_2 (i+1) \rfloor. \end{cases} \tag{7.6}$$

Exp-Golomb coding has been used in both H.264/AVC and HEVC.

Example 7.8. Exp-Golomb encoding and decoding

i) What is the Exp-Golomb code for the symbol index 228_{10}?

ii) Show how the Exp-Golomb codeword 00000100111 would be decoded and compute the value of the corresponding symbol index.

Solution. i) The codeword group in this case is given by

$$\zeta_i = \lfloor \log_2 (i+1) \rfloor = 7,$$

and the residual is given by

$$\nu_i = 1 + i - 2^{\zeta_i} = 101_{10} = 1100101_2.$$

The complete codeword is thus 000000011100101.

ii) The algorithm is straightforward in that, for each new codeword, we count the number of leading zeros (ζ_i), ignore the following 1, and then decode the index i from the next ζ_i bits. In this case there are five leading zeros and the encoded residual $\nu_i = 00111_2 = 7_{10}$. The index can then be calculated from Eq. (7.6) as $i = \nu_i - 1 + 2^{\zeta_i} = 134$.

7.6 Arithmetic coding

Although Huffman coding is still widely employed in standards such as JPEG, an alternative, more flexible method has grown in popularity in recent years. Variants on arithmetic coding are employed as the preferred entropy coding mechanism in most video coding standards with *context-based adaptive binary arithmetic coding* (CABAC) now forming the basis of H.264/AVC, HEVC, and VVC [17]. Arithmetic encoding is based on the use of a cumulative probability distribution as a basis for codeword generation for a sequence of symbols. Some of the earliest references to this approach were made by Shannon [18], by Abraham's in his book on information theory in 1963 [2], and by Jelinek [19]. More recent popularizations of the technique are due to Rissanen [3], Witten [23], and others. Excellent overviews can be found in [4,7,8].

Algorithm 7.7 Arithmetic encoding.

1. Subdivide the half-open interval [0,1) according to the symbol cumulative probabilities: Eq. (7.7);
2. Initialize upper and lower limits: $l(0)=0$; $u(0)=1$;
3. INPUT sequence of symbols $\{x_i\}$;
4. $i=0$;
5. REPEAT
    6. $i=i+1$;
    7. Compute lower interval limit: $l\,(i)=l(i-1)+(u(i-1)-l(i-1))P_x\,(x_i-1)$;
    8. Compute upper interval limit: $u\,(i)=l(i-1)+(u(i-1)-l(i-1))P_x(x_i)$;
9. UNTIL $i=N$;
10. OUTPUT arithmetic codeword, $v\in[l(N),u(N))$;
11. END.

7.6.1 The basic arithmetic encoding algorithm

An arithmetic encoder represents a string of input symbols as a binary fraction. In arithmetic coding, a half-open encoding range [0,1) is subdivided according to the probabilities of the symbols being encoded. The first symbol encoded will fall within a given subdivision; this subdivision forms a new interval which is then subdivided in the same way for encoding the second symbol. This process is repeated iteratively until all symbols in the sequence have been encoded.

Providing some basic additional information is known (i.e., the number of symbols transmitted or the existence of an EOB symbol), the complete sequence of symbols transmitted can thus be uniquely defined by any fractional number (referred to as a *tag*) within the range of the final symbol at the final iteration.

By using binary symbols and approximating the probabilities to powers of two, an arithmetic encoder can be implemented using only shift and add operations. This achieves a computational complexity which is comparable or even better than Huffman coding while providing superior compression performance. The basic arithmetic encoding algorithm is illustrated in Algorithm 7.7.

We saw in Chapter 3 that the average codeword length for a source depends highly on how symbols are created and encoded and that, by combining symbols in groups, we could potentially achieve better performance. Put simply, this is what arithmetic coding does, but it does it in a rather elegant and flexible manner.

Let us assume that we have a set of M symbols that comprise our source alphabet, $S=\{s_0,s_1,s_2\cdots s_{M-1}\}$, where each symbol maps to an integer x in the range $\{0,1,2,3\cdots M\text{-}1\}$ corresponding to its symbol index. The source symbol probabilities are therefore defined as

$$P\,(m)=P\,(x=m)\,,\quad m=0\cdots M-1.$$

The cumulative probability distribution can now be defined in terms of these probabilities as follows:

$$P_x(m) = \sum_{k=0}^{m} P(k), \quad m = 0 \cdots M-1. \tag{7.7}$$

Following our discussion of code optimality in Chapter 3 and in Section 7.3.5, we can see that an optimal coding system will code each symbol in S with an average length of $-\log_2 P_i$ bits.

So now we can proceed to develop the iterative symbol encoding process. Firstly we define a number line as a half-open interval in the range $0 \cdots 1$, i.e., [0,1). We can then define each accumulated probability interval at each iteration by its lower extent and its upper limit (or equivalently, its length). So, if we have a sequence of N input symbols $X = \{x_i\}, i = 1 \cdots N$, and we define $l(i)$ to be the lower limit at stage i of the encoding and $u(i)$ to be the corresponding upper limit, then at the start of encoding,

$$\begin{aligned} l(0) &= P(-1) = 0, \\ u(0) &= P_x(M-1) = 1. \end{aligned}$$

After the first symbol is encoded, i=1,

$$\begin{aligned} l(1) &= P_x(x_1 - 1), \\ u(1) &= P_x(x_1). \end{aligned}$$

Similarly after the second symbol, i=2, we have

$$\begin{aligned} l(2) &= l(1) + (u(1) - l(1))P_x(x_2 - 1), \\ u(2) &= l(1) + (u(1) - l(1))P_x(x_2), \end{aligned}$$

and in general after the ith symbol, x_i, we have

$$\begin{aligned} l(i) &= l(i-1) + (u(i-1) - l(i-1))P_x(x_i - 1), \\ u(i) &= l(i-1) + (u(i-1) - l(i-1))P_x(x_i). \end{aligned} \tag{7.8}$$

This is best illustrated by a simple example (Example 7.9).

Example 7.9. Arithmetic encoding

Consider a sequence of symbols $\{s_2, s_2, s_3\}$ from an alphabet $\{s_1, s_2, s_3\}$ with probabilities $\{P(1), P(2), P(3)\}$. Show the arithmetic encoding process for the three iterations corresponding to this sequence.

Solution. Using our notation above, the sequence maps on to $X = \{x_0, x_1, x_2\} \Rightarrow \{2, 2, 3\}$. The three iterations of the arithmetic coder for this sequence are illustrated below.

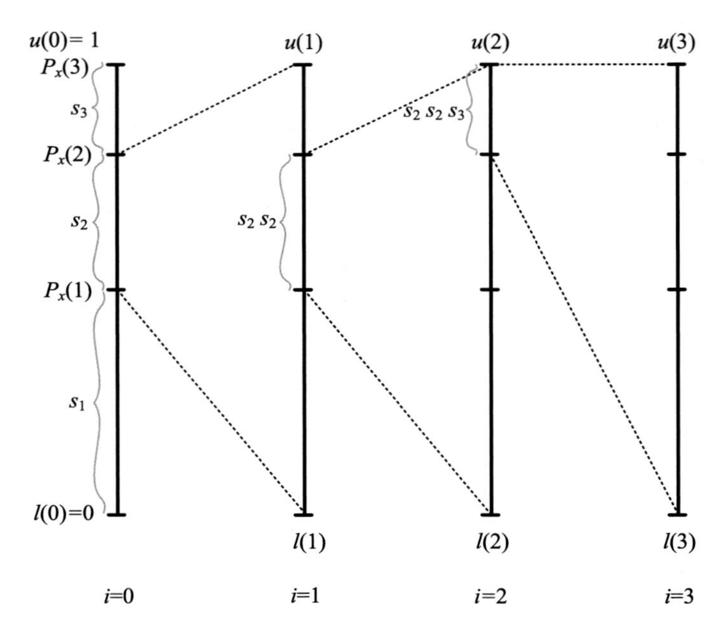

We can see that any value in the range $[l(3), u(3))$ uniquely defines this sequence provided of course that we know when to stop decoding – but more on that later.

Example 7.10. Arithmetic encoding – numerical example

Using the same symbols and sequence as in Example 7.9, assume that the symbol probabilities are $P(1) = 0.5$, $P(2) = 0.25$, $P(3) = 0.25$. Compute the arithmetic code for the sequence $\{s_2, s_2, s_3\}$.

Solution. The three iterations of the arithmetic coder for this sequence are illustrated below, including the tag values for each stage.

As an example of computing the limits for the interim intervals, consider the case when i=2 (i=1 can be worked out by inspection):

$$l(2) = l(1) + (u(1) - l(1))P_x(1),$$
$$l(2) = 0.5 + (0.75 - 0.5)0.5 = 0.625,$$

$$u(2) = l(1) + (u(1) - l(1))P_x(2),$$
$$u(2) = 0.5 + (0.75 - 0.25)0.75 = 0.6875.$$

Repeating this for iteration 3 results in an arithmetic code that can assume any value within the interval $v \epsilon [0.671875, 0.6875) = [0.101011_2, 0.1011_2)$. We might take the lower limit for simplicity or take a value midway between the lower and upper limits. In practice it would be preferable to use the shortest binary representation which in this case is also the lower limit, i.e., 0.101011, or, if we drop the leading zero as it is redundant, $\hat{v}$ =101011.

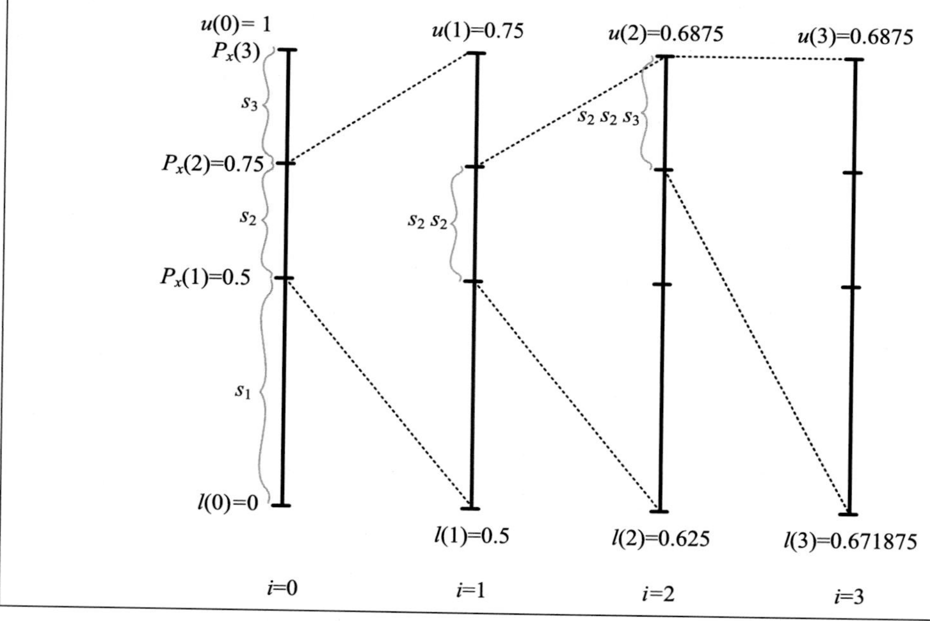

7.6.2 The basic arithmetic decoding algorithm

The encoding algorithm described in Section 7.6.1 creates an arithmetic codeword, $\hat{v} \in [l(N), u(N))$. The purpose of the decoder is to generate the same output sequences, $X = \{x_1, x_2, \ldots x_N\}$, as that processed at the encoder, where $\{x_i = k\} \leftarrow \{s_k\}$. We do this by recursively updating the upper and lower limits of the half-open interval. We initializing these to

$$l\,(0) = P_x(0) = 0,$$
$$u\,(0) = P_x(M-1) = 1.$$

We then compute the sequence of symbol indices x_i using

$$x_1 = \{x : \; P_x(x_1 - 1) \le \hat{v} < P_x(x_1)\}. \tag{7.9}$$

So in general we need to find the next symbol index such that

$$x_i = \{x : l(i) \le \hat{v} < u(i)\}, \; i = 1, 2 \ldots N, \tag{7.10}$$

where, for each value of i, we need to update the lower and upper limits:

$$\begin{aligned} u(i) &= l(i-1) + (u(i-1) - l(i-1))P_x(x_i), \\ l(i) &= l(i-1) + (u(i-1) - l(i-1))P_x(x_i - 1). \end{aligned} \tag{7.11}$$

Algorithm 7.8 Arithmetic decoding.

1. Initialize upper and lower limits: $l(0) = 0$; $u(0) = 1$;
2. INPUT arithmetic code tag, $\hat{v}$;
3. $i = 0$;
4. REPEAT
   5. $i = i + 1$;
   6. Compute symbol index i: $x_i = \{x : l(i) \leq \hat{v} < u(i)\}$ where
   upper interval limit: $u\,(i) = l(i-1) + (u(i-1) - l(i-1))P_x(x_i)$ and
   lower interval limit: $l\,(i) = l(i-1) + (u(i-1) - l(i-1))P_x(x_i - 1)$;
   7. OUTPUT symbol s_{x_i};
8. UNTIL $i = N$;
9. END.

This is formalized in Algorithm 7.8.

Example 7.11. Arithmetic decoding – numerical example

Consider decoding the arithmetic code 0.538, which represents a three-symbol sequence from the alphabet $S = \{s_1, s_2, s_3, s_4\}$, where $P\,(s_1) = 0.6$, $P\,(s_2) = 0.2$, $P\,(s_3) = 0.1$, $P\,(s_4) = 0.1$. What is the symbol sequence represented by this arithmetic code?

Solution. The iterations of the arithmetic decoding process for this codeword are illustrated below.

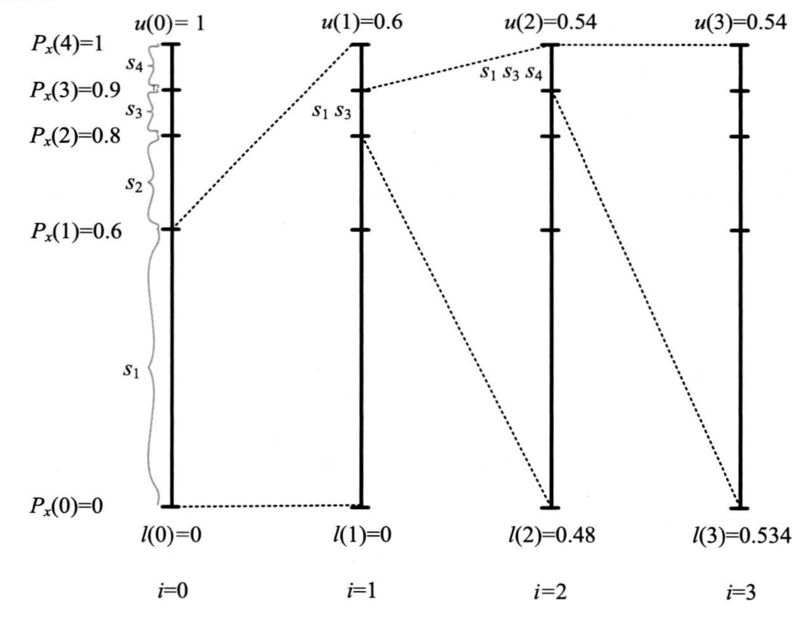

We can see that at stage 1, for decoding symbol 1, the tag lies in the range [0, 0.6), so the symbol output is s_1. Computing the modified cumulative probability values for stage 2 we see that the tag falls in the range [0.48, 0.54), corresponding to the symbol sequence $\{s_1, s_3\}$.

Repeating this for iteration 3 results in the tag falling in the interval [0.534, 0.54), which represents the sequence $\{s_1, s_3, s_4\}$.

7.6.3 Advantages and properties of arithmetic coding

Benefits of arithmetic coding

Arithmetic coding has a number of important properties that make it attractive as a preferred method for entropy encoding. These include:

1. Arithmetic encoding is beneficial with highly skewed probability distributions. Huffman coding can be efficient with large symbol alphabets, where the highest probability for any individual symbol p_{max} is small and the distribution is relatively uniform. However, for smaller alphabets and where the probability distribution is biased, Huffman coding can be inefficient. In arithmetic coding, codewords represent strings of symbols, hence offering the potential for fractional word lengths and a corresponding reduction in overall bit rate.
2. Arithmetic coding separates the modeling process from the coding process. Each individual symbol can have a separate probability (usually referred to as conditioning probability) which may be either predetermined or calculated adaptively according to previous data.
3. Symbol statistics can be simply adapted to changing conditions without major recomputation of coding tables.
4. For the case of binary arithmetic coding, only one probability value is required. It is thus effective to generate numerous probability models with any particular one being invoked according to local context.

Length of arithmetic codes

If a particular sequence from an alphabet with N symbols $S = \{s_i\}$ is encoded using a binary prefix code, then the length of an arithmetic codeword or tag will depend primarily on the distribution of probabilities in the alphabet, on the number of symbols in the alphabet, and on the particular sequence encoded. We saw in Section 7.3.5 that we can compute bounds on the average word length for Huffman codes. We can take a similar approach for arithmetic codes, again based on the Kraft inequality, which gives

$$\left\lceil \log_2 \frac{1}{\prod P(k)} \right\rceil \leq l_\nu \leq \left\lceil \log_2 \frac{1}{\prod P(k)} \right\rceil + 1,$$

where, in this case, the values of $P(k)$ included in the product term are only those associated with the sequence encoded. Hence although the actual length may be lower for specific cases, in order to guarantee an appropriate word length for a unique arithmetic code, we should allow

$$l_\nu = \left\lceil \log_2 \frac{1}{\prod P(k)} \right\rceil + 1.$$

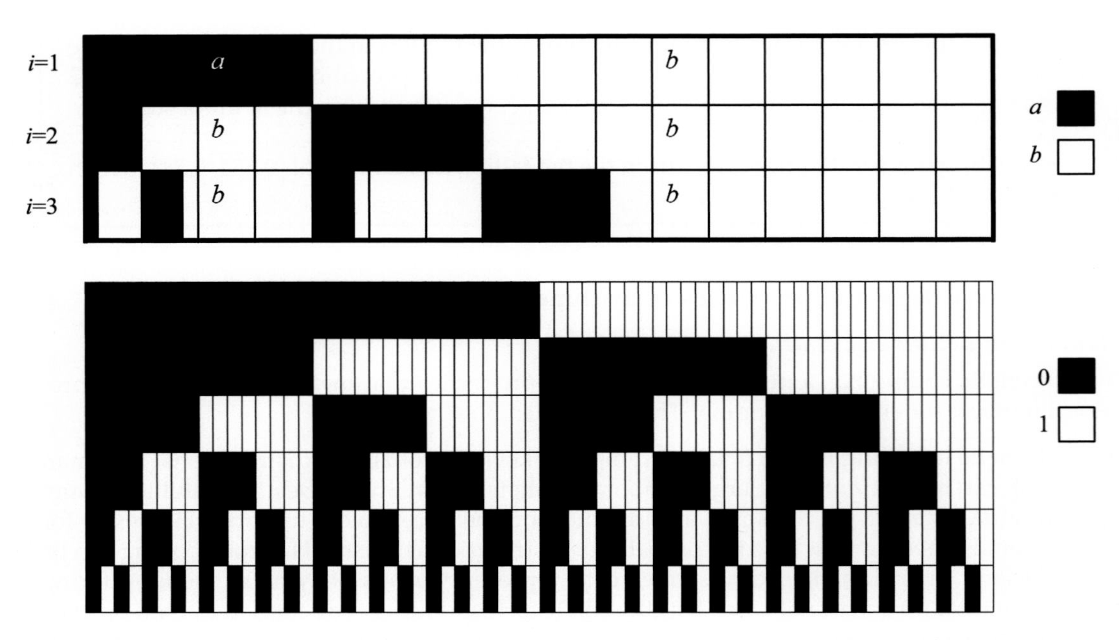

FIGURE 7.12 A graphical illustration of the arithmetic encoding and decoding processes. Top: Probability intervals for three iterations. Bottom: Codeword intervals. Shown are $P(a) = 0.25$ and $P(b) = 0.75$ with example sequences $\{a, b, b\}$ and $\{b, b, b\}$. Adapted from [20].

7.6.4 Binary implementations of arithmetic coding

In practical arithmetic coding systems, binary implementations are essential. This poses some constraints on the operation in terms of finite word length effects and uniqueness, but also presents opportunities in terms of efficient implementations. To help understanding this process, consider the illustration in Fig. 7.12. This shows three iterations of an arithmetic encoder or decoder for the case of a system with two symbols $\{a, b\}$, where $P(a) = 0.25$ and $P(b) = 0.75$. The top diagram shows probability intervals similar to those we have seen before but visualized slightly differently. The bottom diagram shows the corresponding binary codewords. By scanning vertically from a given sequence, we can easily identify the binary codewords that uniquely define that sequence and select the codeword with the minimum length. For example, consider the sequence $\{a, b, b\}$; scanning down to the bottom diagram, we can see that the shortest codeword that uniquely identifies this sequence is 001. Similarly, for $\{b, b, b\}$ it is 11.

Binary encoding implies the restriction (or approximation) of symbol probabilities to integer powers of two, but can provide advantages in sequential bit-wise processing of tag values, both in encoding and decoding stages. This is demonstrated in the following example.

Example 7.12. Binary implementation of arithmetic encoding

Consider a source with four symbols with probabilities $P(s_0) = 0.5$, $P(s_1) = 0.25$, $P(s_2) = 0.125$, $P(s_3) = 0.125$. Perform arithmetic encoding of the sequence $S = \{s_1, s_0, s_2\}$.

Solution. The encoding proceeds as follows.

Stage 1:

The first symbol, s_1, defines the interval $I_1 = [0.5, 0.75)_{10} = [0.10, 0.11)_2$. Examining the upper and lower limits of this interval we can see that the first binary digit after the binary point (1) is invariant and hence can be transmitted or appended to the output file.

Stage 2:

The second symbol, s_0, defines the interval $I_2 = [0.5, 0.625)_{10} = [0.100, 0.101)_2$. Examining the upper and lower limits of this interval we can see that the first two binary digits after the binary point (10) are invariant and hence the 0 can be transmitted or appended to the output file.

Stage 3:

The third symbol, s_2, defines the interval and arithmetic code $I_3 = [0.59375, 0.609375)_{10} = [0.10011, 0.100111)_2$. Examining the upper and lower limits of this interval we can see that the first five binary digits after the binary point (10011) are invariant and hence can be transmitted or appended to the output file. We can transmit any value in this range as our arithmetic code, $\hat{v} \epsilon [0.10011, 0.100111)_2$. In this case we can simply send an extra zero to differentiate the lower limit from the upper limit, i.e., $\hat{v} = 0.100110$.

A simplified diagram of this arithmetic encoding process is shown in the figure below.

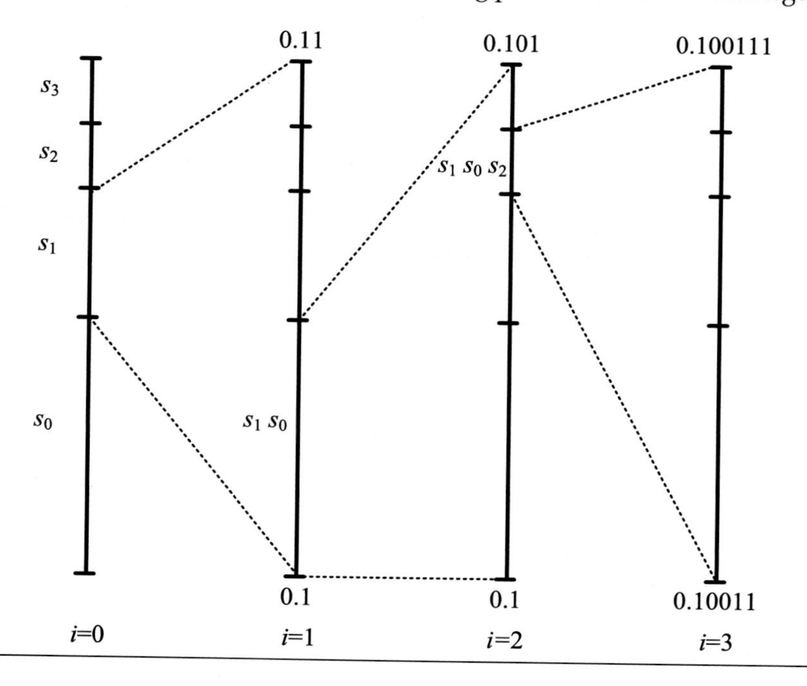

Example 7.13. Binary implementation of arithmetic decoding
Decode the transmitted sequence from Example 7.12.

Solution. The decoding process is summarized in the following table. As each bit is loaded, it can be processed to further differentiate symbols. When the desired number of symbols or an EOB symbol is received, the decoding terminates. Consider the first row of the table: the first 1 does not uniquely define a symbol – it could be s_2 or s_3 in this case. However, when the second bit is processed, it uniquely defines s_1. This is repeated until the complete sequence is decoded.

Rx bit	Interval	Symbol
1	$[0.5, 1)_{10} = [0.1, 1.0)_2$	–
0	$[0.5, 0.75)_{10} = [0.10, 0.11)_2$	s_1
0	$[0.5, 0.625)_{10} = [0.100, 0.101)_2$	s_0
1	$[0.5625, 0.625)_{10} = [0.1001, 0.1010)_2$	–
1	$[0.59375, 0.625)_{10} = [0.10011, 0.10100)_2$	–
0	$[0.59375, 0.609375)_{10} = [0.100110, 0.100111)_2$	s_2

7.6.5 Tag generation with scaling

In cases where a transmitted sequence of symbols is large, the arithmetic encoding process will have to iterate a large number of times and at each iteration the size of the intervals reduces (on average by a factor of 2). To cope with this in practical situations, the word length must be sufficient to ensure that the intervals and tag can be represented with sufficient numerical accuracy. In practice very long word lengths are mitigated by a process of scaling the interval each time a bit of the tag is transmitted. Consider stage 1 in Example 7.12; here $l(1) = 0.5$ and $u(1) = 0.75$. Once the first binary 1 is transmitted, this means that the codeword is forever constrained to the upper half of the number line. So, without any loss of generality, we can scale the interval by a factor of 2 as follows:

$$l'(1) = 2(l(1) - 0.5) = 0,$$
$$u'(1) = 2(u(1) - 0.5) = 0.5.$$

The equations for the next stage are then generated as follows:

$$\begin{aligned} u(i) &= l'(i-1) + (u'(i-1) - l'(i-1))P_x(x_i), \\ l(i) &= l'(i-1) + (u'(i-1) - l'(i-1))P_x(x_i - 1). \end{aligned} \tag{7.12}$$

This process of scaling is repeated every time a new symbol restricts the tag to be completely contained in the upper or lower half-interval. In cases where the tag interval straddles the center of the number line we can perform a similar process, but this time scaling when the interval falls within [0.25,0.75). So we have three types of scaling or renormalization, which can be used after interval subdivision [4,7]:

$$
\begin{aligned}
&E_1 : [0, 0.5) \to [0, 1), \quad E_1(x) = 2x, \quad \text{send } 0 \\
&E_2 : [0.5, 1) \to [0, 1), \quad E_2(x) = 2(x - 0.5), \quad \text{send } 1 \\
&E_3 : [0.25, 0.75) \to [0, 1), \quad E_3(x) = 2(x - 0.25), \quad \text{sc} = \text{sc} + 1.
\end{aligned}
\tag{7.13}
$$

Clearly we must record the scalings used when encoding a sequence as these reflect the final tag produced for that sequence. The case of E3 scaling requires further clarification. We have defined a counter, sc, which records the number of consecutive E3 scalings. These do not directly relate in the transmission of bits but must be compensated for after the next E1 or E2 scaling operation. If the next operation is E1, then a 0 is transmitted followed by sc 1s. If on the other hand it is an E2 operation, then the 1 is appended with sc 0s.

It is not difficult to see how this process can be mimicked at the decoder as the tag bits are processed.

Example 7.14. Binary implementation of arithmetic encoding with interval scaling

The symbols from an alphabet $A = \{a_0, a_1, a_2, a_3\}$ have associated probabilities of occurrence as given in the table below.

Symbol	Probability
a_0	0.375
a_1	0.25
a_2	0.25
a_3	0.125

Compute the arithmetic codeword which represents the sequence $\{a_3, a_1, a_0\}$. Use all three types of interval scaling.

Outline solution

Using interval scaling.

- Giving a codeword of 1110110.

As can be seen, the solution proceeds with three E2 scalings (generating three 1s), followed by two E3 scalings and one E1 scaling (generating a 0). Following the final E1 scaling, we must compensate for the two E3 scalings by appending two 1s after the 0.

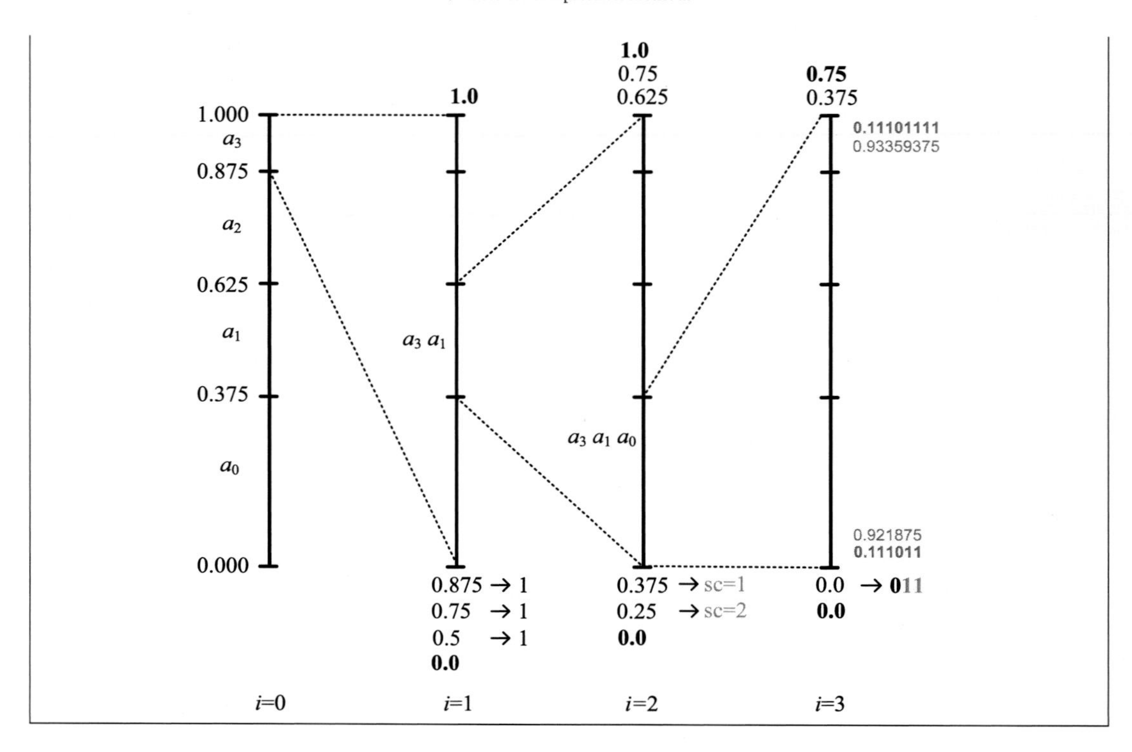

7.6.6 Context-adaptive arithmetic coding

Arithmetic coding has been supported in a number of standards such as JPEG, JPEG2000, H.263, H.264/AVC, HEVC, VVC, VP9, and AV1. However, not until its inclusion in H.264/AVC did it become commonly used in practice. This uptake was due to the fact that CABAC [17] provided significant performance gains over previous methods. CABAC is now the standard entropy encoding method in HEVC and VVC.

By taking account of prevailing context and adapting conditioning probabilities accordingly, significant improvements can be obtained with entropy coders. We saw how to do this with Huffman coding earlier in this chapter and also looked at CAVLC. We will examine it here for the case of arithmetic coding, in particular using CABAC. A good review of adaptive coding methods is provided by Tian et al. [12].

Several other realizations of adaptive arithmetic encoding have been reported in the literature. These include the Q Coder [21], which exploits the idea of renormalization after subdivision (as discussed above) to avoid numerical precision issues (an extension of this, known as the QM Coder, is used in JPEG), and the MQ coder [22]. The latter is used for encoding bit planes of quantized wavelet coefficient codeblocks in the JPEG2000 embedded block coding with optimized truncation coder.

Context-based arithmetic coding

We saw earlier that one of the primary advantages of arithmetic coding is its ability to separate modeling from coding. In context-based arithmetic coding the modeler assigns a probability distribution to the symbols based on the context created by previously processed symbols. For each symbol received, the appropriate probability model is invoked. The context can either be fixed, based on prior knowledge, or modified in real-time, based on the frequency of symbols actually processed. In the former case, the decoder must be aware of the various models in use at the encoder, whereas in the latter case, assuming no channel errors, adaptation at encoder and decoder can be synchronized.

Consider an alphabet with three symbols, s_0, s_1, and s_2, with probabilities $P(s_0)$, $P(s_1)$, and $P(s_2)$. For the initial symbol in a sequence, we would not normally have any context; hence we assume a fixed-probability model. Assuming a first order conditioning model, we would then invoke three additional conditional-probability models for subsequent symbols according to context. This is illustrated in Table 7.7.

Example 7.15. Context-based arithmetic encoding

Consider a source with three symbols with the following precomputed conditional probabilities. Perform arithmetic encoding of the sequence $S = \{s_0, s_1, s_2\}$.

Initial	Context 1	Context 2	Context 3
$P(s_0) = 0.5$	$P(s_0 \mid s_0) = 0.125$	$P(s_0 \mid s_1) = 0.25$	$P(s_0 \mid s_2) = 0.625$
$P(s_1) = 0.375$	$P(s_1 \mid s_0) = 0.75$	$P(s_1 \mid s_1) = 0.25$	$P(s_1 \mid s_2) = 0.125$
$P(s_2) = 0.125$	$P(s_2 \mid s_0) = 0.125$	$P(s_2 \mid s_1) = 0.5$	$P(s_2 \mid s_2) = 0.25$

Solution. The encoding proceeds as follows.

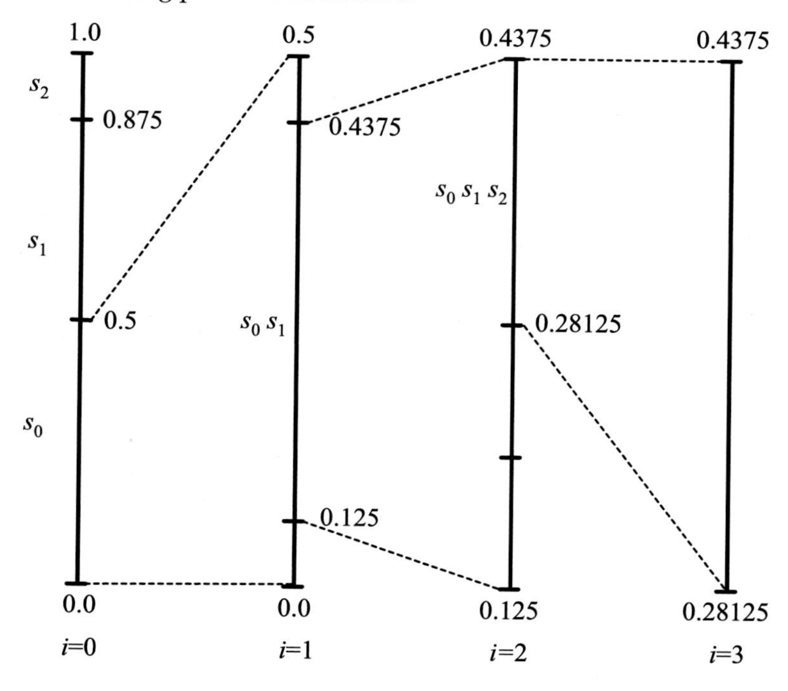

The third symbol, s_2, defines the interval and arithmetic code $I_3 = [0.28125, 0.4375)_{10} = [0.01001, 0.0111)_2$. We can transmit any value in this range as our arithmetic code, $\hat{v}\epsilon[0.01001, 0.0111)_2$; for example, 0110. As an exercise, compare this result using the initial probabilities for each stage.

TABLE 7.7 Conditional probabilities for three different contexts.

Initial	Context 1	Context 2	Context 3
$P(s_0)$	$P(s_0 \mid s_0)$	$P(s_0 \mid s_1)$	$P(s_0 \mid s_2)$
$P(s_1)$	$P(s_1 \mid s_0)$	$P(s_1 \mid s_1)$	$P(s_1 \mid s_2)$
$P(s_2)$	$P(s_2 \mid s_0)$	$P(s_2 \mid s_1)$	$P(s_2 \mid s_2)$

7.6.7 Arithmetic coding with binary symbols

In most contemporary implementations of arithmetic coders, the alphabet is itself binary. This provides benefits, particularly if it exhibits a skewed probability distribution. Because the alphabet for a binary arithmetic coder comprises only two symbols, the probability model is defined by only one probability value. The simplicity of this approach thus lends itself to generating multiple alternative encoders, dependent on local context. For example, take the simple case of encoding a 2D image of black text on a white background where the probability of P(white)= 0.9375 and P(black)=0.0625. Although the entropy of the source is only 0.34 bits/symbol, if we encoded using conventional binary encoding, this would result in 1 bit/symbol.

Encoding only binary decisions (1 or 0) is well suited to encoding many of the SEs associated with signaling modes in modern standards. However, many important symbols are nonbinary (e.g., transform coefficients or motion vectors), and these must be binarized prior to arithmetic coding. This conversion to a variable-length binary string uses an encoding which skews probabilities, such as unary or Exp-Golomb (see Section 7.5). It is thus important to note that binarization not only supports multiplication-free arithmetic, but also enables subsymbol context adaptation, i.e., both at the SE level and within the bin strings.

Context-based adaptive binary arithmetic coding (CABAC)

CABAC has been adopted for the Main and High profiles of H.264/AVC, HEVC, and VVC (although not AV1). It enables the encoder to dynamically adapt to prevailing symbol statistics by exploiting conditioning probabilities, based on the statistics of previously encoded SEs and/or binary symbol strings, and switching between a number of probability models. It offers a fast and accurate means of estimating conditioning probabilities over a short interval and excellent computational efficiency.

CABAC comprises three main stages: binarization, context modeling, and binary arithmetic coding. These are illustrated in Fig. 7.13. The steps are described below. The reader is referred to references [12,17] for a more detailed operational explanation. To summarize:

- *Binarization*: Maps nonbinary-valued SEs to a bin string. This reduces the symbol alphabet by producing a set of intermediate codewords (bin string) for each nonbinary SE. It also

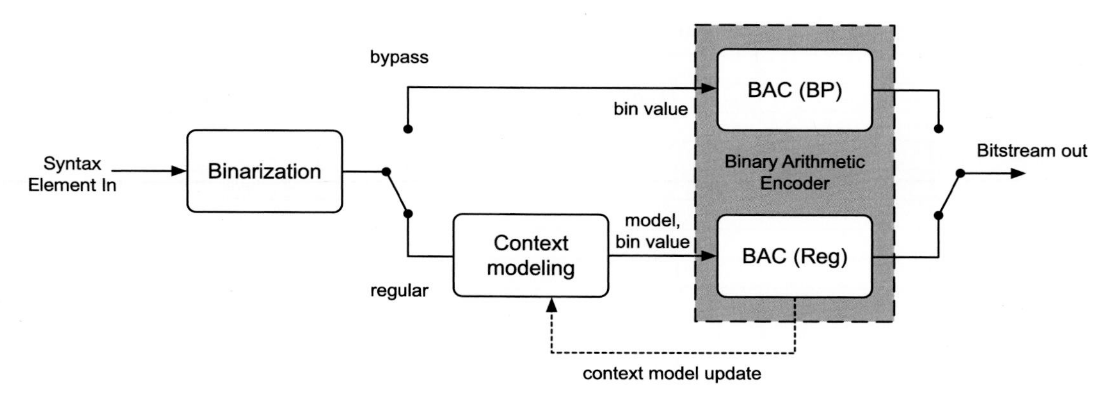

FIGURE 7.13 Context-based adaptive binary arithmetic coding. Adapted from [17].

enables context modeling at a subsymbol level, where conditional probabilities can be used for more common bins and simpler schemes for less frequent bins.

- *Context modeling*: The modeler assigns a probability distribution to the symbols being processed and this is used to initialize the BAC for the SE being processed prior to probability update.
- *Binary arithmetic coding*: CABAC employs a multiplication-free approach to arithmetic encoding and uses recursive interval subdivision and renormalization similar to that described earlier in this chapter. Decoder models are initialized and updates synchronized with the encoder to ensure consistency.

CABAC is used to encode coding parameters (e.g., block type, spatial and temporal prediction mode, and slice and block control information) as well as prediction residuals. Further details can be found in standards documents or overviews [17,24].

Binarization in CABAC

Binarization is implemented using a number of schemes, including: *unary* and *truncated unary* (TU), *fixed-length*, *kth order Exp-Golomb*, and *combinations of these*. To enable encoding of symbols where no prior context can be assumed, a *bypass* operation is performed that assumes $P(0)=P(1)=0.5$ and does not support probability updates. This may at first appear fruitless in terms of coding efficiency compared with straight binary encoding, but it does offer the significant benefit of format and decoding consistency for all SEs. Actually, in HEVC a significant number of bin values use bypass coding because of its enhanced binarization schemes which are already close to optimal.

An example of binarization for transform coefficient residuals in H.264/AVC is given in Table 7.8. This uses of the UEGk scheme, which is often employed for nonbinary SEs, and combines TU and Exp-Golomb (EG). For transform coefficients in H.264 $k=0$, while for motion vector difference (MVD) values $k=3$. The TU code, although simple, is only beneficial for small values. For larger values, the EGk code provides a better fit to the probability distribution. Hence UEG0 uses a TU code for level values up to 14 and then truncates the code without the trailing 0 and adds an EG0 suffix for larger values.

TABLE 7.8 Binarization scheme for transform residual levels in CABAC.

Abs Coeff	Bin string																		
	TU prefix														EG0 suffix				
1	0																		
2	1	0																	
3	1	1	0																
4	1	1	1	0															
...	... ...														... ...				
13	1	1	1	1	1	1	1	1	1	1	1	1	0						
14	1	1	1	1	1	1	1	1	1	1	1	1	1	0					
15	1	1	1	1	1	1	1	1	1	1	1	1	1	1	0	0			
16	1	1	1	1	1	1	1	1	1	1	1	1	1	1	0	1			
17	1	1	1	1	1	1	1	1	1	1	1	1	1	1	1	0	0	0	
18	1	1	1	1	1	1	1	1	1	1	1	1	1	1	1	0	0	1	
...	... ...														... ...				
bin	0	1	2	3	4	5	6	7	8	9	10	11	12	13	14	15	16	17	...

Context models in CABAC

The context models used depend on the SE being coded and are adapted based on a small number of local symbols (to provide a good trade-off between modeling complexity and performance). The context is updated after each bin is processed.

Context models for nonresidual data: For nonresidual data, such as coded block patterns, coding modes, and various flags, the syntax model is normally dependent on the characteristics of neighboring previously encoded blocks – typically above and to the left of the current block. As these selections are standard-dependent and contain large numbers of model choices, we will not consider them further here. For further details please refer to standards documents or overviews [17].

Context models for residual data: In most standards, the context model is selected according to the block type as different block types typically exhibit different distributions. However, for most sequences and coding conditions, similarities between block types can be exploited to reduce the number of context models used.

Using H.264/AVC as an example, there are a total of 399 different contexts, split into ranges according to the type of SE. Each probability model is identified by a context index γ and the context model used is completely determined by its σ_γ together with a value $\overline{\omega}$ which indicates whether the *most probable symbol* (MPS) is a 0 or a 1. For mode or type signaling ($\gamma = 0 \cdots 72$),

$$\gamma = \Gamma_S + \chi_S, \tag{7.14}$$

where Γ_S is the context index offset, i.e., the bottom of the context range for the given SE, and χ_S is the context index increment. For the case of residual coding contexts, the context index is specified by

$$\gamma = \Gamma_S + \Delta_S(\text{cts\_cat}) + \chi_S, \tag{7.15}$$

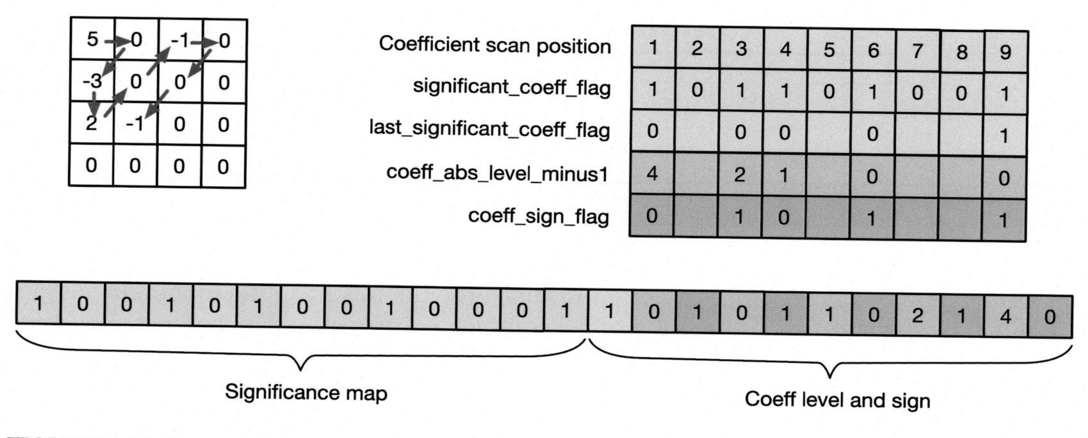

FIGURE 7.14 Formation of significance map and coefficient level and sign for a 4×4 H.264 block.

where cts_cat is an additional offset in the range $0 \cdots 4$ associated with the particular block type being encoded (luma or chroma, inter or intra, and block size). This process is discussed in more detail below for the case of transform coefficient residual coding.

Probability updates in CABAC

In H.264/AVC, 64 representative probability values $P_\sigma \in [0.01875, 0.5]$ are used to represent the least probable symbol (LPS). These are defined as follows:

$$\begin{aligned} P_\sigma &= \alpha \cdot P_{\sigma-1}, & \sigma &= 1, \cdots, 63, \\ \alpha &= \left(\frac{0.01875}{0.5}\right)^{1/63} \simeq 0.95, & \text{and } P_0 &= 0.5. \end{aligned} \tag{7.16}$$

There is no requirement here to explicitly represent these probability values as they are completely defined by σ.

Transform residual coding with CABAC

Using the example of a 4×4 H.264/AVC block shown in Fig. 7.14, the quantized coefficient matrix is scanned in a zig-zag pattern to produce a significance map which determines the locations of all significant quantized coefficients. This is combined with a binary flag that indicates the last significant coefficient in the scan (last_significant_coeff_flag). The levels of these significant coefficients are then captured using a reverse order scan. Levels are encoded using two coding symbols: (i) the coeff_sign_flag (1-bit: 0=positive and 1 =negative) and (ii) coeff_abs_level_minus1, which is binarized using the UEG0 scheme as shown in Table 7.8. An example of this process is shown in Fig. 7.14.

Significance map: For encoding the significance map, the choice of the models and the corresponding context index increments χ_{SIG} and χ_{LAST} in H.264 depend on the scanning position, i.e., for a coefficient coeff[i], which is scanned at the ith position, the context index increments are determined as follows:

$$\chi_{\text{SIG}}(\text{coeff}[i]) = \chi_{\text{LAST}}(\text{coeff}[i]) = i. \tag{7.17}$$

TABLE 7.9 Context updates for the example in Fig. 7.14.

i	1	2	3	4	5	6	7	8	9
AbsCoeff[i]	5	0	3	2	0	1	0	0	1
NumT1[i]	2		2	2		1			0
NumLg1[i]	2		1	0		0			0
$\chi_{\text{AbsCoeff}}(\text{coeff}[i, \text{binIdx} = 0])$	4		4	2		1			0
$\chi_{\text{AbsCoeff}}(\text{coeff}[i, \text{binIdx} > 0])$	7		6	5		5			5

Level information: Levels are scanned in reverse order as this supports more reliable estimation of the statistics, since at the end of the scan there is increased likelihood of successive levels with absolute value equal to 1. Context levels for nonzero significant coefficients are selected based on the number of previously transmitted nonzero levels within the reverse scan.

For encoding coeff_abs_level_minus1, two sets of model are employed: one for the first bin and another one for the other 13 bins as shown in Table 7.9. Any remaining level values and all sign flags are encoded in bypass mode. The selection of the probability model is determined by the accumulated counts of the number of coefficients of absolute value 1 of the number greater than 1 for the scan up to the current coefficient/bin position.

Thus, for encoding the first bin in Table 7.8, the corresponding context index increment at position i is given as

$$\chi_{\text{AbsCoeff}}(\text{coeff}[i, \text{binIdx} = 0]) = \begin{cases} 4, & \text{if NumLgt}[i] > 0, \\ \min(3, \text{NumT1}[i]), & \text{otherwise.} \end{cases} \tag{7.18}$$

For encoding bins 1–13 in Table 7.8 the following expression is used:

$$\chi_{\text{AbsCoeff}}(\text{coeff}[i, \text{binIdx} > 0]) = 5 + \min(4, \text{NumLgt1}[i]), \; 1 \leq \text{binIdx} \leq 13, \tag{7.19}$$

where NumT1[i] is the accumulated number of previously encoded trailing 1s and NumLgt1[i] denotes the accumulated number of previously encoded levels with absolute value greater than 1. Both counters are reset to 0 at the beginning of the block scan. Table 7.9 shows the context index increments for the simple example of Fig. 7.14.

Motion vector residual coding with CABAC

Let us now consider MVD_x (motion vector difference in the x direction) coding H.264/AVC [13]. The bins for MVD_x are shown in Table 7.10 – these use a UEG3 code.[3]

In total, seven context models are used for $|\text{MVD}_x|$ ($\gamma = 40 \cdots 46$). For the first bin, the model is selected based on the previously encoded motion vector difference value, where $e_k = |\text{MVD}_{xA}| + |\text{MVD}_{xB}|$ and A and B represent the blocks to the left of and directly above the current block. This process allows selection of the likely best initial context model according to the motion vector difference magnitudes of previously encoded blocks. The context index offset for the following bins are also shown in Table 7.10.

[3] Larger values of MVD_x are binarized using Exp-Golomb codewords.

TABLE 7.10 Context index offsets for $|MVD_x|$.

e_k	Context index offset χ (bin= 0)
$0 \le e_k < 3$	0
$3 \le e_k < 33$	1
$33 \le e_k$	2
Bin	**Context index offset χ (bin=>0)**
1	3
2	4
3	5
4	6
≥5	6

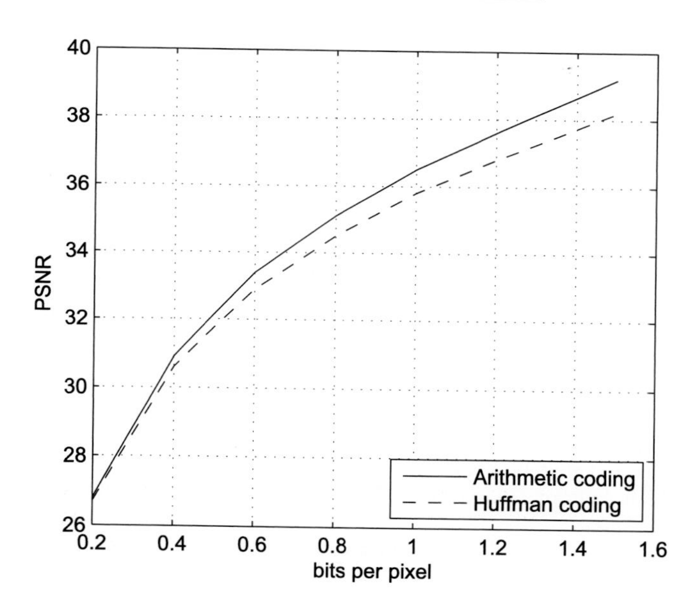

FIGURE 7.15 Huffman vs. Arithmetic coding.

7.7 Performance comparisons

A comparison of the compression performance achieved for basic arithmetic and Huffman coding is shown in Fig. 7.15. These results represent the combination of a simple DCT codec incorporating both types of entropy coder. For the images employed, it shows that arithmetic coding offers advantages (up to 1 dB at lower compression ratios) compared with Huffman coding. As we have seen, further improvements are possible using more sophisticated adaptive and context-based techniques which suit arithmetic coding better than Huffman coding.

For example, CABAC has been reported to offer up to 20% improvement over CAVLC, with typical savings between 9% and 14% in terms of BD-rate [17].

7.8 Summary

This chapter has introduced methods for coding images without loss. We have seen the requirement that useful codes need to have prefix properties so as to avoid the need for explicit synchronization. Huffman codes satisfy this property and are able to code individual symbols from a source at a rate close to its entropy. We then described arithmetic coding which offers performance advantages in that it can represent multiple symbols in a group and hence offer fractional symbol word lengths. We also saw that arithmetic coding is well suited to adaptive encoding – supporting dynamic modification of conditioning probabilities.

References

[1] D.A. Huffman, A method for the construction of minimum-redundancy codes, Proc. IRE 40 (9) (1952) 1098–1101.
[2] N. Abramson, Information Theory and Coding, McGraw-Hill, 1963.
[3] J. Rissanen, Generalized Kraft inequality and arithmetic coding, IBM J. Res. Dev. 20 (1976) 198–203.
[4] A. Said, Introduction to arithmetic coding – theory and practice, in: K. Sayood (Ed.), Lossless Compression Handbook, Academic Press, 2003.
[5] J. Ziv, A. Lempel, A universal algorithm for data compression, IEEE Trans. Inf. Theory 23 (3) (1977) 337–343.
[6] T. Welch, A technique for high-performance data compression, IEEE Comput. (June 1984) 8–19.
[7] K. Sayood, Introduction to Data Compression, 5e, Morgan Kaufmann, 2017.
[8] K. Sayood (Ed.), Lossless Compression Handbook, Academic Press, 2003.
[9] R. Gallagher, Variations on a theme by Huffman, IEEE Trans. Inf. Theory 24 (6) (1978) 668–674.
[10] J. Vitter, Design and analysis of dynamic Huffman codes, J. ACM 34 (4) (1987) 825–845.
[11] ISO/IEC International Standard 10918-1, Information technology – digital and coding of continuous-tone still images – requirements and guidelines, 1992.
[12] X. Tian, T. Le, Y. Lian, Entropy Coders of the H.264/AVD Standard: Algorithms and VLSI Architectures, Springer, 2011.
[13] I. Richardson, The H.264 Advanced Video Coding Standard, 2e, Wiley, 2010.
[14] S. Golomb, Run-length encodings, IEEE Trans. Inf. Theory 12 (3) (1966) 399–401.
[15] D. Taubman, M. Marcelin, JPEG2000 Image Compression Fundamentals, Standards and Practice, Kluwer, 2002.
[16] J. Teuhola, A compression method clustered bit vectors, Inf. Process. Lett. 7 (1978) 308–311.
[17] D. Marpe, H. Schwarz, T. Wiegand, Context-based adaptive binary arithmetic coding in the H.264/AVC video compression standard, IEEE Trans. Circuits Syst. Video Technol. 13 (7) (2003) 620–637.
[18] C. Shannon, A mathematical theory of communication, Bell Syst. Tech. J. 30 (1948) 623–656.
[19] F. Jelinek, Probabilistic Information Theory, McGraw Hill, 1968.
[20] J-R. Ohm, Multimedia Communication Technology, Springer, 2003.
[21] J. Pennebaker, G. Mitchell, G. Langdon, R. Arps, An overview of the basic principles of the Q-coder adaptive binary arithmetic coder, IBM J. Res. Dev. 32 (1988) 717–726.
[22] D. Taubman, E. Ordentlich, M. Weinberger, G. Seroussic, Embedded block coding in JPEG 2000, Signal Process. Image Commun. 17 (1) (2002) 49–72.
[23] I. Witten, R. Neal, J. Cleary, Arithmetic coding for data compression, Commun. ACM 30 (6) (1987) 520–540.
[24] M. Wien, High Efficiency Video Coding, Springer, 2015.

CHAPTER

8

Coding moving pictures: motion prediction

In previous chapters we have seen how to create an appropriate input format and color space in Chapter 4, how to decorrelate and quantize images using transforms such as the discrete cosine transform (DCT) and the discrete wavelet transform, and how to represent the quantized output using various forms of entropy coding. This chapter covers the final component needed to complete our basic video codec – motion prediction. Motion estimation is important as it enables the temporal decorrelation of a signal. So rather than code very similar information repeatedly in each frame, we instead code the residual signal after motion prediction, along with the prediction parameters (the motion vectors).

This chapter firstly introduces the idea of temporal correlation and discusses how it can be exploited to provide additional coding gain. It then compares the merits of various motion models. Block matching, based on a translational motion model, is the approach adopted in all standardized video codecs to date and this is studied in detail with its performance characterized in terms of the effects of block size, search range, and motion vector coding. Alternative approaches to motion estimation, such as those based on phase correlation, are also covered. Although such techniques have rarely been adopted for conventional video compression, they are worthy of mention because of their use in video standards conversion.

One of the primary issues associated with motion estimation, even today, is its complexity – if not optimized, motion estimation can account for 60–70% of the processor loading. Methods of reducing motion estimation complexity have thus received significant attention in recent years and selected low-complexity methods are compared in Section 8.4 in terms of their accuracy–complexity trade-offs. Finally, Section 8.6 describes how we can efficiently code motion information prior to transmission or storage.

8.1 Temporal correlation and exploiting temporal redundancy

8.1.1 Why motion estimation?

Significant spatial redundancies exist in natural images, and we have already seen that these can be exploited by transform, filter-bank, or other coding methods that provide spatial decorrelation. The simplest approach to encoding a sequence of moving images is thus to

Intelligent Image and Video Compression
https://doi.org/10.1016/B978-0-12-820353-8.00017-7

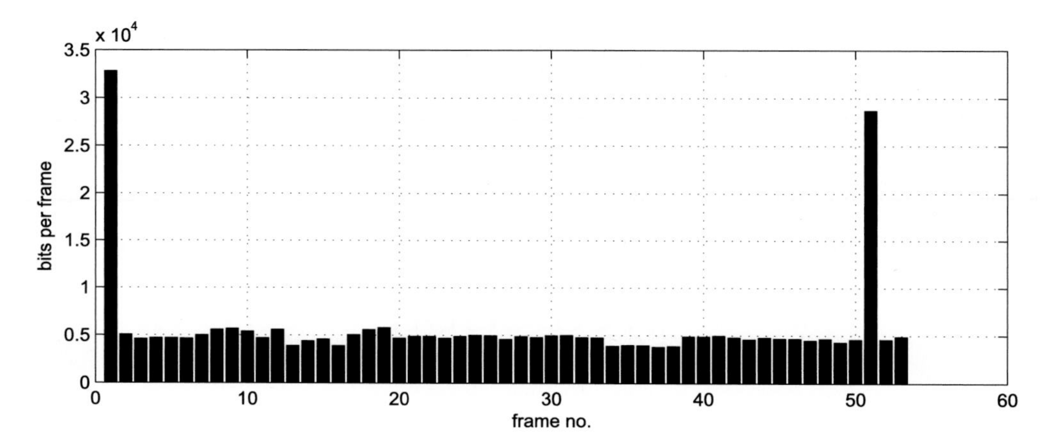

FIGURE 8.1 Variation in the number of bits per frame.

apply an *intraframe* (still image) coding method to each frame (cf. Motion JPEG or Motion JPEG2000). This, as we will see later, does have some benefits, especially in terms of avoiding error propagation caused by temporal prediction. It can also be invoked in cases where a motion model fails to capture the prevailing motion in the scene. However, it generally results in limited compression performance.

We have already seen in Chapter 1, for real-time video transmission over low-bandwidth channels, that there is often insufficient capacity to code each frame in a video sequence independently (25–30 fps is required to avoid flicker). The solution to this problem lies in taking advantage of the correlation that exists between temporally adjacent frames in a video sequence. This *interframe* redundancy can be exploited through motion prediction to further improve coding efficiency. Fig. 8.1 demonstrates the effects of this approach, showing the first 53 frames of the *foreman* CIF sequence, coded with a group of pictures (GOP) length of 50 at approximately 500 kb/s. Frames 1 and 51 are coded as intraframes (I-frames) and the remainder are coded as interframes (also known as unipredicted frames or P-frames).

A number of issues however arise with this approach and these will be discussed further in the following sections:

- Motion estimation yields a residual signal with reduced correlation. Hence transform coding is less efficient on *displaced frame difference* (DFD) signals than on intraframes.
- Motion can be a complex combination of translation, rotation, zoom, and shear. Hence model selection is key in balancing motion prediction accuracy with parameter search complexities.
- Lighting levels can changes from frame to frame – modifying average luminance levels.
- Occluded objects can become uncovered and visible objects can become occluded, confounding motion models and leading to model failure.
- Motion estimation and compensation (MEC) adds complexity to the codec.
- Motion information creates an overhead as it must be transmitted alongside the quantized transform coefficients.

Example 8.1. Comparing intra- and interframe coding

Consider a sequence with 512×512 spatial samples captured at 30 fps with 8 bits per color channel and a GOP length of 100 frames. Compare uncompressed, intra-compressed, and inter-compressed bit rates for this sequence if (for a given quality) the average number of bits per pixel (including motion vector overheads) in intraframe mode is 0.2 bpp per color channel and 0.02 bpp in interframe mode.

Solution. 1. In the uncompressed case we have

$$\text{uncompressed bit rate} = 512 \times 512 \times 30 \times 24 = 189\ \text{Mbps}.$$

2. In the intraframe case we have

$$\text{intra-compressed bit rate} = 512 \times 512 \times 30 \times 0.2 \times 3 = 4.7\ \text{Mbps}.$$

3. In the interframe case, 1 in every 100 frames is an intraframe @ 0.2 bpp per color channel and 99 out of every 100 frames are interframes @ 0.02 bpp per color channel. Thus,

$$\text{inter-compressed bit rate} = 512 \times 512 \times 30 \times \left(\frac{0.2 \times 3 \times 1 + 0.02 \times 3 \times 99}{100}\right) = 514\ \text{kbps}.$$

If these sorts of figures are achievable without compromising quality, then huge coding gains can be achieved by exploiting temporal redundancy. The goal is thus to exploit both temporal and spatial redundancy when coding the video sequence.

8.1.2 Projected motion and apparent motion

While, in the real world, objects move in three spatial dimensions, we are primarily interested here in the projection of their motion onto a 2D image plane as captured by a camera. So, when we use the term "motion estimation" in the context of video compression, this normally refers to the *projected motion* of the object. While it could be argued that tracking an object in 3D might provide a more accurate description of this projected motion, the models and methods associated with doing this are too complex to be of major interest.

In video compression, 2D motion is conventionally estimated by observing the spatio-temporal variations in signal intensity between frames. This approach measures *apparent motion* rather than true projected motion and can lead to ambiguities. Consider for example a disc of uniform luminance rotating about its center. In this case the projected motion is clearly nonzero, whereas the apparent motion is zero. In contrast, an object that is static in the image plane but experiences time-varying illumination changes will have zero projected motion but, because of the intensity changes between frames, will exhibit some apparent motion. The use of apparent motion can be justified since the aim of motion estimation in video compression is not to assess true motion, but instead to minimize the number of bits required to code the prediction residual.

FIGURE 8.2 Temporal correlation between two adjacent video frames.

We will see below that there are many possible motion models that could be adopted for motion estimation. However, the most commonly used is the simple 2D translational model, where the displacement motion vector $\mathbf{d} = [d_x, d_y]^T$. This is not normally computed for a whole frame, but instead for a number of smaller blocks – typically of size 16×16 pixels. The set of all vectors collectively represents the *motion field* for the given frame.[1]

8.1.3 Understanding temporal correlation

We showed in Chapter 3 how natural video data is correlated in time, with the temporal autocorrelation function typically remaining high across a wide range of lag values. This property can be exploited to improve coding efficiency by compensating for the movement of objects in a scene between adjacent frames via MEC. Take, for example, the pair of temporally adjacent frames from a typical 25-fps video sequence (*table*) shown in Fig. 8.2. Although this sequence has relatively high activity – the camera pans from right to left and there are clear movements of the players' hands, bat, and ball – the two frames appear visually very similar. We will consider in the following sections how best to model this motion in order to improve coding gain.

8.1.4 How to form the prediction

Our aim is to estimate the pixel values in the frame currently being encoded (the *current frame*), based on the values in one or more other frames (*reference frames*) in the sequence. These will often be temporally adjacent to the current frame but need not be.

There are two primary ways to form the prediction:

- **Frame differencing**: This method does not employ any motion model (although arguably it could be referred to as a zero order prediction). The reference frame is, without any modification, directly used as the prediction for the current frame. The prediction error is

[1] Formally for apparent motion, the displacement field is referred to as the correspondence field.

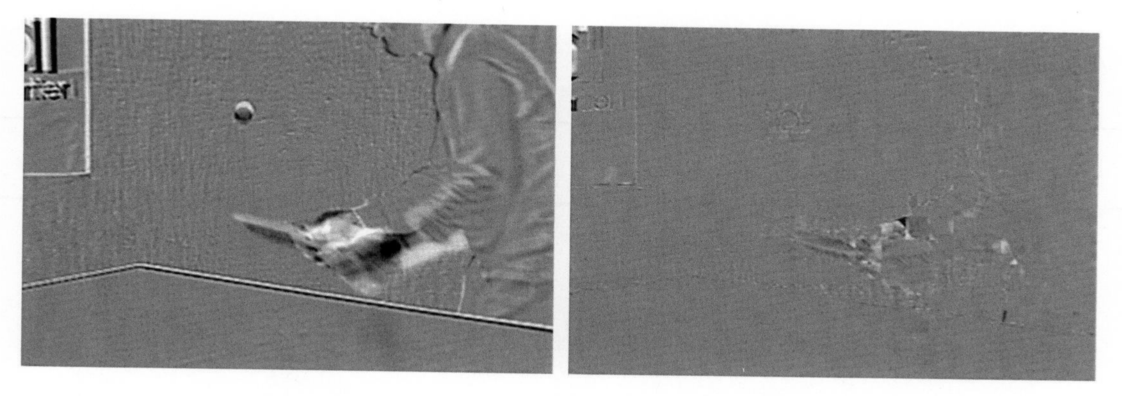

FIGURE 8.3 Frame differencing (left) vs. motion-compensated prediction (right).

thus the difference between the pixel values in the current frame and those colocated in the reference frame and is known as the *frame difference* (FD).
- **Motion-compensated prediction**: In this case a motion model is assumed and motion estimation is used to estimate the motion that occurs between the reference frame and the current frame. Once the motion is estimated, a process known as motion compensation is invoked to use the motion information from motion estimation to modify the contents of the reference frame, according to the motion model, in order to produce a prediction of the current frame. The prediction is called a *motion-compensated prediction* or a *displaced frame* (DF). The prediction error is known as the *displaced frame difference* (DFD) signal.

Fig. 8.3 compares the motion residuals from both of these approaches for those frames from *table* shown in Fig. 8.2. It can be seen that the residual energy is significantly reduced for the case of motion estimation. Fig. 8.4 reinforces this point, showing how the pdf of pixel values is modified for FD and DFD frames, compared to an original frame from the *football* sequence.

Three main classes of prediction are used in video compression:

- **Forward prediction**: The reference frame occurs temporally before the current frame.
- **Backward prediction**: The reference frame occurs temporally after the current frame.
- **Biprediction**: Two (or more) reference frames (forward and/or backward) are employed and the candidate predictions are combined in some way to form the final prediction.

In *forward prediction*, assuming that the motion model used is perfect and based on translation only, the pixel value at location $\mathbf{p} = [x, y]^T$ in the current frame at time k is related to a pixel location in the previous frame at time $k-1$ by[2]

$$s_k[\mathbf{p}] = s_{k-1}[\mathbf{p} + \mathbf{d}], \tag{8.1}$$

where $\mathbf{d}$ is the motion vector corresponding to location $\mathbf{p}$ in the current frame. A similar expression can be generated for *backward prediction*. More sophisticated motion models are considered in the next section.

[2] Without loss of generality we use the adjacent frame here.

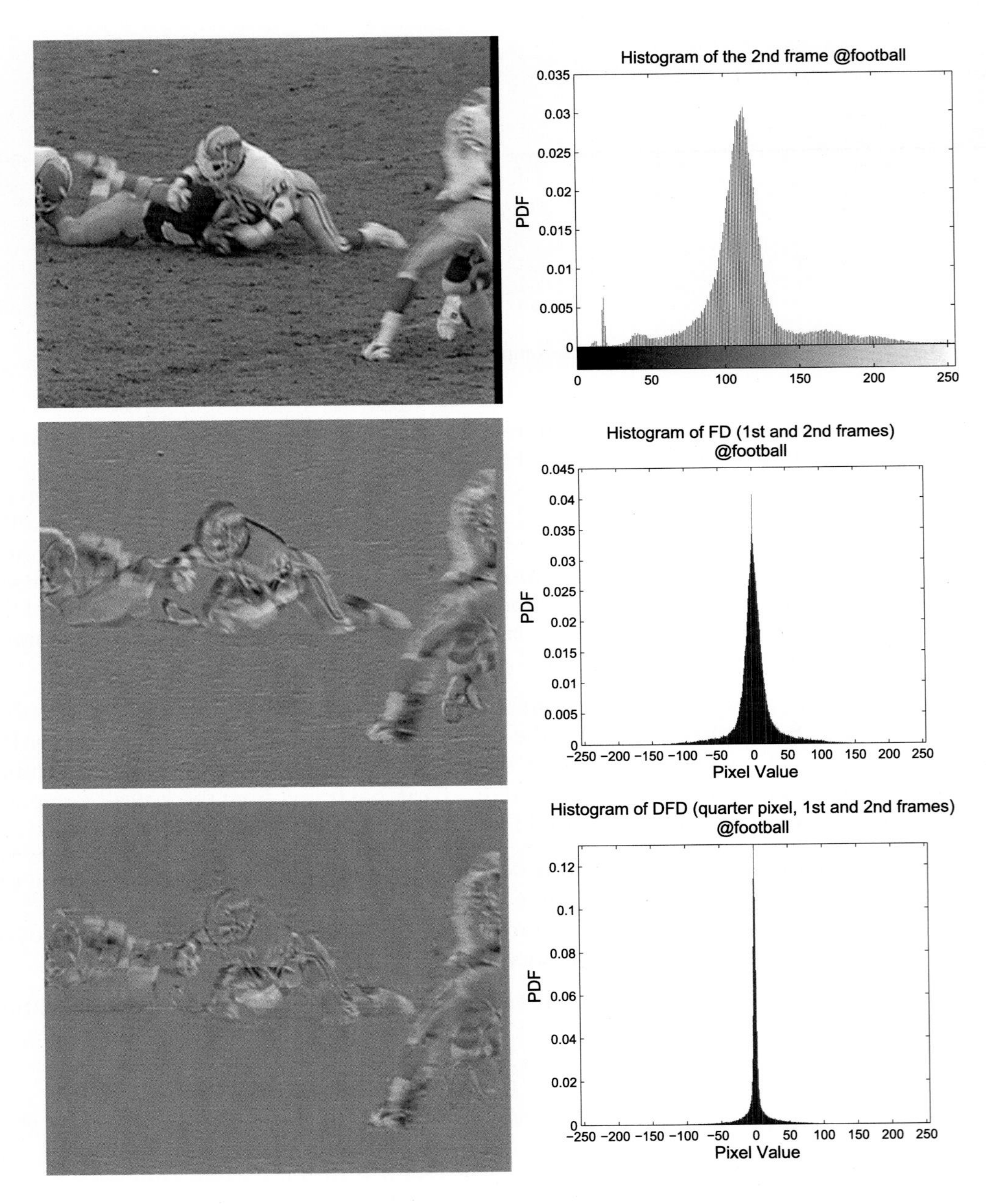

FIGURE 8.4 Probability distributions for an original frame and FD and DFD frames from the *football* sequence.

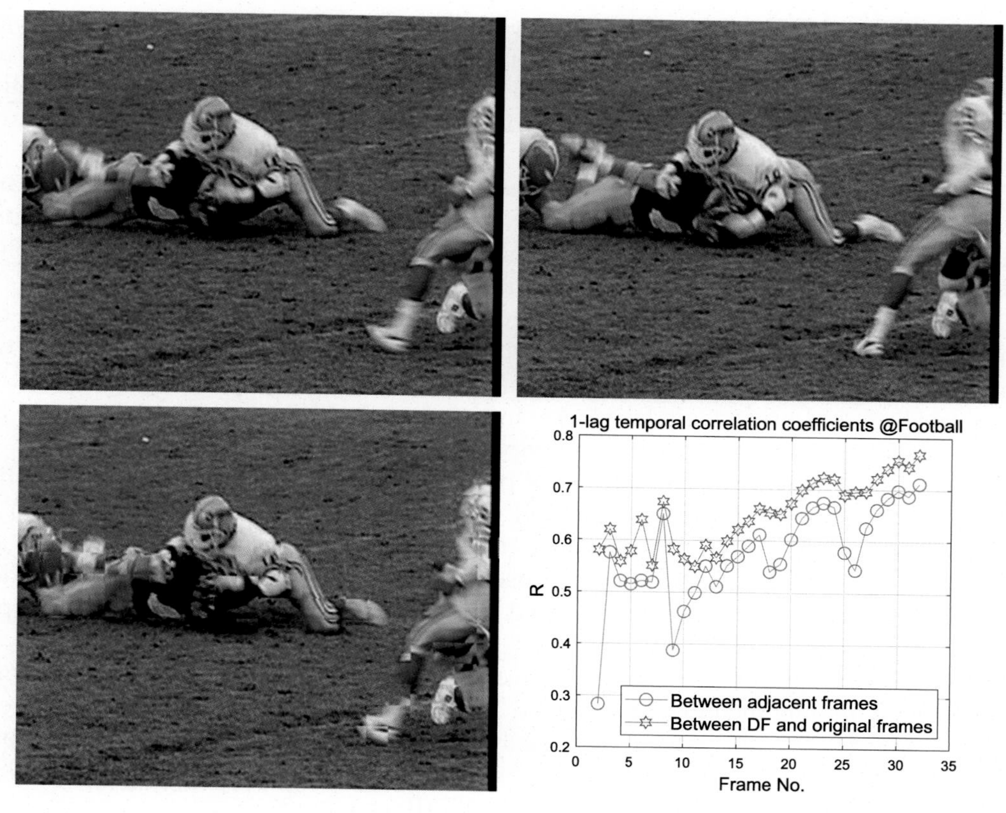

FIGURE 8.5 (Top) Frames 1 and 2 of the *football* sequence. (Bottom left) The motion prediction of frame 2 based on frame 1. (Bottom right) The temporal correlation between adjacent original frames and between each original frame and its motion-compensated prediction.

It should be noted that the motion estimation problem is ill-posed and there is not necessarily a unique solution – the number of independent variables exceeds the number of equations. In fact a solution does not necessarily exist at all, for example in the case of uncovered or occluded pixels. The estimate is also sensitive to the presence of noise.

Let us consider how motion prediction affects temporal correlation. Take, for example, the pair of temporally adjacent frames from a typical 25-fps video sequence (*football*) alongside the motion-compensated prediction of the second frame (using the first frame as a reference) as shown in Fig. 8.5. The correlation between adjacent frames and between each DF and its corresponding original frame is also plotted in this figure. We can see clear improvements in temporal correlation resulting from MEC.

8.1.5 Approaches to motion estimation

The key issues in motion estimation can be summarized as follows:

1. **The motion model**. This can be an accurate, but more complex motion model such as *affine* or *perspective*, or a simple but sufficiently effective model such as *translational*. In practice the translational block matching approach is adopted in most situations. Once the motion model is selected, we must decide how it will be implemented – for example in the spatial or frequency domain.
2. **The matching criterion**. Once the model is selected, we must measure how good the match is for any candidate parameter set. This can be achieved with techniques such as normalized cross-correlation (NCCF), sum of squared differences (SSD), or sum of absolute differences (SAD).
3. **The region of support**. This is the GOP to which the motion model is applied and can range from a whole frame to a single pixel. The region can be arbitrarily shaped in two or three dimensions so as to best reflect the actual motion characteristics of a scene, but is normally based on rectangular blocks. In simple codecs these are typically 16×16 but variable block sizes have become commonplace.
4. **The search range**. This is the window of pixels in one or more reference frames within which the region of support is compared using the matching criterion, in order to estimate the motion and parameters of the model. The search range will depend on the spatio-temporal resolution of the format used and the activity level of the content coded. Normally a rectangular region is employed with a maximum offset between 7 and 63 pixels.
5. **The search strategy**. Once all of the above choices are made, we need to execute a means of efficiently finding the model parameters that best suit the motion. Because of the complexity associated with performing an exhaustive search, fast methods are commonly employed. In the case of translational block matching these include the *N-step search* (NSS), *diamond search* (DS), and *hexagon-based search* (HEXBS) methods.

In all situations, there will exist a trade-off between prediction accuracy and coding overhead and appropriate choices have to be made based on processing power available, video format, and content type. The following sections explore these issues in more detail.

8.2 Motion models and motion estimation

8.2.1 Problem formulation

Apparent motion can be attributed to three main causes:

1. **global motion**, for example camera motion (pan, tilt, zoom, and translation);
2. **local motion**, of objects in the scene;
3. **illumination variations**, due to changes in lighting, shadows, or noise.

The models and methods described in this chapter address the combined motion due to all three of these causes. Other methods exist whereby global motion is extracted initially followed by local refinement (e.g., [1]) but these are not considered further here.

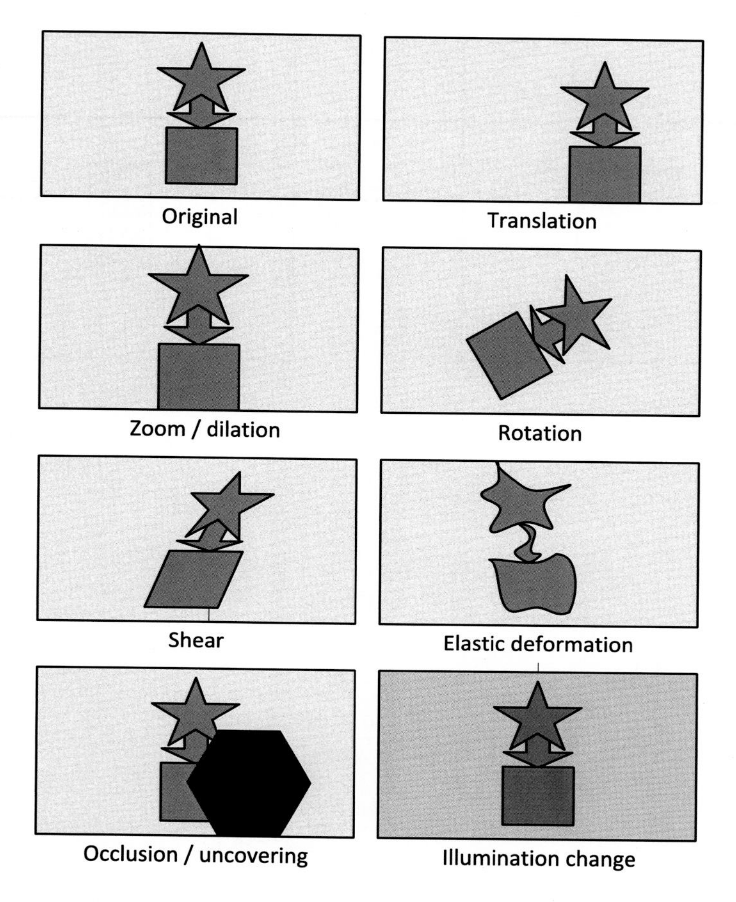

FIGURE 8.6 Motion-based image deformations.

8.2.2 Affine and high order models

The apparent motion in an image or image region can be complex and experience translation, rotation, shear, and dilation as well as suffer from illumination changes between frames and occlusion by other objects (as illustrated in Fig. 8.6). In order to represent such motion characteristics, it is necessary to use an appropriate motion model. Examples of such models are the affine, bilinear, and perspective models as described by Eqs. (8.2), (8.3), and (8.4) respectively.

Affine: We have

$$\begin{aligned} u &= g_x(x, y) = a_1x + a_2y + a_3, \\ v &= g_y(x, y) = a_4x + a_5y + a_6. \end{aligned} \tag{8.2}$$

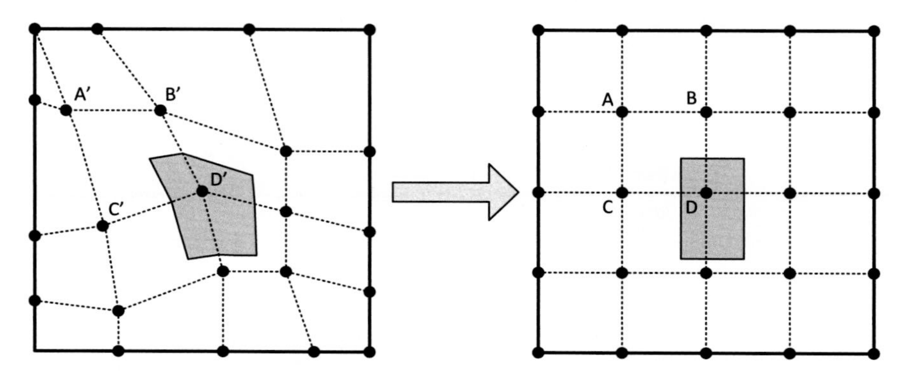

FIGURE 8.7 Mesh-based warping.

Bilinear: We have

$$u = g_x(x, y) = a_1xy + a_2x + a_3y + a_4,$$
$$v = g_y(x, y) = a_5xy + a_6x + a_7y + a_8. \tag{8.3}$$

Perspective: We have

$$u = g_x(x, y) = \frac{a_1x + a_2y + a_3}{a_7x + a_8y + 1},$$
$$v = g_y(x, y) = \frac{a_4x + a_5y + a_6}{a_7x + a_8y + 1}, \tag{8.4}$$

where location (x,y) is mapped to (u,v) using appropriate warping parameters $a_1 \cdots a_8$. An excellent review of image warping transforms is provided by Glasbey in reference [2]. Affine prediction models are now supported in H.266/VVC, standardized in 2020.

Node-based warping

In most cases, the warping parameters are obtained through the use of a 2D deformable mesh formed from an interconnected set of node points, as shown in Fig. 8.7. A method such as *block matching motion estimation* (BMME) is used to estimate the motion of each node through comparisons between a current frame and one or more reference frames and this is in turn used to find the parameters $\{a_i\}$ of the motion model. The model is then used to transform each pixel in the current block region. Meshes can be continuous or discrete, fixed or adaptive, and can adopt a range of approaches to node tracking. A good overview and comparison of these variations is provided in reference [3].

Once the nodes of the mesh have been tracked, the motion parameters can be obtained through solving a set of simultaneous equations, such as those shown in Eq. (8.5) for the bilinear model. Subsequently the transformation can be applied to each pixel in the patch using the appropriate model equation, in this case Eq. (8.3). We have

$$\begin{bmatrix} u_A & u_B & u_C & u_D \\ v_A & v_B & v_C & v_D \end{bmatrix} = \begin{bmatrix} a_1 & a_2 & a_3 & a_4 \\ a_5 & a_6 & a_7 & a_8 \end{bmatrix} \begin{bmatrix} x_A y_A & x_B y_B & x_C y_C & x_D y_D \\ x_A & x_B & x_C & x_D \\ y_A & y_B & y_C & y_D \\ 1 & 1 & 1 & 1 \end{bmatrix}. \tag{8.5}$$

8.2.3 Translation-only models

The translational model is widely used in practice because it is simple to parameterize, it fits well into a regular block-based coding architecture, and importantly, it provides a good fit to the apparent motion in most scenarios. The latter point is worth explaining, since it is obvious that true motion, projected motion, and apparent motion are not usually purely translational.

Recall that our primary purpose is to optimize rate-distortion performance, and this can be achieved with a simple motion model with low coding overhead (i.e., few parameters) and which delivers a low-energy prediction residual. The translational model succeeds in this respect, since at typical video frame rates (24–60 fps) and spatial resolutions, most motion is approximately translational. This can be observed in Figs. 8.2 and 8.3, which show two temporally adjacent frames together with their residual image after block-based translational motion estimation. While small areas exist where the model fails, in most of the frame it works exceptionally well, providing a very low residual signal. Some further examples, where this assumption breaks down, are shown later.

The translational model is a subset of the perspective model, and is given by

$$\begin{aligned} u &= g_x(x, y) = x + a_1 = x + d_x, \\ v &= g_y(x, y) = y + a_2 = y + d_y. \end{aligned} \tag{8.6}$$

It should also be noted that models embracing translation and rotation or translation and dilation are also possible. For example, the latter is given by

$$\begin{aligned} u &= g_x(x, y) = cx + a_1, \\ v &= g_y(x, y) = cy + a_2. \end{aligned} \tag{8.7}$$

Almualla et al. [3] compared warping methods with translational block matching, both based on 16×16 blocks, for low-bit rate coding applications. The results are summarized in Table 8.1. Three algorithms were compared: a block matching algorithm (BMA), a basic integer-pixel block matching algorithm; WBA, a warping-based algorithm based on 16×16 blocks with node warping using an eight-parameter model; and BMA-HO. BMA enhanced with half-pixel accuracy and overlapped block motion estimation. It can be seen that the warping approach does offer gains in prediction quality over the basic BMME algorithm but that this can be offset through enhancements to BMME such as subpixel estimation and the use of overlapped blocks. In terms of complexity (processor execution time), BMA-HO was found to be 50% more complex that BMA, while WBA was found to be 50 times slower than BMA.

TABLE 8.1 Comparison of warping and translational models for three test sequences [3].

PSNR (dB)	Akiyo	Foreman	Table
BMA	39.88	27.81	29.06
WBA	41.45	29.09	29.22
BMA-HO	41.77	29.51	29.87

As mentioned above, the translational motion model is very widely used in practical video compression systems. We will return to it and describe it fully in Section 8.3.

8.2.4 Pixel-recursive methods

Pixel-recursive motion estimation methods were first introduced by Netravali and Robbins [4] and are based on the idea of iteratively minimizing the prediction error using a gradient descent approach. Given a DFD signal for a given frame (or block),

$$\mathrm{DFD}[\mathbf{p},\mathbf{d}] = s_k[\mathbf{p}] - s_{k-1}[\mathbf{p}+\mathbf{d}]. \tag{8.8}$$

The estimate for the motion vector at each location is then formed iteratively using a steepest descent approach; hence,

$$\hat{\mathbf{d}}^{i+1} = \hat{\mathbf{d}}^{i} - \varepsilon \mathrm{DFD}\left[\mathbf{p},\hat{\mathbf{d}}^{i}\right] \nabla s_{k-1}\left[\mathbf{p}+\hat{\mathbf{d}}^{i}\right]. \tag{8.9}$$

It has been observed that the convergence characteristics of pixel-recursive methods can be highly dependent, both on the content (i.e., DFD characteristics) and on the step size parameter ε. They struggle with smooth intensity regions and large displacements; they also inherently produce dense motion fields and this can lead to a high motion overhead unless parameter prediction methods are used. For these reasons, pixel-recursive methods have not been widely adopted in practice.

8.2.5 Frequency domain motion estimation using phase correlation

Principles

The basic idea of phase correlation was introduced by Kuglin et al. [5] and refined by Thomas for broadcast applications [6]; it is illustrated in Fig. 8.8. This method employs the 2D Fourier transform to estimate the correlation between blocks in adjacent frames. Peaks in the correlation surface correspond to the cross-correlation lags which exhibit the highest correlation between the current and reference blocks. One advantage of this approach is that the surface may possess multiple peaks, corresponding to multiple candidate motions. Some form of peak detection would then be required to establish the dominant motion or motion vector. Advantages of the phase correlation approach include:

- It can detect multiple motions in each block, facilitating object-based scene decomposition.
- Block size reduction and/or optimal vector selection for each pixel is possible.

- It can easily cope with large displacements.
- It offers robustness to noise and illumination variations as the Fourier phase is insensitive to mean shifts.
- It facilitates subpixel resolution through increasing the length of the DFT.

Assuming perfect translational block motion, for a block or picture, **S**, each location is related as follows:

$$s_k[x,y] = s_{k-1}\left[x+d_x, y+d_y\right], \tag{8.10}$$

where x and y represent the center pixel locations in block **S** of frame k, and d_x and d_y are the offset parameters which globally translate the block in frame k to align it with that in frame $k-1$. From the 2D DFT,

$$\mathcal{S}_k\left(f_x, f_y\right) = \mathcal{S}_{k-1}\left(f_x, f_y\right) e^{j2\pi(d_x f_x + d_y f_y)}. \tag{8.11}$$

In the case of translational motion, the difference of the 2D Fourier phases in the respective blocks defines a plane in the spatial frequency variables f_x, f_y. The motion vector can be estimated from the orientation of this plane; thus,

$$\arg\left(\mathcal{S}_k\left(f_x, f_y\right)\right) - \arg\left(\mathcal{S}_{k-1}\left(f_x, f_y\right)\right) = 2\pi\left(d_x f_x + d_y f_y\right). \tag{8.12}$$

However, this requires phase unwrapping and it is not easy to identify multiple motions. Instead, the phase correlation method is used to estimate the relative shift between two image blocks by means of a normalized cross-correlation function in the 2D Fourier domain. In this case the correlation between frames k and $k-1$ is defined by

$$C_{k,k-1}[x,y] = s_k[x,y] \star s_{k-1}[-x,-y]. \tag{8.13}$$

Taking the Fourier transform of both sides, we obtain a complex cross-power spectrum,

$$\mathcal{C}_{k,k-1}\left[f_x, f_y\right] = \mathcal{S}_k\left(f_x, f_y\right) \mathcal{S}^*_{k-1}\left(f_x, f_y\right), \tag{8.14}$$

and normalizing this by its magnitude gives the phase of the cross-power spectrum:

$$\angle(\mathcal{C}_{k,k-1}\left[f_x, f_y\right]) = \frac{\mathcal{S}_k\left(f_x, f_y\right) \mathcal{S}^*_{k-1}\left(f_x, f_y\right)}{\left|\mathcal{S}_k\left(f_x, f_y\right) \mathcal{S}^*_{k-1}\left(f_x, f_y\right)\right|}. \tag{8.15}$$

Assuming ideal translational motion by an amount $[d_x, d_y]$, we obtain from Eq. (8.15)

$$\angle(\mathcal{C}_{k,k-1}\left[f_x, f_y\right]) = e^{j2\pi(d_x f_x + d_y f_y)}, \tag{8.16}$$

and taking the inverse Fourier transform of Eq. (8.16), we obtain the phase correlation function,

$$C_{k,k-1}[x,y] = \delta\left[x-d_x, y-d_y\right], \tag{8.17}$$

i.e., the phase correlation function yields an impulse whose location represents the displacement vector.

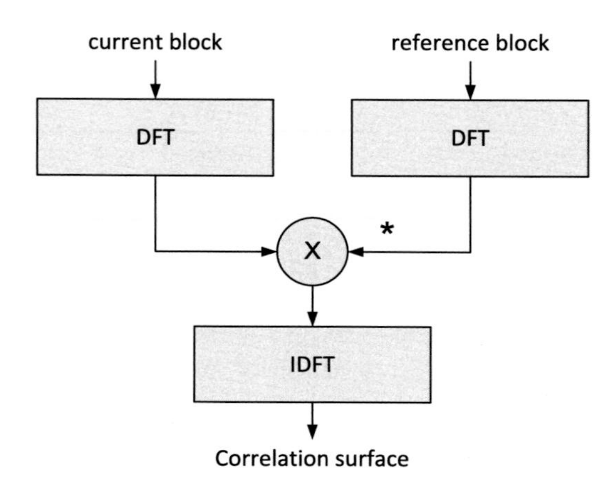

FIGURE 8.8 Phase correlation-based motion estimation.

Applications and performance

Phase correlation is the preferred method for TV standards conversion (e.g., frame rate conversion between US and EU frame rates), primarily due the fact that it produces a much smoother motion field with lower entropy than BMME, significantly reducing motion artifacts. Due to the use of the FFT, phase correlation can be computationally efficient. Phase correlation works well in the presence of large-scale translational motion and it can handle multiple motions. However, the use of variable block size BM is now considered a more flexible approach and is favored for conventional motion estimation in modern video coding standards. An example of phase correlation is given in Fig. 8.9.

8.3 Block matching motion estimation (BMME)

8.3.1 Translational block matching

Motion vector orientation

Fig. 8.10 illustrates the convention we will adopt for motion vector orientation. The motion vector represents the displacement of the closest matching block in the reference frame with respect to the current block location (i.e., reversing the polarity of true motion). We will assume the [0,0] vector corresponds to the center of the search grid and offsets in x and y are positive right and down, respectively.

Region of support – size of the search window

Intuitively, we would wish to search the whole of the reference picture for the best match to the current block. However, this is both impractical and unnecessary in most cases. Consider an 1080p25 sequence with an object traversing the screen in 1 second; with 1920 horizontal pixels, this means that the object moves horizontally by 1920/25=77 pixels per frame. One

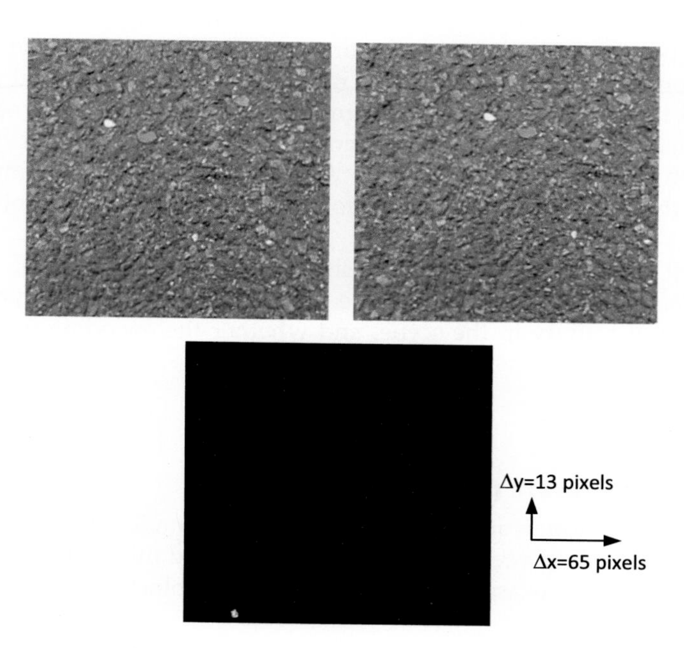

FIGURE 8.9 Example of phase correlation-based motion estimation. Top: Input frames of a tarmac road surface taken with an approximately translating camera. Bottom: Phase correlation surface.

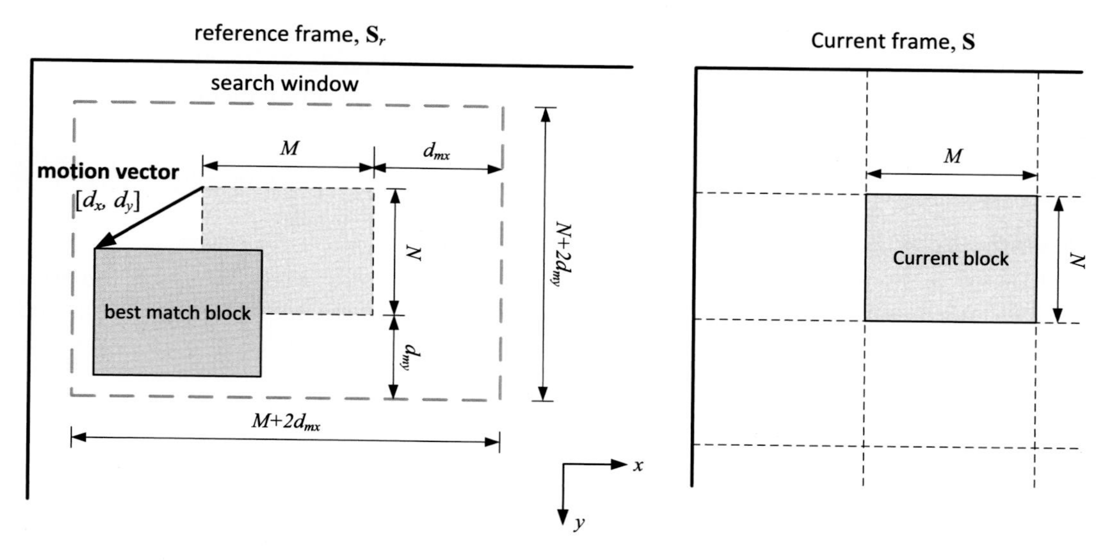

FIGURE 8.10 Translational block matching motion estimation (BMME).

could thus argue that such motion would dictate a search window of ± 77 pixels in order to predict it. In practice, a number of other factors come into play. Firstly, an object traversing a screen in 1 second is very fast and would rarely occur (not in directed content at least); secondly, if it did happen, there would be significant motion blur so the object could never be faithfully depicted anyway; and thirdly, motion masking effects will exist which further diminish the need for faithful reproduction. As such, a compromise window is normally employed where the region of support caters for most motion scenarios while limiting search complexity.

The search grid is normally restricted, as shown in Fig. 8.10, to a $[\pm d_{mx}, \pm d_{my}]$ window and, in most cases, $d_{mx} = d_{my} = d_m$. The value of d_m selected will depend on the image resolution, the temporal activity in the scene, and whether the encoding is performed on- or off-line.

8.3.2 Matching criteria

The block distortion measure (BDM)

For BMME, uniform motion and a translational motion model are normally assumed. We need to select the best motion vector for the current block and this will correspond to the lowest residual distortion. This is assessed with some form of block distortion measure (BDM), formulated as follows:

$$\mathrm{BDM}(i, j) = \sum_{x=0}^{M-1} \sum_{y=0}^{N-1} g(s(x, y) - s_r(x + i, y + j)), \tag{8.18}$$

where $g(.)$ is the function used to compute the BDM, i and j are the offsets for the current motion vector candidate, and x,y are the local coordinates within the current block.

We choose the motion vector that minimizes the BDM for the current block:

$$\mathbf{d} = [d_x, d_y]^{\mathrm{T}} = \arg(\min_{\forall(i,j)}(\mathrm{BDM}(i, j))). \tag{8.19}$$

Absolute difference vs. squared difference measures

There are many possible block distortion functions that could be used to assess the error residual after motion estimation. These include NCCF, mean squared error (or SSD), and SAD. The choice of BDM function is key as it influences both the search complexity and the prediction accuracy.

The two most common functions are SAD (Eq. (8.20)) and SSD (Eq. (8.21))[3]:

$$\mathrm{SAD}(i, j) = \sum_{x=0}^{M-1} \sum_{y=0}^{N-1} |(s(x, y) - s_r(x + i, y + j))|, \tag{8.20}$$

[3] SAD is normally used in preference to mean absolute difference (MAD).

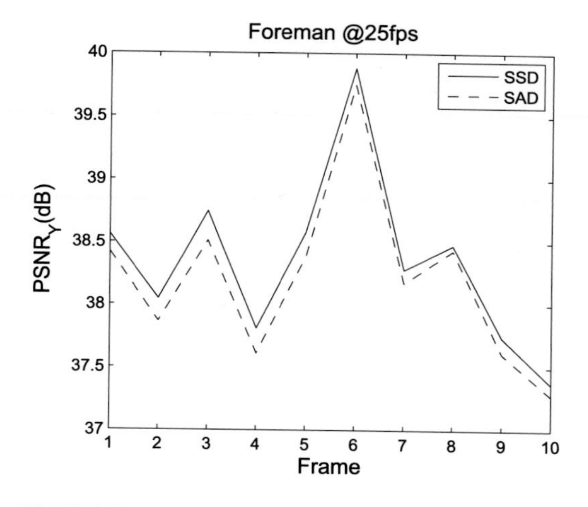

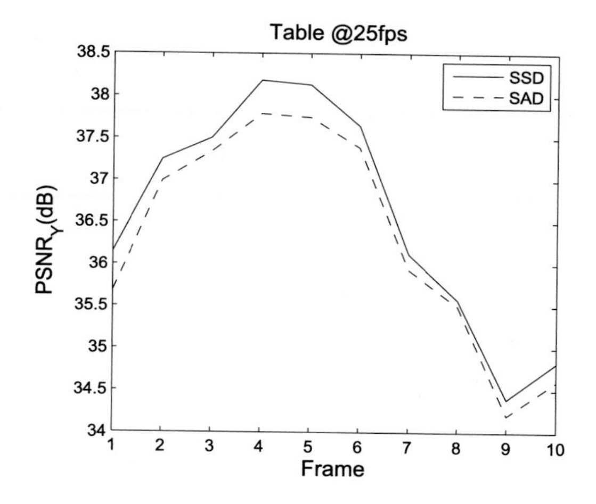

FIGURE 8.11 Effect of matching function: SAD vs. SSD.

$$\mathrm{SSD}(i,j)=\sum_{x=0}^{M-1}\sum_{y=0}^{N-1}(s(x,y)-s_r(x+i,y+j))^2. \tag{8.21}$$

In addition to SSD and SAD measures, as described in Chapter 4, a transform (typically a Hadamard transform) can be applied prior to computing the SAD value. This metric not only serves to measure the overall difference between two images but also, due to the transformation process, factors in an estimate of the coding cost of a residual signal. It is given by

$$\mathrm{SATD}=\sum_{n=1}^{N-1}\sum_{m=0}^{M-1}|r_H(n,m)|, \tag{8.22}$$

where $r_H(n,m)$ are the Hadamard transform matrix values given by

$$\mathbf{R}_{\mathrm{H}}=\mathbf{H}.(\mathbf{S}_1-\mathbf{S}_2).\mathbf{H}^{\mathrm{T}}. \tag{8.23}$$

Fig. 8.11 illustrates the difference in prediction accuracy between these two functions for two different sequences. It can be observed that, although the SSD metric consistently outperforms SAD in terms of quality, there is actually very little difference between the two methods. Furthermore, the SAD metric requires only $3NM$ basic operations (add, sub, abs) per block to compute whereas the SSD requires NM multiplications plus $NM-1$ additions. For this reason, the SAD function is frequently chosen over SSD.

8.3.3 Full search algorithm

The simplest BMME algorithm is based on *full search* or *exhaustive search*. In this case the global minimum of the error surface is found by exhaustively checking all possible $(2d_m+1)^2$ motion vector candidates within the search grid.

$\rho_X = 0.56$ $\rho_Y = 0.33$	$\rho_X = 0.69$ $\rho_Y = 0.49$	$\rho_X = 0.64$ $\rho_Y = 0.46$
$\rho_X = 0.66$ $\rho_Y = 0.48$	**Current Block**	$\rho_X = 0.72$ $\rho_Y = 0.63$
$\rho_X = 0.64$ $\rho_Y = 0.43$	$\rho_X = 0.76$ $\rho_Y = 0.61$	$\rho_X = 0.68$ $\rho_Y = 0.56$

FIGURE 8.12 Autocorrelation values for typical block motion fields (*foreman* at 25 fps) (adapted from [3]).

8.3.4 Properties of block motion fields and error surfaces

The block motion field

For natural scenes, the block motion field is generally smooth and slowly varying. This is because there is normally high correlation between the motion vectors of neighboring blocks. Fig. 8.12 illustrates this for the example of the *foreman* sequence captured at 25 fps.

The effect of block size

Fig. 8.13 shows the effect of different block sizes (8×8 vs. 16×16) in terms of prediction quality and motion overhead. It can be seen that smaller block sizes normally result in better prediction accuracy, but only at the expense of significantly increased bit rates. The improved prediction accuracy is due to the fact that smaller blocks can deal better with multiple motions, complex motions (i.e., provide a better fit to the model), and the different properties of textured and nontextured regions. Because of the conflict between rate and distortion, many modern compression standards employ variable block sizes for motion estimation alongside rate-distortion optimization (RDO) techniques that select the block size that optimizes a rate-distortion trade-off. This is discussed further in Chapters 9 and 10.

The effect of search range

The *search grid* size is dictated by the maximum motion vector offset from the current block location, denoted as d_m. The value of d_m also has an influence on complexity and accuracy, since full search complexity will grow with its square (there are $(2d_m + 1)^2$ candidates to be evaluated) and most fast search methods grow with it linearly. A small value of d_m will mean that fast moving objects will not be captured in the search widow, resulting in poor prediction quality and motion failure. In contrast, a large value of d_m may be inefficient and lead to excessive computational complexity.

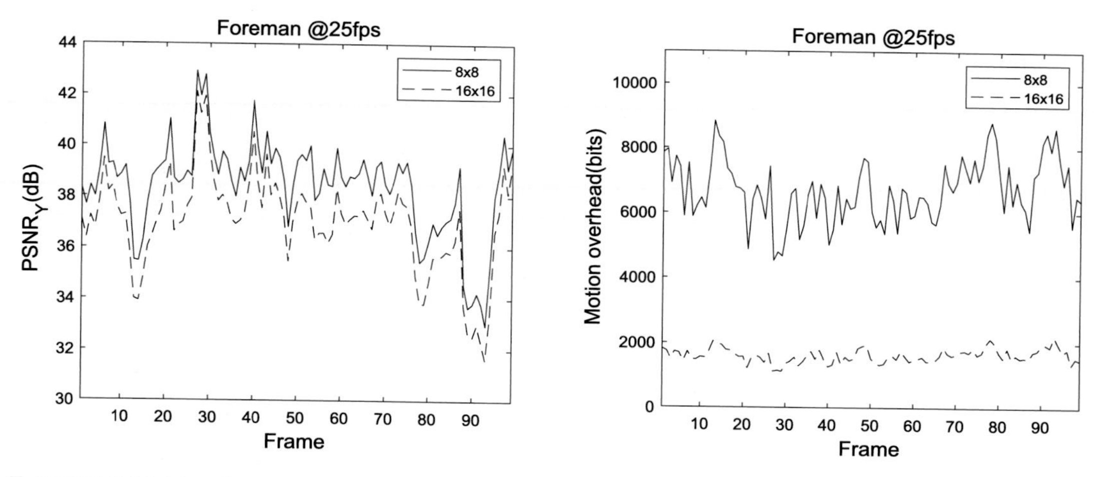

FIGURE 8.13 The effect of block size on motion estimation performance. (Left) The PSNR (luma) values for the predicted frames compared to the original. (Right) The bits consumed by the encoded motion vectors.

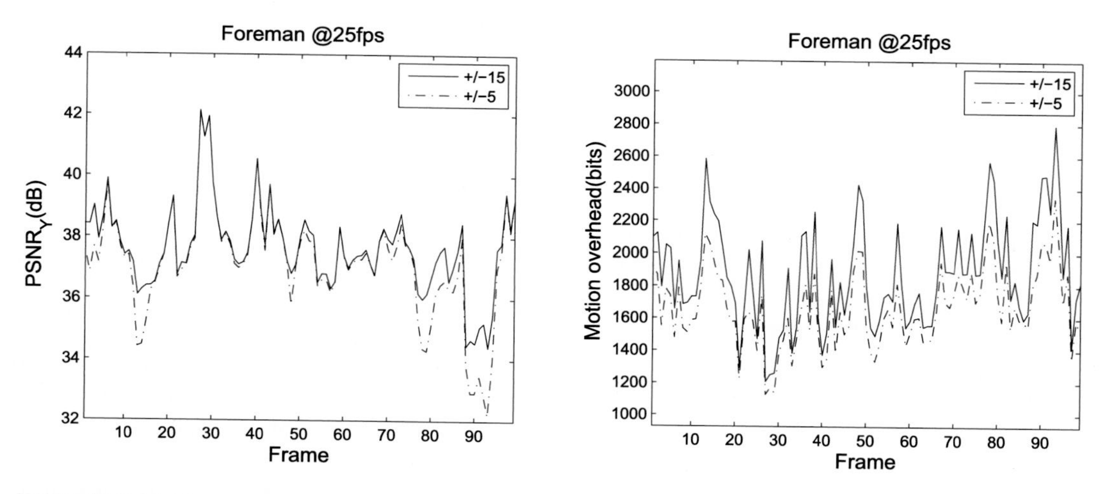

FIGURE 8.14 The influence of search range on prediction accuracy and coding bit rate.

The choice of search window size is dictated by content type and video format. For example sports content will normally require a larger window than talk shows, and HDTV will require a larger window than SDTV for the same frame rate. Fig. 8.14 shows the effect of search window size for the CIF foreman sequence at 25 fps. It can be seen that a small ± 5 window performs satisfactorily most of the time in this case, except for a few temporal regions (around frames 12, 50, 80, and 90) where higher motion causes a significant drop in quality. The larger search window will cope with these fast motions, but does require a small increase in bit rate in order to deal with the increased average motion vector size.

Example 8.2. Full search motion estimation

Given the following 2×2 block in the current frame **S** and the corresponding search window **W**, determine the motion vector, **d**, for this block. Use a full search strategy with an SAD error metric. We have

$$\mathbf{W} = \begin{bmatrix} 1 & 5 & 4 & 9 \\ 6 & 1 & 3 & 8 \\ 5 & 7 & 1 & 3 \\ 2 & 4 & 1 & 7 \end{bmatrix}, \quad \mathbf{S} = \begin{bmatrix} 3 & 9 \\ 1 & 4 \end{bmatrix}.$$

Solution. Consider, for example, the candidate motion vector **d**=[0,0]. In this case the DFD signal is given by

$$\mathbf{E} = \begin{bmatrix} 3 & 9 \\ 1 & 4 \end{bmatrix} - \begin{bmatrix} 1 & 3 \\ 7 & 1 \end{bmatrix} = \begin{bmatrix} 2 & 6 \\ -6 & 3 \end{bmatrix}$$

and the SAD $= 17$. Similarly, for all other $[d_x, d_y]$ offsets, we have the following.

i	j	**SAD**(i, j)	i	j	**SAD**(i, j)
−1	−1	14	0	1	18
−1	0	18	1	−1	7
−1	1	5	1	0	2
0	−1	8	1	1	11
0	0	17			

Hence the motion vector for this block is $\mathbf{d} = [1, 0]$.

The motion residual error surface

The error surface for each block (i.e., the BDM figures for each candidate vector in the search grid) will normally be multimodal, containing one or more local minima. This can be caused by periodicities in local textures, the presence of multiple motions within the block, or just a poor fit between the motion model and the apparent motion. In addition, the global minimum and maximum SAD values can vary enormously, even across a single frame in a sequence. An example of an error surface is shown in Fig. 8.15 together with its corresponding current block (top right) and search window (top left). We can observe that, in the example shown, the SAD surface is indeed multimodal and extends in value from below 1000 to over 11,000.

Motion vector probabilities

The distribution of the block motion field is center-biased, implying that smaller displacements are more probable than larger ones and that the [0,0] vector is (in the absence of global motion) the most likely. This property is illustrated in Fig. 8.16, for a range of sequences coded at 25 fps. The peak in the pdf at [0,0] is clear for the *Akiyo* and *football* sequences, whereas there is an obvious offset due to the camera pan in *coastguard*. The low-activity sequence, *Akiyo*,

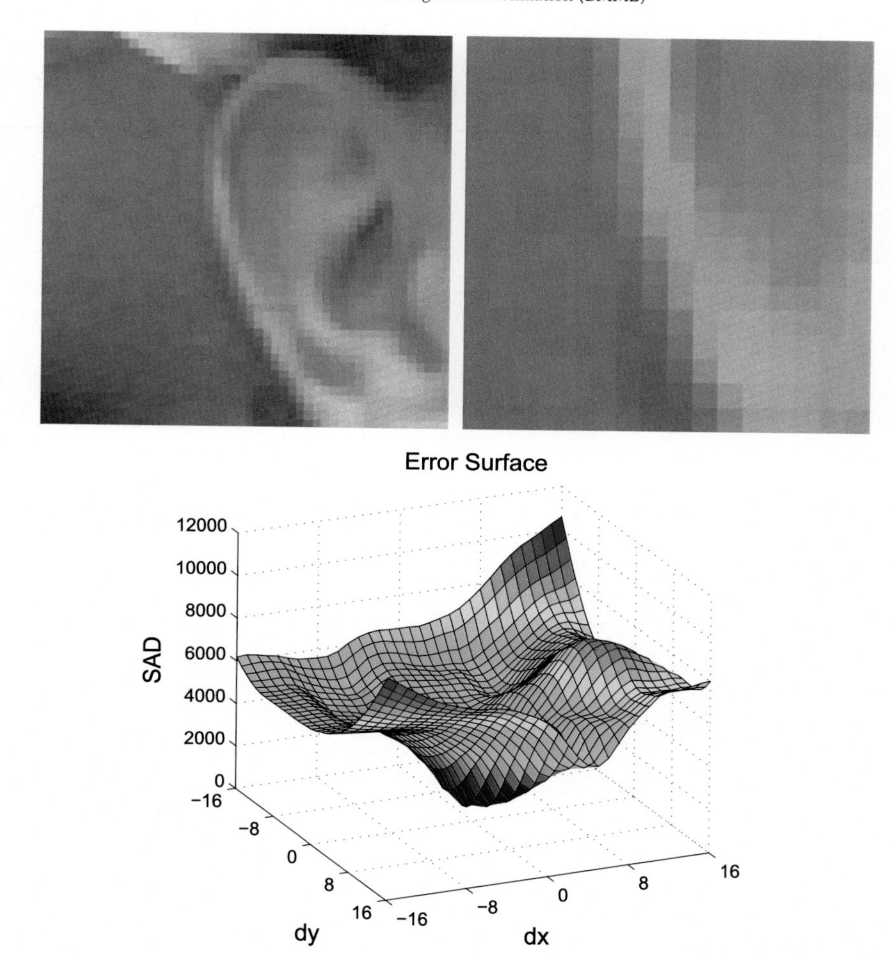

FIGURE 8.15 Typical motion residual error surface for a 16 × 16 block from the *foreman* sequence (CIF at 25 fps, ±16 × ±16 search window). Top left: Search window. Top right: Current block.

shows a large peak in the pdf at [0,0] with very small values elsewhere. The *football* sequence on the other hand shows a much lower peak together with a clear element due to camera panning and a significant pdf energy away from [0,0] due to the complex nature of the scene. The *coastguard* sequence shows a wider main peak, biased due to the camera pan, but also energy away from [0,0], largely due to the difficulty of estimating the random movements of the dynamic water texture.

8.3.5 Motion failure

Although the block-based translational motion model works well in most cases and can provide excellent rate-distortion performance with low complexity, there are situations where

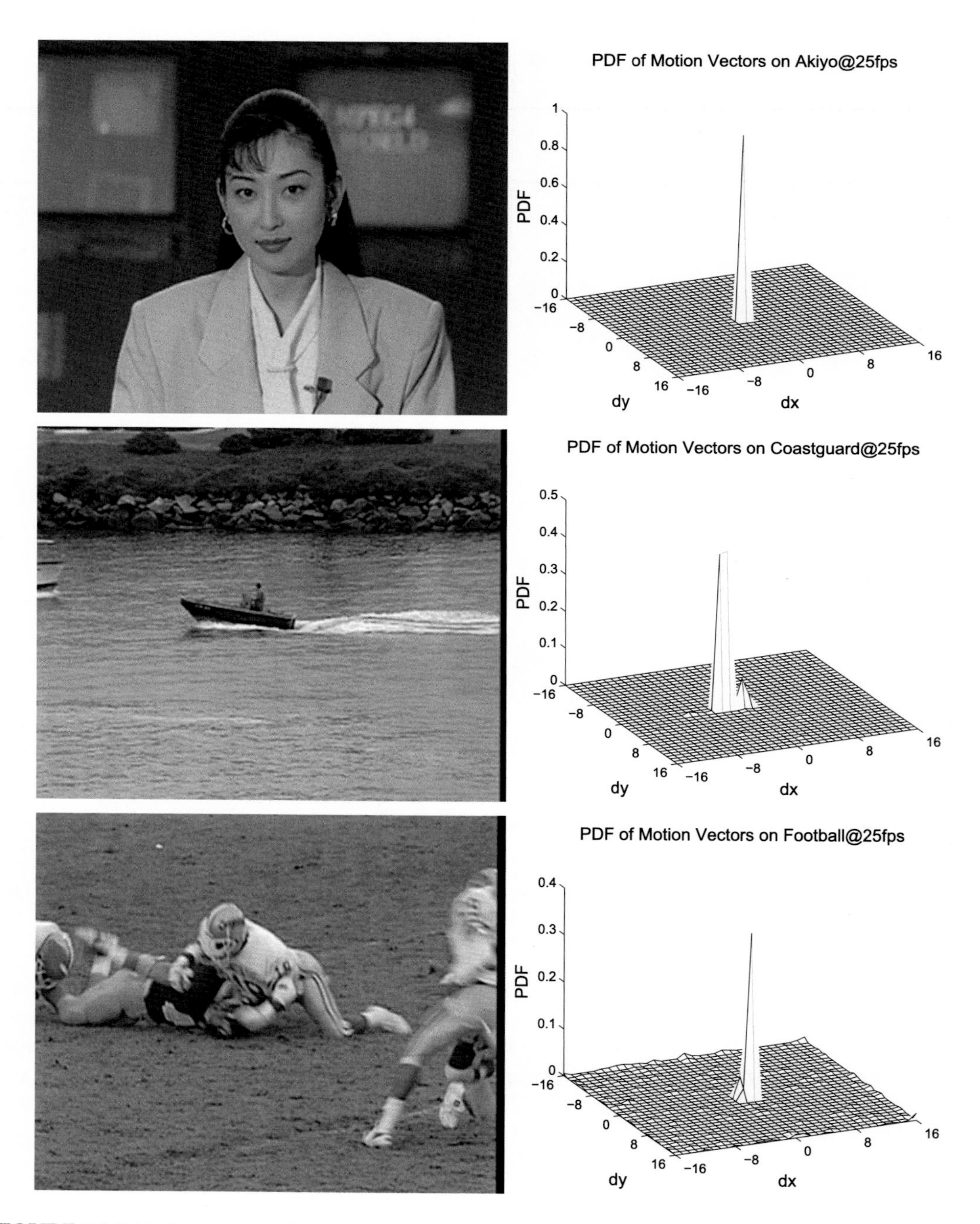

FIGURE 8.16 Motion vector pdfs for *Akiyo*, *coastguard*, and *football* sequences.

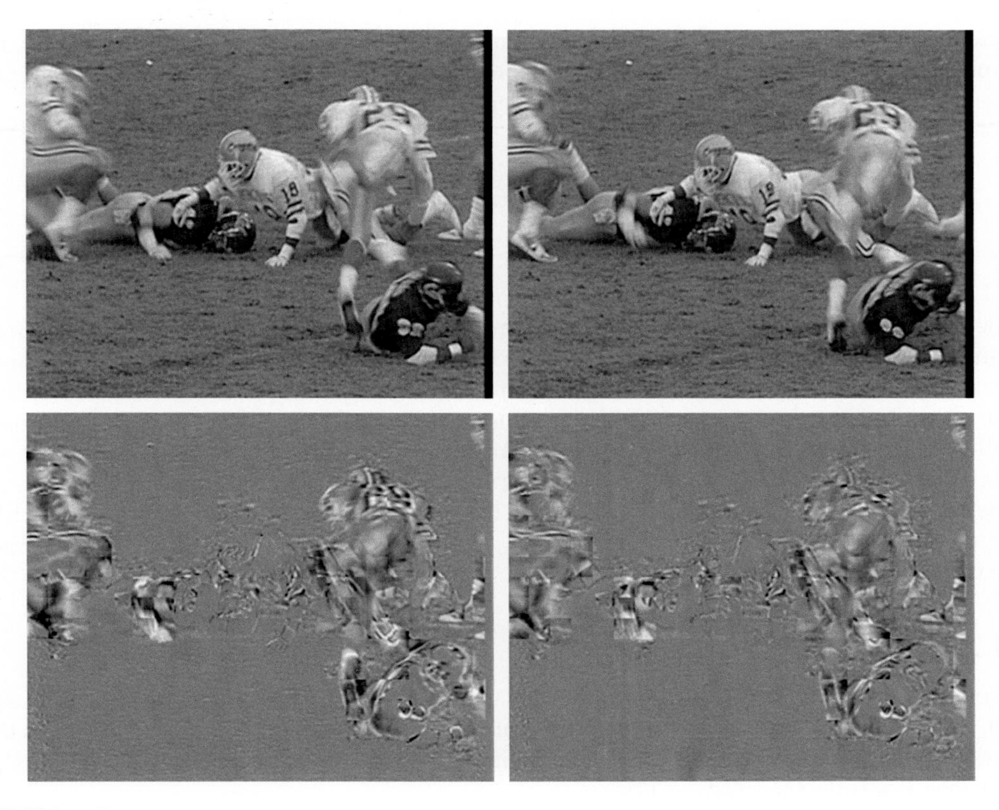

FIGURE 8.17 Example of motion model failure for the *football* sequence. Top left: Current frame. Right: Reference frame. Bottom left: Full-pixel estimation. Right: Half-pixel estimation.

the motion characteristics are such that it fails badly. Such a situation is shown in Fig. 8.17. These two frames, from *football,* capture the high activity typical in such content and the DFD, even after subpixel motion estimation, shows significant artifacts. These artifacts will cause the DFD to have a very low autocorrelation coefficient and hence the decorrelating transform (DCT or otherwise) will perform poorly.

So how do we deal with this situation? Basically two approaches can be used, independently or in combination. The first compares the rate-distortion performance of intra- and interframe coding and chooses the optimal coding mode. The second uses an in-loop deblocking filter to remove introduced edge artifacts in the reference frame. These methods are discussed further in Chapters 9 and 10.

8.3.6 Restricted and unrestricted vectors

In the case of blocks situated at or near the edge of a frame, a portion of the search area will fall beyond the frame border. In standard motion estimation, vectors are restricted to fall within the frame. However benefits can be obtained, especially in the case of camera motion, by artificially extending the search window through border pixel extrapolation, thus supporting the use of *unrestricted motion vectors*. External pixels are normally based on a zero order or symmetric extrapolation of the border values.

8.4 Reduced-complexity motion estimation

8.4.1 Pixel grids and search grids

Before we look at advanced search techniques, it is worth spending a little time differentiating between *pixel grids* and *search grids*. Both represent the region of support for motion search but we will use the latter as the basis for describing the operation of search methods from now on. The pixel grid is simply the image matrix where each element represents the pixel value within the search window. In contrast, the search grid represents the search window in terms of its offsets. Without loss of generality, assuming a square search window, the dimensions of the search grid are $2d_m + 1$ by $2d_m + 1$. This corresponds, for an N by N block size, to a search window with dimensions $2d_m + N$ by $2d_m + N$. These two representations are illustrated in Fig. 8.18.

8.4.2 Complexity of full search

For each motion vector, there are $(2d_m + 1)^2$ motion vector candidates within the search grid. At each search location MN pixels are compared, where each comparison requires three operations (sub, abs, add). The total complexity (in terms of basic arithmetic operations) per macroblock is thus $3MN\,(2d_m + 1)^2$.

Thus, for a picture resolution of $I{\times}J$ and a frame rate of F fps, the overall complexity is (in operations per second) given by

$$C_{FS} = 3IJF\,(2d_m + 1)^2\,. \tag{8.24}$$

Example 8.3. Full search complexity

Consider a video sequence with the following specifications: M=N=16, I=720, J=480, F=30, $d_m = 15$. Compute the complexity of a full search motion estimation algorithm for this sequence.

Solution. As indicated above the full search complexity is given by $3IJF\,(2d_m + 1)^2$ operations per second. Substituting the values for the given sequence gives a processor loading of 29.89×10^9 operations per second!

8.4.3 Reducing search complexity

As we will see shortly, search complexity can be reduced considerably but usually at the expense of suboptimal performance. Many approaches exist and these can be based on:

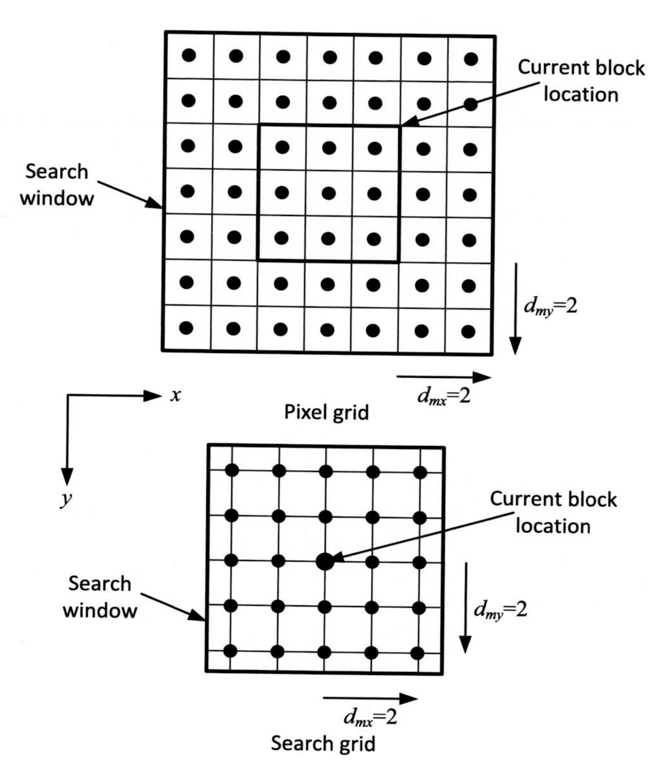

FIGURE 8.18 The pixel grid and the search grid.

1. **A reduced-complexity BDM**: In practice it is difficult to improve upon the use of SAD, which already offers a very good trade-off between complexity and performance. One interesting approach in this category is that of Chan and Siu [7], where the number of pixels used in the BDM evaluation is varied according to the block activity.
2. **A subsampled block motion field**: This approach exploits the smoothness properties of the block motion field. Only a fraction of the block candidates are evaluated and the remainder are interpolated before selecting the minimum BDM. For example, Liu and Zaccarin [8] use a checkerboard pattern to evaluate the motion vectors for half of the blocks. This reduces search complexity by a factor of two with little decrease in prediction quality.
3. **Hierarchical search techniques**: This class of method uses a multiresolution approach to progressively refine motion vector estimates across finer scales. An example of such methods, using a mean pyramid, is presented by Nam et al. in [9]. Such approaches are also reported to provide more homogeneous block motion fields with a better representation of the true motion.
4. **A reduced set of motion vector candidates**: This method restricts the number of blocks evaluated using some form of directed search pattern. Most fast motion estimation algorithms in this category are based on the assumption of a smooth and unimodal error surface (i.e., the error surface increases monotonically with distance from the minimum

Algorithm 8.1 TDL search.

1. Set initial search radius $r = 2^{k-1}$ where $k = \lceil \log_2(d_m) \rceil$;
2. Set checking points to: $\mathbf{\Gamma} = \{[0,0],[0,\pm r],[\pm r,0]\}$;
3. Evaluate the BDM at the 5 candidate locations on the cross and select the grid point with the lowest BDM: $\mathbf{d}' = \arg(\min_{(i,j)\epsilon\Gamma}(\text{BDM}(i,j)))$;
4. IF the minimum remains at the center of the pattern THEN $r = \frac{r}{2}$;
5. IF $r > 1$ THEN $\mathbf{\Gamma} = \mathbf{d}' + \{[0,0],[0,\pm r][\pm r,0]\}$, GO TO (3);
6. IF $r = 1$ THEN invoke 9-point square pattern with $\mathbf{\Gamma} = \mathbf{d}' + \{[0,0],[\pm r,\pm r],[0,\pm r][\pm r,0]\}$, $\mathbf{d} = \arg(\min_{(i,j)\epsilon\Gamma}(\text{BDM}(i,j)))$, STOP.

BDM point). We have already seen that this assumption is not always valid and algorithms can get trapped in local minima (see Fig. 8.15) so the search pattern used must take this into account. The earliest approach of this type was the 2D logarithmic (TDL) search proposed by Jain and Jain [10], and many variants have followed. This class of method is widely considered to offer an excellent compromise between complexity and prediction quality and is commonly used in practice. Selected approaches from this category are described in the following subsections.

5. **Intelligent initialization and termination**: Again, these exploit the properties of the block motion field. Intelligent initialization criteria are used to start the search at the location of a predicted motion vector rather than at the center of the search grid. Such methods can be used with (4) to help avoid local minima.

The reader is directed to [3] for a detailed comparison of these methods. In the following sections we examine some of the more popular techniques in more detail.

8.4.4 2D logarithmic (TDL) search

The TDL search was introduced by Jain and Jain in 1981 [10] as a method of reducing the number of motion vector candidates evaluated. The search patterns used in this method are shown in Fig. 8.19 and are in the shape of a + with five points, except for the final stage where a nine-point square pattern is checked. The largest pattern is initially located at the center of the search grid, all five points are evaluated, and the one with the lowest BDM is selected as the center of the next search. The process is repeated until the minimum point remains at the center, in which case the search pattern radius is halved. This process continues until the radius is equal to 1, when the nine-point square pattern is invoked. This process is described in Algorithm 8.1.

8.4.5 N-step search (NSS)

The NSS, introduced by Koga et al. [11], is similar to the TDL search. It uses a uniform pattern comprising nine evaluation points, as shown in Fig. 8.20, where the search radius decreases by a factor of 2 at each iteration. As for the TDL search, the process terminates when the search radius is equal to 1. The advantage of this approach is that it requires a fixed

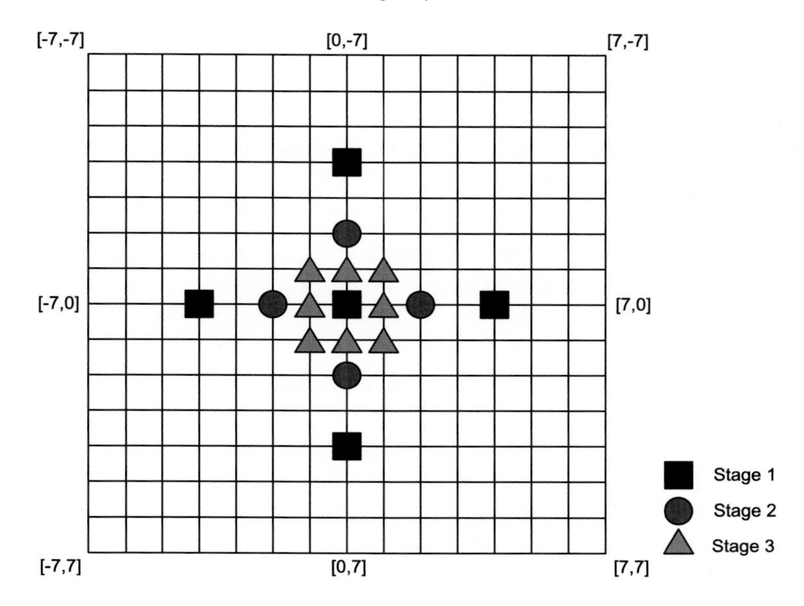

FIGURE 8.19 2D logarithmic search patterns.

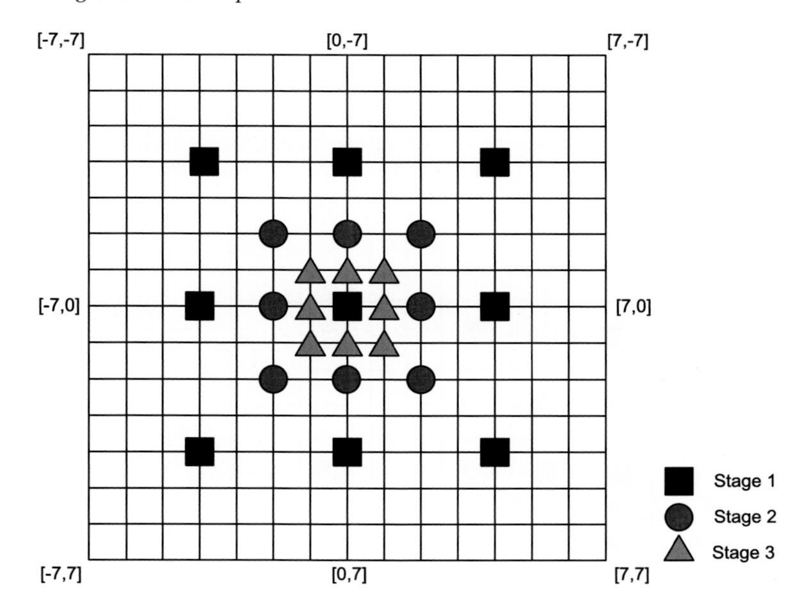

FIGURE 8.20 *N*-step search patterns.

number of iterations, so it offers regularity in terms of hardware or software implementation. The method is described in Algorithm 8.2. A worked example of a three-step search (TSS) is given in Example 8.4.

Algorithm 8.2 NSS search.

1. Set initial search radius $r = 2^{k-1}$ where $k = \lceil \log_2(d_m) \rceil$;
2. Set checking points to: $\mathbf{\Gamma} = \{[0,0], [\pm r, \pm r], [0, \pm r], [\pm r, 0]\}$;
3. Evaluate the BDM at the 9 candidate locations and select the grid point with the lowest BDM: $\mathbf{d}' = \arg(\min_{(i,j)\epsilon\Gamma}(\mathrm{BDM}(i,j)))$;
4. Set $r = \frac{r}{2}$;
5. IF $r < 1$ THEN $\mathbf{d} = \mathbf{d}'$, STOP; ELSE set new search locations: $\mathbf{\Gamma} = \mathbf{d}' + \{[0,0], [\pm r, \pm r], [0, \pm r], [\pm r, 0]\}$; GO TO (3).

The complexity of the NSS algorithm is easy to compute and is

$$C_{NSS} = 3IJF(8k+1), \tag{8.25}$$

where k is the number of steps as indicated in Algorithm 8.2.

Example 8.4. The three-step search

Assuming a 15×15 search grid and a monotonically decreasing error surface with a single minimum, demonstrate how the TSS search algorithm would converge to a motion vector of $\mathbf{d}$=[−3,2].

Solution. The NSS algorithm will always converge in N steps, but the convergence profile will depend on the actual shape of the error surface. One possible solution is shown in the figure below.

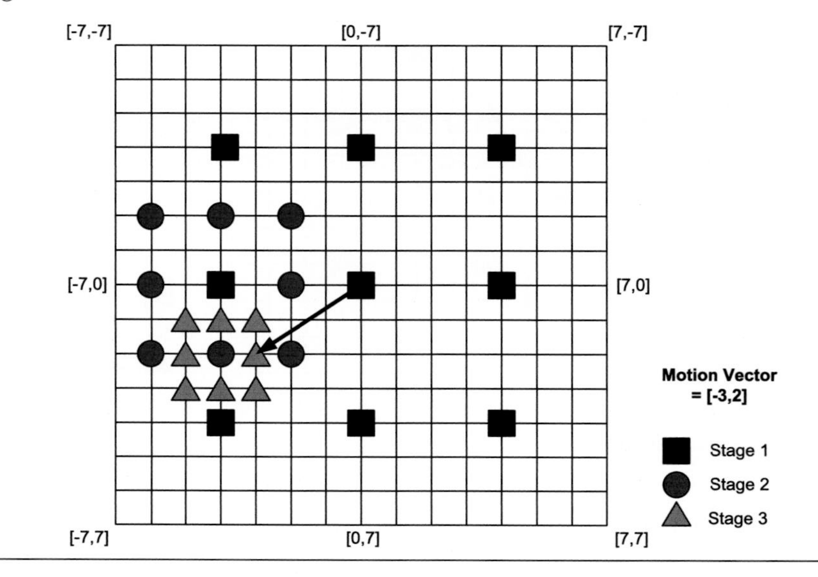

Algorithm 8.3 Diamond search.

1. Initialize motion vector estimate: $\mathbf{d}' = [0, 0]$;
2. Set checking points to: $\mathbf{\Gamma} = \mathbf{d}' + \{[0,0], [0,\pm 2], [\pm 2, 0], [\pm 1, \pm 1]\}$;
3. Evaluate the BDM at the 9 candidate locations on the diamond and select the grid point with the lowest BDM: $\mathbf{d}' = \arg(\min_{(i,j)\in\Gamma}(\text{BDM}(i,j)))$;
4. IF the minimum remains at the center of the pattern THEN $\mathbf{\Gamma} = \mathbf{d}' + \{[0,0], [0,\pm 1], [\pm 1, 0]\}$, $\mathbf{d} = \arg(\min_{(i,j)\in\Gamma}(\text{BDM}(i,j)))$, STOP.
5. GO TO (2);

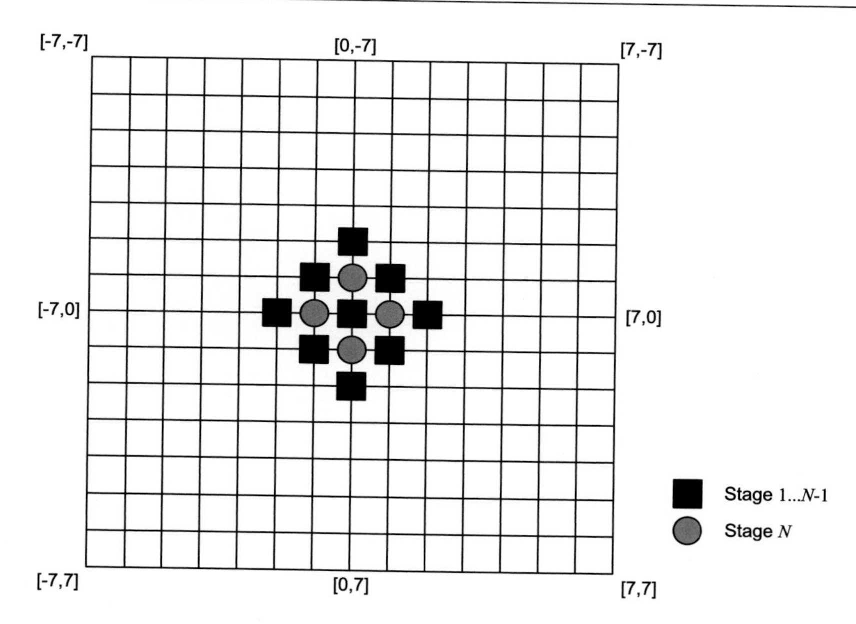

FIGURE 8.21 Diamond search patterns.

8.4.6 Diamond search

The DS was first introduced by Zhu and Ma in 1997 [12] and has demonstrated faster processing, with similar distortion, in comparison with TSS and NTSS approaches. DS adopts two search patterns – one large diamond for general gradient search and a second smaller diamond for final stage refinement. The approach is again similar to the TDL search in that the large diamond pattern at stage k is centered on the point with minimum BDM from stage $k-1$. When the search minimum remains at the center of the large pattern, the small diamond pattern is invoked to refine the motion vector before termination. The two diamond patterns are shown in Fig. 8.21 and the process is described in Algorithm 8.3. An illustration of the DS process is given in Example 8.5. It should be noted that the complexity of DS is dependent on the orientation of the optimal motion vector and so is content-dependent.

Example 8.5. The diamond search

Assuming a 15×15 search grid and a monotonically decreasing error surface with a single minimum, demonstrate how the DS search algorithm would converge to a motion vector of $\mathbf{d}=[-1,-4]$.

Solution. The DS algorithm will not converge in a specific number of steps but will track the gradient of the error surface using the large diamond shape until the minimum stays at the center of the diamond (stages 1–3 in this example). The small diamond shape is then introduced at stage 4 to identify the final motion vector. The convergence profile will of course depend on the actual shape of the error surface. One possible solution is shown in the following figures.

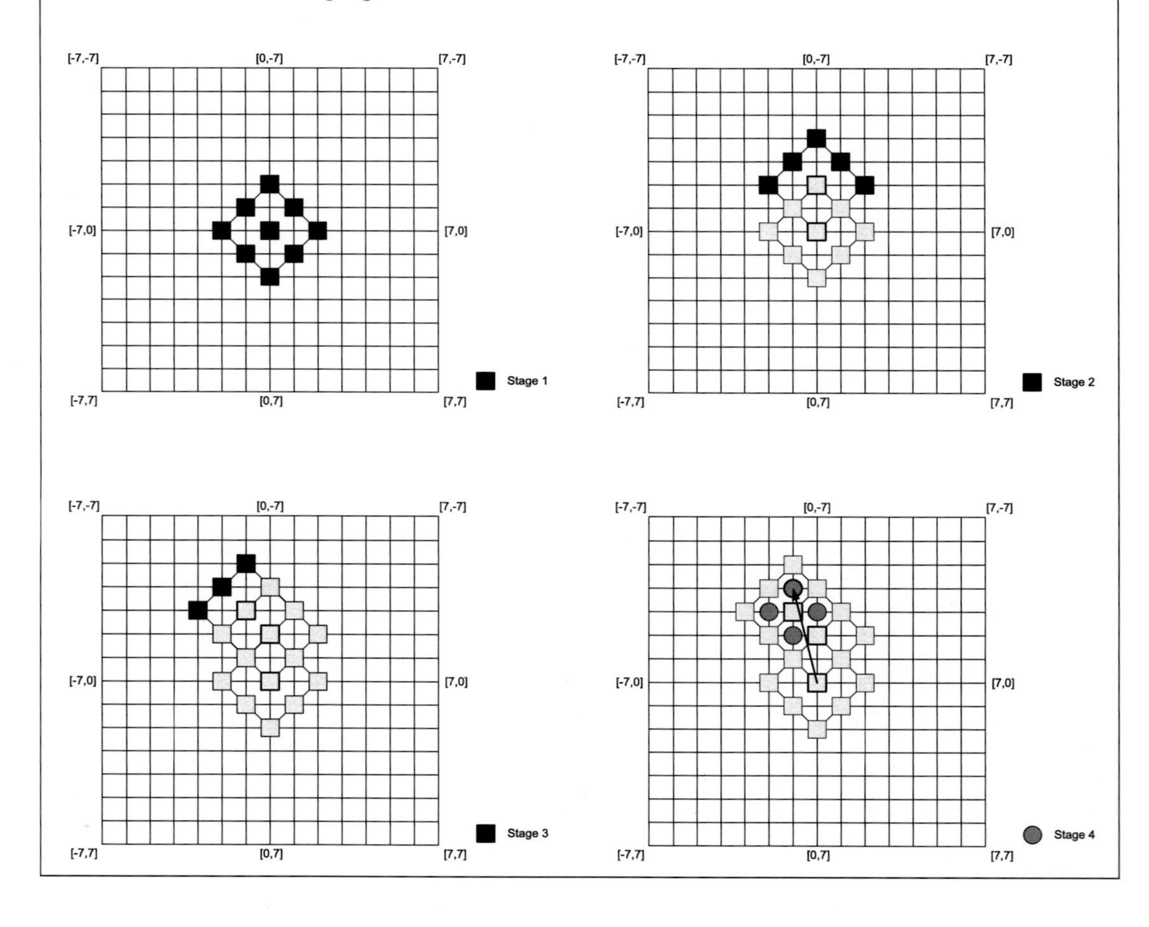

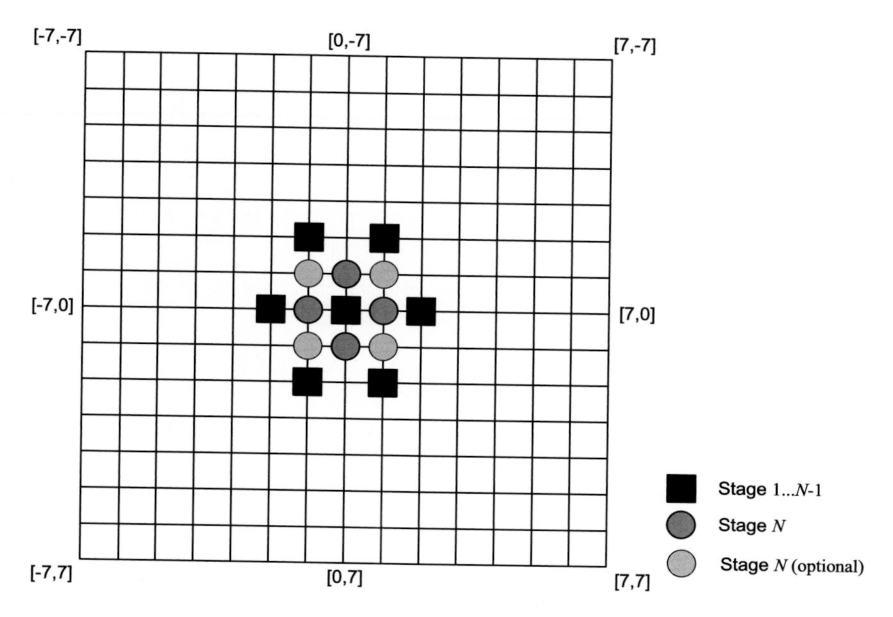

FIGURE 8.22 Hexagonal search patterns.

8.4.7 Hexagonal search

There are many variants on the methods presented above. One worthy of note is the HEXBS, which again uses two pattern types, but this time in the shape of a hexagon (Fig. 8.22). HEXBS was introduced by Zhu et al. [13] and, like DS, the smaller pattern is used only once for refinement at the final iteration. HEXBS has the advantage over DS that it offers greater orientation flexibility with the number of new checking points introduced at each iteration of the gradient descent stage being constant (three new points).

The advancing speed for DS is 2 pixels per step horizontally and vertically, and $\sqrt{2}$ pixels per step diagonally, whereas the HEXBS has horizontal and vertical advancements of 2 pixels per step horizontally and vertically and $\sqrt{5}$ pixels per step in the direction of the off-axis search points. This means that HEXBS is more consistent in its search performance and can outperform DS by up to 80% for some motion vector locations.

The complexity of the HEXBS algorithm can easily be seen to be

$$C_{\text{HEXBS}} = 7 + 3k + 4, \tag{8.26}$$

where k is the number of iterations required to reach the minimum of the error surface.

8.4.8 Test zone search (TZS)

The default fast search algorithm employed in the test models of H.265/HEVC and H.266/VVC is generally referred to as test zone search (TZS). This is a flexible family of hy-

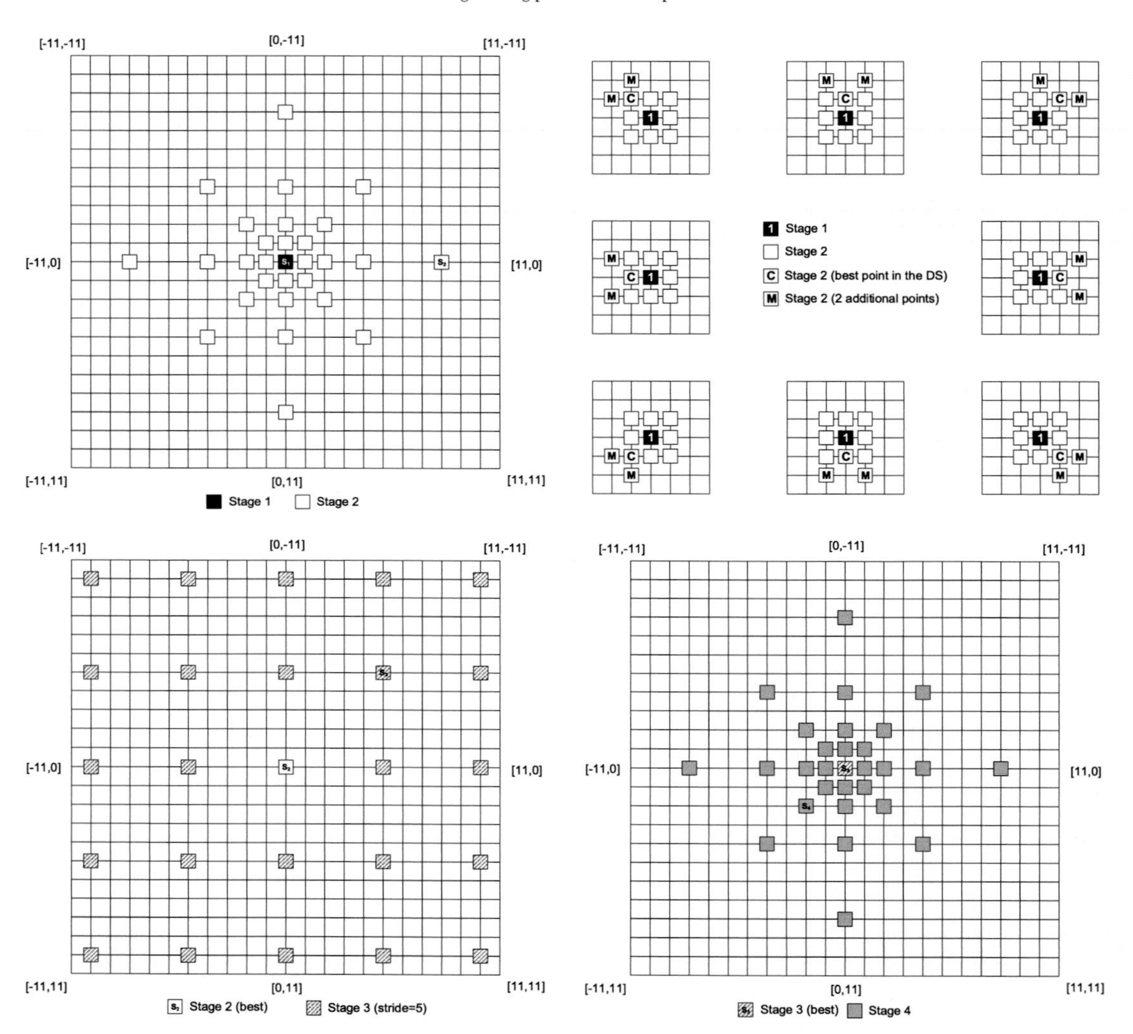

FIGURE 8.23 Illustration of the test zone search (TZS) method.

brid search methods typically combining DS and raster search [14]. It comprises four primary stages, as illustrated in Fig. 8.23. These are:

1. **Motion vector prediction:** The motion vector with the minimum RD cost is predicted from its encoded neighbors (see Section 8.6.1) to provide the initial prediction and search start point. This is shown as 1 in Fig. 8.23.
2. **Initial grid search**: A coarse search is then conducted from the starting point 1 in Stage 1, using a diamond (the default in HEVC) or square pattern. The search points reside within a preconfigured search window and their relative locations $[\Delta x, \Delta y]$ to the starting point

$\boxed{1}$ are given by

$$[\Delta x, \Delta y] \in \{[0,0], [0, \pm 2^n], [\pm 2^n, 0], [\pm 2^{n-1}, \pm 2^{n-1}], \text{ where } n \in 1, 2, 3, \cdots\}. \tag{8.27}$$

If the location of the current best point $\boxed{C}$ ($[\Delta x_c, \Delta y_c]$) relative to the starting point meets the following criterion:

$$|\Delta x_c| = 1 \text{ or } |\Delta y_c| = 1, \tag{8.28}$$

then another two untested neighboring points (examples shown as $\boxed{M}$ in Fig. 8.23 top right) are further compared with $\boxed{C}$. The best match of all the $\boxed{M}$ points is then selected as the start point for Stage 4. Otherwise, after the initial search, the best point $\boxed{2}$ is obtained for further refinement.

3. **Raster search**: If any of the two coordinates of the relative location, $[\Delta x_2, \Delta y_2]$, between the best point obtained in Stage 2 $\boxed{2}$ and the initial starting point $\boxed{1}$, is larger than a predefined threshold (the default threshold is 5 in HEVC), i.e.,

$$|\Delta x_2| > 5 \text{ or } |\Delta y_2| > 5, \tag{8.29}$$

then a raster search with a sparse grid pattern is further applied, centered on point $\boxed{2}$ within the maximum search window and with a fixed stride value (the default stride is 5 in HEVC). This is used to broaden the search space when the best match is far away from the prediction, due to (a) the sparsity of search points and (b) the increased uncertainty over the local motion field. The best point at Stage 3 is denoted as $\boxed{3}$.

4. **Final refinement**: If the current best point obtained in the previous stages satisfies $|\Delta x| < 2$ and $|\Delta y| < 2$, then no further search is performed. Otherwise an iterative refinement is performed using a diamond (the default pattern in HEVC, identical to that used in Stage 2), square, or raster search pattern centered initially on the best point from the previous stage. The two-missing point search is also performed, as described above, when the best point in the DS is close to the starting point of this iteration (based on Eq. (8.28)). The refinement stage terminates when the best match is unchanged, or when the relative location of the best point (to the starting point of this iteration) meets the criterion in Eq. (8.28). Fig. 8.23 (bottom right) shows a DS pattern used to obtain the best point in one refinement iteration.

8.4.9 Initialization and termination criteria

Initialization

The use of intelligent initialization and termination criteria can, in many cases, improve search performance and help to avoid the problems of the search becoming trapped in local minima. Referring to Fig. 8.24, which shows a current block P and its near causal neighbors A–D, a range of candidate initialization vectors can be formulated, based on the local smoothness property of the block motion field. These include

$$\hat{\mathbf{d}}_P = \left\{[0,0], \mathbf{d}_A, \mathbf{d}_B, \mathbf{d}_C, \mathbf{d}_D, \text{med}\left(\mathbf{d}_A, \mathbf{d}_C, \mathbf{d}_D\right), \text{med}\left(\mathbf{d}_A, \mathbf{d}_B, \mathbf{d}_C\right)\right\}. \tag{8.30}$$

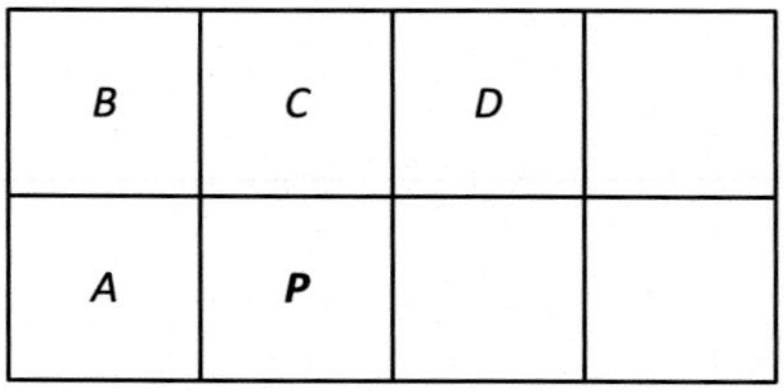

B	C	D	
A	P		

FIGURE 8.24 Predicted motion vectors for initialization.

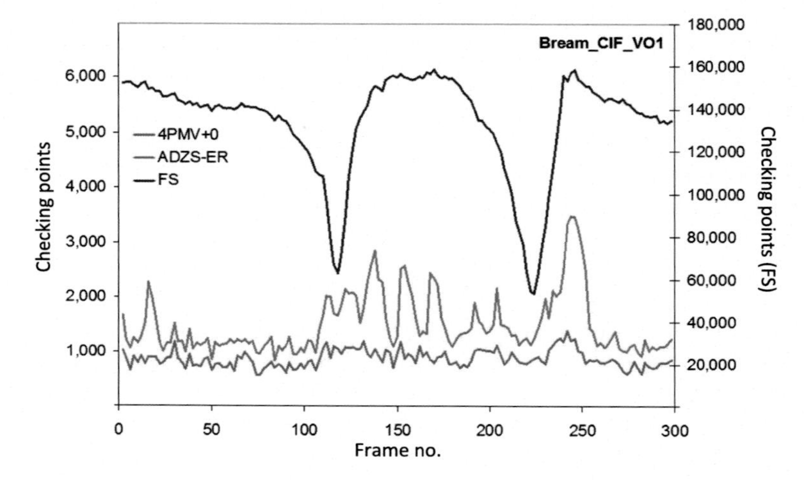

FIGURE 8.25 Fast motion estimation performance comparison for the *bream* CIF sequence at 25 fps: 4PMV+0 vs. ADZS-ER vs. FS.

An initialization algorithm would typically evaluate some or all of the candidate vectors from Eq. (8.30), choose that with the lowest BDM, and start the search at $\hat{\mathbf{d}}_P$ rather than always at [0,0].

Termination

The use of termination criteria is largely dictated by the implementation method used. For example, in the case of a software implementation, a search could be terminated by setting an acceptable block distortion threshold BDM_{Th} and exiting the search when $BDM \leq BDM_{Th}$. Furthermore, inside each SAD evaluation loop, execution could be terminated when the distortion measure for the current offset [*i,j*] exceeds the current minimum, i.e., when $BDM_{i,j} \geq BDM_{opt}$, or when the rate-distortion cost is sufficiently small.

Several methods have been reported that provide significant speed-up using intelligent initialization and termination. These include the work of Tourapis et al. [17,18] on PVMFast and EPZS and the work in the author's lab by Sohm et al. [19]. For example, Fig. 8.25 shows complexity results including the Four predicted motion vectors plus zero vector (4PMV+0) algorithm, DS, and full search.

TABLE 8.2 Reduced-complexity BMME comparisons for the *bream* CIF sequence at 25 fps. Top: Checking points per macroblock. Bottom: SAD error relative to full search.

Algorithm	4PMV+0	ADZS	DS	Simplex	NSS	FS
Minimum	1	1	13	9	32.9	1002.4
Average	4.1	5.9	16.1	14.2	33.0	1018.3
Maximum	9.6	15.3	22.2	21.8	33.1	1024.0
Algorithm	**4PMV+0**	**ADZS**	**DS**	**Simplex**	**NSS**	**FS**
Average	4.248	4.492	4.311	4.078	4.494	3.975
% vs. FS	106.9	113.0	108.5	102.6	113.1	100.0

8.4.10 Reduced-complexity block matching algorithm performance comparisons

Several authors have published tables comparing motion estimation methods in terms of complexity and prediction accuracy (see for example [17]). Table 8.2 compares some of the algorithms presented in this chapter. It can be seen that methods such as the simplex search [20] provide the most accurate results (closest to FS) with a competitive complexity. Techniques adopting intelligent initialization and termination criteria such as 4PMV+0 on the other hand require extremely low numbers of checking points with only a modest reduction in prediction accuracy.

8.5 Skip and merge modes

Video coding standards since H.263+ have supported a special inter-prediction mode, *skip* (referred to as direct mode for B-frames in H.263+, MPEG-4 Visual, and H.264/AVC). This has been found to provide better rate-distortion performance than normal inter-mode coding for content with consistent global motion [16]. In this mode, the motion information for the current block is predicted from its encoded spatial (or temporal) neighbors (similar to the approach described in Section 8.6.1). No prediction residual is encoded. The rate-distortion cost of the skip mode is compared with other prediction modes (e.g., inter and intra) as part of the RDO process. More recently, early skip mode decision algorithms have also been employed to reduce overall computational complexity. In these methods, skip mode will be automatically selected if predefined conditions (e.g., related to block size, motion vector length, or prediction residuals) are satisfied.

The skip mode has been further extended in HEVC and VVC in the form of a *merge* mode, which supports motion vector selection from a configurable list of spatial and temporal candidates, again without encoding a motion vector difference [15]. This feature enhances the coding performance for content with homogeneous motion and helps to overcome potential oversegmentation associated with quadtree partitioning. Fig. 8.26 illustrates the potential locations of the spatial and temporal merge candidates used in HEVC. The spatial candidates are included in the list and processed in the following order: A_1, B_1, B_0, A_0, and B_2. If candidates share the same motion information, they are merged (only one of them is kept in the list). A single temporal candidate can also be added, extracted from the reference pic-

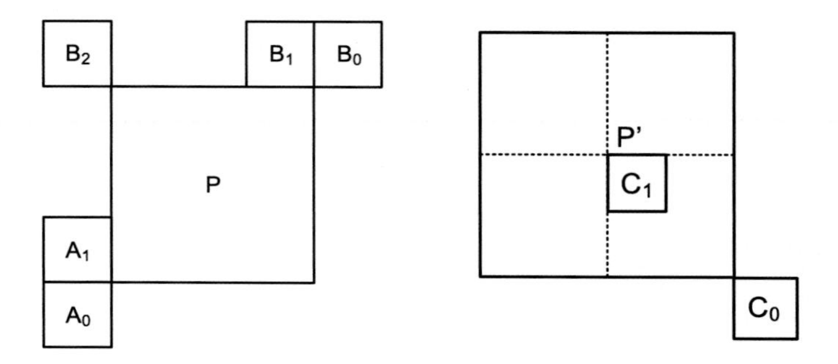

FIGURE 8.26 Locations for spatial (left) and temporal (right) merge candidates.

ture closest to the current picture. Referring to Fig. 8.26, if the candidate C_0 is available, not intra-coded (with motion information), and within the colocated coding tree unit, it will be selected as the temporal merge candidate. Otherwise, the candidate at position C_1 will be included. During the RDO process, if one of these candidates exhibits the best rate-distortion performance (compared to other prediction modes), additional bits will be added to signal its spatio-temporal index.

8.6 Motion vector coding

8.6.1 Motion vector prediction

Standards such as H.264/AVC and HEVC have made substantial progress in improving rate-distortion performance. However, as a consequence, the motion coding overhead has proportionally increased, with over 50% of the bits being dedicated to motion for some sequences with high quantization parameter values.

Coding of motion vectors is normally performed predictively to exploit the smoothness property of the block motion field. Consider coding the current block P in Fig. 8.24. A predicted motion vector can be generated using Eq. (8.31). If block D is not available, then block B can be used instead. We have

$$\hat{\mathbf{d}}_P = \left\{\text{med}\left(\mathbf{d}_A, \mathbf{d}_C, \mathbf{d}_D\right)\right\}. \tag{8.31}$$

The motion vector difference or residual is then computed and entropy coded for transmission:

$$\mathbf{e}_P = \mathbf{d} - \hat{\mathbf{d}}_P. \tag{8.32}$$

8.6.2 Entropy coding of motion vectors

The motion vector prediction process will result in a decorrelated residual signal that will mainly comprise small values. In the same way as we used entropy coding based on Huffman or arithmetic methods in Chapter 7 to code quantized transform coefficients, so too can we apply this to motion vector residuals. Clearly, different VLC tables will be required for Huffman coding and different probability distributions for arithmetic coding. In practice, most current video coding standards employ a form of context-based adaptive binary arithmetic coding as described in Chapter 7.

Example 8.6. Motion vector coding

Motion vectors for four adjacent blocks in two partial rows of a frame are shown below, outlined in bold. Using the prediction scheme $\hat{\mathbf{d}}_P = \text{med}(\mathbf{d}_A, \mathbf{d}_C, \mathbf{d}_D)$, compute the predicted motion vectors for each of these blocks together with their residuals.

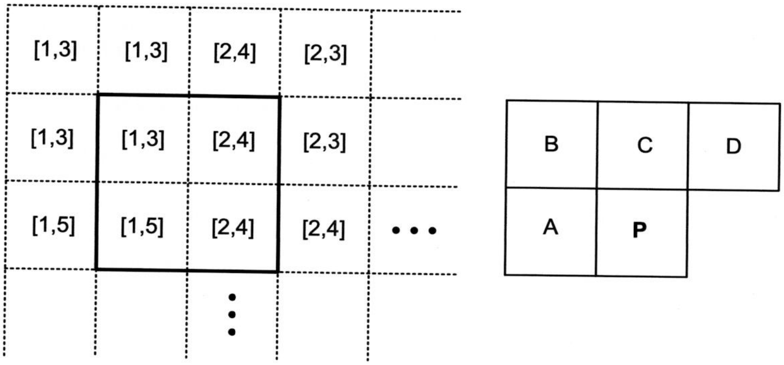

[1,3]	[1,3]	[2,4]	[2,3]
[1,3]	**[1,3]**	**[2,4]**	[2,3]
[1,5]	**[1,5]**	**[2,4]**	[2,4] ...

B	C	D
A	P	

Solution. The predicted motion vectors and residuals for each of the four blocks outlined (labeled top left to bottom right) are given in the following table.

Block, P	1	2	3	4
$\hat{\mathbf{d}}_P$	[1,3]	[2,3]	[1,4]	[2,4]
$\mathbf{e}_P$	[0,0]	[0,1]	[0,1]	[0,0]

Note that the prediction process described here must be modified for boundary blocks. This is described further in Chapter 12.

8.7 Summary

This chapter has addressed the important topic of motion estimation, the primary technique for reducing temporal correlation in an image sequence. We saw how temporal correlations can be exploited to provide additional coding gain. Most of the chapter focused on block matching and block-based motion estimation using a translational motion model. This approach is in widespread use and has been adopted in all standardized video codecs to date. The chapter also described a number of methods for reducing the complexity of motion estimation, comparing a range of fast search methods alongside descriptions of intelligent initialization and termination techniques.

In the next chapter we will show how to integrate the motion estimation algorithm into a DPCM prediction loop at the encoder. We also examine a range of methods for improving still further the coding gain of this block-based hybrid codec.

References

[1] T. Ebrahimi, M. Kunt, Visual data compression for multimedia applications, Proc. IEEE 86 (6) (1998) 1109–1125.
[2] C. Glasbey, K. Mardia, A review of image warping methods, J. Appl. Stat. 25 (2) (1998) 155–171.
[3] M. Al-Mualla, C. Canagarajah, D. Bull, Video Coding for Mobile Communications, Academic Press, 2002.
[4] A. Netravali, J. Robins, Motion compensated television coding: Part 1, Bell Syst. Tech. J. 58 (1979) 631–670.
[5] D. Kuglin, D. Hines, The phase correlation image alignment method, in: Proc. IEEE Intl. Con. on Cybernetics and Society, 1975, pp. 163–165.
[6] G. Thomas, Television motion measurement for DATV and other applications, BBC Research Technical Report 1987/11, 1987.
[7] Y. Chan, W. Siu, New adaptive pixel decimation for block vector motion estimation, IEEE Trans. Circuits Syst. Video Technol. 6 (1) (1996) 113–118.
[8] B. Liu, A. Zaccarin, New fast algorithms for the estimation of motion vectors, IEEE Trans. Circuits Syst. Video Technol. 3 (2) (1993) 148–157.
[9] K. Nam, J. Kim, R. Park, Y. Shim, A fast hierarchical motion vector estimation algorithm using mean pyramid, IEEE Trans. Circuits Syst. Video Technol. 5 (4) (1995) 344–351.
[10] J. Jain, A. Jain, Displacement measurement and its application in interframe image coding, IEEE Trans. Commun. 29 (12) (1981) 1799–1808.
[11] T. Koga, K. Linuma, A. Hirano, T. Ishiguro, Motion compensated interframe coding for video conferencing, in: Proc. National Telecommunications Conference, 1981, pp. G5.3.1–G5.3.5.
[12] S. Zhu, K. Ma, A new diamond search algorithm for fast block matching motion estimation, IEEE Trans. Image Process. 9 (2) (2000) 287–290.
[13] C. Zhu, X. Lin, L. Chau, Hexagon-based search pattern for fast block motion estimation, IEEE Trans. Circuits Syst. Video Technol. 12 (5) (2002) 349–355.
[14] Rosewarne, C. et al., High Efficiency Video Coding (HEVC) Test Model 16 (HM 16), JCT-VC meeting JCTVC-AB1002, ITU-T and ISO/IEC, 2017.
[15] G. Sullivan, et al., Overview of the high efficiency video coding (HEVC) standard, IEEE Trans. Circuits Syst. Video Technol. 22 (12) (2012) 1649–1668.
[16] A. Tourapis, F. Wu, S. Li, Direct mode coding for bipredictive slices in the H. 264 standard, IEEE Trans. Circuits Syst. Video Technol. 15 (1) (2005) 119–126.
[17] A. Tourapis, O. Au, M. Liou, Predictive motion vector field adaptive search technique (PMVFAST): enhancing block-based motion estimation, in: Proc. SPIE 4310, Visual Communications and Image Processing, 2001, pp. 883–893.
[18] A. Tourapis, Enhanced predictive zonal search for single and multiple frame motion estimation, in: Proc. SPIE 4671, Visual Communications and Image Processing, 2002, pp. 1069–1078.
[19] Sohm, O., Method for motion vector estimation, US Patent 7,260,148, 2007.
[20] M. Al-Mualla, C. Canagarajah, D. Bull, Simplex minimization for single- and multiple-reference motion estimation, IEEE Trans. Circuits Syst. Video Technol. 11 (12) (2001) 1209–1220.

CHAPTER

9

The block-based hybrid video codec

In Chapter 8 we saw how motion estimation formed the basis of an efficient video coding system. This chapter describes how it can be integrated into a practical video compression framework based on a hybrid structure, combining block-based motion prediction with block-based transform coding. The architecture of a generic video encoder is presented and the operation of the encoding and decoding stages are described in detail.

The limitations of the basic hybrid architecture are then discussed and a number of practical codec enhancements that have been introduced in recent standards are described. These techniques serve to significantly enhance the performance of the basic codec and have enabled the basic hybrid architecture to provide a doubling of coding gain every 8–10 years over the past 40 years or so. The techniques described are not an exhaustive list of all the features found in modern codecs, but include some of the most significant ones. Standard-specific details are provided in Chapter 12.

We first of all focus on intra-prediction – a means of enhancing performance through spatial prediction of blocks in each frame. We then extend the motion prediction methods introduced in Chapter 8 through subpixel estimation and the use of multiple reference frames. We also examine the influence of variable block sizes and their exploitation in the coding process, in terms of both transforms and motion estimation. Finally we study the important area of in-loop filtering as an approach to reduce edge artifacts.

9.1 The block-based hybrid model for video compression

9.1.1 Picture types and prediction modes

Prediction modes

As we saw in Chapter 8, three main classes of prediction are used in video compression:

- **Forward prediction**: The reference picture occurs temporally before the current picture.
- **Backward prediction**: The reference picture occurs temporally after the current picture.
- **Biprediction**: Two (or more) reference pictures (forward and/or backward) are employed and the candidate predictions are combined in some way to form the final motion vector.

Picture types and coding structures

Three major types of picture (or frame) are employed in most video codecs:

Intelligent Image and Video Compression
https://doi.org/10.1016/B978-0-12-820353-8.00018-9

- **I-pictures or slices**: These are intra-coded (coded without reference to any other pictures).
- **P-pictures or slices**: These are inter-coded with forward (or backward) prediction from one other I- or P-picture (unipredicted).
- **B-pictures or slices**: These are inter-coded with prediction normally from two I-, P-, and/or B-pictures (bipredicted). The resultant prediction can be either one of the source pictures or a weighted combination of them.
- **Instantaneous decoder refresh (IDR) pictures**: I-pictures that are used to trigger the reset of the decoding process – clearing of all decoded picture buffers. These are used to start a new CVS.

Video sequences are also characterized according to the arrangement of picture types. Common terms, related to these structures, used in recent standards include:

- **GOP (group of pictures)**: A repeated pattern of pictures containing one key picture (an I- or P-picture) and zero or more B-pictures.
- **Coded video sequence (CVS):** Used in HEVC and VVC – a sequence of pictures that can be decoded independently of any other CVS.
- **Intra period**: The number of pictures from one I-picture up to, but not including, the next I-picture.
- **Random access (RA)**: A coding mode offering high compression performance, typically used in streaming applications where access to the decoded sequence is limited to specific random access points (random access points pictures) and these must only contain I-slices.
- **Low delay (LD)**: A coding sequence with no structural delay – where the output order and coding order are identical.
- **All intra (AI)**: All pictures in the sequence are intra-coded.

Coded pictures are arranged in a sequence known as a *group of pictures* (GOP).[1] The accepted definition of a GOP is that it will contain one *key picture* (an I- or P-picture) and zero or more B-pictures.[2] Typical GOP structures are shown in Fig. 9.1. In the first case the intra period is 12 frames and the GOP length is 3 frames. The 12-picture sequence in Fig. 9.1 (a) is sometimes referred to as an IBBPBBPBBPBB structure and it is clear, for reasons of causality, that the encoding order is different from that shown in the figure since the P-pictures must be encoded prior to the preceding B-pictures. The hierarchical structure depicted in Fig. 9.1 (b) is typical of GOP structures used in H.264/AVC, H.265/HEVC, and H.266/VVC. Further details of these coding structures can be found in Chapter 12.

The number of bits consumed by different types of picture in a sequence can vary significantly, with I-pictures typically requiring more bits than P-pictures and these more than B-pictures. An example of the bit consumption pattern for the GOP structure shown in Fig. 9.1 (a) is shown in Fig. 9.2.

[1] Standards since HEVC formally no longer use the term GOP. It does however continue to be widely used in practice and so we will continue to use it here.

[2] This differs slightly from earlier definitions (e.g., in MPEG-2 where a GOP described an intra period).

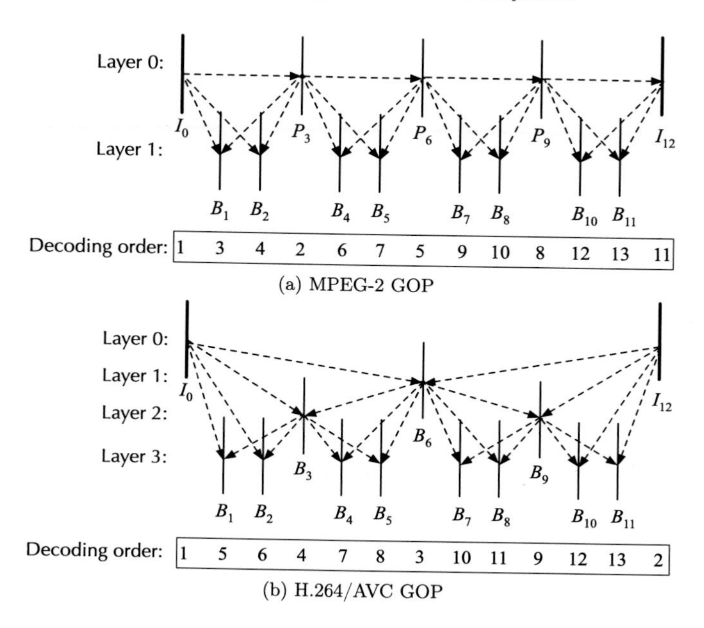

FIGURE 9.1 Typical group of pictures (GOP) structures in (a) MPEG-2 and (b) H.264/AVC.

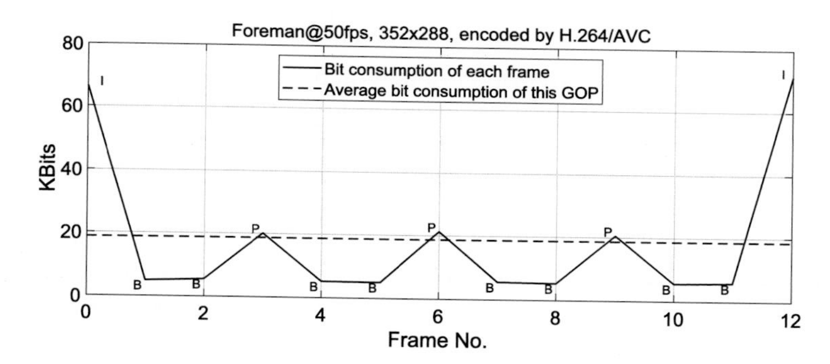

FIGURE 9.2 Typical bit consumption for the GOP structure shown in Fig. 9.1 (a).

9.1.2 Properties of the DFD signal

The DCT is a special case of the Karhunen–Loeve expansion for stationary and first order Markov processes whose autocorrelation coefficient approaches unity. Most still images possess good correlation properties, typically with $\rho \geq 0.9$; this justifies the use of the DCT for intraframe coding as it will provide close to optimal decorrelation. However, in the case of interframe coding, where motion is predicted through motion estimation in a DPCM loop, the autocorrelation coefficient for the DFD signal typically falls to 0.3 to 0.5. Strobach [1] showed that, in many practical cases, the DCT coding gain for a DFD signal is small compared to the intraframe case and that most of the gain is associated with temporal decorrelation. The

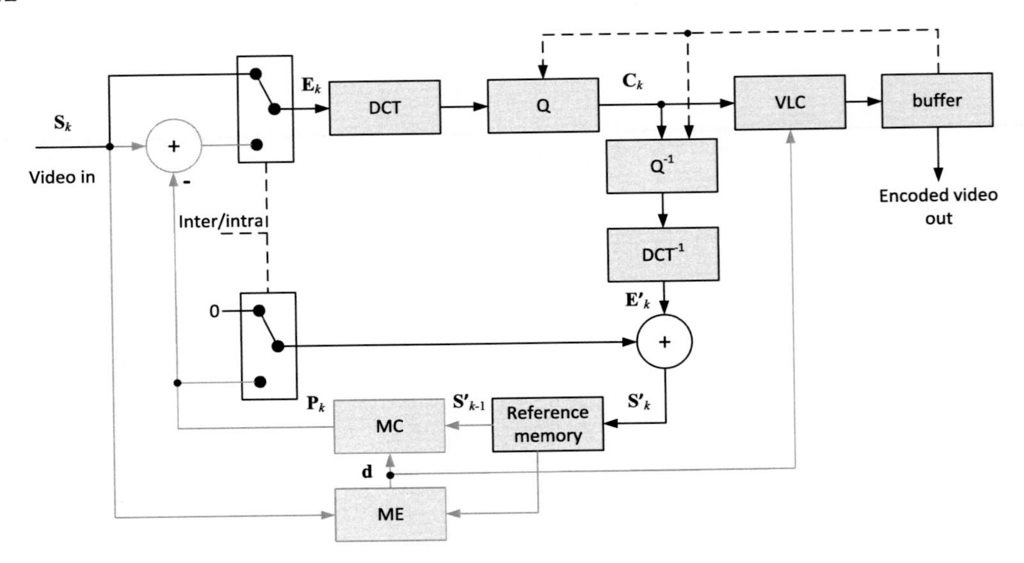

FIGURE 9.3 Video encoder structure (intraframe mode). Dotted lines depict control signals relating to quantized step size, etc.

edges introduced in the DFD signal due to motion failure can also cause ringing artifacts after quantization.

This discussion implies that the hybrid codec, which uses a decorrelating transform such as the DCT to code the DFD residual, is not always optimal. In practice, however, the DFD can be made smoother (i.e., exhibit higher autocorrelation values) if the motion model is more accurate and if any artificial edges are filtered out. These characteristics are promoted by codec enhancements such as subpixel motion estimation, variable block sizes, multiple reference frames, and the use of a deblocking filter in the motion prediction loop. Such enhancements, alongside efficient rate-distortion optimization methods, are largely responsible for the performance, and universal acceptance, of the block-based hybrid codec. They are all discussed in more detail later in this chapter.

9.1.3 Operation of the video encoding loop

Building on the brief overview of the video encoding architecture in Chapter 1, we describe here the operation of the encoding loop more formally and in more detail. A generic structure is given in Fig. 9.3 for the case of intraframe coding and in Fig. 9.4 for the interframe mode. The operation of the encoder in intra-mode is described in Algorithm 9.1. Similarly the operation in inter-mode is described in Algorithm 9.2.

9.1.4 Operation of the video decoder

The operation of the video decoder is illustrated in Fig. 9.5 and described formally in Algorithms 9.3 and 9.4. We can see, by comparing the encoder and decoder architectures, that the

Algorithm 9.1 Intra-mode encoder operation.

1. Inter/intra switch is in the intra position;
2. Compute displaced frame difference (DFD) signal (equal to the video input frame in this case): $\mathbf{E}_k = \mathbf{S}_k$;
3. Perform a forward decorrelating transform (typically a DCT or variant) on the input frame and quantize according to the prevailing rate-distortion criteria: $\mathbf{C}_k = \mathrm{Q}\left(\mathrm{DCT}\left(\mathbf{E}_k\right)\right)$;
4. Entropy code the transformed frame and transmit to the channel;
5. Inverse quantize $\mathbf{C}_k$ and perform an inverse DCT to produce the same decoded frame pixel values as at the decoder: $\mathbf{E}'_k = \mathrm{DCT}^{-1}\left(\mathrm{Q}^{-1}\left(\mathbf{C}_k\right)\right)$;
6. Update the reference memory with the reconstructed frame: $\mathbf{S}'_k = \mathbf{E}'_k + \mathbf{0}$.

Algorithm 9.2 Inter-mode encoder operation.

1. Inter/intra switch is in the inter position;
2. Estimate motion vector for current frame: $\mathbf{d} = \mathrm{ME}\left(\mathbf{S}_k, \mathbf{S}_{k-1}\right)$;
3. Form the motion-compensated prediction frame, $\mathbf{P}_k$: $\mathbf{P}_k = \mathbf{S}'_{k-1}\left[\mathbf{p} + \mathbf{d}\right]$;
4. Compute the DFD signal: $\mathbf{E}_k = \mathbf{S}_k - \mathbf{P}_k$;
5. Perform a forward decorrelating transform on the DFD and quantize according to the prevailing rate-distortion criteria: $\mathbf{C}_k = \mathrm{Q}\left(\mathrm{DCT}\left(\mathbf{E}_k\right)\right)$;
6. Entropy code the transformed DFD, motion vectors and control parameters and transmit to the channel;
7. Inverse quantize $\mathbf{C}_k$ and perform an inverse DCT to produce the same decoded frame pixel values as at the decoder: $\mathbf{E}'_k = \mathrm{DCT}^{-1}\left(\mathrm{Q}^{-1}\left(\mathbf{C}_k\right)\right)$;
8. Update the reference memory with the reconstructed frame: $\mathbf{S}'_k = \mathbf{E}'_k + \mathbf{P}_k$.

Algorithm 9.3 Intra-mode decoder operation.

1. Inter/intra switch is in the intra position;
2. Perform entropy decoding of control parameters and quantized DFD coefficients;
3. Inverse quantize $\mathbf{C}_k$ and perform an inverse DCT to produce the decoded frame pixel values: $\mathbf{E}'_k = \mathrm{DCT}^{-1}\left(\mathrm{Q}^{-1}\left(\mathbf{C}_k\right)\right)$;
4. Update the reference memory with the reconstructed frame and output to file or display: $\mathbf{S}'_k = \mathbf{E}'_k + \mathbf{0}$.

encoder contains a complete replica of the decoder in its prediction feedback loop. This ensures that (in the absence of channel errors) there is no drift between the encoder and decoder operations.

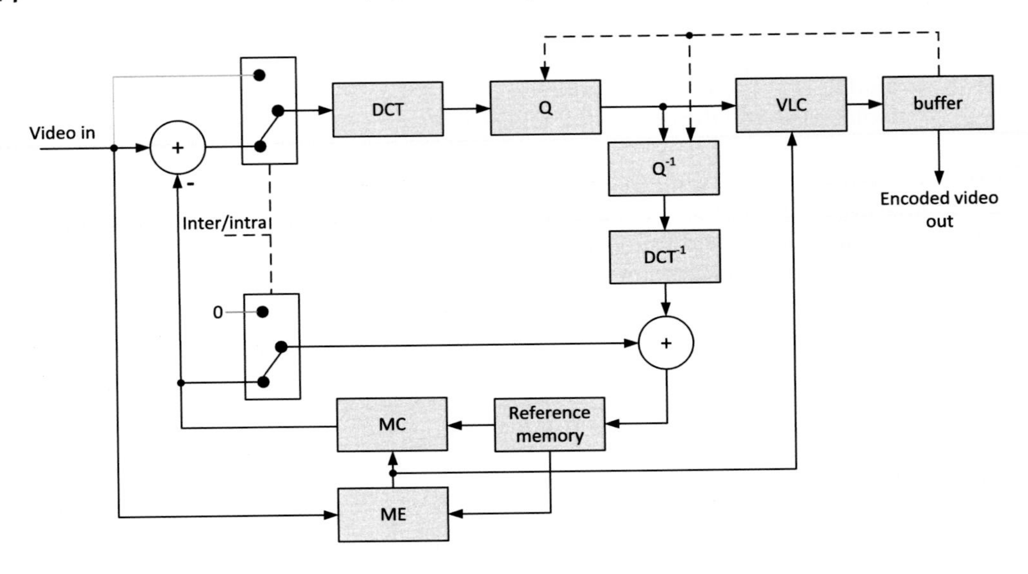

FIGURE 9.4 Video encoder structure (interframe mode).

Algorithm 9.4 Inter-mode decoder operation.

1. Inter/intra switch is in the inter position;
2. Perform entropy decoding of control parameters, quantized DFD coefficients and motion vector;
3. Inverse quantize $\mathbf{C}_k$ and perform an inverse DCT to produce the decoded DFD pixel values: $\mathbf{E}'_k = \mathrm{DCT}^{-1}\left(\mathrm{Q}^{-1}(\mathbf{C}_k)\right)$;
4. Form the motion-compensated prediction frame, $\mathbf{P}_k$: $\mathbf{P}_k = \mathbf{S}'_{k-1}[\mathbf{p}+\mathbf{d}]$;
5. Update the reference memory with the reconstructed frame and output to file or display: $\mathbf{S}'_k = \mathbf{E}'_k + \mathbf{P}_k$.

Example 9.1. Operation of a basic video encoder

Based on the encoder descriptions in Algorithm 9.2 and Fig. 9.4 together with the following data at time k, compute:

1. The current integer-pixel motion vector, $\mathbf{d}$, at time k. Assume the search window is the whole reference frame and that vectors are restricted to point within the frame.
2. The motion-compensated output, $\mathbf{P}_k$, at time k.
3. The DFD for the current input frame, $\mathbf{E}_k$.
4. The transformed and quantized DFD output, $\mathbf{C}_k$.

Assume an image frame size of 12×12 pixels and a macroblock size of 4×4 pixels. The codec is operating in its interframe mode, using block-based translational motion estimation with a block size of 4 × 4 pixels. We have

transform matrix: $\mathbf{A} = \frac{1}{2}\begin{bmatrix} 1 & 1 & 1 & 1 \\ 1 & 1 & -1 & -1 \\ 1 & -1 & -1 & 1 \\ 1 & -1 & 1 & -1 \end{bmatrix}$,

quantization matrix: $\mathbf{Q} = \begin{bmatrix} 2 & 2 & 2 & 2 \\ 2 & 2 & 2 & 2 \\ 2 & 2 & 4 & 4 \\ 2 & 2 & 2 & 4 \end{bmatrix}$,

current input block: $\mathbf{S}_k = \begin{bmatrix} 6 & 6 & 6 & 6 \\ 8 & 8 & 8 & 8 \\ 6 & 6 & 6 & 6 \\ 8 & 8 & 8 & 8 \end{bmatrix}$; assume top left of frame;

reference memory: $\mathbf{S}'_{k-1} = \begin{bmatrix} 0 & 0 & 5 & 6 & 2 & 6 & 0 & 0 & 0 & 0 & 0 & 0 \\ 0 & 0 & 8 & 8 & 8 & 8 & 0 & 0 & 0 & 0 & 0 & 0 \\ 0 & 0 & 6 & 6 & 6 & 6 & 0 & 0 & 0 & 0 & 0 & 0 \\ 0 & 0 & 8 & 8 & 12 & 8 & 0 & 0 & 0 & 0 & 0 & 0 \\ 0 & 0 & 0 & 0 & 0 & 0 & 0 & 0 & 0 & 0 & 0 & 0 \\ 0 & 0 & 0 & 0 & 0 & 0 & 0 & 0 & 0 & 0 & 0 & 0 \\ 0 & 0 & 0 & 0 & 0 & 0 & 0 & 0 & 0 & 7 & 0 & 0 \\ 0 & 0 & 0 & 0 & 0 & 0 & 0 & 3 & 4 & 2 & 3 & 4 \\ 0 & 0 & 0 & 0 & 0 & 0 & 0 & 4 & 4 & 5 & 8 & 0 \\ 7 & 7 & 8 & 9 & 0 & 0 & 0 & 0 & 6 & 7 & 0 & 0 \\ 0 & 0 & 0 & 10 & 3 & 0 & 0 & 0 & 0 & 7 & 0 & 0 \\ 0 & 0 & 2 & 2 & 2 & 0 & 0 & 0 & 0 & 0 & 0 & 0 \end{bmatrix}$

Solution. 1. Current motion vector:
By inspection,

$$\mathbf{d} = [2, 0].$$

2. Motion-compensated output:

$$\mathbf{P}_k = \begin{bmatrix} 5 & 6 & 2 & 6 \\ 8 & 8 & 8 & 8 \\ 6 & 6 & 6 & 6 \\ 8 & 8 & 12 & 8 \end{bmatrix}.$$

3. DFD for current input frame:

$$\mathbf{E}_k = \mathbf{S}_k - \mathbf{P}_k = \begin{bmatrix} 1 & 0 & 4 & 0 \\ 0 & 0 & 0 & 0 \\ 0 & 0 & 0 & 0 \\ 0 & 0 & -4 & 0 \end{bmatrix}.$$

4. Transformed and quantized output:
The transformed output is

$$\mathbf{C}_k = \mathbf{A}\mathbf{E}_k\mathbf{A}^{\mathrm{T}} = \frac{1}{4}\begin{bmatrix} 1 & 1 & 1 & 1 \\ 9 & -7 & -7 & 9 \\ 1 & 1 & 1 & 1 \\ 9 & -7 & -7 & 9 \end{bmatrix}$$

and the quantized output is

$$\mathbf{C}_{k(\mathbf{Q})} = \frac{1}{4}\begin{bmatrix} 1 & 1 & 1 & 1 \\ 9 & -7 & -7 & 9 \\ 1 & 1 & 1 & 1 \\ 9 & -7 & -7 & 9 \end{bmatrix} ./\mathbf{Q} = \begin{bmatrix} 0 & 0 & 0 & 0 \\ 1 & -1 & -1 & 1 \\ 0 & 0 & 0 & 0 \\ 1 & -1 & -1 & 1 \end{bmatrix}.$$

9.2 Intraframe prediction

In Chapter 8 we saw the benefits of temporal prediction as a means of decorrelating between frames prior to transformation. Modern video coding standards (such as AVC, HEVC, and VVC) also support intra-prediction in the spatial domain, as well as PCM modes that enable direct encoding of pixel values. Intra-coding can be especially beneficial in regions which have certain oriented textures or directional edges. The main prediction modes supported in H.264/AVC are intra_4×4, intra_8×8, and intra_16×16. HEVC extends this up to block sizes of 32×32. The basic principles of intra-prediction are described below using H.264/AVC as

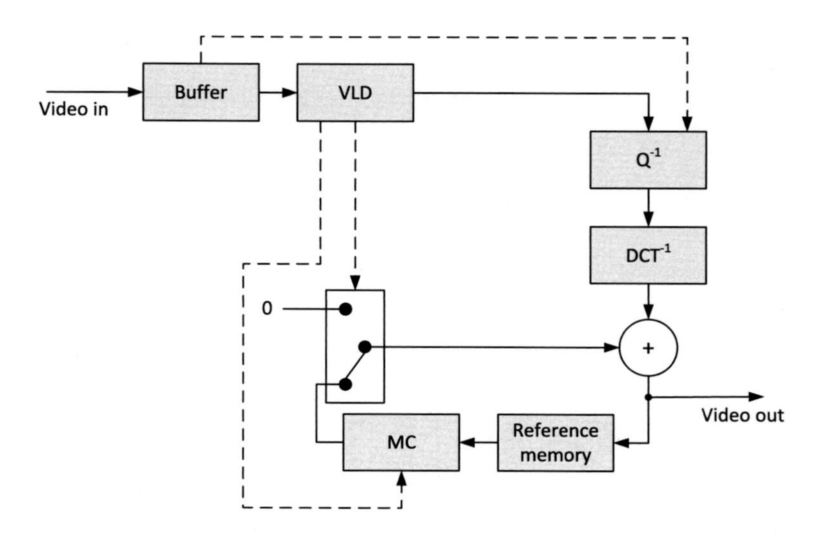

FIGURE 9.5 Video decoder structure.

M	A	B	C	D	E	F	G	H
I	a	b	c	d				
J	e	f	g	h				
K	i	j	k	l				
L	m	n	o	p				

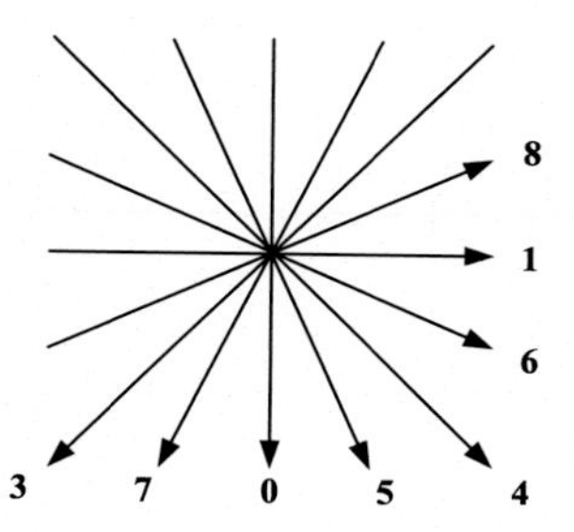

FIGURE 9.6 Intra-prediction modes in H.264/AVC.

an example. For a more detailed coverage of intra-prediction in other standards, the reader is referred to Chapter 10.

9.2.1 Intra-prediction for small luminance blocks

This mode performs well when there is a high level of local detail, particularly with oriented features such as edges. Coding of a given block is performed with reference to a subset of samples from previously coded blocks (available at encoder and decoder) that lie to the left and above the current block.

Fig. 9.6 shows, for example, the intra-coding relationships and orientations used in the 4×4 mode of H.264/AVC. The prediction for each 4×4 block (pixels a–p here) is based on its neighboring (previously decoded) pixels (A–M). In H.264/AVC, the predictor can select from nine prediction orientations plus a DC mode where the average of the neighboring pixels is used as the prediction. These nine modes are shown in Fig. 9.7. Some examples of how the predictions are computed are given below.

Mode 0 (vertical):

$$\{a, e, i, m\} = A. \tag{9.1}$$

Mode 1 (horizontal):

$$\{a, b, c, d\} = I. \tag{9.2}$$

Mode 2 (DC):

$$\{a, b, \cdots, p\} = \left\lfloor (\frac{A+B+C+D+I+J+K+L}{8}) + 0.5 \right\rfloor. \tag{9.3}$$

Mode 4 (down-right):

$$\{a, f, k, p\} = \left\lfloor \left(\frac{A+2M+I}{4}\right) + 0.5 \right\rfloor. \tag{9.4}$$

Other modes are formed by a similar extrapolation approach and a full description can be found in [2]. Standard-specific variations are discussed in Chapter 12.

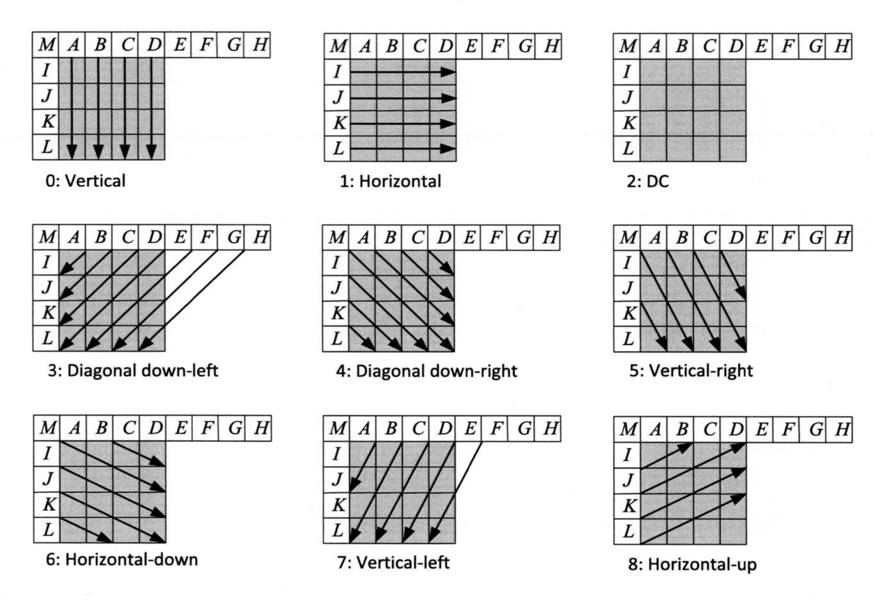

FIGURE 9.7 4×4 intra-prediction modes.

9.2.2 Intra-prediction for larger blocks

For larger smoother image regions it can be beneficial to perform intra-prediction on larger blocks. In the H.264 intra_16×16 mode, the entire 16×16 luma block is predicted simultaneously. The mode supports only four prediction types: Mode 0 (vertical), mode 1 (horizontal), mode 2 (DC), and mode 3 (plane). The first three modes are similar to the 4×4 modes described above except they reference 16 pixels above and to the left of the current block, rather than just four. Details of the plane mode can be found in [2].

Chroma samples can be predicted in a similar manner to that used for luma blocks. Since chroma signals normally exhibit smoother characteristics than luma signals, only larger block sizes tend to be used.

As we will see in Chapter 11, intra-coding can lead to the propagation of errors between frames when it is based on data predicted from previously inter-coded blocks. To avoid this, a *constrained intra-coding mode* can be used that restricts intra-prediction only to be based on intra-coded neighbors.

Example 9.2. Intra-prediction

Compute the sum of absolute differences (SAD) values using the intra_4×4 vertical, horizontal, and DC modes for the 4×4 block of pixels shown below.

1	2	3	4	5	6	7	8	9
1	2	3	4	5				
1	2	2	4	6				
2	2	3	4	3				
2	2	3	4	5				

Solution. The prediction blocks based on vertical, horizontal, and DC modes are given in the figure below.

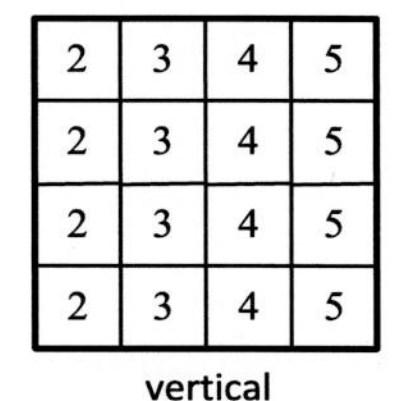

2	3	4	5
2	3	4	5
2	3	4	5
2	3	4	5

vertical

1	1	1	1
1	1	1	1
2	2	2	2
2	2	2	2

horizontal

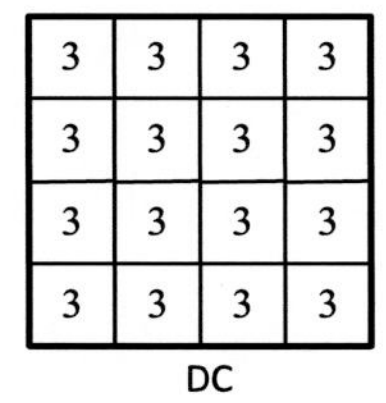

3	3	3	3
3	3	3	3
3	3	3	3
3	3	3	3

DC

The corresponding SAD values are SAD(vertical)=4; SAD(horizontal)=30; SAD(DC)=16. Hence, based on this limited subset, the best match in terms of prediction residual energy is mode 0 (vertical). In this case the residual prior to coding would be as shown below.

0	0	0	0
0	-1	0	1
0	0	0	-2
0	0	0	0

9.3 Subpixel motion estimation

9.3.1 Subpixel matching

In Chapter 8 we introduced the concepts and realization of motion estimation as the basis for efficient video encoding. We limited the resolution of our search and matching to that of the sampling grid used to acquire the images. In reality, the motion in a scene is not related to the sampling grid and the most accurate match might occur with a subpixel displacement. Ericsson [3] introduced the concept of subpixel estimation in 1985 demonstrating, for the sequences used, that up to 2 dB coding gain could be achieved when changing from full-pixel to 1/8-pixel accuracy. This was formalized by Girod in 1993 [4], who showed that limited gains were achieved by decreasing the estimation accuracy beyond 1/8-pixel resolution.

It is clear that the benefits of subpixel estimation will depend on a number of factors, including video format (spatial and temporal resolution) and type of content. However, it is

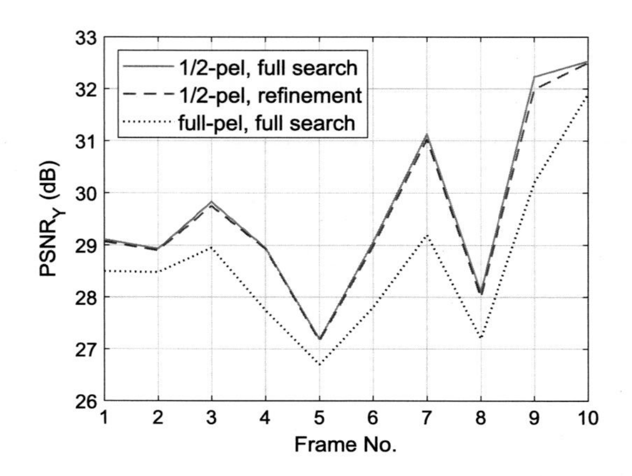

FIGURE 9.8 Subpixel search with local refinement.

acknowledged that subpixel estimation offers significant performance gains and modern coding standards all include the option to predict motion with fractional-pixel accuracy. MPEG-1 was the first international standard to use half-pixel estimation, closely followed by MPEG-2 in 1994. H.263 introduced more flexible block sizes and quarter-pixel estimation in 1996 and its improvement in performance over H.261 has been largely attributed to this innovation [5].

It is worth spending a few moments discussing the interpolation needed to support subpixel motion estimation. We could adopt a number of approaches:

1. Interpolate all of the current frame block and all of the search window and perform the entire motion estimation process at subpixel resolution.
2. Interpolate just the search window and proceed as in (1).
3. Perform integer-pixel search initially, then refine this locally using an interpolated search window and an interpolated current block.
4. Perform integer search initially, then refine this locally just interpolating the search window.

In practice method (4) is adopted, as it offers the lowest complexity with only a minor fall-off in performance (see Fig. 9.8). This approach is illustrated in Fig. 9.9, which shows the black square as the best full-pixel solution. The grid is then interpolated locally at half-pixel resolution and a refinement search is performed to obtain the best half-pixel solution (black circle). The refinement search is normally performed exhaustively and the same matching criterion is used as for the integer-pixel case.

9.3.2 Interpolation methods

Using H.264/AVC as an example, Fig. 9.10 illustrates the interpolation method used for half-pixel and quarter-pixel interpolation. In the case of half-pixel estimation, the search window is interpolated locally around the optimal integer-pixel result. If we assume that the pixels in the frame's search window are locally upsampled by a factor of two (see Chapter 6 for a more detailed discussion of upsampling), then the upsampled result in the x direction

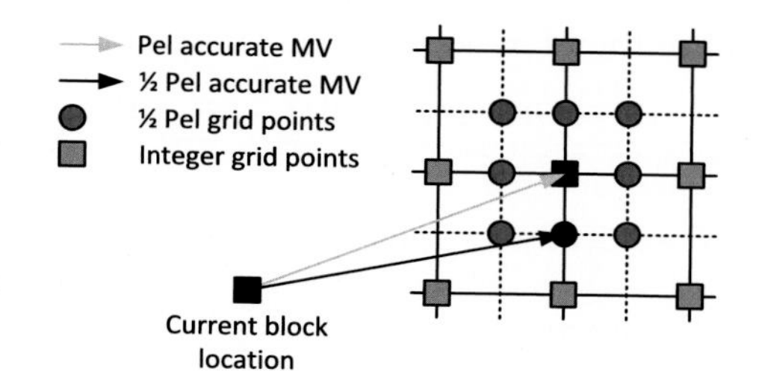

FIGURE 9.9 Motion estimation search grid with half-pixel refinement.

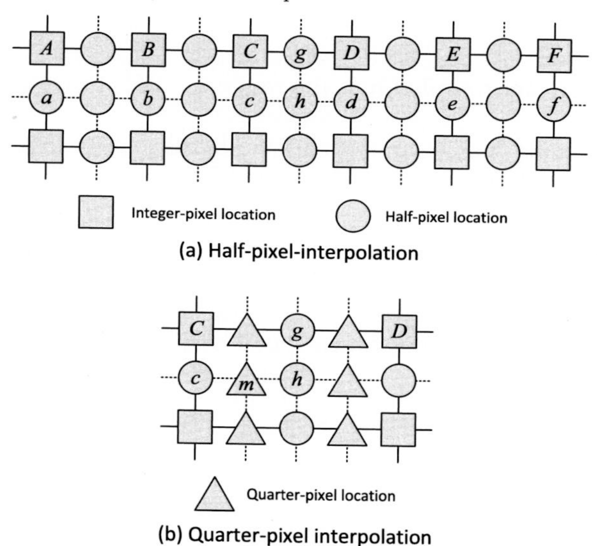

FIGURE 9.10 H.264 subpixel interpolation.

is given by

$$s_u[x] = \begin{cases} s[x/2], & x = 2m, \\ 0, & \text{otherwise}, \end{cases} \tag{9.5}$$

where m is an integer. We can then filter the result to interpolate the half-pixel locations. Normally the filter used is symmetrical and separable to reduce complexity. In the case of H.264/AVC [6], the filter used to create an interpolated half-pixel value, $g[x]$, is as follows:

$$g[x] = \sum_{i=-5}^{+5} h[i] s_u[x-i], \tag{9.6}$$

where the filter coefficients are

$$h[i] = \left\{1, 0, -5, 0, 20, 0, 20, 0, -5, 0, 1\right\}/32. \tag{9.7}$$

In simple terms this means, with reference to the labeling in Fig. 9.10, that

$$g = \frac{(A - 5B + 20C + 20D - 5E + F)}{32}, \tag{9.8}$$

and a similar approach is used vertically and on intermediate pixels such as h, based on pixels $\{a, b, c, d, e, f\}$.

In the case of quarter-pixel refinement, a simpler interpolation is employed; for example, to obtain the interpolated quarter-pixel value for location m in Fig. 9.10, we use the following equation:

$$m = \frac{(c + h)}{2}. \tag{9.9}$$

All interpolated results are rounded into the range $0 \cdots 255$, prior to the execution of the motion search phase.

It should be noted that the choice of interpolation filter is important as it can introduce bias (see for example the work of Bellers and de Hann in [7]).

Example 9.3. Subpixel motion estimation

Implement the full search block matching motion estimation algorithm on the 6×6 search window $\mathbf{S}_{k-1}$, using the current frame template $\mathbf{S}_k$, given below. Use a simple 2-tap interpolation filter to refine the integer-pixel result to half-pixel accuracy. We have

$$\mathbf{S} = \begin{bmatrix} 1 & 5 & 4 & 9 & 6 & 1 \\ 6 & 1 & 3 & 8 & 5 & 1 \\ 5 & 7 & 1 & 3 & 4 & 1 \\ 2 & 4 & 1 & 7 & 6 & 1 \\ 2 & 4 & 1 & 7 & 8 & 1 \\ 1 & 1 & 1 & 1 & 1 & 1 \end{bmatrix}, \quad \mathbf{M} = \begin{bmatrix} 3 & 9 \\ 1 & 4 \end{bmatrix}.$$

Solution. SAD values for selected integer-pixel $[d_x, d_y]$ candidates are as follows.

i	j	SAD(i, j)	i	j	SAD(i, j)	i	j	SAD(i, j)
−1	−1	17	0	1	7	−1	−2	8
−1	0	18	1	−1	11	−2	−2	14
−1	1	15	1	0	13	−2	−1	18
0	−1	2	1	1	17	−2	0	5
0	0	11	0	−2	7	2	−2	18

Clearly, by inspection, all other candidate vectors give SAD values greater than 2. Hence the integer-pixel motion vector for this block is $\mathbf{d} = [0, -1]$.

Next compute the half-pixel accurate motion vector. Use a simple 2-tap filter to interpolate the values of the reference frame only. For example, in the horizontal direction

$$s[x+0.5, y] = \frac{s[x, y] + s[x+1, y]}{2}.$$

Once all values have been interpolated it is conventional to round the result prior to computing the subpixel SAD. We can use $\lceil s[x+0.5] - 0.5 \rceil$ to round down values ending in 0.5. Using this approach, the interpolated half-pixel values in the vicinity of the integer-pixel result are given below.

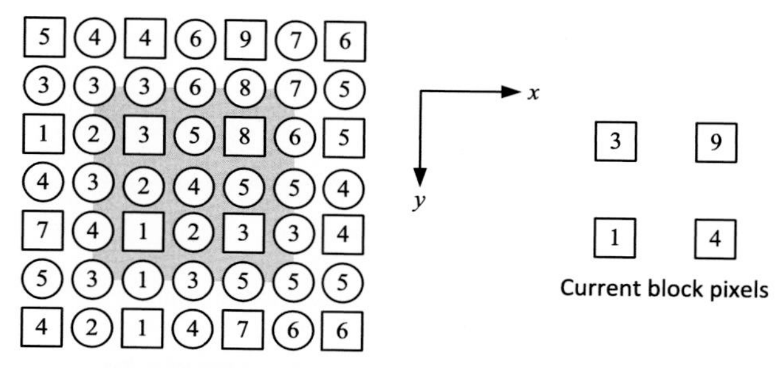

The subpixel SAD values are given in the following table.

i	j	**SAD(i, j)**	i	j	**SAD(i, j)**
0	−1	2	0.5	−1.5	>2
−0.5	−1	10	0	−0.5	>2
0.5	−1	7	−0.5	−0.5	>2
0	−1.5	>2	0.5	−0.5	>2
−0.5	−1.5	>2	–	–	–

Hence the half-pixel motion vector for this block is the same as before, i.e., $\mathbf{d} = [0, -1]$.

9.3.3 Performance

The benefits of fractional-pixel search are shown in Figs. 9.11 and 9.12. Fig. 9.11 shows the PSNR improvements obtained from using half- and quarter-pixel estimation for 10 frames of the *foreman* sequence (CIF at 25 fps). This is based on using the original frame as the reference in each case. It can be seen that up to 1.5 dB gain is achieved in this case. It can also be observed that, while quarter-pixel search does provide additional benefits, the gain is reduced. In most cases search at 1/8-pixel resolution will provide some small additional gain, but going beyond this provides diminishing returns.

Fig. 9.12 shows an example DFD signal for *coastguard*, comparing integer- and half-pixel search methods. The reduction in residual energy is clear, especially in the region of the boat and the riverbank in the background. However, while there is some improvement in the region of the river, the dynamic textures of the water cause severe problems for the trans-

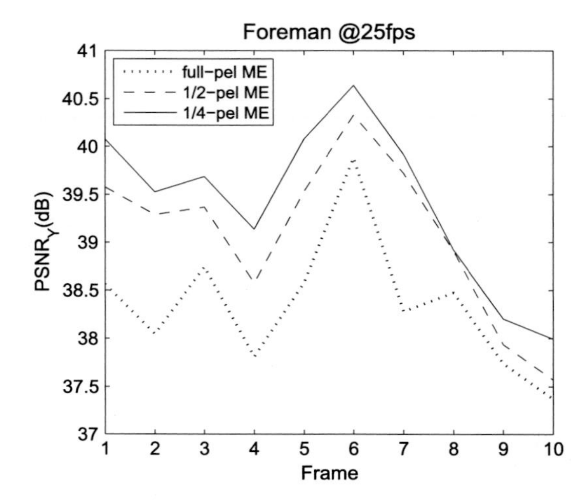

FIGURE 9.11 Comparison of full-, half-, and quarter-pixel motion estimation.

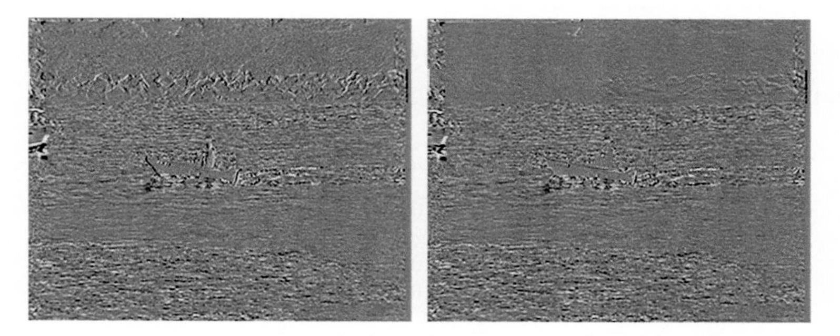

FIGURE 9.12 Influence of subpixel motion estimation on residual error. Left: Integer-pixel motion estimation. Right: Half-pixel motion estimation.

lational motion model. Techniques which model such textures using dynamic models will be discussed further in Chapter 13. It has been found that the use of quarter-pixel accurate motion prediction contributes between 10% and 20% saving in bit rate, compared with the integer-pixel case.

9.3.4 Interpolation-free methods

We have seen that quarter-pixel motion estimation can provide a significant contribution to the rate-distortion performance of a video codec. By using subpixel estimation, based on interpolated search values, the residual energy for each predicted macroblock can be significantly reduced, but only at the cost of increased computational complexity. This latter point is particularly relevant for codecs such as H.264/AVC and HEVC, where the cost of an exhaustive set of macroblock segmentations needs to be estimated for optimal mode selection.

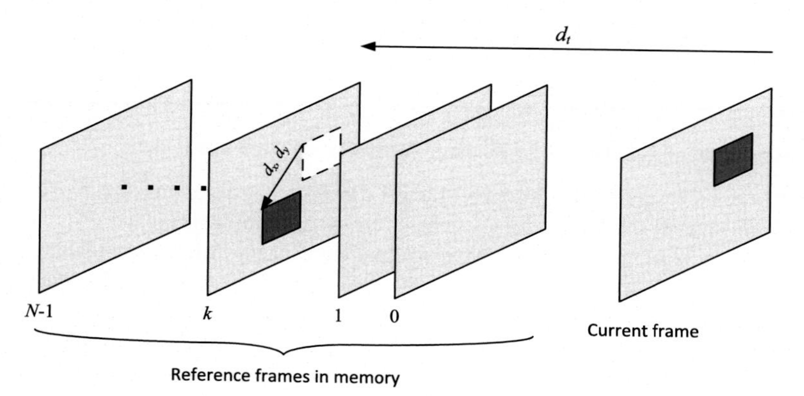

FIGURE 9.13 Multiple-reference frame motion estimation.

To mitigate the cost of interpolation, novel schemes for subpixel motion estimation based solely on the whole-pixel SAD distribution have been proposed. The approach of Hill et al. [8] enables both half-pixel and quarter-pixel searches to be guided by an intelligent model-free estimation of the SAD surface. They use a method based on a 2D parameterized parabolic interpolation between the estimated ambiguity samples, approximating the actual behavior in the presence of pure translations. They also include a fallback strategy that improves the rate-distortion performance to a level close to full interpolation. This approach is faster (in terms of fps) than a fully interpolated system by a factor of between 12% and 20%.

Hill and Bull [9] extended the method of [8] using a 2D kernel approach. This reduces the number of quarter-pixel search positions by a factor of 30%, giving an overall speed-up of approximately 10% compared to the EPZS quarter-pixel method.

9.4 Multiple-reference frame motion estimation

9.4.1 Justification

In conventional motion estimation, only one reference frame is employed. However, it was observed by Wiegand and Girod [10] that such approaches do not exploit long-term statistical dependencies in video sequences and that *multiple-reference frame motion estimation* (MRF-ME) can provide many benefits. Multiple reference frames were first standardized in H.263 and have been a feature in most codecs since that time [6]. Normally the frames employed as references are simply past decoded frames; however, they may also include warped versions to account for various camera motions.

Reference frames are assembled in a reference picture buffer at both the encoder and the decoder, and synchrony of contents must be maintained at both ends. This is illustrated in Fig. 9.13. This is simple when a sliding window update is employed but requires signaling when other, more intelligent updates are used.

9.4.2 Properties, complexity, and performance of MRF-ME

Properties

A good description of the properties of long-term memory is provided by Almualla et al. in [12]. To summarize (assuming a sliding window update method):

1. The distribution of long-term memory spatial displacements is center-biased. This means that $\mathbf{d} = [0, 0, d_z]$ has the highest relative frequency of occurrence.
2. The distribution of long-term memory temporal displacements is zero-biased. This means that the most recent reference frame is the most commonly adopted candidate, i.e., $\mathbf{d} = [0, 0, 0]$ is the most common vector.
3. The long-term memory motion field is smooth and slowly varying. Hence search methods that work well for single-frame search can also offer solutions for the multiple frame case.

Performance and complexity

MRF-ME exploits the long-term statistical dependencies in an image sequence to provide coding gain. However, in order to achieve this, the motion estimation algorithm must not only search across spatial locations in the search grid, but also temporally across multiple candidate frames. With increases in processor power and the increasing integration and lowering cost of memory, this approach is now feasible. Nonetheless it can still present a major computational overhead and reduced-complexity approaches are desirable. Fig. 9.14 (left) illustrates the performance gains possible for the case of *foreman* (QCIF @25 fps). The figure shows the benefits of long-term memory up to 50 reference frames, indicating an increase in PSNR from 33 dB for one frame up to 35.5 dB for 50 frames.

This improvement in coding performance is, however, offset by the additional complexity required to search for the optimal motion vector. This is shown in Fig. 9.14 (right), which demonstrates, for the same sequence, that the number of locations searched per frame (for full search) increases from 80,000 for the case of one reference frame to over 4,000,000 for 50 reference frames. Such an increase in complexity, even with today's technology, could make MRF-ME intractable.

9.4.3 Reduced-complexity MRF-ME

Several methods for improving the efficiency of MRF-ME without compromising its associated coding gain have been proposed. Su and Sun [11] exploit the smoothness and correlations within the block motion field to provide similar coding performance to the H.264/AVC JM but at around 20% of the complexity. Al-Mualla et al. [13] showed that their simplex method can provide benefits for long-term memory motion estimation. The simplex-based approach has a number of useful properties that make it attractive in such situations. For example, it is not based inherently on a unimodal error surface and it offers diversity in its candidate search locations. Table 9.1 presents coding performance and complexity results for this algorithm; this is based on 300 frames of *Foreman* (QCIF @25 fps), using an SAD distortion measure with a block size of 16×16 pixels and a maximum motion displacement of 15 pixels. The search was performed to full-pixel accuracy over 50 reference frames.

It can be seen that significant gains in performance are possible using this approach; a coding gain decrease of only 0.1 dB is accompanied by a speed-up of some 33 times.

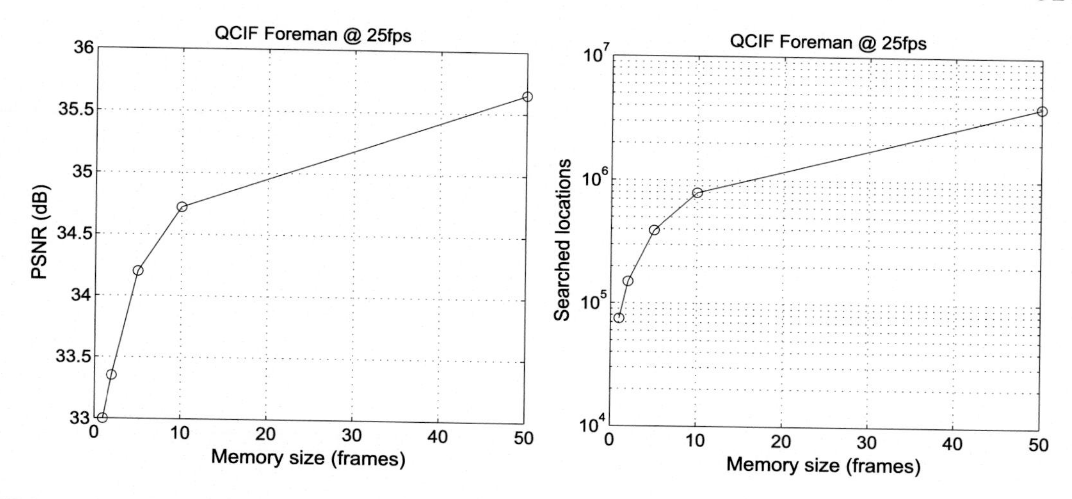

FIGURE 9.14 MRF-ME performance. Left: PSNR. Right: Complexity.

TABLE 9.1 Performance of simplex-based (SMS) multiple-reference frame motion estimation. Results from [13].

Algorithm	FS	MR-FS	MR-SMS
PSNR (dB)	32.2	33.97	33.87
Search locations/frame	77,439	3,554,700	106,830

9.4.4 The use of multiple reference frames in current standards

In current standards such as HEVC and VVC, the reference pictures for a specific frame in a GOP can be predefined in the configuration file which is readable by the encoder. Depending on the GOP structure in use, the default reference frame configurations for HEVC and VVC test models use up to five reference frames, which can be I-, P-, and/or B-pictures. An example of a multiple-reference frame configuration based on the low delay mode (see Section 10.5) is shown in Fig. 9.15, in which the indices of possible reference frames for encoding each frame in the GOP are given. This configuration was optimized alongside frame level quantization parameter (QP) values and rate-distortion optimization parameters.

9.5 Variable block sizes for motion estimation

9.5.1 Influence of block size

One of the main reasons for the performance gains of recent standards has been the introduction of variable block sizes, both for transforms and prediction. This enables the shape and size of the block to be optimized in a rate-distortion sense, taking account of content

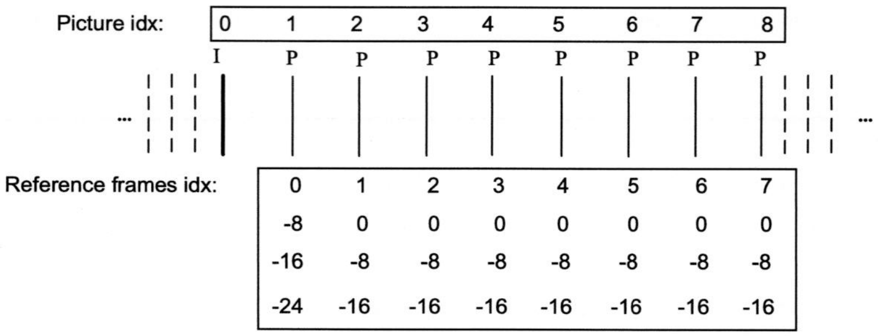

Picture idx:
0 1 2 3 4 5 6 7 8
I P P P P P P P P
...
...
Reference frames idx:
0 1 2 3 4 5 6 7
-8 0 0 0 0 0 0 0
-16 -8 -8 -8 -8 -8 -8 -8
-24 -16 -16 -16 -16 -16 -16 -16

FIGURE 9.15 Multiple-reference frame configuration for the low delay configuration in HEVC and VVC test models.

0
0 1
0
1
0 1
2 3
16x16 modes
0
0 1
0
1
0 1
2 3
8x8 modes

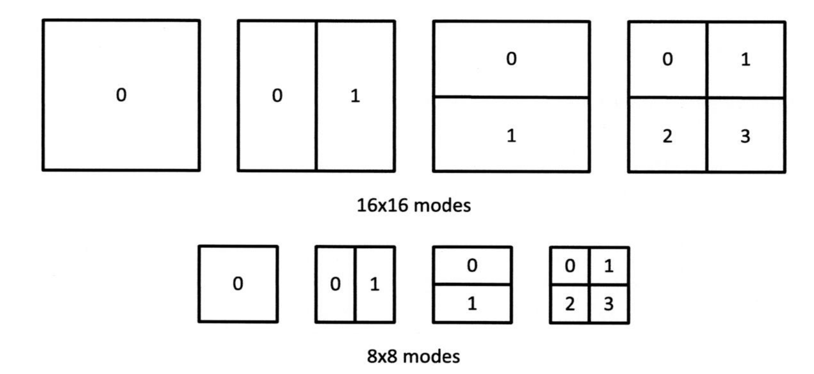

FIGURE 9.16 Variable block sizes supported by H.264/AVC. Top: 16×16 modes 1 to 4. Bottom: 8×8 modes 1 to 4.

type, spatial and temporal characteristics, and bit rate constraints. We saw in Chapter 8 that block size can influence both DFD accuracy and motion bit rate overhead. It thus seems perfectly sensible that block sizes should not be fixed, but instead should be allowed to adapt to content type within a rate-distortion framework.

9.5.2 Variable block sizes in practice

H.264/AVC was the first standard to introduce variable block sizes and this concept has been extended in subsequent standards. The block sizes supported by H.264/AVC are:

- 16×16 **modes**: 16×16, 16×8, 8×16, 8×8,
- 8×8 **modes**: 8×8, 8×4, 4×8, 4×4.

These are shown in Fig. 9.16 and have been found to offer approximately 15% saving over the use of fixed block sizes. The distribution of block sizes in an actual coded frame are shown in Fig. 9.17 for the *foreman* sequence, produced using the Elecard Streameye analyzer [14].

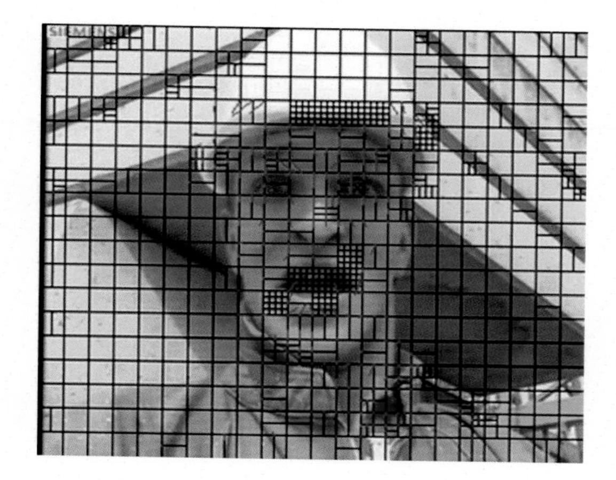

FIGURE 9.17 Block size distribution (*foreman* sequence coded using H.264/AVC). Produced using [14].

9.6 Variable-sized transforms

9.6.1 Integer transforms

Recent standards such as H.264/AVC [6] and HEVC [15] have employed variable-sized transforms rather than the fixed size used previously. For example in H.264/AVC, the smallest block size is 4×4 and a separable integer transform which is closely related to the4×4 DCT is used. This integer transform was introduced by Malvar et al. [16] and offers the following benefits:

- The transform block size is matched to the basic motion estimation block, so it performs better in the presence of complex local motions and regions with random textures.
- Because of the improvements resulting from variable block sizes in HEVC and H.264/AVC, the residual signal after motion compensation has reduced spatial correlation. As discussed previously, this implies that the transform is less efficient and hence a 4×4 integer transform can often be as good as a larger transform in terms of decorrelation.
- It offers the benefits of reducing noise around edges.
- The inverse transform is matched to the forward transform with exact integer operations, and hence encoder–decoder mismatches are eliminated.
- The integer transform is computationally efficient, requiring a smaller processing word length, and can be implemented with just addition and shift operations.

In Chapter 5 we saw that the 1D 4×4 DCT is given by

$$\mathbf{A} = \begin{bmatrix} \frac{1}{2} & \frac{1}{2} & \frac{1}{2} & \frac{1}{2} \\ \sqrt{\frac{1}{2}}\cos(\frac{\pi}{8}) & \sqrt{\frac{1}{2}}\cos(\frac{3\pi}{8}) & -\sqrt{\frac{1}{2}}\cos(\frac{3\pi}{8}) & -\sqrt{\frac{1}{2}}\cos(\frac{\pi}{8}) \\ \frac{1}{2} & -\frac{1}{2} & -\frac{1}{2} & \frac{1}{2} \\ \sqrt{\frac{1}{2}}\cos(\frac{3\pi}{8}) & -\sqrt{\frac{1}{2}}\cos(\frac{\pi}{8}) & \sqrt{\frac{1}{2}}\cos(\frac{\pi}{8}) & -\sqrt{\frac{1}{2}}\cos(\frac{3\pi}{8}) \end{bmatrix} \tag{9.10}$$

$$= \begin{bmatrix} a & a & a & a \\ b & c & -c & -b \\ a & -a & -a & a \\ c & -b & b & -c \end{bmatrix}, \quad \text{where} \quad \begin{aligned} a &= \tfrac{1}{2}, \\ b &= \sqrt{\tfrac{1}{2}}\cos(\tfrac{\pi}{8}), \\ c &= \sqrt{\tfrac{1}{2}}\cos(\tfrac{3\pi}{8}). \end{aligned}$$

Letting $d = \frac{c}{b}$, the integer version of the transform can be written as follows:

$$\mathbf{A} = \begin{bmatrix} 1 & 1 & 1 & 1 \\ 1 & d & -d & -1 \\ 1 & -1 & -1 & 1 \\ d & -1 & 1 & -d \end{bmatrix} \otimes \begin{bmatrix} a & a & a & a \\ b & b & b & b \\ a & a & a & a \\ b & b & b & b \end{bmatrix},$$

where $\otimes$ represents a scalar multiplication. Since $d \approx 0.414$, the next step is to approximate d by a value of 0.5, to give

$$\mathbf{A} = \begin{bmatrix} 1 & 1 & 1 & 1 \\ 2 & 1 & -1 & -2 \\ 1 & -1 & -1 & 1 \\ 1 & -2 & 2 & -1 \end{bmatrix} \otimes \begin{bmatrix} a & a & a & a \\ \frac{b}{2} & \frac{b}{2} & \frac{b}{2} & \frac{b}{2} \\ a & a & a & a \\ \frac{b}{2} & \frac{b}{2} & \frac{b}{2} & \frac{b}{2} \end{bmatrix}. \tag{9.11}$$

Since for all row vectors, $\mathbf{a}_i \mathbf{a}_j^{\mathrm{T}} = 0$, the basis vectors remain orthogonal. To make the matrix orthonormal, we must ensure that $\|\mathbf{a}_i\| = 1$. Hence,

$$a = \frac{1}{2}, \quad b = \sqrt{\frac{2}{5}}, \quad d = \frac{1}{2}.$$

Extending this separable transform to process 2D signals gives

$$\begin{aligned} \mathbf{C} &= \mathbf{A}\mathbf{S}\mathbf{A}^{\mathrm{T}} \\ &= \begin{bmatrix} 1 & 1 & 1 & 1 \\ 2 & 1 & -1 & -2 \\ 1 & -1 & -1 & 1 \\ 1 & -2 & 2 & -1 \end{bmatrix} [\mathbf{S}] \begin{bmatrix} 1 & 2 & 1 & 1 \\ 1 & 1 & -1 & -2 \\ 1 & -1 & -1 & 2 \\ 1 & -2 & 1 & -1 \end{bmatrix} \otimes \mathbf{E}, \end{aligned} \tag{9.12}$$

where

$$\mathbf{E} = \begin{bmatrix} a^2 & \frac{ab}{2} & a^2 & \frac{ab}{2} \\ \frac{ab}{2} & \frac{b^2}{4} & \frac{ab}{2} & \frac{b^2}{4} \\ a^2 & \frac{ab}{2} & a^2 & \frac{ab}{2} \\ \frac{ab}{2} & \frac{b^2}{4} & \frac{ab}{2} & \frac{b^2}{4} \end{bmatrix}.$$

Thus we can perform the forward and inverse transform operations as $\mathbf{S} = \mathbf{A}\mathbf{C}\mathbf{A}^{\mathrm{T}}$, with the scalar multiplication by $\mathbf{E}$ absorbed into the quantization process.

9.6.2 DC coefficient transforms

We saw earlier that intra-prediction for luma and chroma signals can provide significant benefits for coding nontextured regions. If a smoothly varying region extends across the whole coding block, then additional benefits can be obtained by performing a further transform on the DC coefficients from the 4×4 transforms. In H.264, a 2×2 transform is also applied to the DC coefficients of the four 4×4 blocks of each chroma component. This provides benefits because, for smoothly varying regions, the autocorrelation coefficient will approach unity and the reconstruction accuracy is proportional to the inverse of the size of the transform. For nontextured regions the reconstruction error for an 8×8 transform is therefore half of that for a 4×4 transform.

Example 9.4. Reduced-complexity integer transform

Consider the forward integer 4×4 transform given in Eq. (9.11). Show that this can be implemented using only shift and add operations and compute its complexity.

Solution. The architecture for the fast integer transform is given below.

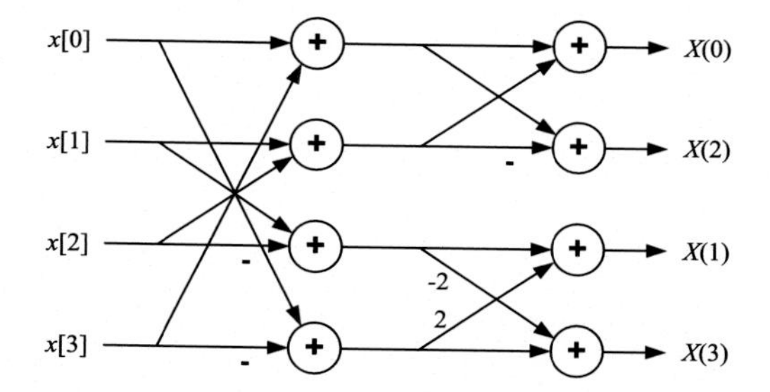

It can be observed that the transform requires only eight additions/subtractions and two 1-bit shifts.

9.7 In-loop deblocking operations

Blocking artifacts can arise in the hybrid video codec because of motion estimation and transform coding of the DFD signal. Because of this, deblocking filters have been used in various forms in standardized video codecs since H.261. They have been proven to improve both objective and subjective performance and have become progressively more sophisticated over time. Wiegand et al. [6] and List et al. [17] provide excellent overviews of the deblocking filter used in H.264/AVC; a summary is given below.

Deblocking filters perform best when they are integrated into the motion prediction loop, as this eliminates drift and enables benefits for all frame types. The idea of the deblocking operation is illustrated in Fig. 9.18. The edge samples are filtered according to a set of criteria that relate to the prevailing quantization conditions coupled with estimates of whether the edge is real or induced by coding. The absolute differences of pixels near a block edge are first computed and then referenced against the likely size of any quantization-induced artifact. If the edge is larger than any that might have been induced by coding, then it is most likely to be an actual edge and should not be filtered. It has been reported [6] that the incorporation of the deblocking filter saves as much as 10% in bit rate for equivalent quality.

Algorithm 9.5 Deblocking operation.

1. Set thresholds α (QP) and β (QP);
2. Filter a_0 and b_0 iff: $|a_0 - b_0| < \alpha$ (QP) AND $|a_1 - a_0| < \beta$ (QP) AND $|b_1 - b_0| < \beta$ (QP);
3. Filter a_1 and b_1 iff: $|a_2 - a_0| < \beta$ (QP) OR $|b_2 b_0| < \beta$ (QP).

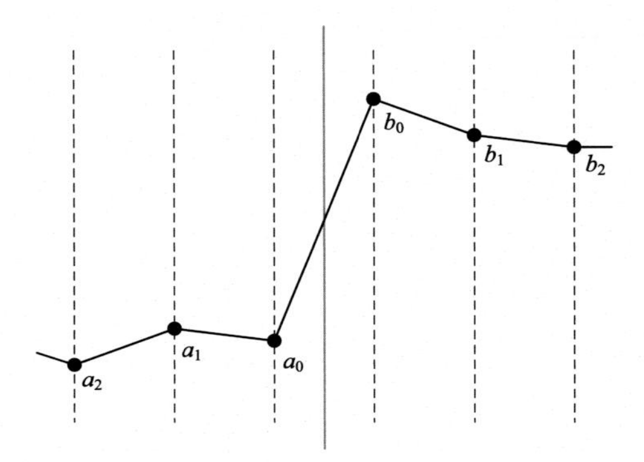

FIGURE 9.18 1D edge example.

FIGURE 9.19 Illustration of the effect of a deblocking filter. Left: Without deblocking. Right: With deblocking filter applied.

The conditions for filtering to be invoked depend on two thresholds, α (QP) and β (QP), where the second threshold is much smaller than the first. The deblocking process is described in Algorithm 9.5.

Example frames from *foreman* with, and without, a deblocking filter applied are shown in Fig. 9.19. The subjective benefits of the filtering operation can be clearly seen. What cannot be seen directly are the benefits of the smoothed and better correlated DFD signal which will enable more efficient transform coding and hence reduced bit rate.

9.8 Summary

This chapter has integrated the concepts from Chapters 4, 5, 7, and 8 to form what is referred to as a block-based hybrid coder. This architecture has been universally adopted through standardization processes and is the work horse of all of our TV, storage, and streaming applications. This basic architecture has benefited from several enhancements by way of advancing international standards. Some of the key enhancements have been described in this chapter and it is these that are responsible for providing the excellent performance offered by recent standards. Further standard-specific details are provided in Chapter 12.

For interactive demonstrations of video codec operation, the reader is referred to the MATLAB® code associated with this text, which can be downloaded from https://fan-aaron-zhang.github.io/Intelligent-Image-and-Video-Compression.htm.

References

[1] P. Strobach, Tree-structured scene adaptive coder, IEEE Trans. Commun. 38 (4) (1990) 477–486.
[2] Recommendation ITU-T H.264, International Standard ISO/IEC 14496-10, ITU-T, 2013.
[3] S. Ericsson, Fixed and adaptive predictors for hybrid predictive/transform coding, IEEE Trans. Commun. 33 (12) (1985) 1291–1302.
[4] B. Girod, Motion-compensated prediction with fractional-pel accuracy, IEEE Trans. Commun. 41 (4) (1993) 604–612.
[5] B. Girod, E. Steinbach, N. Farber, Performance of the H.263 compression standard, J. VLSI Signal Process. 17 (1997) 101–111.
[6] T. Wiegand, G. Sullivan, G. Bjøntegaard, A. Luthra, Overview of the H.264/AVC video coding standard, IEEE Trans. Circuits Syst. Video Technol. 7 (13) (2003) 560–576.
[7] E. Bellers, G. de Haan, Analysis of sub-pixel motion estimation, Proc. SPIE Vis. Commun. Image Process. (1999) 1452–1463.
[8] P. Hill, T. Chiew, D. Bull, C. Canagarajah, Interpolation-free subpixel accuracy motion estimation, IEEE Trans. Circuits Syst. Video Technol. 12 (16) (2006) 1519–1526.
[9] P. Hill, D. Bull, Sub-pixel motion estimation using kernel methods, Signal Process. Image Commun. 25 (4) (2010) 268–275.
[10] T. Wiegand, X. Zhang, B. Girod, Long term memory motion-compensated prediction, IEEE Trans. Circuits Syst. Video Technol. 1 (9) (1999) 70–84.
[11] Y. Su, M-T. Sun, Fast multiple reference frame motion estimation for H.264/AVC, IEEE Trans. Circuits Syst. Video Technol. 3 (16) (2006) 447–452.
[12] M. Al-Mualla, C. Canagarajah, D. Bull, Video Coding for Mobile Communications, Academic Press, 2002.
[13] M. Al-Mualla, C. Canagarajah, D. Bull, Simplex minimization for single- and multiple-reference motion estimation, IEEE Trans. Circuits Syst. Video Technol. 11 (12) (2001) 1209–1220.
[14] http://www.elecard.com/en/products/professional/analysis/streameye.html.
[15] G. Sullivan, J-R. Ohm, W.J. Han, T. Wiegand, Overview of the high efficiency video coding (HEVC) standard, IEEE Trans. Circuits Syst. Video Technol. 22 (12) (2012) 1649–1668.
[16] H. Malvar, A. Hallapuro, M. Karczewicz, L. Kerofsky, Low-complexity transform and quantization in H.264/AVC, IEEE Trans. Circuits Syst. Video Technol. 13 (7) (2003) 598–603.
[17] P. List, A. Joch, J. Lainema, G. Bjøntegaard, M. Karczewicz, Adaptive deblocking filter, IEEE Trans. Circuits Syst. Video Technol. 13 (7) (2003) 614–619.

CHAPTER

10

Measuring and managing picture quality

Assessing perceptual quality is one of the most critical yet challenging tasks in image and video processing. Visual perception is highly complex, influenced by many confounding factors, not fully understood, and difficult to model. For these reasons, the characterization of compression algorithm performance has invariably been based on subjective assessments where a group of viewers are asked their opinions on quality under a range of test conditions. Methods for conducting such trials and for analyzing the results from them are described in Section 10.2. A discussion of the properties of some publicly available subjective test databases is given in Section 10.3.

Objective measures of video quality have conventionally been computed using the absolute or squared difference between the coded version of a picture and its original version. It is however well known that the perceptual distortion experienced by the human viewer cannot be fully characterized using such simple mathematical differences. Because of the limitations of distortion-based measures, perception-based metrics have begun to replace them. These offer the potential for enhanced correlation with subjective opinions, thus enabling more accurate estimates of visual quality. However, many still have shortcomings. The most promising of these are reviewed in Section 10.4, including the structural similarity image metric (SSIM), multiscale SSIM (MS-SSIM), visual information fidelity (VIF), the video quality metric (VQM), the visual signal to noise ratio (VSNR), spatial and temporal most apparent distortion (MAD), motion-tuned spatio-temporal quality assessment (a.k.a. MOVIE), and the perception-based video metric (PVM). We also describe and compare a machine learning-based video quality metric, video multimethod assessment fusion (VMAF), which combines multiple conventional assessment methods using a regression model.

Video metrics are not just used for comparing different coding algorithms. They also play an important role in the picture compression and delivery processes, for example enabling in-loop rate-quality optimization (RQO). Section 10.5 addresses this issue, describing some of the most common techniques that enable us to select the optimal coding parameters for each spatio-temporal region of a video. Finally, methods for controlling the bit rate generated by a codec in the context of time-varying channel conditions (rate control) are described in Section 10.6.

Intelligent Image and Video Compression
https://doi.org/10.1016/B978-0-12-820353-8.00019-0

10.1 General considerations and influences

Firstly it is worth stating that assessing the quality of impaired image or video content, whether due to transmission losses or compression artifacts, is not straightforward. It can be achieved in one of two ways:

1. **Subjectively:** Requiring many observers and many presentations of a representative range of impairment conditions and content types. Subjective testing conditions must be closely controlled, with appropriate screening of observers and postprocessing of the results to ensure consistency and statistical significance. They are costly and time consuming, but generally effective.
2. **Objectively:** Using metrics that attempt to capture the perceptual mechanisms of the human visual system (HVS). The main issue here is that simple metrics bear little relevance to the HVS and generally do not correlate well with subjective results, especially at lower bit rates when distortions are higher. More complex, perceptually inspired metrics, although improving significantly in recent years, can still be inconsistent under certain test conditions. The outcome of this is that mean squared error (MSE)-based metrics are still the most commonly used assessment methods, both for in-loop optimization and for external performance comparisons.

10.1.1 What do we want to assess?

Good overviews of the reasons and processes for evaluating visual quality are provided in [1] and [2]. The primary motivations are summarized below:

1. To compare the performance of different video codecs across a range of bit rates and content types.
2. To compare the performance of different video codecs across a range of channel impairments.
3. To compare the influence of various parameters and coding options for a given type of codec.

The latter is of particular relevance to in-loop rate-distortion optimization (RDO) (or RQO) as we will see later.

10.1.2 Noise, distortion, and quality

Noise or artifacts are introduced into image and video content throughout the whole processing workflow. These include sensor noise introduced during content acquisition, artifacts caused by compression during coding, transmission errors due to noisy communication channels, and sample aliasing due to resolution adaptation (if applied). Noise can be defined from different perspectives:

- **Mathematical distortion**: An objective measure of difference between the original acquired content and the processed version.

- **Perceptual quality**: A measure of the perceptual impact of the unwanted signal and how it detracts from the enjoyment or interpretation of an image or video.

In most application scenarios, the aim of picture coding is, with a limited bit rate, to reconstruct image or video content with optimal visual quality rather than simply to minimize its mathematical distortion with respect to the original version. It should however be noted that significant correlation often exists between these two measures.

10.1.3 Influences on perceived quality

Human visual perception

More than a century ago, vision scientists began to pay attention to our perceptual sensitivity to image and video distortions. We saw in Chapter 2 that this sensitivity varies with screen brightness, local spatial and temporal frequency characteristics, types of motion, eye movements, various types of artifact, and of course the viewing environment. In order to ensure validity of subjective tests and consistency in the performance of objective metrics, the influence of these sensitivities must be, as far as possible, represented in the content used for evaluations and captured in the structure of any metric employed.

It should also be noted that the performance of the HVS varies significantly across subjects, depending on age, illness, fatigue, or visual system defects. It is also possible that biases can influence the opinions of viewers through personal preferences or even boredom.

Viewing environment

In order to provide consistent subjective assessments of picture quality it is essential that the details and parameters of the viewing conditions are recorded and kept as consistent as possible between tests. This is particularly important when comparing results across different laboratories. Several factors about the viewing environment will influence the perception of picture quality. These include:

- **Display size**: The more the display fills peripheral vision, the more engaging it will appear.
- **Display brightness and dynamic range**: Flicker and temporal CSF will depend on display brightness and higher dynamic range displays impart a greater sense of depth.
- **Display resolution**: The spatial and temporal update rates will influence the types of artifacts perceived.
- **Ambient lighting**: High ambient lighting levels and reflections from the display will reduce perceived contrast levels and introduce additional artifacts.
- **Viewing distance**: This parameter interacts with spatio-temporal resolution and will have a significant impact on perceived quality.
- **Audio quality**: It has been clearly demonstrated that variations in audio quality will influence the perception of video quality.
- **Sensory misalignment**: As we have described in Chapter 2, certain viewing environments can cause viewer discomfort, for example the sensory conflict between depth cues in 3D and head-mounted displays.

Content type

We have seen in Chapter 2 how content type can influence perceived quality. For example, a plain scene with little motion will be much easier to code than a spatio-temporally busy sequence with complex textures and motions. If codec A is used to code easy content and a similar codec, B, is used to code more difficult content, then it would appear that codec A is superior in terms of rate-distortion performance to codec B, when in reality this is not necessarily the case.

Tests should therefore be based on representative and reasonably challenging content offering a range of spatial and temporal activity levels (see Section 10.2.2). To ensure that the focus of the assessor remains on picture quality, rather than on the narrative, short sequences (typically 10 s) are normally selected and these are not usually accompanied by an audio track. All codecs under test should be evaluated on all sequences and at all impairment levels.

Artifact types

We will not list all of the artifact types produced by all types of codec here, but hopefully the reader is by now aware of these from previous chapters. In general they can be classified as follows:

- **Edges**: Due to blocking from transforms or motion estimation, or to contouring effects associated with quantization.
- **Blurring**: Due to the loss of high-frequency detail during quantization of block transform or wavelet coefficients.
- **Ringing**: Quantization artifacts in filter-bank synthesis stages can cause ringing effects.
- **Dissimilarity**: Due for example to inconsistencies in geometric transformations of planar regions in certain codecs, errors in warping rigid textures, or synthesizing dynamic textures (see Chapter 13 for further details).

10.2 Subjective testing

10.2.1 Justification

Despite recent advances in the performance of objective metrics and the large number of them available, none are yet universally accepted as a definitive measure of quality.[1] As a consequence it is necessary to use controlled subjective testing. Subjective assessment methods are employed widely to characterize, compare, or validate the performance of video compression algorithms. Based on a representative set of test content and impairment conditions, they are intended to provide a robust indication of the reactions of those who might view the systems tested.

Most subjective testing experiments conducted are based on recommendations from the ITU and other organizations that have been developed over many years, based on the col-

[1] It could be argued that VQM [49], as standardized by ANSI, is more widely accepted than others although it is not the best performing.

lective experiences of many organizations. The primary reference documents are ITU-R Rec. BT.500 [4] and ITU-T Rec. P.910 [5]. The discussion that follows is primarily based on these recommendations.

10.2.2 Test sequences and conditions

Test material

In general, test material should be selected that is appropriate to the problem or application area being addressed. It will normally be defined in terms of its fixed parameters, which include the number of sequences used, sequence duration, sequence content, spatial resolution, temporal resolution, and bit depth.

For most general assessments, the test material will be "critical, but not unduly so" [4]. This means that it should contain content that is difficult to code but should remain representative of typical viewing. The sequences selected should always include critical material since results for noncritical material cannot usually be extrapolated. However, if the tests are intended to characterize performance for specific difficult cases, then it would be expected that the material used would be selected to reflect these cases.

In general it is usual to select at least four types of sequence for testing. This number will provide a minimum coverage of activity levels while not boring the assessors with an overlong test session.

Activity or information levels

The amount of spatial and temporal activity present in a clip has a high impact on the degree of compression possible and the quality of the resulting reconstruction. In general test sequences should be selected that are consistent with the range of channel conditions prevailing in the application areas of interest. It is useful, prior to final selection of test material, to formally assess the spatio-temporal activity levels of the candidate clips to ensure that these cover the appropriate spatio-temporal information space [3,5].

Following the approach recommended in recommendation ITU-T Rec. BT.910 [5], the spatial information (SI) measure is based on the standard deviation of each frame $\mathbf{S}_z$ in the sequence after Sobel filtering. The maximum value of the standard deviation, over all frames, is selected as the SI metric:

$$\mathrm{SI} = \max_{\forall z}\left\{\sigma_{\forall(x,y)}\left(\mathrm{Sobel}\left(\mathbf{S}_z(x,y)\right)\right)\right\}. \tag{10.1}$$

Alternative SI measures also include the mean and root mean square of each Sobel filtered frame [6].

The temporal information (TI) measure used in BT.910 is based on the standard deviation of the difference, over consecutive frames, between colocated luma pixel values. The TI measure is computed, as above, as the maximum value of the standard deviation over all frames in the sequence. Hence,

$$\mathrm{TI} = \max_{\forall z}\left\{\sigma_{\forall(x,y)}\left(\mathbf{S}_z(x,y) - \mathbf{S}_{z-1}(x,y)\right)\right\}. \tag{10.2}$$

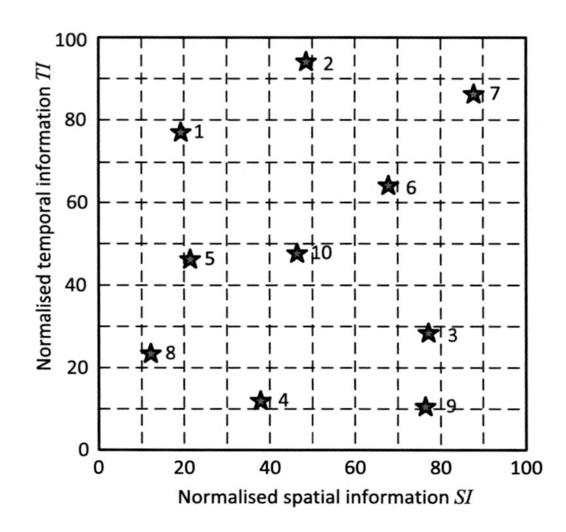

FIGURE 10.1 Spatial and temporal information coverage for a range of test sequences.

It should be noted that an alternative measure of TI is used by Winkler in [3], based on the root mean square of motion vector magnitudes. It is usual to produce an SI vs. TI plot to ensure that the sequences selected for testing provide adequate coverage of the SI-TI space. An indicative plot, with a good coverage, is shown in Fig. 10.1. It also includes an information measure based on color.

Test conditions

The *test conditions* refer to those parameters that will change during the test session. These might include codec types, codec parameters, and bit rates, and a typical test would compare two or more codecs or codec configurations for a range of sequences and coding rates.

It is normal to evaluate codec performance over a representative, well-distributed set of test conditions. Most assessment methods are sensitive to the range and distribution of conditions. The sensitivity can be reduced by restricting the range of conditions but also by including some explicit (direct anchoring) extreme cases or by distributing these throughout the test without explicit identification (indirect anchoring). Furthermore, the larger the number of test conditions, the longer the test session becomes. Tests must avoid viewer fatigue and this imposes constraints on the number of sequences employed and the number of test conditions evaluated.

The number of presentations is equal to $N = N_s \times N_c \times N_r$, where N_s is the number of source sequences, N_c is the number of test conditions ($N_c = N_{c_1} N_{c_2}$, where N_{c_1} is the number of codecs under test and N_{c_2} is the number of bit rates tested [usually including uncompressed]), and N_r is the redundancy factor (i.e., the number of times each condition is repeated). If each presentation takes T seconds, then the time for each test is $N \times T$ seconds, and if there are K observers, then the total time for the complete trial is $N \times T \times K$ seconds (not including dead time for change-overs and breaks).

TABLE 10.1 Selected subjective testing viewing environment parameters. From [4,5].

Viewing condition	Home	Laboratory
Ratio of inactive to peak screen luminance	≤ 0.02	≤ 0.02
Display peak luminance	200 cd/m^2	See recommendations BT.814 and BT.815
Background chromaticity	–	D65
Maximum observer angle	30°	30°
Viewing distance	3–6 H	3–6 H
Room illumination	200 lux	Low ($\leq$ 20 lux)

10.2.3 Choosing subjects

Depending on the nature of the test, observers may be expert or nonexpert. Studies have found that systematic differences can occur between different laboratories conducting similar tests [4]. The reasons for this are not fully understood, but it is clear that expert observers will view material differently from nonexperts. Other explanations that have been suggested include gender, age, and occupation [4]. In most cases, for consumer applications, it is expected that the majority of observers should be nonexpert and that none should have been directly involved in the development of the system under test.

Before final selection of the assessors, all candidates should be screened to ensure that they possess normal visual acuity (with or without corrective lenses). This can be done using a Snellen chart for visual acuity and an Ishihara chart to check for color vision. The number of assessors used depends on the scale of the test and the sensitivity and reliability of the methodology adopted. Normally it is recommended that at least 15 subjects are employed. This author recommends that the number is slightly higher to allow for outlier removal during results processing.

10.2.4 Testing environment

Testing environments are normally specified according to the requirements of the test – usually meaning either a realistic consumer environment or laboratory conditions:

- **Laboratory conditions**: These are intended to provide test conditions that enable maximum detectability of impairments.
- **Home conditions**: These are intended to enable assessment in viewing conditions closer to typical consumer-end environments.

Selected parameters associated with these two environments are listed in Table 10.1.

10.2.5 Testing methodology and recording of results

A generic diagram of a subjective testing methodology is shown in Fig. 10.2. The source delivers the presentation clips, either directly to the subject or via a system under test that introduces impairments dependent on the test conditions set at the time. The subject or sub-

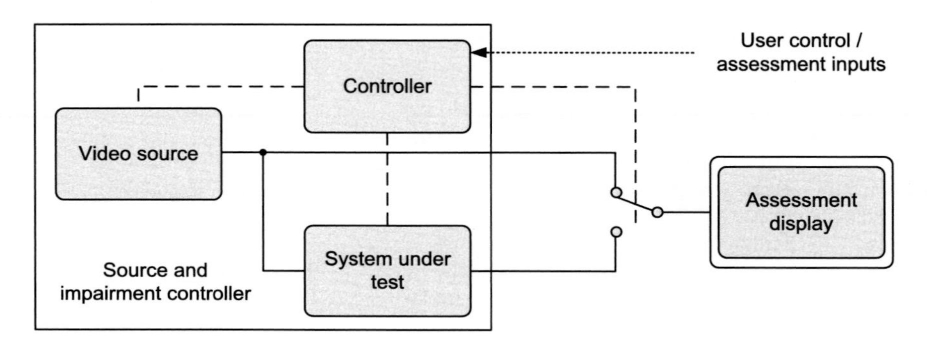

FIGURE 10.2 Generic arrangement for subjective video evaluation.

jects will view the content on an assessment display which shows the impaired video clip and in some cases also the unimpaired clip, either simultaneously or sequentially. The controller will control the timing and switching between different test conditions. In some cases the user may be able to influence the order, repetition, and timing of clips.

Testing methodologies broadly fall into two categories, based on their use of single or multiple stimuli. Before we consider these, there are a few general points that are common to all methods.

General principles of subjective testing

- **Duration of test**: To avoid fatigue, BT.500 recommends that the duration of the test session should be limited to 30 minutes. In practice, it has been the author's experience that tests often do consume more time and 40–45 minutes is not uncommon, especially in cases where the observer is permitted to control the display and revisit sequences multiple times (e.g., the Subjective Assessment Methodology for Video Quality (SAMVIQ) [2]).
- **Preparing assessors**: Prior to the test, assessors should receive full instructions on the reason for the test, its content, the types of impairment that will occur, and the method of recording results. The first part of any test should include dummy runs which familiarize the assessor(s) with the methodology. These also help to stabilize the observer's opinions.
- **Presentation order**: It is normal during the test session to randomize the presentations – within and across tests. This ensures that any influences of fatigue or adaptation are balanced out. It is also normal to include anchors (extreme cases). In some cases the assessor will know which presentations contain anchors, but this is usually not the case.
- **Recording opinions**: The opinions of each assessor must be recorded for posttest analysis. This is often done using a line bisection process based on gradings such as those shown for the double-stimulus continuous quality scale (DSCQS) test in Fig. 10.3. The observer places a mark on the line for each presentation assessed and this typically falls into one of five quality scale bins as shown in the figure. The final grade is normally scaled to the range 0–100. In other cases, an automated user interface will be provided, such as that shown in Fig. 10.4 for the SAMVIQ methodology [2].
- **Recording test conditions**: Because of the inconsistencies that can exist between subjective trials, it is vital to record all salient information about the test and the parameters and

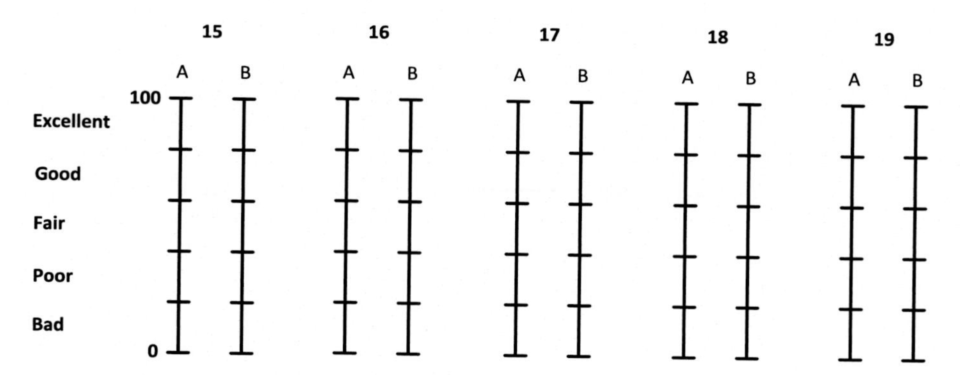

FIGURE 10.3 A section of the DSCQS assessment form.

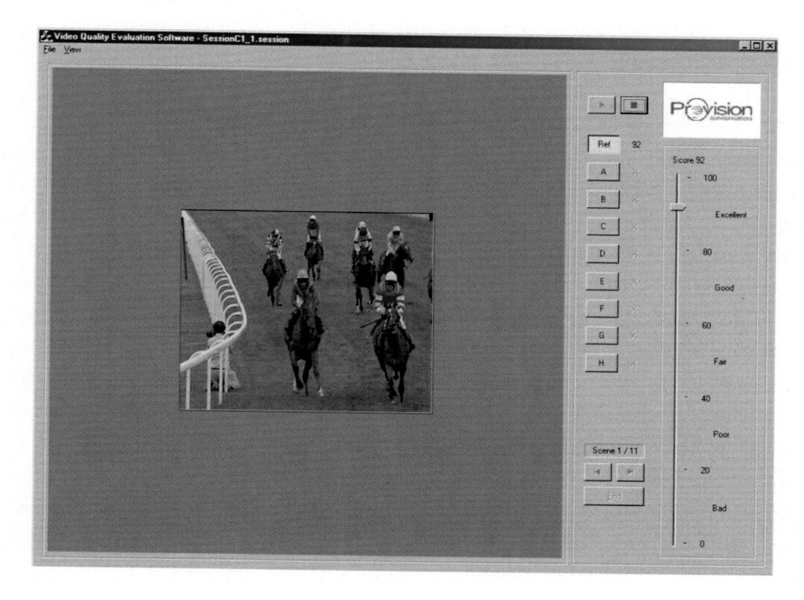

FIGURE 10.4 Interface for SAMVIQ testing (courtesy of ProVision Communications).

equipment used. Good examples are provided by Ebrahimi's group in the context of HEVC – for example, see [7,8].

Double-stimulus methods

Double-stimulus evaluations remain the most popular means of evaluating compressed video quality. The most commonly used procedure is ITU-R Rec. BT.500, which, although originally intended for television applications, is now used more widely. ITU-T Rec. P.910 was targeted specifically at multimedia applications, but shares many similarities with BT.500. ITU-R Rec. BT.500 describes two main double-stimulus methods:

- **Double-stimulus continuous quality scale (DSCQS)**: The DSCQS methodology is best suited for cases where the qualities of the test material and the original are similar and the aim is to assess how well the system under test performs relative to the original. The test is arranged as a sequence of paired clips (A and B), one of which is the original (anchor) and the other is the original impaired by the system under test. The assessor does not know which clip is the original as the order is randomized. This is shown in Fig. 10.5. The pair is shown twice, the first time for familiarization and the second for voting.
- **Double-stimulus impairment scale (DSIS)**: DSIS is similar to DSCQS except that the pair is presented only once and the assessor knows which clip is the anchor as it is always shown before the impaired version. DSIS is generally better suited to assess the robustness of a system or the effects of more noticeable distortions such as those imparted by transmission errors. This method is very similar to the degradation category rating (DCR) method in ITU-T Rec. P.910. The DSIS grading scale again comprises five grades but this time they are labeled: imperceptible, perceptible but not annoying, slightly annoying, annoying, and very annoying.

Single-stimulus methods

In single-stimulus methods, there is no explicit anchor and the assessor is presented with a randomized sequence of test conditions, normally including the original. The absolute category rating (ACR) in ITU-T Rec. P.910 does this using 10-s clips across a range of test conditions, with voting on the same scale as in Fig. 10.3. A single-stimulus continuous quality evaluation method is proposed as part of ITU-R Rec. BT.500. This is intended to better capture the time and content variations that happen in practical content delivery. Here it is recommended that the test session is organized with longer program segments (e.g., sport) of around 5 minutes duration. These are concatenated into a test session of duration 30–60 minutes comprising a range of program segments with various quality parameters. One of the main advantages of single-stimulus methods is reduced testing time.

SAMVIQ [2] is a single-stimulus method where the assessor has some control over viewing order and repetitions. The test is organized as a series of test sequences which are assessed in a given order and where the assessor cannot proceed to the next sequence until the previous one has been completely assessed. Within each sequence a number of algorithms and test conditions can be presented in any order as selected by the assessor (however, the selection buttons are randomized for each new sequence evaluated). An example interface with a slider for score entering is shown in Fig. 10.4.

Triple-stimulus methods

Some authors have proposed that triple-stimulus methods provide increased consistency in results, particularly for example when comparing interlaced and progressive scanning methods for HDTV. The triple-stimulus continuous evaluation scale (TSCES) methodology was proposed by Hoffman et al. [9], and simultaneously displays the video processed by the system under test alongside (usually they are stacked vertically) two extreme anchors – the original and a low-quality coded version. The conventional five-grade scale is also used and the assessor is asked to score the system under test in the context of both anchors.

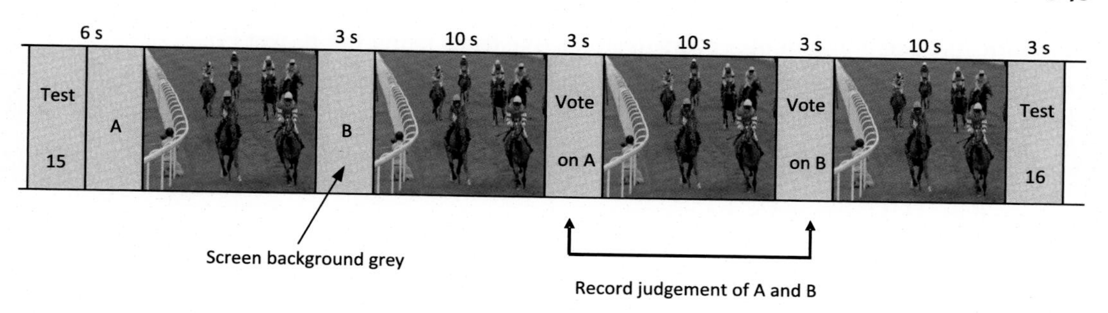

FIGURE 10.5 Typical timing of a DSCQS subjective test.

Pair comparison methods

Pair comparison methods are also double-stimulus, but this time they are based on comparisons between two systems under test for the same test conditions.

10.2.6 Statistical analysis and significance testing

It is essential that the data collected is analyzed and presented in a robust and consistent fashion. Most test results are scaled to the range 0–100 and that is assumed here. Generally, following the process described in BT.500 [4], let us assume that there are N presentations, where each consists of one of N_c test conditions.

Calculation of mean scores

The mean score across all observers for each presentation is given by

$$\bar{u}_{csr} = \frac{1}{K}\sum_{k=1}^{K} u_{kcsr}, \tag{10.3}$$

where u_{kcsr} is the score for observer k in response to sequence s under test condition c for repetition r, and K is the total number of observers. This is then similarly processed after screening of observers to produce $\bar{u}_{sc}$, the mean opinion score (MOS) for each sequence under a given test condition.

It is common in many tests to compensate scores relative to the reference content. This is particularly important for single-stimulus methods where a process of hidden reference removal is applied to produce difference MOSs (DMOSs).

Confidence interval

When the results of a study are presented, it is good practice to also include the confidence interval [10]. The confidence interval for a mean is the interval that will contain the population mean a specified proportion of the time, typically 95%. Confidence intervals are based on the size of each sample and its standard deviation and provide more information than point

Algorithm 10.1 Screening of observers in subjective trials using the β_2 test [4].

1. REPEAT for each observer, k:
   2. $k = k + 1$
   3. FOR $c, s, r = 1, 1, 1$ to C, S, R DO:
      4. IF $2 \leq \beta_{2csr} \leq 4$ THEN: IF $u_{kcsr} \geq \bar{u}_{csr} + 2\sigma_{csr}$ THEN $P_k = P_k + 1$; IF $u_{kcsr} \leq \bar{u}_{csr} - 2\sigma_{csr}$ THEN $Q_k = Q_k + 1$;
      5. ELSE: IF $u_{kcsr} \geq \bar{u}_{csr} + \sqrt{20}\sigma_{csr}$ THEN $P_k = P_k + 1$; IF $u_{kcsr} \leq \bar{u}_{csr} - \sqrt{20}\sigma_{csr}$ THEN $Q_k = Q_k + 1$;
      6. IF $\frac{P_k+Q_k}{N_c N_s N_r} > 0.5$ and $\left|\frac{P_k-Q_k}{P_k+Q_k}\right| < 0.3$ THEN reject observer k;
   7. END FOR;
8. UNTIL $k = K$.

estimates. The 95% confidence interval is given by

$$[\bar{u}_{csr} - \delta_{csr}, \bar{u}_{csr} + \delta_{csr}], \tag{10.4}$$

where

$$\delta_{csr} = 1.96\frac{\sigma_{csr}}{\sqrt{K}}.$$

The standard deviation for each presentation is given by

$$\sigma_{csr} = \sqrt{\sum_{k=1}^{K} \frac{(\bar{u}_{csr} - u_{kcsr})^2}{(K-1)}}. \tag{10.5}$$

The absolute difference between the experimental mean and the true mean (based on an infinite number of observers) is smaller than the 95% confidence interval of Eq. (10.4) with a probability of 95% (on the condition that the scores are normally distributed).

Screening of observers

The screening (used in DSIS and DSCQS) for a normally distributed set of scores conventionally uses the β_2 test, based on the kurtosis coefficient. When β_2 is between 2 and 4, the distribution can be assumed to be normal. The kurtosis coefficient is defined as the ratio of the fourth order moment to the square of the second order moment. Thus,

$$\beta_{2csr} = \frac{m_4}{(m_2)^2} \quad \text{with } m_x = \frac{\sum_{k=1}^{K} (u_{kcsr} - \bar{u}_{kcsr})^x}{K}. \tag{10.6}$$

We now compute P_k and Q_k, as given in Algorithm 10.1, in order to determine whether any observers should be rejected in forming the distribution.

Algorithm 10.2 Subjective testing methodology.

1. Select the codecs under test and the parameter settings to be used (GOP structures, filtering, RDO modes, ME modes, etc.);
2. Define the coding conditions to be evaluated (spatial and temporal resolutions, bit rates etc.);
3. Identify or acquire the test sequences that will be used, ensuring that they are sufficiently varied and challenging, representative of the applications of interest, and of appropriate duration;
4. Select an appropriate evaluation methodology (dependent on conditions being tested, impairment levels, etc.);
5. Design the test session and environment (complying with recommendations of whichever methodology is being adopted);
6. Dry run the session to remove any bugs or inconsistencies;
7. Organize the full test, inviting sufficient subjects (assessors), with prescreening to eliminate those with incompatible attributes;
8. Run the tests (including assessor briefing and dummy runs) and collect the results;
9. Perform postprocessing to remove outliers and assess significance;
10. Produce test report fully documenting laboratory conditions, test conditions, and analysis/screening methods employed.

10.2.7 The complete evaluation process

The preceding sections have provided a general overview of subjective testing methods. Both double- and single-stimulus methods have been formalized through recommendations such as ITU-R Rec. BT.500-13, ITU-T Rec. P.910, and SAMVIQ [2,4,5]. The reader is advised to refer to the relevant standard or procedure documents before implementing any actual subjective testing experiments. As a guide, however, a general methodology for subjective testing is described in Algorithm 10.2.

10.2.8 Subjective testing through crowdsourcing

Laboratory-based experiments with fully controlled viewing conditions can produce accurate ground truth data for perceptual video quality. However, there are many cases in practice where lab-based tests cannot be conducted. For example, when:

- Participants are difficult to source locally.
- A laboratory environment is not readily accessible.
- There are a large number of test videos to be evaluated.
- Social distancing must be enforced.

The last point has become particularly pertinent during the COVID-19 global pandemic, where social lockdowns were enforced across the world. In such situations, crowdsourcing-based experiments offer an alternative solution to conventional subjective testing. These allow researchers to recruit subjects online and efficiently collect subjective data for large-scale studies. In crowdsourcing experiments, participants cannot be easily controlled and monitored; they will view the test content on a range of different display devices with various viewing conditions. This in turn will result in much higher measurement variability than in conventional subjective testing. This can however be mitigated by employing an increased number of subjects. Ghadiyaram and Bovik [11] compared results collected from an online crowdsourced subjective test and a lab-based experiment for the same test material, and reported high consistency and correlation between these two experimental approaches.

10.2.9 Measuring immersion in visual content

The need to assess the impact on user experience of new media formats, such as VR, AR, high dynamic range (HDR), or higher spatio-temporal resolution, goes well beyond the capability of visual quality metrics. We not only need to measure the quality of the visual content, but also the degree to which the viewer engages with the content's narrative and/or its presentation format. Traditionally this information has been captured via questionnaires and audience feedback after trial screenings – primarily to make qualitative judgments about the likely value of the film or program rather than its format.

Analysis of a subject's experience has been assessed using proxies such as physiological or behavioral cues including electroencephalography, functional magnetic resonance imaging, heart rate, respiration, and skin temperature [12]. Most of these methods are however highly intrusive, sensitive to external factors, or not applicable to large-scale deployment. One method that supports continuous quantitative measurement of immersion is the dual-task method presented by Hinde, Smith, and Gilchrist [13]. It is based on the principle that the brain is a linear machine and that the time taken to switch from one task to a second task is modulated by the subject's level of engagement in the first task. Fluctuations in performance in the secondary task are taken as evidence that attentional resources are being focused on the primary task to the detriment of the secondary task. Recent experimental work by Hinde, Gilchrist, and Bull at Bristol Vision Institute in collaboration with Noland and Thomas at BBC R&D has confirmed the power of reaction time as a proxy for immersion. We demonstrated that this approach can assess not only engagement with the narrative but also the impact of audiovisual parameters such as spatial resolution and dynamic range. The study used professional HDR content including episodes from the BBC natural history series Planet Earth II and Blue Planet II. Four different formats were used, at standard and high dynamic range, and at high and ultrahigh definition. The secondary task involved detecting and classifying a brief high- or low-frequency audio tone presented periodically during the program. A significant effect of dynamic range was observed, with HDR content found to be more immersive than standard dynamic range.

10.3 Test datasets and how to use them

A reliable subjective database has three primary uses. Firstly it will produce a robust comparison of the selected compression methods in the context of the test conditions employed. Secondly it provides a very useful basis for validating objective metrics, based on well-established statistical methods. Finally it can be utilized to characterize HVS properties in the context of compression and hence provides a valuable tool for refining objective quality metrics.

10.3.1 Databases

A brief overview of the primary publicly available databases is presented below. The reader is referred to the appropriate reports associated with each trial for further details. An excellent critique and comparison of some of these databases was provided by Winkler in [3].

VQEG FRTV

The earliest major subjective database for objective video quality assessment was generated via the Video Quality Experts Group (VQEG) FRTV Phase 1 program [14] (followed by phase 2 in 2003 [15]). The phase I database was constructed in 2000 to address quality issues associated with the introduction of digital TV standards worldwide. The database used 287 assessors (in four groups) and 20 video sequences with 16 distortion types to generate over 26,000 subjective opinion scores. FRTV-I employed two resolutions: 720 × 486 and 720 × 576, all interlaced. The testing methodology used was the DSCQS method described earlier.

Although the FRTV-I database exhibits good coverage and uniformity, it does have limitations in the context of contemporary coding requirements:

1. The artifacts presented do not fully reflect recent advances in video compression as coding is primarily based on MPEG-2.
2. Its assessments are based on standard-definition formats. It does not include high-definition material.
3. Its content is interlaced, whereas the current trend is toward progressive formats, especially for streaming applications.

However, despite these limitations, because of the large number of sequences, the large number of subjects, its good coverage, and its good uniformity [3], VQEG FRTV Phase I is still one of the more commonly used databases for objective quality metric validation.

LIVE

An equally important subjective test database was developed by the Laboratory for Image and Video Engineering (LIVE) [16,17,21]. The LIVE database presents impairments based on both MPEG-2 and H.264/AVC compression algorithms and also includes results for simulated wired and wireless transmission errors. Ten reference sequences are used with a resolution of 768×432, where each reference has 15 distorted versions. A single-stimulus continuous test procedure was adopted for this dataset and final scores were again scaled between 0 and 100. Compared to the VQEG FRTV-I database, only 38 assessors were employed in the LIVE viewing trials. According to the analysis of Winkler in [3], LIVE also provides a narrower range of source material (in terms of SI and TI) and distortions (in terms of PSNR and MOS values).

BVI-HD

Both VQEG FRTV and LIVE databases present the results of subjective studies using standard-definition (SD) sequences based on compression algorithms developed over 15 years ago. To further investigate subjective quality of compressed high-definition content using state-of-the-art coding approaches, Zhang et al. [18] developed an HD video quality database, BVI-HD, which contains 384 test sequences generated from 32 diverse and representative sources based on HEVC compression and a less conventional synthesis-based codec. A double-stimulus test methodology was employed to collect subjective data from a total of 86 subjects. The content diversity and large number of test sequences make BVI-HD a robust test database for video quality assessment.

Netflix public database

Another contribution to HD video quality assessment is the Netflix public dataset, which was initially used to evaluate the performance of VMAF, a machine learning-based quality metric (see Section 10.4.4) [20]. It includes 70 test videos created from nine source sequences using H.264/AVC. Spatial resolution resampling, which is a commonly used tool together with compression for internet streaming, was also applied (with four sampling ratios) when processing some of the test sequences.

UHD subjective databases

More recently, subjective quality databases including ultrahigh-definition content have been publicly released. One notable example is the SJTU 4K Video Subjective Quality Database [22]. This contains 60 HEVC compressed test sequences from 10 reference videos alongside subjective opinions collected from 42 subjects. Another widely used database, the MCML 4K database, was developed by the researchers from Yonsei university [28]. They used three different video codecs, H.264/AVC, H.265/HEVC, and Google's VP9, to generate 240 distorted video sequences from 10 original UHD sources. Resolution resampling was also applied to produce more distortion variations. Twenty-four subjects were employed to rate all reference and test sequences based on a single-stimulus methodology.

Subjective databases based on crowdsourcing

When subjective databases are used to train objective quality assessment methods, especially those based on machine learning models, much larger volumes of video data are often required than for quality metric evaluation. Although it may be possible to collect sufficient source content and generate their extensive distorted versions, these sequences cannot be efficiently assessed by subjects through conventional lab-based psychophysical experiments. Crowdsourced evaluation has therefore become increasingly popular in recent years, and it is now feasible to collect subjective data online from voluntary or paid users in a less controlled experimental environment. Yim et al. have recently published a subjective database containing YouTube user-generated content, which includes approximately 1500 video clips at five spatial resolutions (without real original sources) [30]. Each of these was assessed by more than 100 subjects on a crowdsourcing platform. This database provides a valuable resource for artifact analysis and no-reference quality metric development.

Others

Other examples of subjective databases include the IRCCyN/IVC [19], VQEG HDTV [23], BVI-Texture database [32], VQEG multimedia Phase I [25], BVI-HFR [24], and BVI-SR [34]. Most of these are targeted at specific research purposes and are less suitable for general validation of image and video quality metrics. A further comparison of some of these can be found in [3].

10.3.2 The relationship between mean opinion score and an objective metric

Subjective results are frequently used as a benchmark for establishing a relationship between DMOS (or MOS) scores and a specific objective picture quality metric. The scores produced by the objective video quality metric must be correlated with the viewer scores in a predictable and repeatable fashion. The relationship between predicted and DMOS need not be linear as subjective testing can have nonlinear quality rating compression at the extremes of the test range. The linearity of the relationship is thus not so critical, but rather the stability of the relationship and a dataset's error variance determine predictive usefulness.

BT.500 [4] describes a method of finding a simple continuous relationship between $\bar{u}$ (the mean score) and the metric based on a logistic function. Firstly the range of mean score values is normalized as follows (after screening):

$$p = \frac{(\bar{u} - u_{\min})}{(u_{\max} - u_{\min})}. \tag{10.7}$$

Typical relationships between p and a given distortion measure D generally exhibit a skew-symmetric sigmoid form. Hence the function $p = f(D)$ can be approximated by a logistic function of the form

$$\tilde{p} = \frac{1}{1 + e^{(D - D_M)G}}, \tag{10.8}$$

where D_M and G are constants that can be quite simply derived from the experimental data [4].

10.3.3 Evaluating metrics using public (or private) databases

Judgements of the performance of a particular objective metric relative to a body of MOS values associated with a specific database are normally based on certain statistical attributes. These are conventionally related to measures of prediction accuracy, monotonicity, and consistency of its fit. The following measures are commonly used.

Linear correlation

The Pearson linear correlation coefficient (LCC) is used as a measure of the accuracy of fit of the metric to the subjective scores. It characterizes how well the metric under test can predict the subjective quality ratings. The general form is defined for a set of N measurement–

prediction pairs (x_i, y_i) as

$$r_p = \frac{\sum_{i=0}^{N-1}(x_i - \bar{x})(y_i - \bar{y})}{\sqrt{\sum_{i=0}^{N-1}(x_i - \bar{x})^2}\sqrt{\sum_{i=0}^{N-1}(y_i - \bar{y})^2}}, \tag{10.9}$$

where x_i would represent the actual MOS (or DMOS) score and y_i would represent the predicted MOS (or DMOS) score.

Rank order correlation

The degree to which the model's predictions correlate with the relative magnitudes of subjective quality ratings is assessed using a rank order metric. This characterizes the prediction monotonicity, i.e., to what degree the sign of differences across tests correlate between the subjective scores and the metric's prediction of them. Conventionally the Spearman rank order correlation coefficient is used for this purpose:

$$r_s = \left| \frac{\sum_{i=0}^{N-1}(\mathcal{X}_i - \bar{\mathcal{X}})(\mathcal{Z}_i - \bar{\mathcal{Z}})}{\sqrt{\sum_{i=0}^{N-1}(\mathcal{X}_i - \bar{\mathcal{X}})^2}\sqrt{\sum_{i=0}^{N-1}(\mathcal{Z}_i - \bar{\mathcal{Z}})^2}} \right|, \tag{10.10}$$

where $\mathcal{X}_i$ is the rank order of x_i, $\mathcal{Z}_i$ is the rank order of the quality indices, z_i, and $\bar{\mathcal{X}}$ and $\bar{\mathcal{Z}}$ are their median values.

Outlier ratio

The outlier ratio, r_o, effectively measures prediction consistency – how well the metric predicts the subjective scores over the range of content and impairments. An outlier is normally classed as a predicted data point that is greater than a threshold distance from the corresponding MOS point. Conventionally a threshold of twice the standard deviation error of the MOS values is used. So if the number of data points that satisfy Eq. (10.11) is N_o, then the outlier ratio is simply given in Eq. (10.12):

$$|x_i - y_i| > 2\sigma_{y_i}, \tag{10.11}$$

$$r_o = \frac{N_o}{N}. \tag{10.12}$$

Prediction error

The prediction error, e, is a statistical parameter which measures the prediction accuracy of objective models. It is based on the root mean square error between predicted MOS (or DMOS) and subjective MOS (DMOS) values, and its definition is given as follows:

$$e = \sqrt{\sum_{i=0}^{N}(y_i - x_i)^2}. \tag{10.13}$$

It should be noted that the calculation of prediction error, linear correlation, and outlier ratio requires predicted subjective scores, which are dependent on the nonlinear regression (fitting) described in Section 10.3.2. In some cases where the regression accuracy is poor, these statistical measurements may become less robust. However rank order correlation estimates the ranking correlation between predicted and subjective scores and thus does not rely on the fitting process. This explains why the rank order correlation is more frequently used than the other three.

Significance test

Based on a subjective database, the statistical parameters described above can expose differences in terms of prediction performance between various objective quality metrics. A significance test can be conducted to identify whether this difference is statistically significant. Based on statistical theory, the F-test is a popular model that can be used to identify the best fitting model for a certain dataset by assessing the equality of variances between two models at a 95% significance level [26]. For use in quality assessment, an approach is proposed in [16] that allows an F-test to be performed between the prediction residual sets R^a and R^b of two objective quality metrics, a and b. Each prediction residual r_i^a for metric a is the difference between a predicted MOS (or DMOS) and its corresponding actual MOS (or DMOS):

$$R^a = \{r_i^a\} = \{y_i^a - x_i\}. \tag{10.14}$$

In this approach, a hypothesis is proposed with regard to the two-tailed F-test: that the variances of residuals from the two tested objective metrics are statistically equivalent. The F-statistic is hence computed as

$$F = \frac{\sigma_a^2}{\sigma_b^2}, \tag{10.15}$$

where σ_a^2 and σ_b^2 represent the variances of R^a and R^b.

10.4 Objective quality metrics

10.4.1 Why do we need quality metrics?

Quality assessment plays a crucial role in many aspects of image and video processing, in particular related to video coding and communications. In the context of communicating pictures, they have three primary uses:

1. **Algorithm development and benchmarking**: We have seen previously that subjective evaluations, while very effective in characterizing the strengths and weaknesses of competing coding algorithms, are expensive and time consuming. The existence of reliable and consistent objective metrics provides a much simpler means of comparison. Furthermore they can provide guidance regarding the benefits or otherwise of various algorithmic modifications to a given codec (for example the benefits of adding a loop filter or using multiple reference frames).

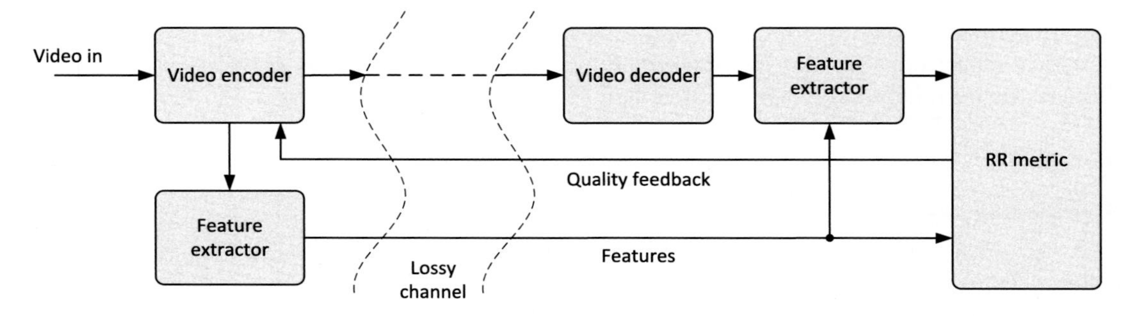

FIGURE 10.6 Reduced-reference video quality assessment.

2. **Rate-quality optimization**: Quality assessments are increasingly needed in the encoding loop to make instantaneous RQO decisions about which coding modes and parameter settings to use for optimal performance given certain content and rate constraints.
3. **Streaming control**: In the case of network delivery of video content, it is beneficial for the encoder and transmitter to be aware of the quality of the signal at the receiver after decoding. This enables the encoder to be informed of the prevailing channel conditions and hence to make appropriate decisions in terms of rate and error control.

Objective quality assessment methods are generally classified as either full-reference (FR), reduced-reference (RR), or no-reference (NR).

- **FR methods:** These are widely used in applications where the original material is available, such as when assessing image and video coding algorithm performance or when making RQO decisions.
- **NR methods:** These are normally only employed where reference content is not available [27]. Scenarios could include when evaluating the influence of a lossy communication system at the receiver. It is extremely difficult to produce "blind" metrics and hence the use of NR methods is generally restricted to specific operating scenarios and distortion types. They do not generalize well and reduced-reference metrics are preferable if possible.
- **RR methods [29]:** These use only partial information about the source for predicting quality. They find application in lossy communication networks where quality predictions can be informed by partial knowledge of the original content and possibly also the channel state. Fig. 10.6 depicts a possible scenario for an RR metric, where sparse features are extracted at the encoder and transmitted through a protected (low data rate) side channel to the decoder. At the decoder, similar features are extracted from the reconstructed signal and compared in the RR metric. An indication of the reconstruction quality at the decoder can then be fed back to the encoder so that it can make informed coding decisions based on the prevailing channel state. Clearly any additional side information places an overhead on the bit rate of the coded information and this must be assessed in the context of the quality gains achieved.

10.4.2 A characterization of PSNR

For certain types of visual signal under certain test conditions, PSNR can provide a simple and efficient approach to distortion estimation. For example, Huynh-Thu and Ghanbari [31] showed that PSNR can offer consistent results when used to compare between similar codecs or codec enhancements based on the same test data. However, it is equally clear that MSE measures will fail badly for certain impairments that in reality have little perceptual impact. For example, a small spatial or temporal shift, an illumination change, or a small variation in a contextual texture [33] will all cause a large increase in MSE. Girod [35] provides a comprehensive overview of the limitations of MSE-based measures and an excellent analysis is also presented by Wang and Bovik in [33]. They sum up the situation nicely, listing the assumptions that underpin the use of MSE:

1. Signal quality is independent of temporal or spatial relationships between samples.
2. Signal quality is independent of any relationship between the original signal and the error signal.
3. Signal quality is independent of the signs of the error signal.
4. All samples contribute equally to signal quality.

Based on results from the VQEG FRTV Phase I database, Zhang and Bull [36] analyze the correlation between PSNR quality predictions (after nonlinear regression) and subjective DMOSs for a range of different conditions. The primary results are illustrated in Fig. 10.7, where it can be observed that in general the points fit only loosely and that the fit worsens at lower quality levels (higher DMOSs).

Zhang and Bull investigated how content influences PSNR-based predictions by subdividing the FRTV-I dataset into five groups based on the dominant content type (spatial texture, structural elements, plain luminance, dynamic motion, or mixed). This grouping was determined subjectively because conventional SI and TI measures do not provide sufficient granularity. The correlations for these content classes are presented in Fig. 10.7. It can be observed that, for videos with significant texture content, most subjective DMOS scatter points fall below the PSNR predictions. This illustrates the existence of HVS masking effects for static and dynamic textured content. In contrast, for sequences with significant structural content, the PSNR-predicted DMOS values tend to fall below the subjective scores, indicating that (despite edge masking effects) the HVS is very sensitive to errors around structural elements. The scatter plot for plainer content with little high-frequency energy similarly indicates little masking protection.

Zhang and Bull [59] also present PSNR-predicted DMOS vs. DMOS scatter plots based on different types of coding distortion: compressed content, compressed with transmission errors, and compressed with interpolation. These results are shown in the final three subplots in Fig. 10.7. PSNR can be seen to perform reasonably well for pure coding distortion, where errors are widely distributed, but provides much poorer predictions for the cases with interpolation errors and transmission failures. It is clear that in the case of highly distorted content, PSNR is not effective and alternative methods are needed.

This analysis confirms that:

1. Visual masking exists and is more evident for content with spatial and temporal textures than for plain luminance areas.

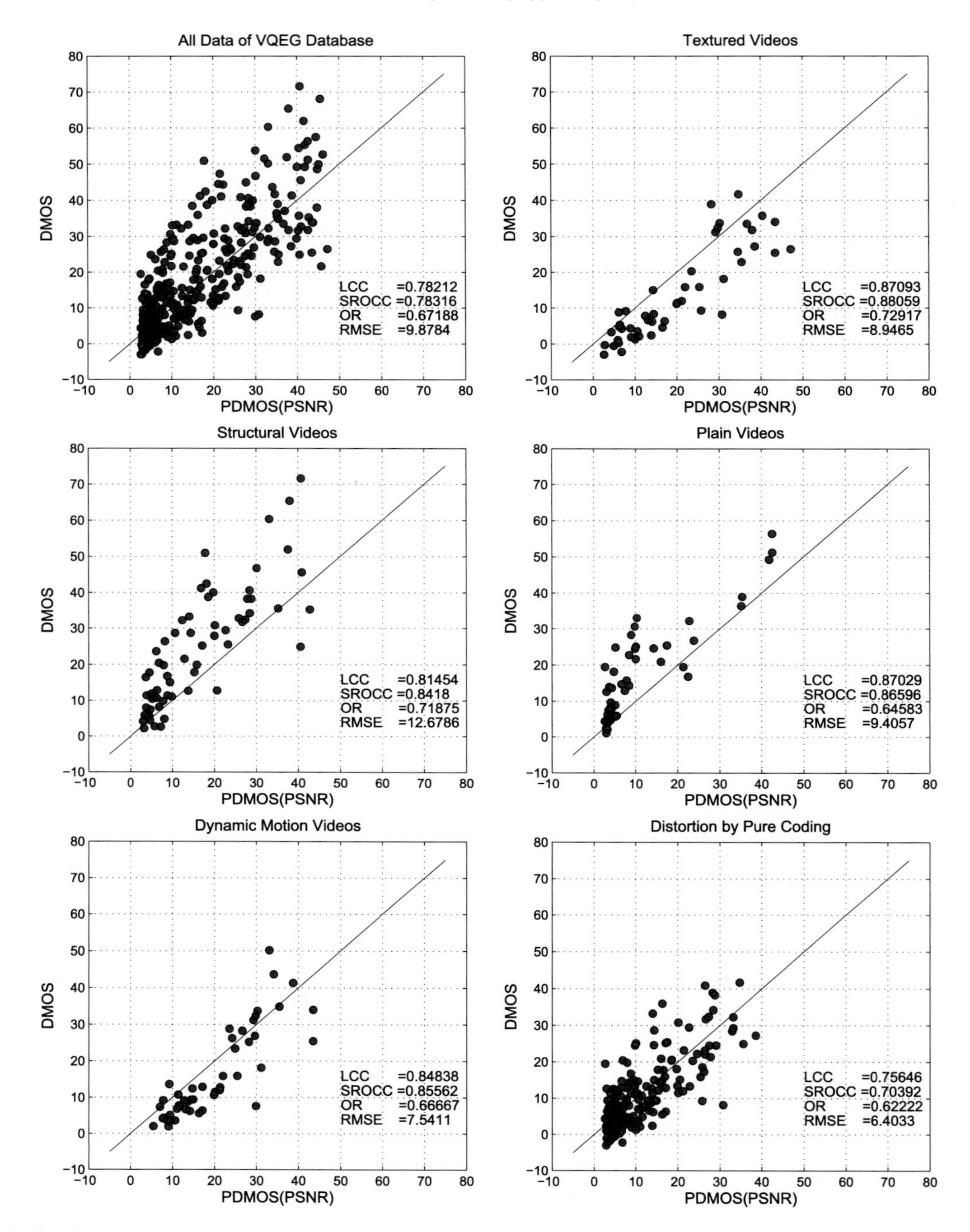

FIGURE 10.7 Scatter plots of DMOS- vs. PSNR-predicted DMOS for various classes of sequence from the VQEG database.

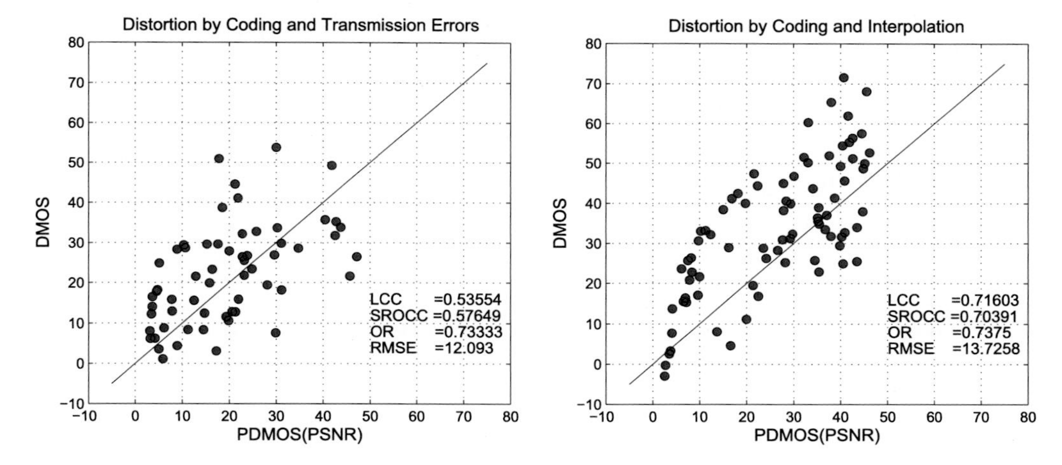

FIGURE 10.7 *(continued)*

2. Highly visible artifacts such as those in plain areas or due to transmission loss tend to cause the HVS to overestimate distortion.
3. Two distinct perceptual strategies are utilized by the HVS – *near-threshold* and *suprathreshold*. It is clear that artifact detection is more important for cases of lower perceptual quality.

Example 10.1. Texture masking and PSNR

Consider the following original 256 × 256 image (left) and its two distorted versions. In the middle image, the foreground is corrupted by additive Gaussian white noise (AGWN) of variance 0.003. The right-hand image has the background corrupted, again with AGWN of variance 0.003. The PSNR values for the two corrupted images are identical and the foreground and background regions are of equal area.

Compare these images visually and comment on the visibility of the noise.

Solution. The PSNR value for the distorted images is 28 dB, so both have identical distortions in the MSE sense. Most observers will find the distortions to the background more noticeable than those to the foreground and this is particularly the case where textured content is contextual in the scene and where the content is viewed in the absence of a reference image. This shows that distortions with different perceptual impacts are not always differentiated by PSNR measures.

10.4.3 A perceptual basis for metric development

The principle aim of any video quality metric is to correlate well with visual perception under a wide range of conditions and impairments. Secondary requirements might also include metric complexity and metric latency, particularly relevant for on-line decision making within a codec or transmission system.

HVS characteristics such as contrast sensitivity and visual masking have been exploited, both in coding and in quality assessment. For example, CSF weighting has been employed to reflect the HVS sensitivity across the range of spatial and temporal frequencies, and visual masking – the reduction of perceived distortions at certain luminances and in spatio-temporal textures – has also been exploited. Masking effects are more evident in spatially textured regions, where the HVS can tolerate greater distortions than in smoother areas. A similar phenomenon exists with dynamic textures.

When HVS properties are exploited, it is generally found that the metric provides enhanced correlation with subjective judgments, compared to conventional distortion measures such as MSE. It has been noted that this improvement is more significant when the distorted content is similar to the original [38].

When distortion becomes significant, visual attention is more likely to be attracted by visible artifacts. Under these circumstances, it is often more efficient to characterize the artifacts rather than to purely compute distortions. Based on this concept, quality metrics have been developed which emulate the perception process using a two-step strategy: near-threshold and suprathreshold. Examples of this class of metric include those reported by Kayargadde et al. [40], Karunasekera et al. [41], Carnec et al. [42], Chandler and Hemami (VSNR) [43], and Larson and Chandler (MAD) [44].

10.4.4 Perception-based image and video quality metrics

Over the past two decades or so, several perceptually inspired objective quality assessment metrics have been proposed. Most of these quality metrics adopted a similar high-level structure as shown in Fig. 10.8, which includes four key components (or a subset): preprocessing, similarity comparison, artifact detection, and pooling. A brief description of each component is provided below.

- **Preprocessing**: The test pictures and their references are preprocessed using various filters, transformed (e.g., into the frequency domain), or decomposed into multiple subbands. Their spatial and temporal features may also be extracted for subsequent analysis.
- **Similarity comparison**: The similarity between the processed test and reference signals is measured based on various visual statistics at different spatial and temporal scales. Sim-

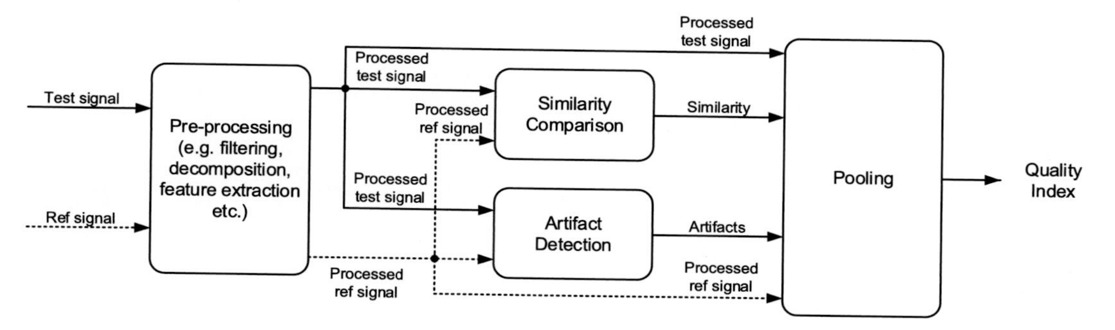

FIGURE 10.8 A high-level architecture of a generic perception-based video quality metric.

ilarity maps are generated at various scales (e.g., pixel, block, or frame) as inputs to the pooling operation.

- **Artifact detection**: Artifacts are commonly defined as unwanted signals introduced into the test content which are not present in the corresponding reference picture or sequence. At this stage, different classes of artifact may be detected at various scales to provide inputs to the pooling process alongside the similarity maps.
- **Pooling**: This final stage combines the similarity maps and the detected artifacts together with other extracted spatial and temporal features using a linear or nonlinear model to obtain a final sequence level quality index. The combining model can be as simple as a spatio-temporal mean or as complex as a (deep) neural network.

We continue this section by briefly describing notable examples of perception-based quality metrics, emphasizing how HVS characteristics can be exploited at different stages in these approaches. For further details the reader is referred to references [1,45–48].

SSIM

The integrity of structural information in an image or video is an important cue for visual perception. Wang et al. [50] developed an image quality assessment approach, SSIM, which estimates the degradation of structural similarity based on the statistical properties of local information between a reference and a distorted image. This is an improved version of the previous universal image quality index and combines three local similarity measures based on luminance, contrast, and structure. These three terms can be rearranged as in Eq. (10.16) to give the structural similarity between an impaired image $\mathbf{p}$ and its reference $\mathbf{q}$:

$$\mathrm{SSIM}(\mathbf{p},\mathbf{q}) = \frac{1}{N}\sum_{i=1}^{N}\left(l^{\alpha}(p_i,q_i)\cdot c^{\beta}(p_i,q_i)\cdot s^{\gamma}(p_i,q_i)\right), \tag{10.16}$$

in which α, β, and γ are combination parameters which equal 1 by default; $l(p_i,q_i)$, $c(p_i,q_i)$, and $s(p_i,q_i)$ are luminance, contrast, and structure comparison functions for the local patches in the test (p_i) and reference (q_i) images, which are calculated using the following equations:

$$l(p_i, q_i) = \frac{(2\mu_{p_i}\mu_{q_i} + C_1)}{\left(\mu_{p_i}^2 + \mu_{q_i}{}^2 + C_1\right)}, \tag{10.17}$$

$$c(p_i, q_i) = \frac{(2\sigma_{p_i}\sigma_{q_i} + C_2)}{\left(\sigma_{p_i}^2 + \sigma_{q_i}^2 + C_2\right)}, \tag{10.18}$$

$$s(p_i, q_i) = \frac{(2\sigma_{p_i,q_i} + C_2)}{(2\sigma_{p_i}\sigma_{q_i} + C_2)}, \tag{10.19}$$

where μ and σ are the local mean and standard deviation of p_i and q_i, $C_1 = (k_1L)^2$ and $C_2 = (k_2L)^2$ are constants used to stabilize the equation in the presence of weak denominators, and σ_{p_i,q_i} is the sample cross-correlation (zero mean); $k_1 = 0.01$, $k_2 = 0.03$, and L is the dynamic range of the pixel values. It is conventional for SSIM to be calculated using a sliding window, typically of size 11×11, using a circularly symmetric Gaussian weighting function. The SSIM value for the whole image is then computed as the average across all individual windowed results. In Eq. (10.16), N represents the total number of local windows in each image.

The range of SSIM values extends between -1 and $+1$ and only equals 1 if the two images are identical. An advantage of SSIM is that it offers superior performance to PSNR in many cases and that it is relatively simple to implement. It is interesting to note in the context of Example 10.1 that, although the PSNR values are identical for both distorted images, the SSIM scores are 0.81 for the case of the right-hand image and 0.95 for the middle image. This indicates that SSIM better reflects subjective opinions in the context of texture masking.

It has however been recognized that SSIM suffers from a number of problems, particularly that it is sensitive to relative scalings, translations, and rotations. A complex wavelet-based approach, CW-SSIM, has been developed to address these issues [51] as well as an enhanced multiscale version (MS-SSIM) [52] (described in more detail below). A further extension to SSIM called V-SSIM which also takes account of TI [53] incorporates weighting of the SSIM indices of all frames. This metric has demonstrated improved performance compared to PSNR on the VQEG FRTV Phase I database [17].

MS-SSIM

MS-SSIM extends the original SSIM from a single-scale metric to a multiple-scale solution. This accounts for the influence of viewing conditions, such as viewing distance and display setup. Based on the same notations used for SSIM,

$$\text{MS-SSIM}(\mathbf{p}, \mathbf{q}) = \frac{1}{N}\sum_{i=1}^{N}\left(l_M^{\alpha_M}(p_i, q_i) \cdot \prod_{j=1}^{M}\left(c_j^{\beta_j}(p_i, q_i) \cdot s_j^{\gamma_j}(p_i, q_i)\right)\right), \tag{10.20}$$

in which $l_j(p_i, q_i)$, $c_j(p_i, q_i)$, and $s_j(p_i, q_i)$ are three comparison functions for local patches at scale j ($j = 1, \cdots, M$). Here a low-pass filter is employed to obtain downsampled images with a sampling factor of 2. In the original literature, the default value for the maximum decomposition level (scale) M is 5, and α_j, β_j, and γ_j are weight parameters which are $\beta_1 = \gamma_1 = 0.0448$, $\beta_2 = \gamma_2 = 0.2856$, $\beta_3 = \gamma_3 = 0.3001$, $\beta_4 = \gamma_4 = 0.2363$, and $\alpha_5 = \beta_5 = \gamma_5 = 0.1333$.

VIF

Based on the different models of distortions and the HVS, Sheikh and Bovik proposed a VIF measurement method for image quality assessment [54]. This approach models image distortion in terms of signal attenuation and additive noise in the wavelet domain (Eq. (10.21)), and also considers the visual perception process as a stationary, zero-mean additive white Gaussian noise model, as shown in Eqs. (10.22) and (10.23):

$$\mathcal{I}_t = \mathcal{G}\mathcal{I}_r + \mathcal{V}, \tag{10.21}$$

$$\mathcal{O}_r = \mathcal{I}_r + N, \tag{10.22}$$

$$\mathcal{O}_t = \mathcal{I}_t + N, \tag{10.23}$$

where $\mathcal{I}_r$ and $\mathcal{I}_t$ represent the random field (RF) from a subband in the reference and test images, respectively, $\mathcal{V}$ is a stationary additive zero-mean Gaussian white noise RF, $\mathcal{G}$ is a deterministic scalar gain field, $\mathcal{O}_r$ and $\mathcal{O}_t$ denote the visual signal at the output of the HVS model from the subbands of the reference and test images, respectively, and $\mathcal{N}$ denotes the HVS (zero-mean Gaussian) noise in the wavelet domain.

VIF calculates two mutual information quantities for multiple image subbands: (i) the reference image information, $I(\mathcal{I}_r; O_r)$, between the input ($\mathcal{I}_r$) and the output ($\mathcal{O}_r$) of the HVS channel for a reference image subband, and (ii) the test image information, $I(\mathcal{I}_r; O_t)$, between and $\mathcal{I}_r$ and the output ($\mathcal{O}_t$) of the HVS channel for a test image subband. We have

$$\text{VIF} = \frac{\sum_{j=1}^{N} I(\mathcal{I}_r; O_r)}{\sum_{j=1}^{N} I(\mathcal{I}_r; O_t)}, \tag{10.24}$$

where N stands for the total number of subbands.

This metric has been evaluated on both VQEG FRTV Phase I and Phase II databases, and produces evident improvements compared to PSNR and SSIM [54].

VQM

Pinson and Wolf's VQM [49] is an objective method for video quality assessment that closely predicts subjective quality ratings. It is based on impairment filters that combine measures of blurring, jerkiness, global noise, block distortion, and color distortion. VQM computes seven parameters based on different quality features. These are obtained by filtering the impaired and reference videos to extract the property of interest. Spatio-temporal features are then extracted and a quality parameter is obtained for the feature by comparing the statistics of the filtered original and impaired video regions.

The parameters used in VQM are si_{loss} (blurring), hv_{loss} (edge shifts), hv_{gain} (blocking), si_{gain} (edge sharpening), $\text{chroma}_{\text{spread}}$ (color distribution), $\text{chroma}_{\text{extreme}}$ (localized color impairments), and $\text{ct-ati}_{\text{gain}}$ (spatio-temporal impairments). VQM forms a linear combination of these to yield the following final quality assessment:

$$\begin{aligned}\mathrm{VQM} = & -0.2097\mathrm{si}_{\mathrm{loss}} + 0.5969\mathrm{hv}_{\mathrm{loss}} + 0.2483\mathrm{hv}_{\mathrm{gain}} + 0.0192\mathrm{chroma}_{\mathrm{spread}} \\ & -2.3416\mathrm{si}_{\mathrm{gain}} + 0.0431\mathrm{ct\text{-}ati}_{\mathrm{gain}} + 0.0076\mathrm{chroma}_{\mathrm{extreme}}. \end{aligned} \tag{10.25}$$

In VQEG tests in 2004, VQM demonstrated superior performance in predicting MOS scores compared to all other algorithms evaluated. It has subsequently been adopted as an ANSI and ITU standard.

MOVIE

Scene motion is a key component in influencing visual perception and it can provide a considerable degree of artifact masking. However, perceived quality is also dependent on viewing behavior; for example whether or not a moving object is being tracked by the observer. Seshadrinathan et al. [17] introduced a motion-tuned spatio-temporal quality assessment method (MOVIE). MOVIE analyzes both distorted and reference content using a spatio-temporal Gabor filter family, and the quality index consists of a spatial quality component (inspired by SSIM) and a temporal quality component based on motion information.

The accuracy of MOVIE-based predictions has been evaluated using the VQEG FRTV Phase I database, where it was demonstrated to offer significant correlation improvements compared to both PSNR and SSIM. It also demonstrates excellent performance on the LIVE video database. One issue with MOVIE however is its high computational complexity, due to the large number of Gabor filters used and the temporal extent required for its calculation.

VSNR

The near-threshold and suprathreshold properties of the HVS were exploited by Chandler and Hemami in their VSNR still image metric [43]. This emulates the cortical decomposition of the HVS using a wavelet transform. A two-stage approach is then applied to assess the detectability of distortions and determine a final measure of VSNR. VSNR has been evaluated using the LIVE image database with very good results.

MAD and STMAD

Building on the approach used in VSNR, Larson and Chandler [44] developed the MAD model. MAD models both near-threshold distortions and appearance-based distortions. It employs different approaches for high-quality images (near-threshold distortions) and low-quality images (suprathreshold distortion). These are combined using a nonlinear model to a obtain final quality index.

MAD was extended to cope with temporal components in [55] where spatio-temporal slices are processed based on the spatial MAD and the results are then weighted using motion information. The temporal MAD index is combined with spatial MAD (computed from individual frames) to obtain spatio-temporal MAD (ST-MAD) for video. Excellent correlation performance with subjective results is reported based on the LIVE video database.

PVM

The perception-based video metric (PVM) was proposed by Zhang and Bull [57,59], building on their artifact-based video metric (AVM) model. PVM simulates the HVS perception processes by adaptively combining noticeable distortion and blurring artifacts (both exploiting the shift invariance and orientation selectivity properties of the dual-tree complex wavelet transform) using an enhanced nonlinear model. Noticeable distortion is defined by thresholding absolute differences using spatial and temporal masks which characterize texture masking effects, and this makes a significant contribution to quality assessment when the distorted video is similar to the original. Blurring artifacts, estimated by computing high-frequency energy variations and weighted with motion speed, are found to improve the overall metric performance in low-quality cases when it is combined with noticeable distortion.

Importantly, PVM, as with its predecessor, AVM, was designed for use with synthesized as well as conventionally coded material. Early results indicated that these were the only metrics capable of robust performance in this respect.

VMAF

In contrast to the approaches described above, Video Multimethod Assessment Fusion (VMAF) [20] is a machine learning-based video quality metric. Developed by researchers from Netflix, it predicts subjective quality by combining several other metrics: the detail loss metric (DLM) [58], VIF (at four different scales), and the averaged temporal frame difference (as shown in Section 10.2.2) using a support vector machine (SVM) regressor. Consistent performance improvements have been reported for VMAF over conventional quality assessment methods on various subjective databases. The original VMAF model was trained on HD video content, while various extensions have also been developed for UHDTV and cellular phone viewing scenarios.

VDP and VDP-2

In VDP-2 Mantiuk et al. [56] built on the previous VDP metric to create a metric that correlates well with subjective opinions for content with extended or diminished intensity ranges. The VDP-2 metric predicts both error visibility and quality (MOS) and was based on new contrast sensitivity measurements. The model was validated using LIVE (image) and TID2008 image databases and demonstrated improved performance compared to its predecessor HDR-VDP and VDP metrics, especially for low luminance conditions. It also compared favorably with the MS-SSIM metric.

Reduced-complexity metrics and in-loop assessment

Current image and video quality assessment methods have been developed based on various HVS characteristics in order to achieve quality prediction close to subjective opinions. These approaches provide enhanced correlation with subjective scores but are often disadvantaged by high complexity or high latency, and hence are not appropriate for real-time operation.

Recently, Soundararajan and Bovik presented a reduced-reference video quality metric [29] based on spatio-temporal entropic differencing. This performs well and supports low-complexity operation through the use of reduced spatio-temporal content.

In the context of perceptual video coding, Zhang and Bull proposed an AVM [36] using the DT-CWT as the basis for assessment of both conventionally compressed and synthesized content. AVM correlates well with subjective VQEG scores and has the advantage that it can

TABLE 10.2 Performance comparison of PSNR, SSIM, MS-SSIM, VIF, VQM, PVM, MOVIE, STMAD, and VMAF on the BVI-HD database [59].

Metric	PSNR	SSIM	MS-SSIM	VIF	VQM	PVM	MOVIE	STMAD	VMAF
LCC	0.6009	0.5784	0.7554	0.7658	0.7654	0.6830	0.7450	0.7536	0.7830
SROCC	0.5923	0.5753	0.7518	0.7700	0.7584	0.6723	0.7295	0.7541	0.7733
OR	0.5625	0.5703	0.4349	0.4609	0.4401	0.4896	0.4401	0.4141	0.4453
RMSE	13.8420	14.1642	11.2943	11.1051	11.0933	12.6410	11.5051	11.3895	10.7361

be easily integrated into a synthesis-based framework because of its high flexibility and low complexity due to extensive parameter reuse.

Comparing results

The performance of PSNR, SSIM, MS-SSIM, VIF, VQM, PVM, MOVIE, ST-MAD, and VMAF has been evaluated on the BVI-HD database, and their performance is summarized in Table 10.2 with scatter plots of the best performers shown alongside PSNR in Fig. 10.9. It can be seen that improved correlation performance has been achieved over PSNR by almost all the perception-based quality metrics described above. However, BVI-HD is acknowledged to be a challenging database as it contains diverse content and a diverse range of coding regimes. It is evident that none of these metrics perform particularly well. Interestingly the fusion-based metric, VMAF, is the best performer but nonetheless only with a relatively low SROCC value of 0.77. It is also interesting that VMAF has been specifically tuned, via its machine learning process, to work with HD content, whereas others, such as PVM, were originally tuned on SD content. In contrast, Fig. 10.10 shows earlier results for some of these metrics on SD content using the VQEG FRTV database (which includes interlaced content). This demonstrates the importance of the training data but also the challenges that remain for the development of generic and reliable perceptual metrics.

10.4.5 The future of metrics

While it is clear that perceptual metrics have advanced significantly in the past decade or so, they are still not widely accepted for general picture quality assessment. There are many complex reasons for this. For example, they mostly measure video fidelity (the closeness of the impaired version to the original), they generally measure degradation (i.e., the impaired version is assumed to be worse than the original), and they do not generally take account of viewing patterns (viewers tend to focus on regions of interest [ROIs]).

One of the most challenging problems for the future is to create a reliable, yet low-complexity, in-loop quality assessment measure with which to precisely estimate subjective quality and detect any noticeable coding artifacts. MSE methods are used extensively for this at present, but will be inappropriate as synthesis-based coding becomes more commonplace.

For real-time or in-loop processing, quality metrics should be able to:

1. perform assessment at different spatial levels (such as GOP, picture, region, or coding unit),
2. offer manageable computational complexity,

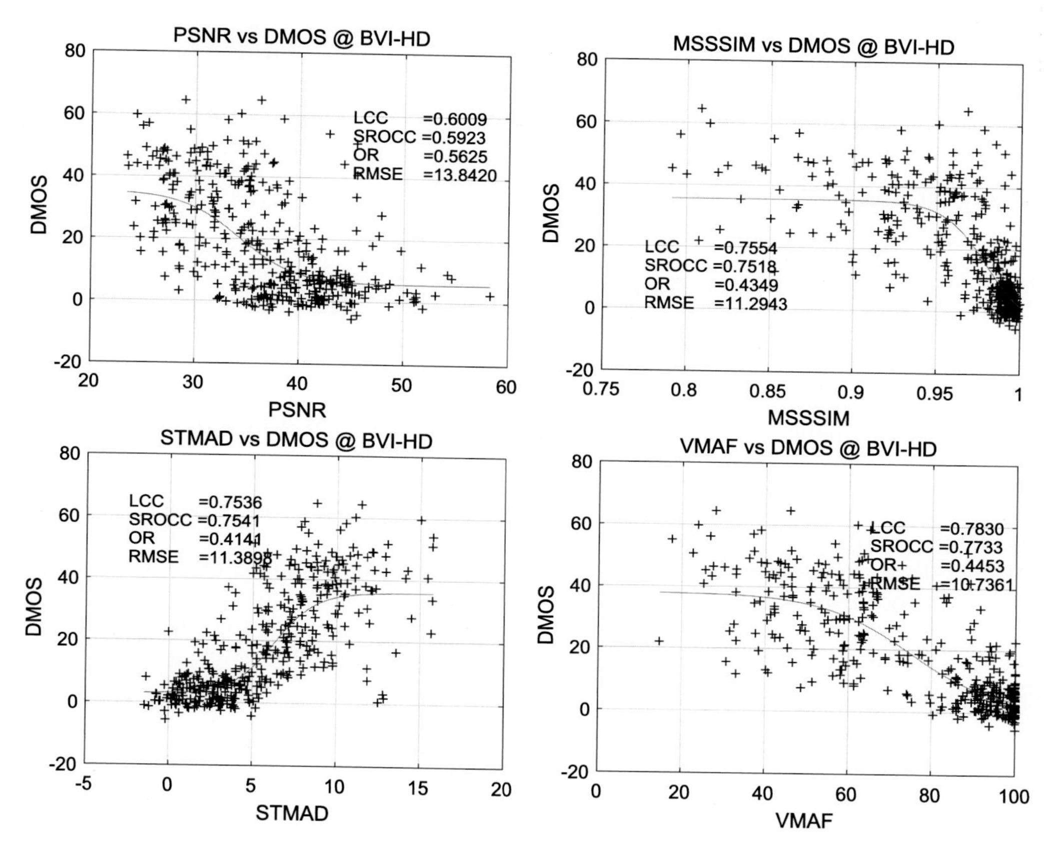

FIGURE 10.9 Performance comparison of PSNR, MS-SSIM, STMAD, and VMAF on the BVI-HD database [59].

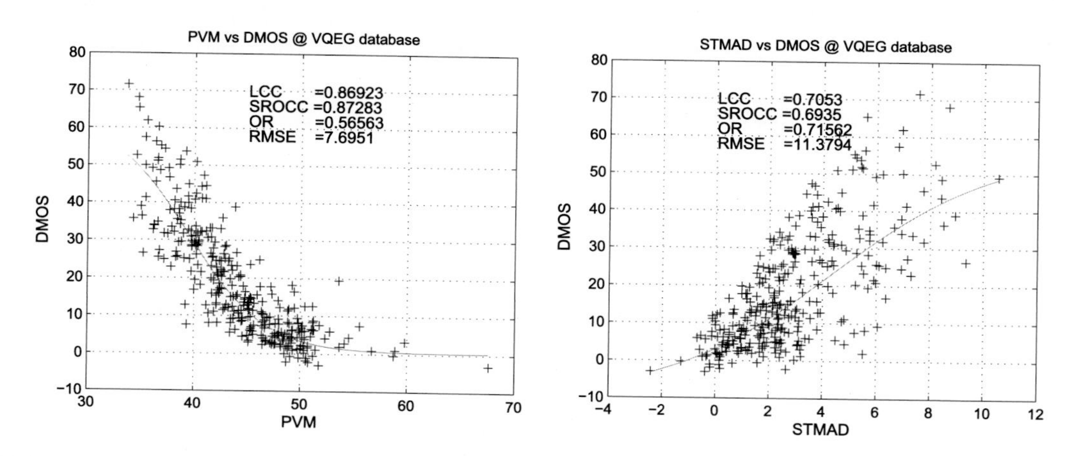

FIGURE 10.10 Performance of PVM and STMAD on the VQEQ FRTV database [59].

3. differentiate the effects of an extended video parameter space, including higher dynamic range, frame rate, and resolution, and also take account of different environmental conditions, and
4. ideally provide compatibility with emerging perceptual coding and analysis–synthesis compression methods (see Chapter 13).

Few if any existing metrics meet these criteria.

10.5 Rate-distortion optimization

Building on our introduction to rate and distortion in Chapter 4, we will consider this important area of compression in more detail here. As outlined previously, the bit rate for any compressed sequence will depend on a number of factors. These include:

- **The video content**: High spatio-temporal activity will in general require more bits to code.
- **The encoding algorithm used**: For example, wavelet- or discrete cosine transform (DCT)-based, intra-only or motion-compensated.
- **The encoding parameters selected**: At the coarsest level this includes things such as spatial resolution and frame rate. At a finer granularity, issues such as quantizer control, intra- vs. inter-modes, and block size choices will be key. The difference in performance between an encoding based on good parameter choices and one based on poor choices can be very significant.

The rate vs. distortion (or rate vs. quality) characteristic for a given codec–sequence combination provides a convenient means of comparing codecs and assessing how parameter selections can influence performance. A plot of operating points might, for a range of different parameter values, look like Fig. 10.11. The aim is to ensure that the parameters selected at any time are those that yield the lowest possible distortion for any given coding rate. To achieve this we must perform RDO or RQO.

For further reading on this topic, excellent reviews of rate-distortion theory and practice are provided by Ortega and Ramchandran in [60] and Sullivan and Wiegand in [61]. Wang et al. [62] also provide a good overview of R-D bounds. For more recent work on RDO related to current standards, the reader is referred to [63,65].

10.5.1 Classical rate-distortion theory

Rate-distortion theory lies at the heart of source coding. It refers to the trade-off between source fidelity and coding rate and can be stated simply as either:

1. representing a source with the minimum number of bits given a quality constraint, or
2. achieving the optimal quality for a given bit rate.

Shannon's *separation principle* states that, for the case of transmission of data over a noisy channel (under certain assumptions), the optimal use of bandwidth results if source and channel coding are treated independently. In other words, we should first obtain the most efficient data representation for the source in an error-free environment and then add an appropriate

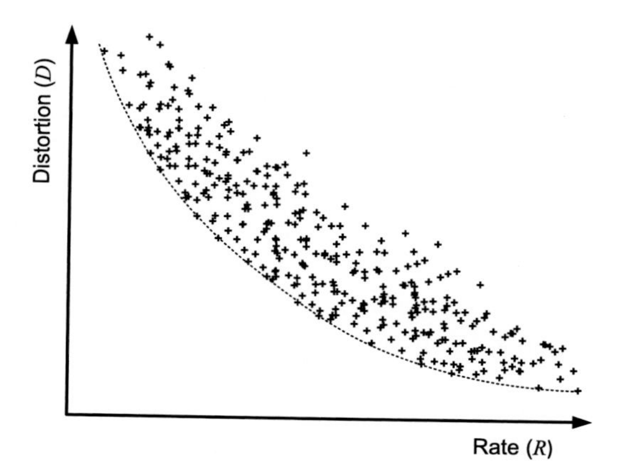

FIGURE 10.11 Rate-distortion plot for various coding parameter choices. The associated convex hull provides the operational R-D characteristic for the given codec–source combination.

level of protection to ensure error-free delivery over a lossy channel. As we have already discussed in Chapter 9, the underlying assumptions do not always apply but in general it is not a bad starting point.

Rate-distortion theory has conventionally been used to establish performance bounds (in terms of bit rate and quality) for given types of source, usually characterized in terms of their signal statistics. One of the problems faced in this respect is that simple statistical characterizations do not usually fit well with complex nonstationary sources like video.

Even in the simple case of optimizing a scalar quantizer, we still need to select an appropriate statistical model for the source. Only then can we investigate practical algorithms and assess how closely they perform to the bounds for that distribution. The problem with this approach is twofold:

1. If the model is poor, then the bounds are meaningless and practical or heuristic designs may well outperform them. Ortega and Ramchandran [60] provide an example where a low-complexity wavelet image encoder outperforms an encoder based on an i.i.d. Gaussian model by over 3 dB.
2. Even if the model is good, a practical solution may be unrealistically complex or require huge latency. For the example given in [60], the wavelet coder is relatively low-complexity, whereas the "R-D optimal" encoder is of infinite complexity.

Distortion measures

Let the distortion between a symbol of a coded source, $\tilde{s}$, and its original version, s, be denoted as $d(s,\tilde{s})$. In the case where $s=\tilde{s}$, $d(s,\tilde{s})=0$. The average distortion across the source is given by

$$D = d(\mathbf{s},\tilde{\mathbf{s}}) = E\{d(s,\tilde{s})\} = \sum_{s}\sum_{\tilde{s}} P(s,\tilde{s})\, d(s,\tilde{s}). \tag{10.26}$$

Normally the distortion measure used is either squared error, $d(s,\tilde{s}) = (s-\tilde{s})^2$, or absolute error, $d(s,\tilde{s}) = |s-\tilde{s}|$. The average distortion for a source over N samples, for the case of a squared error distortion, is the MSE:

$$d(\mathbf{s},\tilde{\mathbf{s}}) = \frac{1}{N}\sum_{n=1}^{N} d_n(s,\tilde{s}) = \frac{1}{N}\sum_{n=1}^{N} (s[n]-\tilde{s}[n])^2. \tag{10.27}$$

More often, in practical RDO applications the sum of squared differences (SSD) is used, which is computed as in Eq. (10.27), but omitting the $1/N$ term.

The memoryless Gaussian source

As we have discussed above, closed-form solutions for R-D bounds are in general difficult to obtain. However, for the case of a Gaussian i.i.d. source with variance σ^2, the bound can be computed (assuming high bit rates) and is given by

$$\bar{R}(D) = \begin{cases} \frac{1}{2}\log_2\left(\frac{\sigma^2}{D}\right), & 0 \le D < \sigma^2, \\ 0, & D \ge \sigma^2, \end{cases} \tag{10.28}$$

where $R(D)$ denotes the rate-distortion function (which gives the bit rate R needed to deliver a distortion, D, for a given source) and $\bar{R}(D)$ is the rate-distortion bound (i.e., the minimum rate among all possible coders for an infinite-length vector).

In practice, for more general and complex sources such as natural video, theoretical R-D bounds serve little purpose and more pragmatic approaches must be adopted. These are considered next.

10.5.2 Practical rate-distortion optimization

Operational control of a source encoder is a major issue in video compression and the aim of practical RDO is to select the best coding modes and parameter settings for the prevailing operating conditions. The available bit budget must thus be allocated in a manner such that the overall distortion is kept as low as possible in accordance with a given rate constraint. Parameter settings for modern codecs will cover a wide range of encoder attributes, such as block size, motion estimation references, inter- or intra-mode selection, and quantization step size. The optimization problem is made particularly difficult due to the nonstationary nature of video content, where the best encoder settings can vary significantly for different spatio-temporal regions. Furthermore the rate-distortion costs are not generally independent for all coding units (for example due to the effects of prediction).

It should be noted that, especially for lower bit rates, it is not practical to consider quantization and entropy coding independently. Unlike higher bit rates and quality settings, where the number of alphabet symbols is large, for the case of lower bit rates it may for example be advantageous to choose a quantization point that delivers slightly higher distortion because it leads to significantly lower bit rates.

From source statistics to a parameterisable codec

Practical RDO methods do not normally start with a source model. Instead they are based on the assumption of a given parameterizable encode-decode combination which is known to perform well if the right parameters are chosen. In such cases, parameter choices are not based on source statistics but rather on some transformation of the source data (e.g., DCT coefficients or wavelet subband coefficients). This has the benefit of making the modeling more robust as the statistics of transformed data tend to be easier to model. For example, subband coefficients can be roughly approximated to a memoryless source modeled with an i.i.d. Laplacian process.

However, if the subband coefficients were indeed memoryless, we could entropy encode them based on this assumption using first order entropy. In practice the assumption is not sufficiently good as the coefficients, when scanned after quantization, typically present long chains of zeros. Hence the need for tricks such as zig-zag run-value coding that we explored in Chapter 7.

RDO methods must find the best operating points based on the optimal parameter vector, $\mathbf{p}_{opt}$. If we evaluate all parameter combinations for a given realization of a source (e.g., an image) over a range of rates, then we can obtain the best rate-distortion curve for that coder–sequence combination. The convex hull, as shown in Fig. 10.11, represents the operational bound against which any practical RDO heuristic algorithm can be measured.

RDO complexity

It is worth quickly mentioning the complexity of the RDO process. In off-line cases, where a single encode will be reused multiple times, the value proposition will justify more complex or multipass encoding (a good example is encoding for a Blu-ray disc), while in other cases it may not.

Lagrangian optimization

Lagrangian methods have become widely accepted as the preferred approach for RDO in recent standards, primarily due to their effectiveness and simplicity [63,66]. Initially we will consider the case where each coding unit, i (the basic optimization building block), can be optimized independently of all others and where the only optimization parameter is a quantization index, j. While this assumption clearly breaks down in cases where context dependence or prediction are invoked, it can provide a useful and tractable solution.

Normally a discrete Lagrangian approach is adopted for RDO, as introduced by Shoham and Gersho [60,67], in order to find the maxima or minima of a function, subject to external constraints, where a closed-form solution does not exist. The Lagrangian cost function for a coding unit i is given by

$$J_{ij}(\lambda) = D_{ij} + \lambda R_{ij}, \tag{10.29}$$

where the quantization index j dictates the trade-off between rate and distortion and the Lagrange multiplier λ controls the slope of lines in the R-D plane that intersects the R-D characteristic to select specific operating points. This is shown graphically in Fig. 10.12.

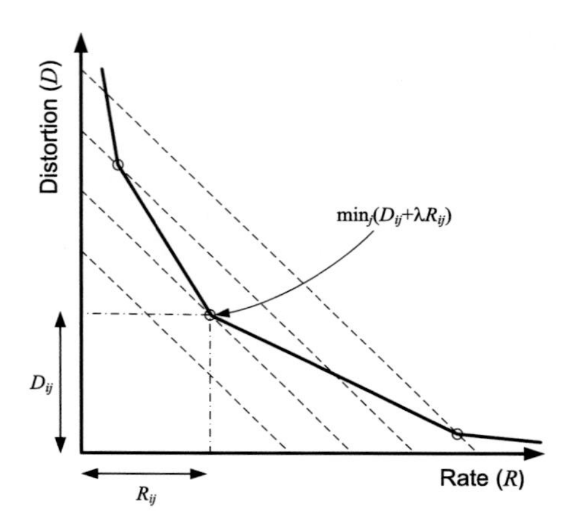

FIGURE 10.12 Lagrangian optimization for a coding unit, i.

If we minimize Eq. (10.29), then we obtain

$$\frac{\partial(J_{ij}(\lambda))}{\partial R_{ij}} = \frac{\partial(D_{ij} + \lambda R_{ij})}{\partial R_{ij}} = \frac{\partial D_{ij}}{\partial R_{ij}} + \lambda_{\text{opt}} = 0. \tag{10.30}$$

This implies that the optimal value of λ is given by the negative slope of the distortion function:

$$\lambda_{\text{opt}} = -\frac{\partial D_{ij}}{\partial R_{ij}}. \tag{10.31}$$

Clearly, if we minimize J for the case of $\lambda = 0$ we are minimizing the distortion and if we do the same when $\lambda = \infty$ we minimize the rate. However in general, the slope is not usually known so we must estimate it using operational techniques as described below. It should be noted that Lagrangian approaches are constrained to select operating points on the convex hull of the R-D characteristic. In some cases, preferable nonoptimal points may exist. This can be addressed using alternative, but potentially more complex techniques such as dynamic programming [60].

10.5.3 The influence of additional coding modes and parameters

Modern coding standards offer high coding efficiency, not just because of their underlying architecture, but also because of their adaptability to different content types and channel constraints. We will examine this in more detail in Chapter 12, but the adaptation is typically based on some or all of the following decisions:

- quantizer control,
- when to use block or picture skip modes,
- determination of motion vectors and multiple reference frame indices,
- the use of integer-, half-, or quarter-pixel motion estimation,

- the determination of intra- or inter-modes and the associated block partitions. For example in H.264/AVC the standard uses a 16×16 luma macroblock with the following options:
  - intra-modes (for all frames): nine 4×4 modes and four 16×16 modes;
  - inter-modes (only for P- and B-frames): macroblock partitions of 16×16, 16×8, 8×16, 8×8, 8×4, 4×8, 4×4.

Each mode has an associated rate-distortion cost and the encoder must select the mode having the lowest cost. This is of course complicated by the fact that the optional modes will have different efficiencies for different content types and at different bit rates. The optimization problem is complicated even further because the rate-distortion costs are not generally independent for all coding units, for example, because of the spatial and/or temporal dependencies introduced through intra- and inter-prediction. This means that the Lagrangian optimization of Eq. (10.29) should theoretically be performed jointly over all coding units in the video sequence, something that in most cases would be prohibitively complex. A number of simplification techniques have therefore been proposed.

Perhaps the most obvious method of determining parameter choices or mode decisions is to use an exhaustive search. By comparing all possible options and parameter combinations, the optimum can be selected. This can be prohibitively expensive, so alternatives are generally used. Much simpler strategies for decision making can be made based on indicative thresholds. For example, the decision as to whether an intra- or inter-mode should be selected could be based on a comparison between the SAD of a 16×16 macroblock with respect to its mean and the minimum SAD after integer motion search. Similar strategies have been reported for other decisions. However, the primary method of choice for RDO is based on the use of Lagrange multiplier methods and these are normally applied separately to the problems of mode selection and motion estimation.

Lagrangian multipliers revisited

Following the approach above and as described in [63], the aim is to solve the constrained problem

$$\min_{\mathbf{p}} D(\mathbf{p}) \text{ s.t. } R(\mathbf{p}) \leq R_T, \tag{10.32}$$

where R_T is the target bit rate and $\mathbf{p}$ is the vector of coding parameters. As before, this can be represented as an unconstrained Lagrangian formulation based on the following cost function:

$$\mathbf{p}_{\text{opt}} = \arg\min_{\mathbf{p}} \left\{ D(\mathbf{p}) + \lambda R(\mathbf{p}) \right\}. \tag{10.33}$$

Intra-modes: Let us first consider the problem of mode selection for intra-coding modes. If we assume that the quantizer value Q is known and that the Lagrange parameter λ_{MODE} is given, then the parameter (or coding mode) selection is given by

$$J_{\text{MODE}}(\mathbf{S}_k, \mathbf{p}_k \mid Q, \lambda_{\text{MODE}}) = D(\mathbf{S}_k, \mathbf{p}_k \mid Q, \lambda_{\text{MODE}}) + \lambda_{\text{MOD}E} \cdot R(\mathbf{S}_k, \mathbf{p}_k \mid Q, \lambda_{\text{MODE}}), \tag{10.34}$$

where the parameter vector $\mathbf{p}$ is varied over all possible coding modes for the given coding unit (or subimage), $\mathbf{S}_k$. As discussed previously, the distortion measure D is normally based on SSD between the original block and its reconstructed version and the rate R is normally measured after entropy coding.

Skip modes: For the case of computing *skip* modes, the distortion and rate calculations do not depend on the current quantizer value, but simply on the SSD between the current coding unit and that produced by prediction using the inferred motion vector. The rate for a skipped block is approximately 1 bit per coding unit in H.264.
Inter-modes: For the case of *inter*-modes, let us assume we know the Lagrange parameter $\lambda_{\mathrm{MOTION}}$ and the decoded reference picture $\tilde{\mathbf{S}}$. Rate-constrained motion estimation can be performed by minimizing the cost function as follows:

$$\mathbf{p}_{opt} = \arg\min_{\mathbf{p}} \left\{ D_{\mathrm{DFD}}(\mathbf{S}_k, \mathbf{p}_k) + \lambda_{\mathrm{MOTION}} \cdot R_{\mathrm{MOTION}}(\mathbf{S}_k, \mathbf{p}_k) \right\}. \tag{10.35}$$

The distortion measure used in this case is normally either the SAD or the SSD, as defined by Eq. (10.36) with l=1 or l=2, respectively:

$$D_{\mathrm{DFD}}(\mathbf{S}_k, \mathbf{p}_k) = \sum \left| s\left[x, y, z\right] - \tilde{s}[x - d_x, y - d_y, z - d_z] \right|^l, \tag{10.36}$$

where **d** is the motion vector for the current parameter set. The rate calculation for motion must include all the bits required to code the motion information, normally based on entropy coded predictions.

RDO in H.264/AVC and H.265/HEVC reference encoders

As shown in Eq. (10.28), based on a high rate approximation, rate R can be formulated as the logarithmic function of D for entropy-constrained quantization [61],

$$R\left(D\right) = a \log_2 \left(\frac{b}{D}\right), \tag{10.37}$$

where a and b are two parameters that characterize the relationship between R and D. According to the high rate approximation, D can be further modeled using the quantization interval Q as

$$D = \frac{Q^2}{3}, \tag{10.38}$$

where Q can be obtained from the quantization parameter (QP) in H.264/AVC and H.265/HEVC using

$$Q^2 = 2^{(QP-12)/3}. \tag{10.39}$$

If Eqs. (10.37)–(10.39) are substituted into (10.31), it follows that

$$\lambda = c \cdot 2^{(QP-12)/3}, \tag{10.40}$$

where $c = ln2/(3a)$.

In order to determine the value of c, Wiegand et al. [61,63] conducted a Lagrange multiplier selection experiment in the context of H.263. The experimental results show that c is approximately independent of video content, with a fixed value of 0.85 for mode selection (interframes):

$$\lambda_{\mathrm{MODE}} = 0.85 Q_{\mathrm{H.263}}^2. \tag{10.41}$$

The Lagrangian approach has been used widely since its introduction in H.263 and has been adopted in the test models for H.264/AVC and H.265/HEVC. For the case of the H.264/AVC test model JM 19, a different expression has been determined empirically. For I-, P-, and B-frames,

$$\begin{cases} \lambda_I &= 0.57 \cdot 2^{(QP-12)/3}, \\ \lambda_P &= 0.85 \cdot 2^{(QP-12)/3}, \\ \lambda_B &= 0.68 \cdot \max(2, \min(4, (QP-12)/6)) \cdot 2^{(QP-12)/3}. \end{cases} \tag{10.42}$$

A simpler model has been employed in the HEVC test model HM 16.21 [64]:

$$\lambda_{I,P,B} = 0.57 \cdot 2^{(QP-12)/3}. \tag{10.43}$$

The relationship between λ_{MODE} and λ_{MOTION} is given by

$$\begin{cases} \lambda_{MOTION} = \lambda_{MODE}, & \text{for SSD}, \\ \lambda_{MOTION} = \sqrt{\lambda_{MODE}}, & \text{for SAD}. \end{cases} \tag{10.44}$$

Example 10.2 shows the calculation of the frame level Lagrangian multiplier λ values for three preconfigured coding modes in the HEVC reference encoder (HM 16.21).

Example 10.2. Lagrangian multiplier calculation in HEVC HM 16.21

Given preconfigured coding parameters for three coding modes in HEVC HM 16.21 below [64], compute the value of the Lagrangian multiplier λ for each frame.

Assume an all intra (AI)-mode (GOP size equals 1):

Frame no.	0
Frame type	I
QP	22

Assume a low delay (LD) mode (GOP size equals 8):

Frame no.	0	1	2	3	4	5	6	7	8
Frame type	I	P	P	P	P	P	P	P	P
QP	21	27	26	27	26	27	26	27	23

And assume a random access (RA) mode (GOP size equals 16):

Frame no.	0	1	2	3	4	5	6	7	8
Frame type	I	B	B	B	B	B	B	B	B
QP	19	29	27	29	26	29	27	29	23
Frame no.	9	10	11	12	13	14	15	16	
Frame type	B	B	B	B	B	B	B	P	
QP	29	27	29	26	29	27	29	23	

Solution. According to Eq. (10.43), Lagrangian multiplier λ values (for mode selection) are calculated as in the following tables:

AI mode:

Frame no.	0
Frame type	I
QP	22
λ	5.7

LD mode:

Frame no.	0	1	2	3	4	5	6	7	8
Frame type	I	P	P	P	P	P	P	P	P
QP	21	27	26	27	26	27	26	27	23
λ	4.6	18.2	14.5	18.2	14.5	18.2	14.5	18.2	7.2

RA mode:

Frame no.	0	1	2	3	4	5	6	7	8
Frame type	I	B	B	B	B	B	B	B	B
QP	19	29	27	29	26	29	27	29	23
λ	2.9	29.0	18.2	29.0	14.5	29.0	18.2	29.0	7.2
Frame no.	9	10	11	12	13	14	15	16	
Frame type	B	B	B	B	B	B	B	P	
QP	29	27	29	26	29	27	29	23	
λ	29.0	18.2	29.0	14.5	29.0	18.2	29.0	7.2	

Example 10.3. Rate-distortion optimization

For the case presented in Example 10.2, when encoding a 64 × 64 block (a coding tree unit in HEVC) of frame 16 in the *foreman* sequence using the HEVC RA configuration, the rate (bits) and distortion (SSD) figures are obtained for three sets of coding parameters (using different modes, partitions, etc.):

Set 1: $R_1 = 1564$ and $D_1 = 30317$,

Set 2: $R_2 = 1585$ and $D_2 = 28620$,

Set 3: $R_3 = 6021$ and $D_3 = 32610$.

Compare the RD performance of these three parameter sets using the Lagrangian multiplier method.

Solution. According to Example 10.2, the value of the Lagrangian multiplier λ for this frame is 7.2. Based on Eq. (10.33), the cost for each set is calculated as

$$\text{Set 1}: \; J_1 = D_1 + \lambda R_1 = 30317 + 7.2 \times 1564 = 41577.8,$$

$$\text{Set 2}: \; J_2 = D_2 + \lambda R_2 = 28620 + 7.2 \times 1585 = 40032.0,$$

$$\text{Set 3}: \; J_3 = D_3 + \lambda R_3 = 32610 + 7.2 \times 6021 = 75961.2.$$

The coding parameter set with the best RD performance (the minimum cost) is thus Set 2.

10.5.4 From rate-distortion optimization to rate-quality optimization

The techniques described throughout this section have been based on two assumptions: (i) that the aim of the encoding process is to represent the picture sample values as closely as possible and (ii) that MSE (or other L_p norms) accurately reflects the perceptual quality and importance of the spatio-temporal region being coded. In practice we have seen that, although the second assumption is clearly violated, MSE can provide monotonic and consistent indicators for parameter variations within a given coding strategy for given content [31].

In the context of the first assumption, new perceptually inspired coding strategies have been reported based on analysis–synthesis models (see Chapter 13). The aim of these is not to minimize the distance between the original and coded versions, but instead to obtain a perceptually plausible solution. In this case, MSE is unlikely to provide any meaningful indication of perceived quality and will no longer be a valid objective function. Emphasis must therefore shift from RDO to RQO, demanding new embedded perceptually driven yet low-complexity quality metrics [36].

Among existing perceptual video compression metrics, SSIM is one of the most commonly used quality assessment methods for in-loop RQO due to its accuracy and simplicity. Recent work has demonstrated improvements in rate-quality performance with SSIM-based RQO when compared to conventional RDO approaches in the context of H.264 and HEVC [37,39]. It should be emphasized that RQO-based video compression is still in its infancy and, as shown in Section 10.4.4, quality metrics with lower computational complexity (e.g., SSIM) do not always correlate well with subjective quality opinions. In contrast, more advanced methods, such as MOVIE and STMAD, are not appropriate for in-loop application due to their high complexity and/or latency characteristics.

10.6 Rate control

Rate control is an important component in a practical video codec since it ensures that the coded video is delivered to the channel at a rate compatible with the prevailing content complexity and channel capacity. Putting to one side for the moment the influences of specific coding modes, the primary issues that influence the bit rate of an encoded video are:

1. the different requirements for I-pictures and P- or B-pictures (I-pictures can consume 5–10 times the number of bits of a P- or B-picture) and
2. the influence of source complexity. Content with high spatio-temporal activity will normally require many more bits than for the case of low-activity content. This is depicted in Fig. 10.13.

Hence if the quantizer is kept fixed, then the peak to mean ratio of the output bitstream could be excessively high, translating to an inefficient use of bandwidth. The peak output bit rate could also exceed the maximum channel capacity .

Even when rate control is used, bit rate fluctuations are inevitable and buffering is invariably employed to smooth out these variations. A simple depiction of the interface between an encoder and the channel and the channel and a decoder via buffers is shown in Fig. 10.14. This figure explicitly includes a rate controller that is used to moderate the flow of coded

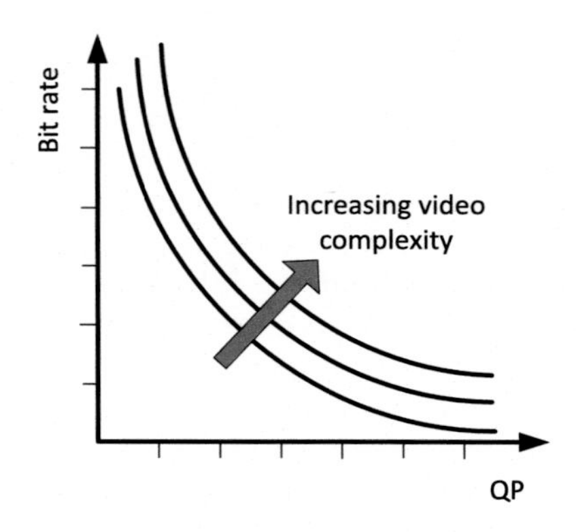

FIGURE 10.13 Managing bit rate in the context of changing source complexity.

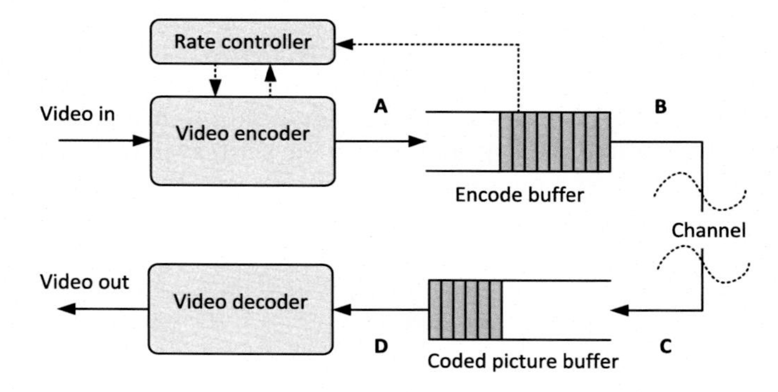

FIGURE 10.14 Transmission buffering at encoder and decoder.

data into the encode buffer. This is normally achieved by means of a rate-quantization model that delivers QP to the encoder based on dynamic measures of buffer occupancy and content complexity. If we consider the delays due to the buffering operations and the channel, we can represent the playout schedule for this type of configuration, as shown in Fig. 10.15.

The buffers described above and the associated rate control mechanisms are nonnormative in most standards. However, to ensure that rate control is performed correctly at the encoder, recent standards have incorporated the concept of a *hypothetical reference decoder* (HRD). This is used to emulate idealized decoder behavior in terms of buffering and bit consumption. Without rate control, decoder buffers would regularly underflow or overflow, resulting in playout jitter or loss.

Rate control and RDO jointly present a dilemma: in order to perform RDO for a coding unit, the QP must be known, and this is usually based on SAD or a variance measure. However, the actual SAD of the coding unit is not available until after the RDO process is complete; it is thus necessary to estimate the SAD of the current coding unit. Alongside this we also need to compute target bits for the current frame, and this is complicated by the fact that header information such as coding unit modes and motion vector information are not available before the RDO process completes.

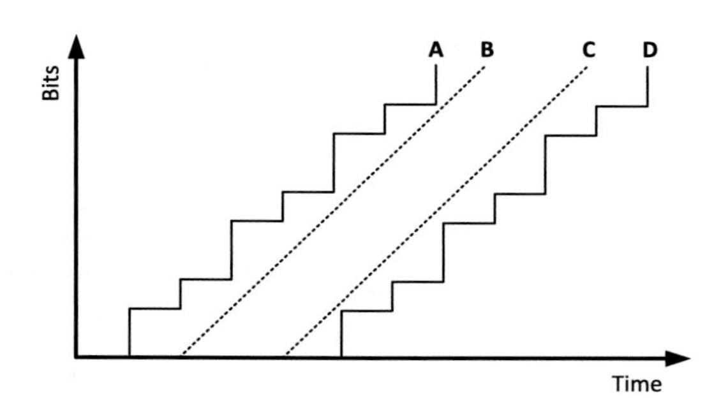

FIGURE 10.15 Streaming playout schedule on a picture by picture basis, corresponding to the points labeled in Fig. 10.14.

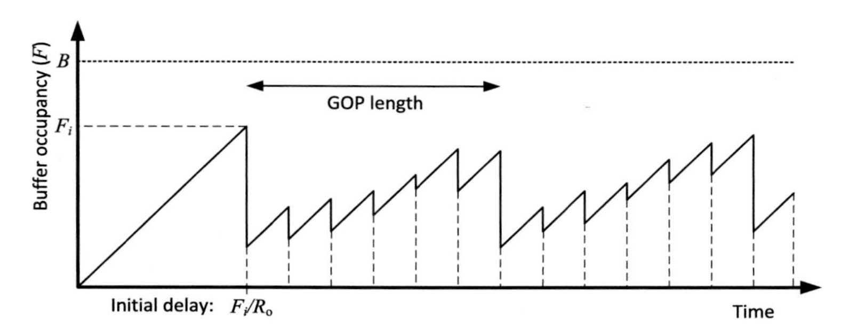

FIGURE 10.16 Example decoder buffer timing, indicating HRD parameters.

10.6.1 Buffering and HRD

Bit rate allocation is normally associated with a buffer model that is specified as part of an HRD in the video coding standard. The HRD is usually a normative component of a standard since it is important in dictating the requirements on compliant bitstreams.

The HRD is an idealized representation of a decoder that is conceptually connected to the output of the encoder. It comprises a decoder buffer as well as the decoder itself. The buffer is normally modeled using a leaky bucket approach which simply provides a constant rate flow of bits to and from the channel. Bits are assumed to flow into the decoder buffer (sometimes called the coded picture buffer [CPB]) at a constant rate and are assumed to be extracted from it by the decoder instantaneously in picture-sized blocks. This is the scenario depicted in Fig. 10.14 and Fig. 10.15 where, in the absence of loss in the channel, the content of the buffer at the receiver simply mirrors that at the encoder. An HRD-compliant bitstream must thus be processed by the CPB without overflow and underflow. This requirement must be satisfied by the rate control implemented in the encoder.

The HRD buffer model is normally described by three parameters: (i) the output rate R_o (assumed constant), (ii) the initial occupancy F_i (which is the occupancy level at which decoding is assumed to start), and (iii) the buffer size B. The rate controller must track the occupancy F of the hypothetical buffer and adjust the encoder parameters within its scope, in order to avoid buffer underflow and overflow (see Fig. 10.16).

Example 10.4. Buffering and HRD parameters

Consider the operation of a constant-bit rate video transmission system at 24 fps with the following parameters:

$$R_o = 500 \text{ kbps}, \; B = 128 \text{ kb}, \; F_i = 100 \text{ kb}, \text{ GOP} = 6 \text{ frames}.$$

If the initial pictures transmitted have the following sizes, compute the occupancy of the decoder buffer and determine whether underflow or overflow occurs.

Frame no.	Picture size (kbits)
1	60
2–6	20
7	30
8–12	10
13	30
14–18	10
19	50
20–24	30

Solution. The ramp-up time for the buffer is $F_i/R_o = 0.2s$. Plotting the buffer occupancy over the 1-s period of these initial pictures, we can observe that the decoder buffer overflows when pictures 18 and 19 are transmitted.

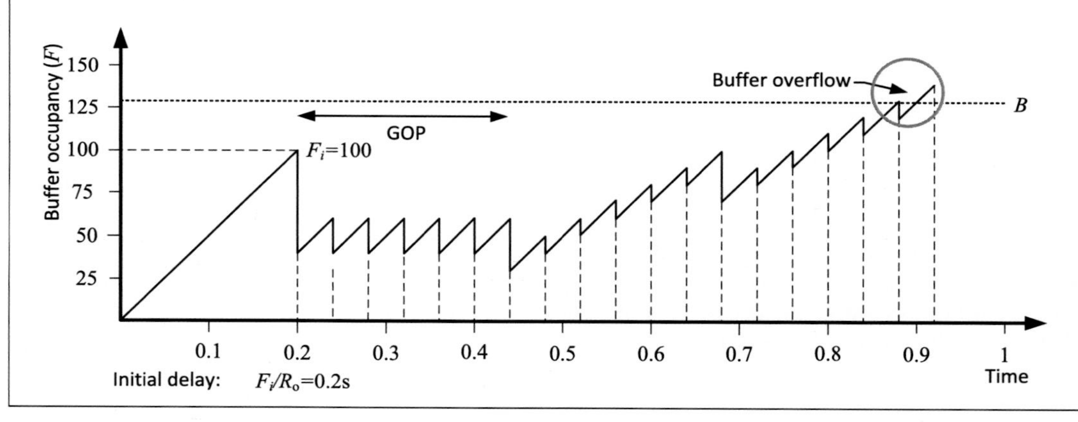

10.6.2 Rate control in practice

Fig. 10.17 shows a generic rate control architecture. While simplified, this has many similarities to those used in MPEG-2, H.264/AVC, and HEVC. The basic elements are described below.

Buffer model

The buffer model represents the hypothetical reference decoder as described in Section 10.6.1. It is parameterized by its size and initial fullness, and takes a further input

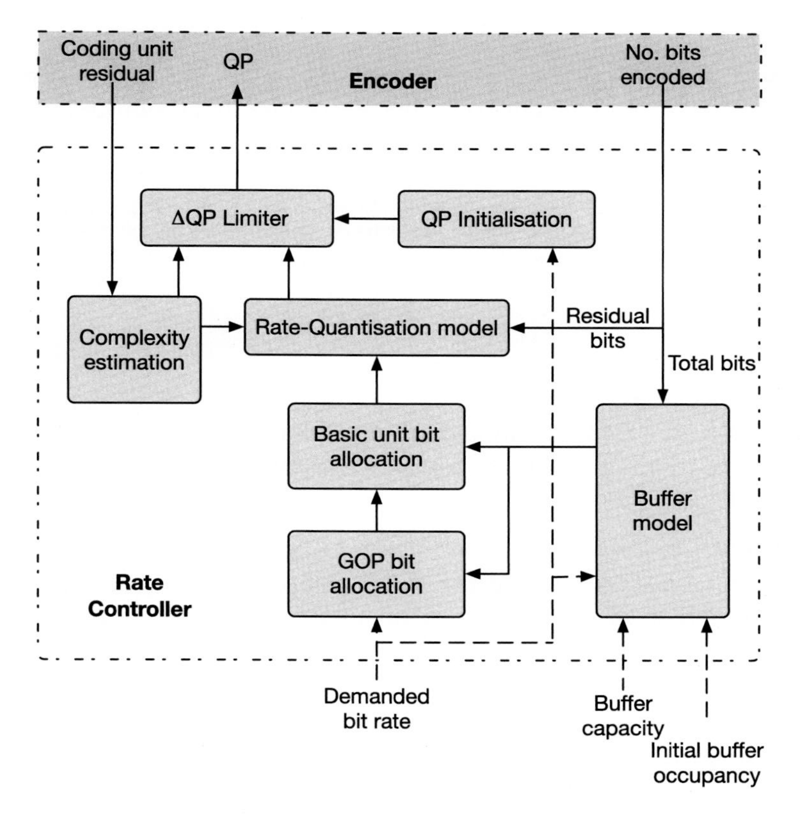

FIGURE 10.17 Generic rate controller.

according to the bit rate of the channel. It provides an output of its current occupancy, F, to the bit allocation blocks described below.

Complexity estimation

In order to control the bit allocation process it is essential to estimate the complexity of the coded residual. The estimation of source complexity is normally based on the residual data (since this is what the quantizer influences). A measure such as MAD, sum of absolute values (SAD), SSD, or sum of absolute transform differences [68] of the residual is often used for this purpose.

Rate-quantization model

Reflecting the variability indicated in Fig. 10.13, the rate-quantization model describes how the available bits are related to quantizer values when content of varying complexity is encoded.

ΔQP limiter

In order to avoid excessive changes in QP values that might introduce noticeable artifacts or cause oscillatory behavior in the rate controller, the change in QP value is normally limited to a small increment, typically ± 2.

QP initialization

The QP value must be set to some value prior to coding. An average value for the bit rate per pixel can be easily computed, based on the bit rate demanded and the video format; thus,

$$\text{bpp} = \frac{R_1(1)}{f \cdot X \cdot Y}, \tag{10.45}$$

where $R_i(k)$ is the instantaneous available bit rate at the time of coding frame k of GOP i, f is the frame rate, and X and Y are the number of horizontal and vertical pixels in the picture. This can then be related to QP values by way of a look-up table such as that recommended for H.264/AVC in [69].

GOP bit allocation

Based on the demanded bit rate and the buffer occupancy, a target bit allocation for the GOP is computed. GOP level bit allocation normally initializes the IDR picture and the first P-picture. A scheme used for this in H.264/AVC is as follows [70]:

$$QP_i(1) = \max\left\{\mathrm{QP}_{i-1}(1) - 2, \min\left\{\mathrm{QP}_{i-1}(1) + 2, \frac{\sum_{k=2}^{N_{i-1}} \mathrm{QP}_{i-1}(k)}{N_{i-1} - 1} - \min\left\{2, \frac{N_{i-1}}{15}\right\}\right\}\right\}, \tag{10.46}$$

where $\mathrm{QP}_i(k)$ is the QP value assigned to picture k of GOP i and N_i is the total number of pictures in GOP i. If $\mathrm{QP}_i(1) > \mathrm{QP}_{i-1}(N_{i-1}) - 2$, this is further adjusted as follows:

$$\mathrm{QP}_i(1) \leftarrow \mathrm{QP}_i(1) - 1, \tag{10.47}$$

where $\mathrm{QP}_{i-1}(N_{i-1})$ represents the QP value of the previous P-picture.

Coding unit bit allocation

This refers to the process of allocating bits to smaller coding elements, such as individual pictures, slices, or smaller coding units. The bit allocation to each picture is based on the target buffer level, the actual buffer occupancy, the bit rate demanded, and of course the frame rate. QP values for each of the remaining P- and B-frames are thus assigned based on the number of remaining bits available for the GOP. The allocation process should also take account of whether the current picture is of P or B type since these may be provided with different bit allocations. The scheme used for this in H.264/AVC, based on a quadratic model derived assuming a Laplacian source distribution, is as follows [70].

Firstly, the MAD of the current stored picture is estimated, $\tilde{\xi}_i(k)$, based on the corresponding value for the previous picture, as given by

$$\tilde{\xi}_i(k) = a_1 \cdot \xi_i(k-1) + a_2, \tag{10.48}$$

where a_1 and a_2 are the coefficients of the prediction model, initialized to 1 and 0 and updated after each unit is coded. Next, the quantizer step size is computed for each of the remaining P-pictures, using the following equation:

$$T_i(k) = c_1 \cdot \frac{\tilde{\xi}_i(k)}{Q_{step,i}(k)} + c_2 \cdot \frac{\tilde{\xi}_i(k)}{Q^2_{step,i}(k)} + m_{h,i}(k), \tag{10.49}$$

where $m_{h,i}(k)$ is the total number of header bits and motion vector bits, c_1 and c_2 are the coefficients of the quadratic model, updated after each picture is processed, and $Q_{step,i}(k)$ is the quantizer step size for picture k, corresponding to the target number of bits. In order to compute the quantizer step size in Eq. (10.49) we first need to calculate a value for the target number of bits for frame k, $T_i(k)$. This is usually computed based on a weighted combination of the buffer occupancy and the number of bits remaining for the rest of the GOP. Further details on this scheme can be found in [70].

Several other schemes have been proposed for rate control that improve prediction accuracy for both texture and header bits, for example, the contribution of Kwon et al. [68]. In the context of HEVC, several modifications to the above approach have been proposed and the reader is referred to the contributions by Sanz-Rodriguez and Schierl [71] and by Li et al. [72].

10.6.3 Regions of interest and rate control

In certain applications, and for certain types of content, a priori information about the scene, or areas within it, can be exploited to improve coding efficiency or better manage the bit rate. For example, in many surveillance videos, there may be one or more *regions of interest* that would benefit from prioritized bit allocation. A good discussion of the trade-offs involved in this is provided by Sadka in [73].

Agrafiotis et al. [74] presented a modified rate control method that allows the definition of multiple priority regions, the quality of which varies, based on the ROI characteristics and the target bit rate. The results presented demonstrate increased coding flexibility, which can lead to significant quality improvements for the ROI and a perceptually pleasing variation in quality for the rest of the frame, without violating the target bit rate. This work was presented in the context of low-bit rate sign language coding for mobile terminals, but it can also be of benefit in many other applications, including surveillance and healthcare.

10.7 Summary

This chapter has covered a range of methods for measuring and controlling visual quality. We first described the most common subjective testing methods and showed how these can provide a consistent means of comparing codec performance. We then went on to discuss a number of objective measures that can be used as an efficient substitute for subjective evaluation in certain cases. MSE-based methods were reviewed briefly and their strengths and weaknesses were highlighted. We demonstrated cases where the perceptual distortion experienced by the human viewer cannot however be characterized using such simple mathematical differences. PVMs have therefore emerged and a number of these were reviewed and compared, showing improved correlation with subjective scores.

The second part of this chapter addressed the issue of RDO, describing some of the most common techniques that enable us to select the optimal coding parameters for each spatio-temporal region of a video. After reviewing conventional rate-distortion theory, we focused on practical solutions such as those employing Lagrange multipliers, showing that these can provide a tractable solution for modern codecs. Related to this, the final part of the chapter studied the requirements of, and methods for, rate control as a means of adapting the output bit rate of an encoder according to the capacity of a given channel.

References

[1] S. Winkler, Digital Video Quality: Vision Models and Metrics, Wiley, 2005.
[2] F. Kozamernik, P. Sunna, E. Wycken, D. Pettersen, Subjective Quality Assessment of Internet Video Codecs – Phase 2 Evaluations Using SAMVIQ, EBU Technical Review, January 2005, pp. 1–22.
[3] S. Winkler, Analysis of public image and video database for quality assessment, IEEE J. Sel. Top. Signal Process. 6 (6) (2012) 1–10.
[4] Recommendation ITU-R BT.500-13: Methodology for the subjective assessment of the quality of television pictures, ITU-R, 2012.
[5] Recommendation ITU-T P.910, Subjective video quality assessment methods for multimedia applications, ITU-T, 1999.
[6] Digital transport of one-way video signals-parameters for objective performance assessment, ANSI Standard ATIS-0100801.03, 2003.
[7] F. De Simone, L. Goldmann, J-S. Lee, T. Ebrahimi, Towards high efficiency video coding: subjective evaluation of potential coding methodologies, J. Vis. Commun. Image Represent. 22 (2011) 734–748.
[8] P. Hanhart, M. Rerabek, F. De Simone, T. Ebrahimi, Subjective quality evaluation of the upcoming HEVC video compression standard, in: Applications of Digital Image Processing XXXV, 2012, 8499.
[9] H. Hoffmann, T. Itagaki, D. Wood, T. Hinz, T. Wiegand, A novel method for subjective picture quality assessment and further studies of HDTV formats, IEEE Trans. Broadcast. 54 (1) (2008) 1–13.
[10] D. Lane, Introduction to statistics online edition, Available from http://onlinestatbook.com/, Rice University.
[11] D. Ghadiyaram, A. Bovik, Massive online crowdsourced study of subjective and objective picture quality, IEEE Trans. Image Process. 25 (1) (2015) 372–387.
[12] S. Moon, J. Lee, Implicit analysis of perceptual multimedia experience based on physiological response: a review, IEEE Trans. Multimed. 19 (2) (2017) 340–353.
[13] S. Hinde, T. Smith, I. Gilchrist, Does narrative drive dynamic attention to a prolonged stimulus?, Cognitive Research: Principles and Implications 3 (45) (2018).
[14] Video Quality Experts Group, Final report from the video quality experts group on the validation of objective quality metrics for video quality assessment, VQEG, Tech. Rep., 2000 [Online]. Available http://www.its.bldrdoc.gov/vqeg/projects/frtv, phaseI.
[15] Video Quality Experts Group, Final VQEG report on the validation of objective models of video quality assessment, VQEG, Tech. Rep., 2003 [Online]. Available http://www.its.bldrdoc.gov/vqeg/projects/frtv, phaseII.
[16] H. Sheikh, Z. Wang, L. Cormack, A. Bovik, LIVE image quality assessment database release 2, http://live.ece.utexas.edu/research/quality.
[17] K. Seshadrinathan, A. Bovik, Motion tuned spatio-temporal quality assessment of natural videos, IEEE Trans. Image Process. 19 (2010) 335–350.
[18] F. Zhang, F. Mercer Moss, R. Baddeley, D. Bull, BVI-HD: a video quality database for HEVC compressed and texture synthesised content, IEEE Trans. Multimed. 20 (10) (2018) 2620–2630.
[19] P. Le Callet, F. Autrusseau, Subjective quality assessment irccyn/ivc database, http://www.irccyn.ec-nantes.fr/ivcdb/, 2005.
[20] Z. Li, et al., Toward a practical perceptual video quality metric, Netflix Tech Blog, 2016.

[21] K. Seshadrinathan, et al., Study of subjective and objective quality assessment of video, IEEE Trans. Image Process. 19 (6) (2010) 335–350.
[22] L. Song, et al., The SJTU 4k video sequence dataset, in: Fifth IEEE Intl. Workshop on Quality of Multimedia Experience (QoMEX), 2013, pp. 34–35.
[23] Video Quality Experts Group, Experts Group, Report on the validation of video quality models for high definition video content, VQEG, Tech. Rep., 2010 [Online]. Available, http://www.its.bldrdoc.gov/vqeg/projects/hdtv/hdtv.aspx.
[24] A. Mackin, F. Zhang, D. Bull, A study of high frame rate video formats, IEEE Trans. Multimed. 21 (6) (2019) 1499–1512.
[25] A. Mackin, F. Zhang, D. Bull, Final report from the video quality experts group on the validation of objective models of multimedia quality assessment, VQEG, Tech. Rep., 2008 [Online]. Available http://www.its.bldrdoc.gov/vqeg/projects/multimedia-phase-i/multimedia-phase-i.aspx.
[26] D.C. Howell, Statistical Methods for Psychology, 7th ed., Cengage Wadsworth, 2010.
[27] P. Marziliano, F. Dufaux, S. Winkler, T. Ebrahimi, A no-reference perceptual blur metric, in: Proc. IEEE Int Conf. on Image Processing, vol. 3, 2002, pp. 57–60.
[28] M. Cheon, J. Lee, Subjective and objective quality assessment of compressed 4K UHD videos for immersive experience, IEEE Trans. Circuits Syst. Video Technol. 28 (7) (2018) 1467–1480.
[29] R. Soundararajan, A. Bovic, Video quality assessment by reduced reference spatio-temporal entropic differencing, IEEE Trans. Circuits Syst. Video Technol. 23 (4) (2013) 684–694.
[30] J. Yim, et al., Subjective quality assessment for YouTube UGC dataset, arXiv preprint, arXiv:2002.12275, 2020.
[31] D. Huynh-Thu, M. Ghanbari, Scope of validity of PSNR in image/video quality assessment, Electron. Lett. 44 (13) (2008) 800–801.
[32] M. Papadopoulos, F. Zhang, D. Agrafiotis, D. Bull, A video texture database for perceptual compression and quality assessment, in: IEEE Intl. Conf. on Image Processing (ICIP), 2015, pp. 2781–2785.
[33] Z. Wang, A. Bovik, Mean squared error: love it or leave it?, IEEE Signal Process. Mag. 21 (1) (2009) 98–117.
[34] A. Mackin, M. Afonso, F. Zhang, D. Bull, A study of subjective video quality at various spatial resolutions, in: IEEE Intl. Conf. on Image Processing (ICIP), 2018, pp. 2830–2834.
[35] B. Girod, What's wrong with mean squared error?, in: A. Watson (Ed.), Digital Images and Human Vision, MIT Press, 1998.
[36] F. Zhang, D. Bull, A parametric framework for video compression using region-based texture models, IEEE J. Sel. Top. Signal Process. 5 (7) (2011) 1378–1392.
[37] S. Wang, A. Rehman, Z. Wang, S. Ma, W. Gao, SSIM-motivated rate-distortion optimization for video coding, IEEE Trans. Circuits Syst. Video Technol. 22 (4) (2012) 516–529.
[38] T. Pappas, T. Michel, R. Hinds, Supra-threshold perceptual image coding, in: Proc. IEEE Int Conf. on Image Processing, 1996, pp. 234–240.
[39] A. Rehman, Z. Wang, SSIM-inspired perceptual video coding for HEVC, in: IEEE Int. Conf. on Multimedia and Expo, 2012, pp. 497–502.
[40] V. Kayargadde, J. Martens, Perceptual characterization of images degraded by blur and noise: experiments, J. Opt. Soc. Am. 13 (1996) 1166–1177.
[41] S. Karunasekera, N. Kingsbury, A distortion measure for blocking artifacts in images based on human visual sensitivity, IEEE Trans. Image Process. 4 (6) (1995) 713–724.
[42] M. Carnec, P. LeCallet, D. Barba, An image quality assessment method based on perception of structural information, in: Proc. IEEE Int Conf. Image Processing, vol. 2, 2003, pp. 185–188.
[43] D. Chandler, S. Hemami, VSNR: a wavelet-based visual signal to noise ratio for natural images, IEEE Trans. Image Process. 16 (9) (2007) 2284–2298.
[44] E. Larson, D. Chandler, Most apparent distortion: full reference image quality assessment and the role of strategy, J. Electron. Imaging 19 (1) (2010), 011 006(1–21).
[45] S. Chikkerur, V. Sundaram, M. Reisslein, L. Karam, Objective video quality assessment methods: a classification, review and performance comparison, IEEE Trans. Broadcast. 57 (2) (2005) 17–26.
[46] W. Lin, C. Kuo, Perceptual visual quality metrics: a survey, J. Vis. Commun. Image Represent. 22 (4) (2011) 297–312.
[47] S. Sheikh, A. Bovik, Image information and visual quality, IEEE Trans. Image Process. 15 (2) (2006) 430–444.
[48] S. Winkler, P. Mohandas, The evolution of video quality measurement: from PSNR to hybrid metrics, IEEE Trans. Broadcast. 54 (3) (2008) 660–668.

[49] M. Pinson, S. Wolf, A new standardized method for objectively measuring video quality, IEEE Trans. Broadcast. 50 (3) (2004) 312–322.
[50] Z. Wang, A. Bovik, H. Sheikh, E. Simoncelli, Image quality assessment: from error visibility to structural similarity, IEEE Trans. Image Process. 13 (4) (2004) 600–612.
[51] Z. Wang, E. Simoncelli, Translation insensitive image similarity in complex wavelet domain, in: IEEE Intl. Conf. Acoustics, Speech and Signal Processing, 2005, pp. 573–576.
[52] Z. Wang, E. Simoncelli, A. Bovik, Multi-scale structural similarity for image quality assessment, in: Proc. IEEE Asilomar Conf. Signals, Syst., Comput., 2003, pp. 1398–1402.
[53] Z. Wang, L. Lu, A. Bovik, Video quality assessment based on structural distortion measurement, Signal Process. Image Commun. 19 (2) (2004) 121–132.
[54] H.R. Sheikh, A.C. Bovik, G. de Veciana, An information fidelity criterion for image quality assessment using natural scene statistics, IEEE Trans. Image Process. 14 (12) (2005) 2117–2128.
[55] P. Vu, C. Vu, D. Chandler, A spatiotemporal most apparent distortion model for video quality assessment, in: Proc. IEEE Int Conf. on Image Processing, 2011, pp. 2505–2508.
[56] R. Mantiuk, et al., HDR-VDP-2: a calibrated visual metric for visibility and quality predictions in all luminance conditions, ACM Trans. Graph. 30 (4) (2011) 40 (Proc. SIGGRAPH'11).
[57] F. Zhang, D. Bull, Quality assessment method for perceptual video compression, in: Proc. IEEE Intl. Conf. on Image Processing, 2013, pp. 39–43.
[58] S. Li, F. Zhang, L. Ma, K.N. Ngan, Image quality assessment by separately evaluating detail losses and additive impairments, IEEE Trans. Multimed. 13 (5) (2011) 935–949.
[59] F. Zhang, D. Bull, A perception-based hybrid model for video quality assessment, IEEE Trans. Circuits Syst. Video Technol. 26 (6) (2016) 1017–1028.
[60] A. Ortega, K. Ramchandran, Rate-distortion methods for image and video compression, IEEE Signal Process. Mag. 15 (6) (1998) 23–50.
[61] G. Sullivan, T. Wiegand, Rate-distortion optimization for video compression, IEEE Signal Process. Mag. 15 (6) (1998) 74–90.
[62] Y. Wang, J. Ostermann, Y. Zhang, Video Processing and Communications, Prentice Hall, 2002.
[63] T. Wiegand, H. Schwarz, A. Joch, F. Kossentini, G. Sullivan, Rate-constrained coder control and comparison of video coding standards, IEEE Trans. Circuits Syst. Video Technol. 13 (7) (2003) 688–703.
[64] K. Andersson, et al., Joint Video Team (JVT) of ISO/IEC MPEG and ITU-T VCEG Document JVT-O0038, AHG 3 Recommended settings for HM, 2016.
[65] E-H. Yang, X. Yu, Rate distortion optimization for H.264 interframe coding: a general framework and algorithms, IEEE Trans. Image Process. 16 (7) (2007) 1774–1784.
[66] H. Everett, Generalised Lagrange multiplier method for solving problems of optimum allocation of resources, Oper. Res. 11 (1963) 399–417.
[67] Y. Shoham, A. Gersho, Efficient bit allocation for an arbitrary set of quantizers, IEEE Trans. Acoust. Speech Signal Process. 36 (1988) 1445–1453.
[68] D. Kwon, M. Shen, C. Kuo, Rate control for H.264 video with enhanced rate and distortion models, IEEE Trans. Circuits Syst. Video Technol. 17 (5) (2007) 517–529.
[69] G. Sullivan, T. Wiegand, K. Lim, Joint Model Reference Encoding Methods and Decoding Concealment Methods, JVT-I049, San Diego, 2003.
[70] K. Keng-Pang Lim, G. Sullivan, T. Wiegand, Joint Video Team, (JVT) of ISO/IEC MPEG and ITU-T VCEG Document JVT-O079, Text Description of Joint Model Reference Encoding Methods and Decoding Concealment Methods, 2005.
[71] S. Sergio Sanz-Rodrıguez, T. Schierl, A rate control algorithm for HEVC with hierarchical GOP structures, in: Proc. IEEE Intl. Conf. on Acoustics Speech and Signal Processing, 2013, pp. 1719–1723.
[72] B. Li, H. Li, L. Li, J. Zhang, Rate control by R-lambda model for HEVC, in: JCTVC-K0103, 11th JCTVC Meeting, China, October 2012.
[73] A. Sadka, Compressed Video Communications, Wiley, 2002.
[74] D. Agrafiotis, D. Bull, C. Canagarajah, N. Kamnoonwatana, Multiple priority region of interest coding with H.264, in: Proc. IEEE Intl. Conf. on Image Processing, 2006, pp. 53–56.

CHAPTER

11

Communicating pictures: delivery across networks

Video compression algorithms rely on spatio-temporal prediction combined with variable-length entropy encoding to achieve high compression ratios but, as a consequence, they produce an encoded bitstream that is inherently sensitive to channel errors. This becomes a major problem when video information is transmitted over unreliable networks, since any errors introduced into the bitstream during transmission will rapidly propagate to other regions in the image sequence.

In order to promote reliable delivery over lossy channels, it is usual to invoke various error detection and correction methods. This chapter firstly introduces the requirements for an effective error-resilient video encoding system and then goes on to explain how errors arise and how they propagate spatially and temporally. We then examine a range of techniques in Sections 11.4, 11.5, and 11.6 that can be employed to mitigate the effects of errors and error propagation. We initially consider methods that rely on the manipulation of network parameters or the exploitation of network features to achieve this; we then go on to consider methods where the bitstream generated by the codec is made inherently robust to errors (Section 11.7). We present decoder-only methods that conceal rather than correct bitstream errors in Section 11.8. These deliver improved subjective quality without adding transmission overhead. Finally in Section 11.9, we describe congestion management techniques, in particular HTTP adaptive streaming (HAS), that are widely employed to support reliable streaming of video under dynamic network conditions.

11.1 The operating environment

11.1.1 Characteristics of modern networks

The IP network

The *internet* is essentially a collection of independent packet switched networks operating under the same protocol and connected via routers. Each packet has a header that identifies its source and destination IP addresses, where packet delivery to the destination is conducted using TCP/IP protocols. The *Internet Protocol* (IP) provides a connectionless basis for packet delivery and the *transmission control protocol* (TCP) provides a connection-oriented delivery

Intelligent Image and Video Compression
https://doi.org/10.1016/B978-0-12-820353-8.00020-7

mechanism using *automatic repeat request* (ARQ) techniques, but with no bounds on delivery time. It is thus a best-effort service with no guarantee of the actual delivery of packets.

For less reliable networks and real-time delay-constrained applications, the *user datagram protocol* (UDP) is often used. The *real-time transport protocol* (RTP) operates on top of UDP and contains essential timing and sequence information. Alongside RTP, the *real-time control protocol* (RTCP) enables quality of service feedback from the receiver to the sender to enable it to adjust its transmission characteristics to better suit the prevailing network conditions. This is often used in conjunction with the *real-time streaming protocol* (RTSP) for control of streaming media servers and for establishing and controlling media sessions.

To support the large-scale streaming activities of organizations such as Netflix, YouTube, Facebook, Amazon, and others, content delivery (or distribution) networks now form an essential piece in the delivery jigsaw. Content delivery networks are geographically distributed networks of proxy servers and associated data centers that are configured and located close to clusters of end users in order to provide high service availability and performance.

The wireless network edge

Most modern wireless networks also operate under IP protocols to support seamless integration with the internet backbone. The bandlimited and error-prone nature of the wireless environment, however, needs additional mechanisms for adaption. These are normally based around a suite of operating modes combining different error control and modulation schemes (MCS modes) that can be selected according to prevailing channel conditions. They are implemented in the *physical* (PHY) and *medium access control* (MAC) layers of the modem.

There are two main classes of wireless network and these converged under 4G standards and continue to be integrated under the emerging 5G standard. *Wireless local area networks* (WLANs) are invariably based on IEEE 802.11 standards and offer local high-speed data transfer via hotspots. The *digital cellular network* was originally designed to support voice and data traffic but now, under 3G, 4G, and 5G standards, provides the primary means of video downloading and conversational video services.

We can see the effect of a typical wireless channel on system performance in Fig. 11.1. This shows data logged on a wireless video research platform at the University of Bristol. The data logging traces show how an 802.11 WiFi link adapts as a user moves away from the access point (time: 1–70 s) and then back towards it (time: 70–140 s). This illustrates how the radio environment changes and the effect that this has on packet delays, packet delay variability, linkspeed (modulation and coding modes), and the bit rate at the application layer. Clearly, source coding methods need to adapt to these variable bandwidth conditions, while the addition of channel coding or other means of error resilience is needed to cope with the errors introduced during transmission.

11.1.2 Transmission types

Downloads and streaming

Access to a video file or stream can be achieved in many different ways. In the simplest case this can be in the form of a file transfer where data is stored on the local machine and then opened in an appropriate player. *Progressive downloads* are also file transfers, typically using the HTTP protocol, and the files are also stored locally, but they are more intimately linked

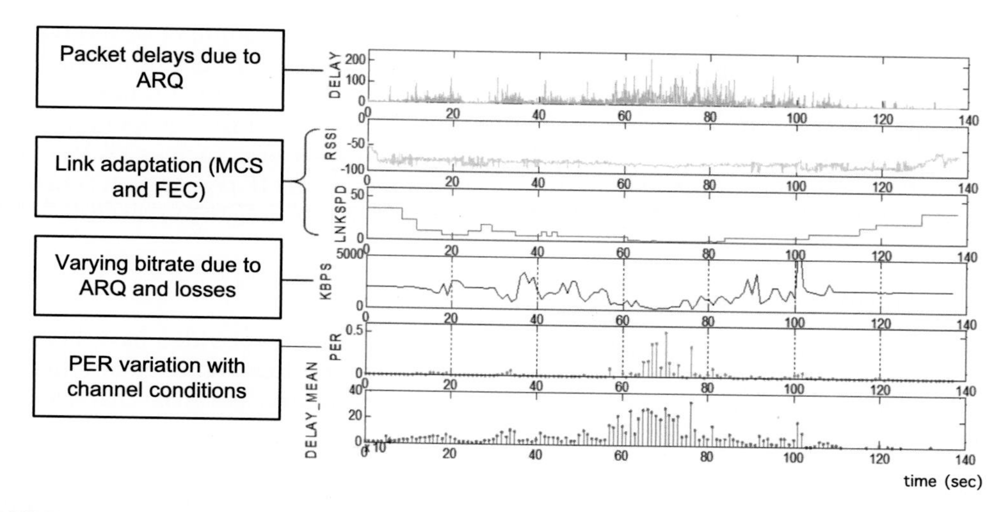

FIGURE 11.1 Logged radio link parameter variations with channel conditions for a wireless video trial.

to the player. The player can start playback before the download is complete but there is generally no interactive support. As the file is downloaded from the server, it will be buffered, and when sufficient data is available for continuous display, it will begin to play.

In the case of streaming, the video is not usually stored on the user's device but is played directly from a buffer. A streaming server will interact with the user's player to enable interactive navigation through the video (e.g., FFWD, RWD, etc.).

Interactive communication

With streamed video, playout will commence when the receiver buffer is sufficiently full to ensure sustained playout. Clearly, if conditions dictate that the network cannot support the encoded video bit rate, then the buffer will progressively empty and eventually the video will stutter and freeze. In cases where interaction is required, the demands on delay are much more stringent; 200–300 ms is typically cited as the maximum tolerable round trip delay for conversational services in order to avoid unnatural hesitations.

Unicast transmission

In unicast transmission, one sender is connected to one receiver. This is mandatory for conversational services, but can be inefficient for other services since every client demands its own bandwidth. For the case of a single transmitter unicasting a video at B bps to N receivers, the total bandwidth required is NB bps. The advantage of unicast transmission is that there is an intimate link between the sender and receiver and hence a feedback channel can be established to signal errors and thus support retransmission and error control. Unicast streaming is used by operators such as YouTube and Skype.

Multicast or broadcast transmission

In the case of multicast transmission, many receivers join a single stream. This is much more efficient than the unicast case, as the N receivers now only consume B bps. However, all receivers will experience different channel conditions, and if the number of receivers is large, then feedback mechanisms become prohibitive and other error control mechanisms must be invoked. Multicasting thus offers much less service flexibility than unicasting, but provides much better bandwidth flexibility. Multicast transmission is used by operators such as Ustream. Broadcasting is typically used for over-the-air TV services such as DVB as well as in LANs. It differs from multicasting in the sense that no specific IP registration is needed to join the stream and packets can be received by anyone who connects to the service.

11.1.3 Operating constraints

As discussed above, IP streaming, conferencing, and entertainment video have been the primary drivers for the growth in multimedia communication services over fixed internet and mobile networks. The channels associated with these networked services can be characterized by the following three attributes.

1. **The bandwidth of the wireless channel is limited**: This necessitates the use of sophisticated source coding techniques to match the bit rate for the signal to the capacity of the channel.
2. **The channel quality for wireless networks can be highly variable**: This is due to signal fading and interference giving rise to erroneous transmission and the need for error protection, error correction, and retransmissions.
3. **Congestion can occur in packet switched networks**: This can lead to dropped packets and variable delays. These demand intelligent buffering and rate control strategies and impose strict bounds on retransmission.

We have seen how attribute (1) can be addressed by exploiting data correlations in a video sequence – using interframe prediction (motion estimation), intraframe prediction, and variable-length entropy encoding. However, as a consequence these methods produce an encoded bitstream that is inherently sensitive to channel errors, and this presents a particular problem in the context of attributes (2) and (3).

Clearly we want our encoder to exhibit excellent rate-distortion (or rate-quality) performance, but what the user is really interested in is the quality at the decoder, as this is what he or she experiences. In fact, the rate-distortion performance at the encoder is irrelevant if all packets are lost during transmission. In order to promote reliable delivery over lossy channels, it is usual to invoke various additional error detection and correction methods. In summary, the encoded video bitstream needs to be error-resilient as well as efficient in a rate-distortion sense.

11.1.4 Error characteristics

Types of errors

Errors that occur during transmission can be broadly divided into two categories: *random bit errors* and *erasure errors*. The former are normally model-led, often with a uniform pdf, whereas the latter are bursty in nature. Bit errors are normally characterized by their *bit error rate* (BER) and can be caused by thermal noise or other environmental factors during transmission. Erasure errors affect longer segments of data and can be caused by congestion in a

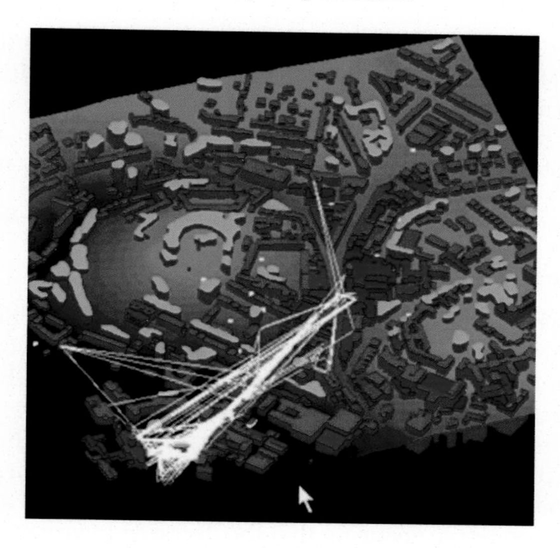

FIGURE 11.2 Radiowave propagation modeling using the Prophecy ray tracing tool (courtesy of A. Nix).

packet network or fading, shadowing, and dropout in a wireless network. They will often exceed the correcting capability of any channel coding system and will demand retransmission or lead to the loss of a whole packet (an erasure at the *application layer*).

Test data

In order to model the effects of errors on a video signal, we need to ensure that the encoding method used is evaluated with representative test data. This can be obtained through data logging over prolonged periods using a real system, such as illustrated in Fig. 11.1, or by emulating system performance using wireless propagation modeling or by statistical channel modeling. We will not discuss Fig. 11.1 in detail here, but we will return to some of the things it depicts later in the chapter.

In practice, while it is useful to characterize the performance of a system using more generic models, for real-world usefulness it is preferable to use propagation models combined with appropriate network simulation tools to generate realistic bit error patterns at the application layer. Such methods are described in [1]. An example output from a ray tracing analyzer that provides accurate representations of multipath effects in a given environment for a specific wireless configuration is illustrated in Fig. 11.2 and is described in reference [2].

Types of encoding

The encoding mode will have a significant influence on the way that errors manifest themselves and propagate spatially and temporally. For example, transform coding in combination with variable-length coding can cause spatial error propagation within a block if symbol synchronization is lost. Propagation across multiple blocks will occur if an end of block (EOB) codeword is incorrectly decoded; in this case any errors within a transform block will propagate spatially to adjacent regions of pixels. The situation is similar for wavelet subbands.

Spatial or temporal propagation of errors can also be caused by predictive coding. Errors within any pixel or transform coefficient used as the basis for prediction at the decoder will lead to incorrect reconstructions and these will propagate around the prediction loop, leading to errors in subsequent pixels or transform coefficients. For example, errors in a reference frame used for motion estimation will be propagated temporally by the motion compensation process.

11.1.5 The challenges and a solution framework

Up until this point we have attempted to achieve the best rate-distortion performance from our compression system. We have removed as much redundancy as possible from the video signal, without compromising quality. However, we are now faced with the dilemma that, in order to make the compressed bitstream robust in the presence of errors, we need to introduce redundancy back into the bitstream in the form of channel coding, thus offsetting some or possibly all of the gains made during compression. Clearly this is not a good situation!

The challenge of error-resilient coding is therefore to maintain the coding gain obtained through source compression while ensuring that the decoded bitstream retains integrity during transmission, so that it can be faithfully decoded. In order to achieve this we need to:

1. understand the causes of errors and their effects (Section 11.2),
2. characterize these from a psychovisual point of view to understand the impact of loss artifacts on the human visual system (HVS) (Chapter 2 and Section 11.3),
3. exploit appropriate channel coding and retransmission solutions but use these, where possible, in a manner suited to video content (Sections 11.3 and 11.4),
4. exploit the adaptive mechanisms that exist in many wireless or internet protocols to work jointly across the application, access control, and physical layers to produce joint source-channel coding or cross-layer solutions (Sections 11.4 and 11.6),
5. exploit the structure of the video coding process to make the encoded bitstream inherently more robust to errors (Sections 11.5 and 11.7),
6. understand where the errors occur in the reconstructed video and take steps to conceal rather than correct them (Section 11.8).

Fig. 11.3 shows a generic architecture that captures the essence of this approach. In practice, an error-resilient solution combines aspects of transport layer solutions and application layer solutions in a sympathetic manner. That is, where possible, network parameters should be adapted to suit video requirements and video parameters should be managed to match the network constraints. These issues are discussed in more detail in the following sections.

11.2 The effects of loss

11.2.1 Synchronization failure

As we have already seen in Chapter 7, a coded video bitstream normally comprises a sequence of variable-length codewords (VLCs). VLC methods such as Huffman or arithmetic coding are employed in all practical image and video coding standards. Fixed-length code-

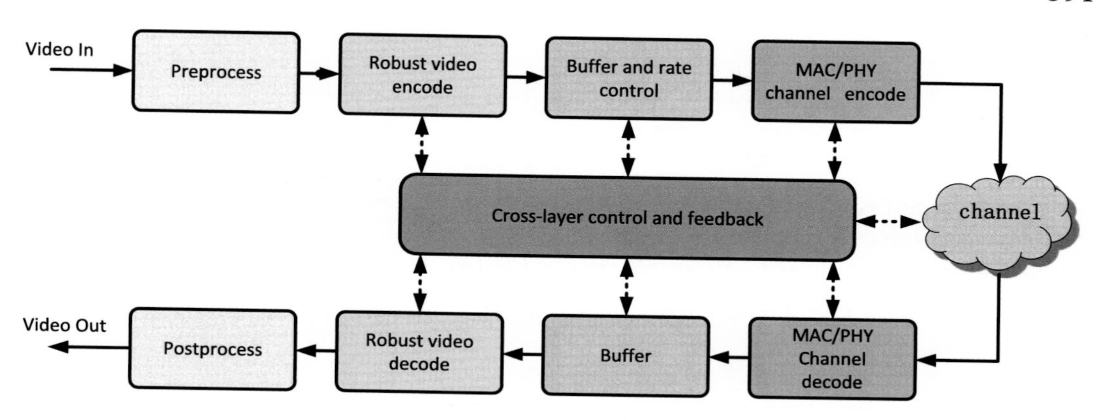

FIGURE 11.3 Generic error-resilient video transmission architecture.

words (FLCs) are rarely used in practice except when symbol probabilities do not justify the use of VLC (e.g., in some header information). A further exception to this is the class of error-resilient methods, including pyramid vector quantization (PVQ), which we will study in Section 11.7.2.

As we will see in the following subsections, the extent of the problem caused will be related to the location where the error occurs and the type of encoding employed.

The effect of a single bit error

With FLC, any codeword affected by a single bit error would be decoded incorrectly but, because of the implicit synchronization of FLC, all other codewords remain unaffected. This means that there is no loss of bitstream or symbol synchronization. In contrast to VLC, single bit errors can cause the decoder to decode a longer or shorter code word, thus meaning that subsequent symbols, although delivered correctly, will most likely be decoded incorrectly due to the loss of synchronization at the decoder. This loss of synchronization can continue until the next explicit resynchronization point and will often result in the wrong number of symbols being decoded. This might mean that important implicit synchronization symbols such as EOB will be missed. In practice, most Huffman codes will resynchronize themselves, whereas arithmetic codes rarely do. The two scenarios in Example 11.1 illustrate the problems of synchronization loss in VLC, demonstrating the subtle differences between synchronization at bitstream, symbol, and coding unit (e.g., transform block) levels.

Example 11.1. Loss of VLC synchronization due to a single bit error

Given the following set of symbols and their corresponding Huffman codewords,

Alphabet	A	B	C	D	E
Huffman code	0	10	110	1110	1111

with the following transmitted message,

Message	B	A	D	E	C	B	A
Encoded bitstream	10	0	1110	1111	110	10	0

a) Compute and comment on the decoded sequence of symbols if there is a single bit error in the sixth bit.

b) Repeat (a) for the case of a single bit error at the tenth bit position.

Solution. a) The table of decoded bits and symbols is given below with the errored bit in bold font:

Encoded message	B	A	D	E	C	B	A	–
Received bitstream	10	0	11**0**0	1111	110	10	0	
Parsed bitstream	10	0	110	0	1111	110	10	0
Decoded message	B	A	C	A	E	C	B	A

In this case the symbol D is incorrectly decoded as a C followed by an A and then bitstream resynchronization occurs. However, the fact that an extra symbol is decoded means that the subsequent symbols are displaced and hence symbol synchronization is lost.

b) The table of decoded bits and symbols is given below with the errored bit in bold font.

Encoded message	B	A	D	E	C	B	A
Received bitstream	10	0	1110	11**0**1	110	10	0
Parsed bitstream	10	0	1110	110	1110	10	0
Decoded message	B	A	D	C	D	B	A

In this case, because of the lengths and encodings of the adjacent symbols, the error is constrained to only two symbols. After this, both bitstream and symbol resynchronization are achieved. Superficially this appears a lot better than the first case. However, consider the situation where the symbols represent {run/size} encodings of transform coefficients. The fact that there are errors in the fourth and fifth symbols means that, after the third symbol, the whole block will be decoded incorrectly because the runs of zeros corresponding to symbols 4 and 5 are incorrect. However, because symbol resynchronization is regained, any subsequent EOB symbol would be correctly detected and block level synchronization will be maintained. See Example 11.2 for more details.

11.2.2 Header loss

Errors in header information can cause catastrophic errors or error propagation. If any bit within a header is corrupted, then the result is often a complete failure of the decoder since

FIGURE 11.4 The effect of a single bit error on the DCT/Huffman encoded *puppet* image.

the basic parameters needed to decode the bitstream are incorrect. There is therefore some justification for header information to be protected more strongly than other data as, without it, little else matters.

11.2.3 Spatial error propagation

Errors can propagate spatially during decoding for a number of reasons:

1. **DPCM prediction**: For example as used in the encoding of DC transform coefficients in JPEG. This will normally result in a luminance or chrominance shift in the image blocks decoded after the error.
2. **Intra-prediction**: Modern compression standards employ sophisticated intra-prediction modes that can result in the propagation of errors.
3. **Loss of VLC symbol synchronization within a block**: For example in a discrete cosine transform (DCT)-based system, this will result in the appearance of artifacts relating to superpositions of incorrect DCT basis functions.
4. **Loss of block synchronization**: Errors will propagate across blocks if EOB symbols are incorrectly decoded. This normally leads to a loss of positional synchronization.

The effect of a single error on a simple JPEG DCT/Huffman encoded image is illustrated in Fig. 11.4. This shows the effects described in Examples 11.1 and 11.2. We can clearly see the block where the error has occurred as the corrupted DCT basis functions are evident. The figure also illustrates the effects of corrupted DC levels propagated by the DPCM method used. Also shown is the effect of a spatial shift in the reconstructed image due to EOB symbols being missed in blocks adjacent to the errored block. Eventually symbol synchronization is regained but the effects of the spatial displacement remain. The effect of higher error rates and the impact these have on subjective image quality are shown in Fig. 11.5.

FIGURE 11.5 Impact of increased error rates on a DCT image codec. Left: 0.01% BER. Middle: 0.1% BER. Right: 1% BER.

Example 11.2. Spatial error propagation due to VLC errors

Given the set of symbols and their corresponding Huffman codewords from Example 11.1, let us now assume that the symbols correspond to {run/size} values for an intra-coded DCT block in an image. If in this case, symbol B corresponds to EOB, then comment on the spatial error propagation associated with the following transmitted sequence, when a single error is introduced at bit position 12 in the first block of the image:

Message	**A**	**D**	**E**	**C**	**B**	**D**	**A**	**E**	**C**	**B**
Encoded bitstream	0	1110	1111	110	10	1110	0	1111	110	10

Solution. The table of decoded bits and symbols is given below with the errored bit in bold font.

Message	**A**	**D**	**E**	**C**	**EOB**	**D**	**A**	**E**	**C**	**EOB**
Received bits	0	1110	1111	11**1**	10	1110	0	1111	110	10
Parsed bits	0	1110	1111	1111	0	1110	0	1111	110	10
Decoded	A	D	E	E	A	D	A	E	C	EOB

In this case, symbols C and B are incorrectly decoded as E and A, and hence the first EOB symbol is missed. So although bitstream and symbol resynchronization occurs, block synchronization is lost. All subsequently decoded blocks will thus be displaced by one block position to the left. So unless some explicit form of synchronization is introduced, the whole image after the second block will be corrupted.

11.2.4 Temporal error propagation

Temporal error propagation occurs when corrupted spatial regions are used as the basis for prediction of future frames. As shown in Fig. 11.6, a single corrupted block in the reference

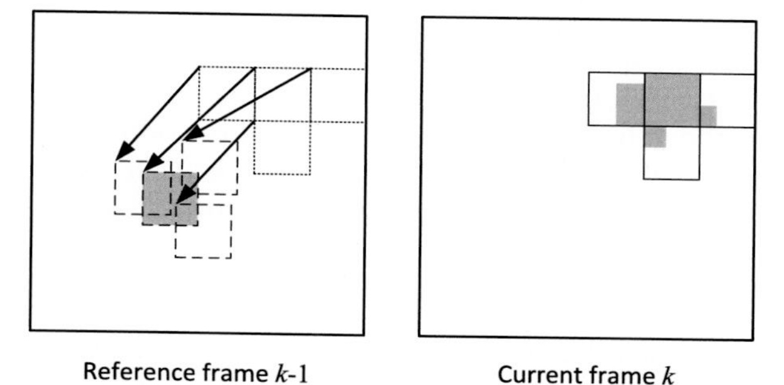

FIGURE 11.6 Temporal error propagation from a single corrupted block in frame $k-1$ to multiple corrupted blocks in frame k.

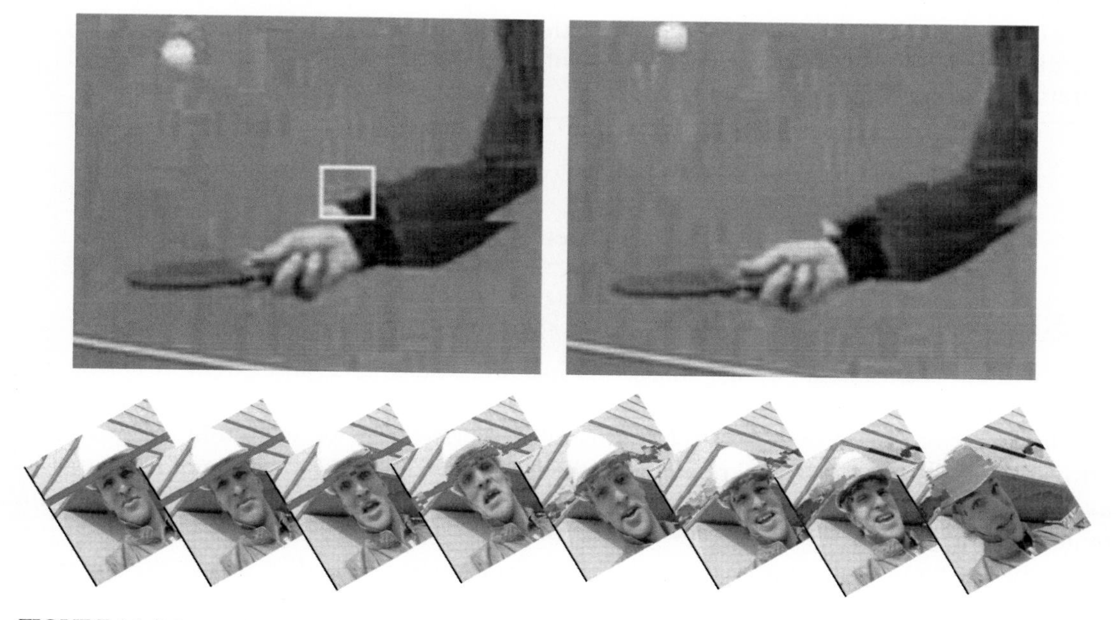

FIGURE 11.7 Example of temporal propagation. Top: Across four frames of the *table tennis* sequence. Bottom: Across 190 frames of *foreman*.

frame can propagate to multiple (four in this case) blocks in the predicted current frame. The spatial spread of the error is dependent on the local distribution of motion vectors and more active regions will tend to spread errors more widely than regions of low motion. Fig. 11.7 shows two examples of temporal propagation in practice. Here, the error introduced locally due to spatial data loss still persists and has spread, several frames later.

It should be noted that corrupted motion vectors can also provide a source of errors due to incorrect motion compensation at the decoder.

Example 11.3. Temporal propagation of errors due to prediction
Consider the same symbol encoding and transmitted sequence as given in Example 11.2. Assume that the symbols represent {run/size} encodings related to the transform coefficients in two DFD blocks and that these are used as the basis for future motion-compensated predictions. Comment on the temporal error propagation that would result from this single bit error in bit position 12.

Solution. As with Example 11.2, because the EOB symbol is missed, block synchronization is lost, and all blocks decoded after the errored block will be displaced.

In this case, however, because we are dealing with inter-coded DFD blocks, there is a second issue. The DFD blocks will be used as the basis of temporal prediction at the decoder, and since in the decoded version will contain spatial errors, these will propagate temporally to subsequent frames because of motion prediction. Because all blocks after the errored block are displaced and because the errored block occurs early in the frame, if there is no explicit resynchronization point, almost all the data in future predicted frames are likely to be corrupted.

11.3 Mitigating the effect of bitstream errors

11.3.1 Video is not the same as data!

This might seem like a strange subsection heading but, to a certain extent, it is true. If we consider, for example, the case of a numerical database, then in order to guarantee its effectiveness we would expect it to be transmitted perfectly. If we were told it contained errors (especially if we did not know where they were), we probably would not trust it and avoid using it. However, in the case of an image or video sequence, the situation is, in general, quite different.

On the one hand we have seen that a single bit error can be highly destructive, due to error propagation. On the other hand we know that the HVS is tolerant to certain types of artifact and can accept loss if it is handled correctly. In fact the compression process itself is nothing but a managed process of introducing noise into an image and we are normally quite happy with the resulting approximations of the original content. Consider, for example, the video frames shown in Fig. 11.8. On the left is the original compressed image (at 30 Mb/s coded using PVQ – see Section 11.7.2) while the one on the right has been reconstructed in the presence of 25% *packet error rate* (PER) (i.e., 1/4 of all packets contain uncorrectable errors). Artifacts do exist, as shown in the bottom subfigure, but are in this case well masked and tolerable.

The process of managing bitstream errors is not straightforward since they are unpredictable. We can however model the impact of loss on the bitstream and take steps to avoid it, correct it, or conceal it; these approaches are all addressed in the following sections.

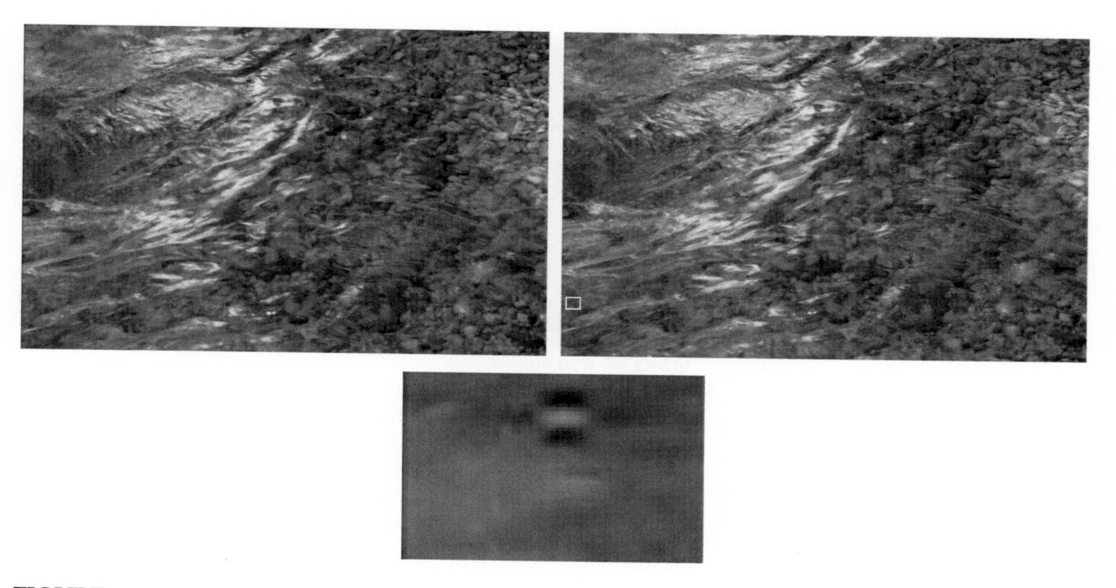

FIGURE 11.8 Subjective appearance of loss. PVQ encoded *riverbed* sequence at 30 Mb/s. Left: Encoded at 30 Mb/s (intra). Right: Reconstructed after 25% PER. Bottom: Magnified artifact from box shown in top right image.

TABLE 11.1 FLC vs. VLC encoding for various channel conditions. 256 × 256 *Lena* image at 2 bpp.

PSNR (dB)	DCT-only	DCT + DC pred.	DCT +DC pred. + VLC
Error-free	33.7	33.9	38.5
0.1% BER	24.6	21.0	6.4

11.3.2 Error-resilient solutions

We have seen that the more redundancy we remove from the source, the more fragile our resulting bitstream becomes. We thus need to trade some rate-distortion performance in a clean channel for improved error resilience in the presence of errors.

Let us first consider the case of transmission of an encoded image over a noisy channel with varying degrees of loss. Table 11.1 compares three cases: (i) a DCT-only transform with fixed-length encoding, (ii) a DCT transform-based system with DC prediction, and (iii) a DCT-based system with DC prediction and Huffman-based VLC. As can be seen, the VLC-based system provides excellent coding performance in a clean channel, yet it falls off badly as channel conditions deteriorate. On the other hand, the FLC scheme provides poor performance in a clean channel yet significantly outperforms the VLC case in lossy channels.

This trade-off between clean channel performance and error resilience is one of the challenges we face in designing our system. Ideally we want the best error resilience without compromising clean channel quality. In practice, however, it is necessary to strike a compromise and sacrifice some clean channel performance for improved error resilience in the

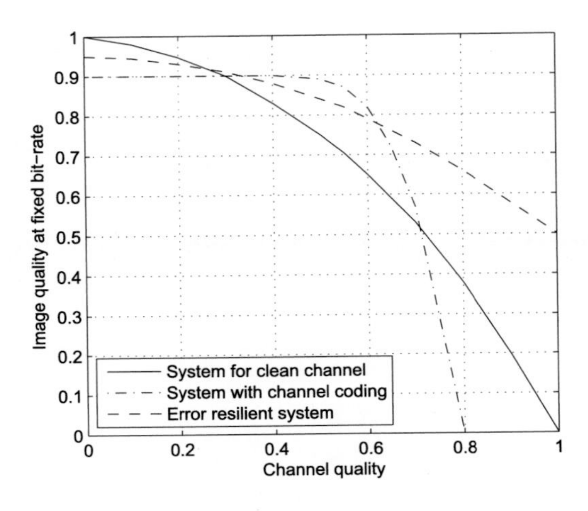

FIGURE 11.9 Idealized comparison of performance.

presence of loss. These characteristics are illustrated in Fig. 11.9, and methods for achieving this error resilience with graceful degradation are described in the following sections.

11.4 Transport layer solutions

In this section we use the term transport layer (a little loosely) to describe all aspects of the video transmission process that are not intimately associated with the application layer (i.e., the video codec). In many cases, the error control mechanisms will be implemented at the transport layer in the ISO 7 layer model. However, they may also be implemented in MAC and PHY layers. Many excellent references exist on this well-researched topic, for example [3–6].

11.4.1 Automatic repeat request (ARQ)

ARQ is an interactive technique that is widely used in many communication protocols, including TCP/IP and 802.11. It relies on feedback from the decoder to determine whether a message (packet) has been correctly received. The detection of an error is normally performed through the addition of some form of *cyclic redundancy check* or through *forward error correction* (FEC) coding. When an error is detected, the decoder returns a NACK (Negative ACKnowledge) signal which instructs the encoder to retransmit the data. There are a number of obvious advantages and problems with this method:

Advantages

- ARQ is simple and efficient. The redundancy introduced through retransmission is efficient as it is targeted specifically at lost data.
- It can be effective for coping with packet loss or large burst errors.

Problems

- A delay is introduced while the decoder waits for data to be retransmitted, and in real-time video applications this may imply that frames have to be dropped (video frames have to be rendered at specific time intervals).
- ARQ requires a feedback channel which is not always available, for example when multicasting or broadcasting.

Delay-constrained retransmission

In practice, ARQ is a simple yet effective technique for ensuring reliable transmission of data but, in the context of real-time video, it has a usefulness-lifetime, limited by the maximum tolerable latency between encoding and decoding. To address this problem, *delay-constrained retransmission* is normally used and this can be implemented at both receiver and transmitter. In the former case, the decoder will only request a retransmission of packet n if the following condition is satisfied:

$$T_c + T_{RT} + T_x < T_d(n), \tag{11.1}$$

where T_c is the current time, T_{RT} is the round trip delay for the network, $T_d(n)$ is the playout time for packet n, and T_x is a variable that accounts for other factors such as the tolerance on delay estimates and other uncertainties. Similarly, this decision could be made at the encoder if the following condition is satisfied:

$$T_c + T_{RT}/2 + T_x < T_d'(n), \tag{11.2}$$

where $T_d'(n)$ is in this case the estimated playout time for the packet.

11.4.2 FEC channel coding

In FEC methods, additional parity data is appended to the compressed signal, which allows the decoder to correct a certain number of errors. FEC can be combined with layered coding (or data partitioning) to provide unequal error protection, where different parts of the compressed bitstream are protected by different strength codes, according to importance.

The use of FEC increases the total data rate required for transmission and clearly offsets some of the benefits gained from source compression – redundancy can reach 100% for a ½ rate code. Furthermore, in general, FEC must be chosen with a particular worst-case channel scenario in mind. For channels that have a highly variable quality, this worst-case situation may imply the need for a very powerful FEC code which can severely reduce the compression performance. Furthermore, such a system will fail catastrophically whenever this worst-case threshold is exceeded. One method that is frequently used to help mitigate the effects of burst errors is to use interleaving. However, this is suboptimal and can introduce large delays.

Erasure codes

FEC can be very effective in cases of packet erasures as usually the positions of the lost packets are known. Efficient erasure codes include the Reed–Solomon erasure (RSE) correcting code. With an RSE(n, k) code, k data packets are used to construct r parity packets, where

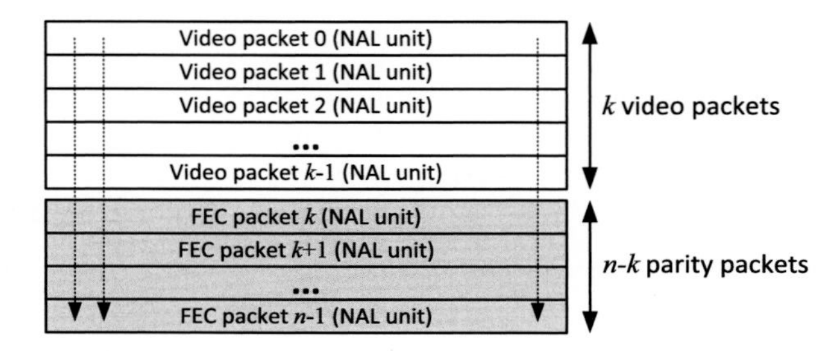

FIGURE 11.10 Cross-packet FEC.

$r = n - k$, resulting in a total of n packets to be transmitted. The k source packets can be reconstructed from any k packets out of the n transmitted packets. This provides for error-free decoding for up to r lost packets out of n. The main considerations in choosing values of r and k are:

- **Encoding/decoding delay**: In the event of a packet loss, the decoder has to wait until at least k packets have been received before decoding can be performed. So, in order to minimize decoding delay, k must not be too large.
- **Robustness to burst losses**: A higher value of k means that, for the same amount of redundancy, the FEC will be able to correct a larger number of consecutive lost packets, or bursts. For example, an RSE(4, 2) code and an RSE(6, 4) code can both protect against a burst error length of 2, but the RSE(6, 4) has only 50% overhead whereas RSE(4, 2) has 100% overhead.

Cross-packet FEC

If FEC is applied within a packet or appended to individual packets, in cases where packets are not just erroneous but are lost completely during transmission (e.g., over a congested internet connection), the correction capability of the code is lost. Instead, it is beneficial to apply FEC across a number of packets, as shown in Fig. 11.10. One problem with using FEC is that all the k data packets need to be of the same length, which can be an issue if GOB fragmentation is not allowed. The performance of cross-packet FEC for the case of different coding depths (8 and 32) is shown in Fig. 11.11. This clearly demonstrates the compromise between clean channel performance and error resilience for different coding rates.

Unequal error protection and data partitioning

We saw in Section 11.2 that not all types of video data are equally important. For example, header information is normally most important, as without it little else is possible. Motion information plays a particularly important role, especially since the relatively small fraction of the total bit rate it occupies (approximately 10%). Faithful reconstruction of the encoded sequence depends heavily on the accuracy of the motion vectors. In order to reflect these priorities, the video data can be partitioned into separate data segments, each being allotted a protection level appropriate to its importance.

Prioritization of data may also be implemented in conjunction with layered coding, as described in Section 11.9.

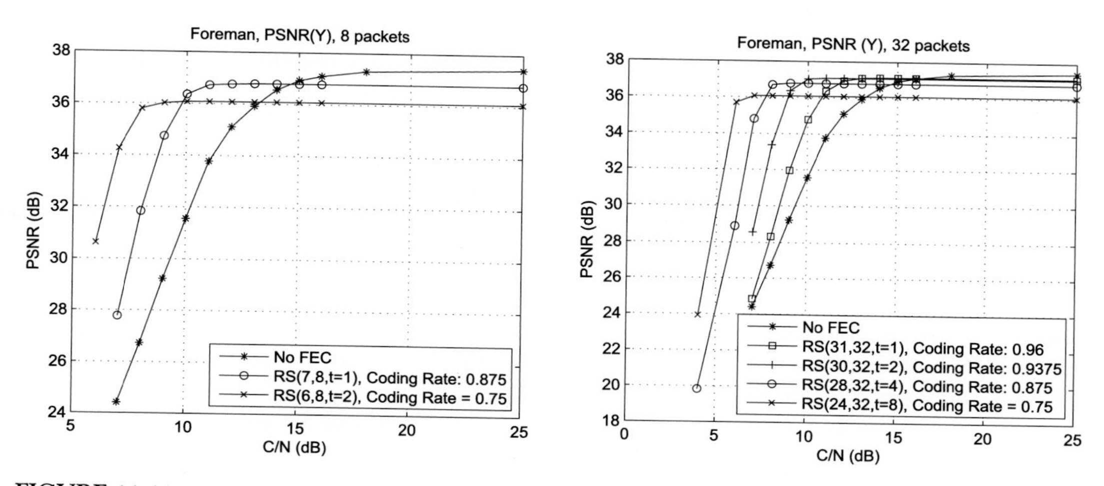

FIGURE 11.11 Performance of cross-packet FEC for a coding depth of 8 (left) and 32 (right) for various coding rates.

Rateless codes

Rateless codes, as the name suggests, do not have a fixed code rate. They are based on the principle of Raptor coding and offer a flexible and adaptive mechanism for dealing with variable wireless channel conditions. Raptor codes, proposed by Shokrollahi in [7], have gained significant traction as an effective means of channel coding for multimedia information. They are part of the family of codes known as fountain codes (first introduced by Luby in [8] as LT codes) that combine low-density parity check (LDPC) and LT codes. They offer the property of linear, $O(k)$, time encoding and decoding. Fountain codes encode k symbols into a (potentially) limitless sequence of encoding symbols, where the probability that the message can be recovered increases with the number of symbols received. For example, in the case of Raptor codes, the chance of decoding success is 99% once k symbols have been received. The recent 5G new radio (5G-NR) standard employs quasicyclic LDPC (QC-LDPC) codes as the channel coding for data transmission.

11.4.3 Hybrid ARQ (HARQ)

Hybrid ARQ (HARQ) is a commonly used tool in modern mobile broadband networks such as 4G and 5G. HARQ [9] combines FEC coding (error detection and correction) with ARQ. It suffers from some of the basic limitations of ARQ, but offers some efficiency gains. In HARQ, the data is encoded with an appropriate FEC code, but the parity bits are not automatically sent with the data. Only when an error is detected at the decoder are these additional parity bits transmitted. If the strength of the FEC code is sufficient to correct the error, then no further action is taken. If however this is not the case, then the system reverts to a full ARQ retransmission. Typically a system will alternate between data and FEC packets during retransmission.

HARQ is a compromise between conventional ARQ and conventional FEC. It operates as well as ARQ in a clean channel and as good as FEC in a lossy channel. In practice, with multiple retransmissions of the same data, a receiver would store all transmissions and use these multiple copies in combination. This is often referred to as *soft combining*.

11.4.4 Packetization strategies

The packetization strategy [10] can have an influence on performance in the presence of loss. The basic principles can be summarized as:

- Longer packets lead to improved throughput for clean channels (lower overhead).
- Shorter packets facilitate better error resilience as their loss will have less impact on the video sequence and error concealment will be easier.
- To support error resilience, video packet fragmentation should be avoided. Every network abstraction layer (NAL) unit (see Section 11.5.1) should thus be transmitted in one and only one UDP frame.

To illustrate the third bullet point, if a video packet containing encoded VLC symbols is split across two UDP packets, then the loss of one of these would render the other useless because of VLC dependencies.

11.5 Application layer solutions

11.5.1 Network abstraction

Standards since H.264/AVC have adopted the principle of a NAL that provides a common syntax, applicable to a wide range of networks. A high-level parameter set is also used to decouple the information relevant to more than one slice from the media bitstream. H.264/AVC, for example, describes both a sequence parameter set (SPS) and a picture parameter set (PPS). The SPS applies to a whole sequence and the PPS applies to a whole frame. These describe parameters such as frame size, coding modes, and slice structuring. Further details on the structure of standards such as H.264 are provided in Chapter 12.

11.5.2 The influence of frame type

I-frames, P-frames, and B-Frames

The type of frame employed has a direct influence on the way errors propagate temporally. The main characteristics of I-frames, P-frames, and B-frames are listed below:

- **I-frames:** I-frames are not predicted with reference to any other frame. They therefore provide a barrier to temporal error propagation. They do, however, in H.264/AVC and HEVC, exploit intra-prediction and hence they can propagate errors spatially. Also I-frames form the basis for prediction P- and B-frames, so any spatial errors occurring in an I-frame can propagate temporally to subsequent frames.

- **P-frames:** P-frames are predicted from other P-frames or from I-frames. They will thus both propagate errors contained in their reference frames and also introduce new spatial errors that can propagate spatially and temporally.
- **B-frames:** B-frames are predicted from two or more frames. In the case of earlier standards such as MPEG-2, they are not used as a reference for predicting other frames and hence do not propagate errors temporally – any spatial errors introduced in this type of B-frame are constrained to that frame. However this is no longer the case for modern standards where B-frames are used as reference frames.

Intra-refresh

We have seen that the group of pictures (GOP) structure for a video sequence defines the relationship between I-, P-, and B-frames. All encoded sequences will have at least one I-frame and most will have them inserted periodically (the intra period typically depends on compression ratio with fewer I-frames for more highly compressed sequences). Many codecs support intra-coding of blocks in P- or B-frames. This provides a more flexible approach to rate-distortion optimization and also helps to limit temporal error propagation. Blocks that are intra-coded in this way can be selected according to rate-distortion criteria, according to their concealment difficulty, or randomly.

Some codecs such as x264 take this further and support periodic intra-refresh instead of key frames. This effectively spreads the I-frame across several frames using a column of intra-coded blocks that progressively steps across the video frame. In this case, motion vectors are restricted in the same way as in slice structuring, so that blocks on one side of the refresh column do not reference blocks on the other side. This approach has the additional benefit that the peak to mean bit rate ratio is constrained (unlike the situation where large I-frames are used in isolation), offering better streaming and lower latency.

Reference picture selection

In simple motion-compensated hybrid codecs, the most recent previously coded frame is used as a reference for prediction of the current frame. More recent standards (from H.263+ onwards) support the use of multiple reference frames and reference picture selection (RPS) modes. These allow, with some constraints, the choice of any previously coded frame as a reference. This can be beneficial in limiting error propagation in error-prone environments.

In RPS, both the encoder and the decoder store multiple reference frames. If, through feedback from the decoder, the encoder is instructed that the most recent reference frame has been received erroneously, then it will switch to another (error-free) reference picture for future predictions. The RPS selection mode was originally adopted in H.263+ as Annex N. This is shown in Fig. 11.12, where a NACK(4) signal is fed back from the decoder and received during the encoding of frame 5. For frame 6, the encoder switches to use frame 3, which was received error-free, as the new reference frame.

Periodic reference frames

In terms of coding efficiency, it is normally more efficient to predict a picture from a frame several frames displaced from it, than to use intraframe coding. Techniques such as periodic RPS [11] exploit this by encoding every Nth frame in a sequence using the Nth previous frame as reference. All the other frames are coded as usual. This scheme is illustrated in Fig. 11.13.

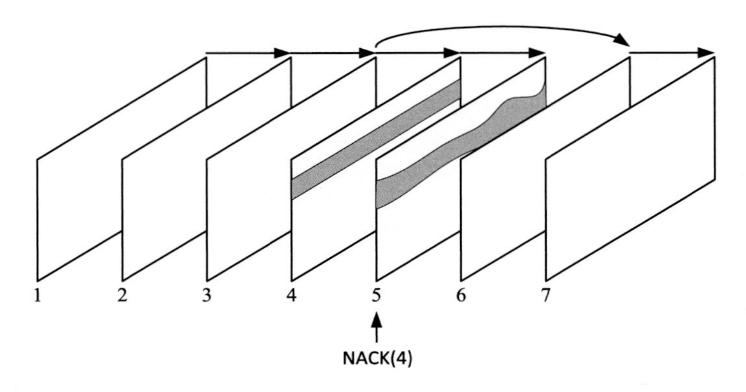

FIGURE 11.12 Error resilience based on reference picture selection.

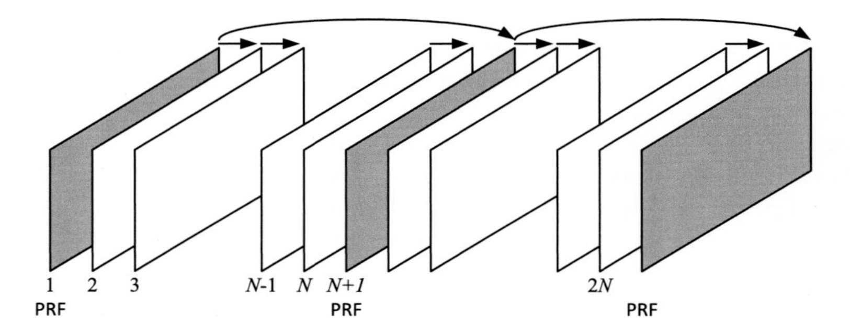

FIGURE 11.13 Periodic reference frames.

Its advantage is that if any errors in a PR frame can be corrected through the use of ARQ or FEC before it is referenced by the next PR frame, then this will effectively limit the maximum temporal error propagation to the number of frames between PR frames.

Since PR frames require far fewer bits to code than intraframes at the same quality, extra bits can be used to protect the PR frames, thus providing better performance under packet losses than intraframe coding. Because this technique does not depend on feedback, it is applicable to broadcast and multicast scenarios. This type of approach has been adopted in coding standards such as H.264/AVC, referred to as *inter-prediction with subsequences*.

11.5.3 Synchronization codewords

Periodic resynchronization of a bitstream, and its associated symbols and data structures, can be achieved by the insertion of explicit synchronization codewords. For example, these might be inserted at the end of each row of blocks, slice, or frame. Resynchronization codewords are uniquely identifiable and are distinct from all other codes or concatenations of codes. They offer the following advantages and disadvantages:

- **Advantages**: If an error is encountered, the maximum loss is limited by the distance between synchronization markers. Once a new synchronization marker is encountered, decoding can proceed again correctly.
- **Disadvantages**: Resynchronization codewords incur a significant overhead, especially if used frequently. In the case of a JPEG encoded image, resynchronization markers at the end of each row typically incur a 1–2% overhead, whereas if they were located after each block, they would typically incur over 30% overhead.[1]

11.5.4 Reversible VLC

When an error is encountered in conventional VLC, there is a strong likelihood that the error will propagate spatially until the next resynchronization point. In the case of *reversible VLC* (RVLC), backward decoding as well as forward decoding can be performed. Hence some of the data lost with conventional VLC can be faithfully decoded with RVLC. RVLC methods initially found favor in H.263+ and MPEG-4, and can be used effectively in combination with synchronization markers to provide enhanced error resilience. They were introduced by Takishima et al. [12], who showed how both symmetric and asymmetric codes can be generated.

RVLC codes [12,13] should exhibit both prefix and suffix properties so that the codewords are decodable in both directions. The types of code that can offer this so-called biprefix property are limited and few are optimal. Codes such as Golomb–Rice codes or Exp-Golomb codes can be mapped into an equally efficient biprefix code, but in most case the resulting biprefix code is less efficient. Girod [14] introduced an alternative method for generating bidirectionally decodable VLC bitstreams, based on modifications to prefix codewords. This elegant approach results in a coding efficiency equivalent to that of Huffman coding.

It is known that VLC data comprising DCT coefficients or differential motion vectors exhibit a *generalized Gaussian distribution*. As discussed in Chapter 7, these can be efficiently entropy coded using Exp-Golomb codes which can be implemented without look-up tables and reversible versions are easy to design.

A simple example of RVLC is provided in Example 11.4.

Example 11.4. Reversible VLC

Consider the following symbol-codeword mappings:

$$a \leftrightarrow 0, \quad b \leftrightarrow 11, \quad c \leftrightarrow 101 = \text{EOB}.$$

The sequence $\{a, b, c, b, a, a, b, c, b, a, c\}$ is transmitted followed by a SYNC codeword. If the 9th and 11th bits are corrupted, show how this encoding can minimize errors through bidirectional decoding.

[1] Overhead will vary according to compression ratio and content.

Solution.

Message	a	b	c	b	a	a	b	c	b	a	c	SYNC
Rx. bitstream	0	11	101	11	**1**	0	**01**	101	11	0	101	SYNC
Fwd. decode	a	b	c	b	–	–	X	X	X	X	X	SYNC
Rev. decode	X	X	X	X	X	X	–	c	b	a	c	SYNC

Note that the codewords are symmetrical, giving us a clue that they should be bidirectionally decodable. The table shows the forward and reverse decoding operations. Through bidirectional decoding we have successfully recovered most of the symbols, having lost only three symbols where the errors occur. If we assume that codeword *c* represents an EOB, then we can see that the RVLC method recovers two of the three blocks of data transmitted.

11.5.5 Slice structuring

Most recent standards support the idea of decomposing a frame into slices prior to coding. A slice comprises a slice header followed by an integer number of macroblocks or coding units. Since all forms of prediction are constrained to within the slice boundary, slice structuring is effective in limiting error propagation. It therefore provides a simple basis for error-resilient coding. Some typical generic slice structures are shown in Fig. 11.14.

Flexible macroblock ordering (FMO)

Flexible macroblock ordering (FMO) is a slice structuring approach that is attractive for error resilience and for supporting concealment. It forms part of the H.264/AVC standard. The video frame is divided into several independently decodable slice groups (Fig. 11.14) and these can assume checkerboard, scattered, and interleaved patterns as shown. For example, in the checkerboard pattern, each frame can be represented as two slice groups with macroblocks interleaved as shown in the figure. It is easy to see that if a slice in one group is lost, then the remaining one can be employed to provide a basis for the interpolation of the missing slice. Providing that only one of the two slices is lost, FMO distributes errors across a larger region of the frame.

A disadvantage of slice structuring is that coding gain is reduced. This is especially true in FMO, since neighboring macroblocks cannot be used as a basis for prediction. It has been reported that the efficiency drop or bit rate increase is approximately 5% at QP = 16 and up to 20% for QP = 28 [15].

An example of using dispersed mode slice groups is shown in Fig. 11.15. Here, each slice corresponds to two rows of macroblocks and the figure shows the case where slices 2 and 3 are lost from one of the slice groups. It can be clearly seen that data from the remaining slice group is distributed in such a way as to facilitate interpolation of the lost data. This is the basis of the error concealment approach presented in Section 11.8.4.

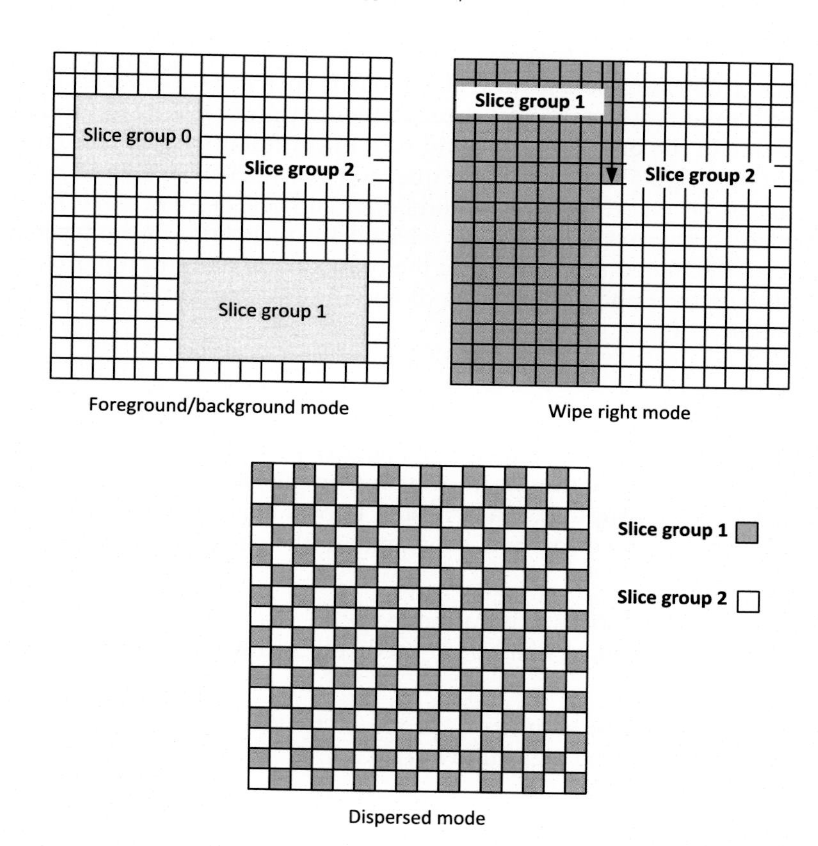

FIGURE 11.14 Example slice structures.

FIGURE 11.15 Use of dispersed mode slice groups as a basis for error concealment.

Redundant slices

Redundant slices (RSs) are also a feature of standards such as H.264/AVC. These support the inclusion of redundant information for error resilience purposes. The redundant slice can be a simple duplicate of the original or it could be a lower-fidelity copy. This copy can then be used in place of the original if it is lost. The question is, which slices should be included as redundancy? This has been addressed by a number of authors. Ferre et al. [16], for example, present a method whereby macroblocks are selected for inclusion on the basis of an end-to-end distortion model aimed at maximizing the reduction in distortion per redundant bit. This shows advantages over alternative methods such as cross-packet FEC and loss adaptive rate-distortion optimization.

11.5.6 Error tracking

This technique also relies on the existence of a feedback channel, not only to indicate that an error has occurred, but also to provide information on the location of the error (e.g., which slices or blocks have been corrupted). If an encoder knows which blocks have been received in error, it can predict how the errors will propagate to the currently encoded frame; it can then take one of two actions:

1. encode the corresponding blocks in intra-mode,
2. encode the corresponding blocks based on predictions only from blocks that have been correctly received.

This so-called *selective recovery* technique was proposed by Wada et al. in [17]. Steinbach et al. propose an efficient error tracking algorithm in [18].

11.5.7 Redundant motion vectors

Many of the error resilience tools present in H.264/AVC have been omitted in subsequent standards. Carreira et al. [19] exposed the increased susceptibility of H.265/HEVC coded streams to network errors and proposed a two-stage approach which limited temporal error propagation following frame loss. Firstly, at the encoder, reference pictures are selected dynamically, using Lagrangian optimization to ensure that the number of prediction units that depend on a single reference is reduced. Secondly, they employ motion vector prioritization, based on spatial dependencies, to select an optimal subset of redundant motion vectors to transmit as side information. These results indicate significant reductions in temporal error propagation, achieving total quality gains up to 5 dB for 10% of packet loss ratio for only a small increase in motion overhead.

11.6 Cross-layer solutions

We have seen that compressed video (in the context of the HVS) is tolerant to a certain level of errors, provided the associated artifacts are managed effectively. However, the aim of almost all networks is to deliver perfect data; anything else is deemed useless and is normally discarded by lower network layers without being sent to the application. Resource management and protection mechanisms that are traditionally implemented in the lower layers of the OSI stack (PHY, MAC, and transport) are, by intent, isolated from the application layer. We have seen however how the error characteristics of these lower layers are intimately coupled with their manifestations at the application layer.

There has thus recently been significant interest in performing optimization across the communication layers in order for the demands of the application to be reflected in the parameterization of the network layers and vice versa. This is referred to as *cross-layer optimization*. Cross-layer optimization differs from *joint source channel coding* in that the aim of the latter is to optimize channel coding, dependent on the rate constraints of the channel. The former however takes a much more holistic approach, taking account of packetization strategies, retransmission strategies, delay constraints, and the loss tolerance of the data being encoded.

An excellent overview of the various aspects of cross-layer optimization is given by van der Scharr in [4,20].

11.6.1 Link adaptation

One form of cross-layer optimization that is realizable in practice is link adaptation, based on end-to-end distortion metrics and modeling. WLANs such as IEEE 802.11a/g/n support numerous modulation and coding schemes (MCSs), each providing different throughputs and reliability levels. Conventional algorithms for adapting MCS attempt to maximize the error-free throughput, based on RF signal measures such as residual signal strength indication (RSSI) and BERs or PERs. They do not take account of the content of the data stream and they rely heavily on the use of ARQ and FEC to correct erroneous transmissions. In contrast, authors such as Ferre et al. [21] have presented ways of achieving this by minimizing the video distortion of the received sequence.

Ferre et al. use simple, local rate distortion measures and end-to-end distortion models at the encoder to estimate the received video distortion at the current transmission rate, as well as at the adjacent lower and higher link speeds (MCS modes). This allows the system to select the mode which offers the lowest distortion, adapting to the channel conditions to provide the best video quality. Simulation results, based on H.264/AVC over IEEE 802.11g, show that the proposed system closely follows the optimal theoretic solution. Example results are provided in Figs. 11.16, 11.17, 11.18, and 11.19. The first three figures illustrate the significant difference in switching characteristics between the cases of a throughput-based metric and a video quality-based metric. Fig. 11.19 shows the result of the metric applied to the *foreman* sequence plotted against channel carrier to noise ratio. It can be clearly seen that the switching characteristic closely tracks the optimal trajectory between MCS modes.

11.7 Inherently robust coding strategies

We have seen above that one of the contributing causes of error propagation in compressed video is variable-length entropy coding. In this section we consider two methods from the re-

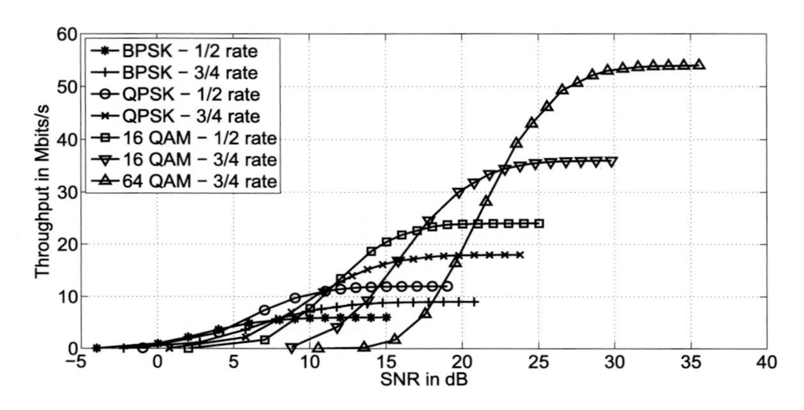

FIGURE 11.16 802.11g MCS switching characteristics based on throughput. From [21].

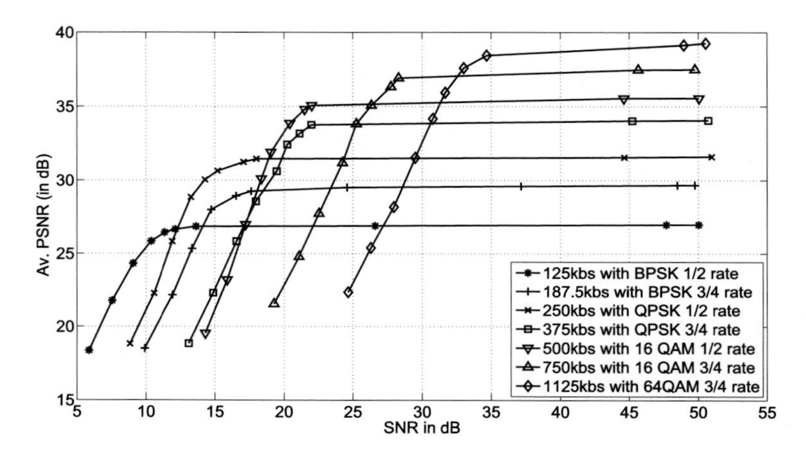

FIGURE 11.17 802.11g MCS switching characteristics based on video distortion. From [21].

search literature that offer the benefits of FLC while preserving the compression performance offered by VLC. The first, *error-resilient entropy coding* (EREC), is a pseudofixed-length scheme where inherent synchronization is introduced through bitstream restructuring. The second, PVQ, is a true fixed-length scheme which exploits subband statistics to create a model that supports FLC.

11.7.1 Error-resilient entropy coding (EREC)

Principle of operation

EREC [22,23] is a simple yet elegant approach to the coding of variable-length blocks of data. It was introduced by Redmill and Kingsbury in [22]. Based on a form of bin packing, using a fixed-length slotted structure, it enables the introduction of implicit synchro-

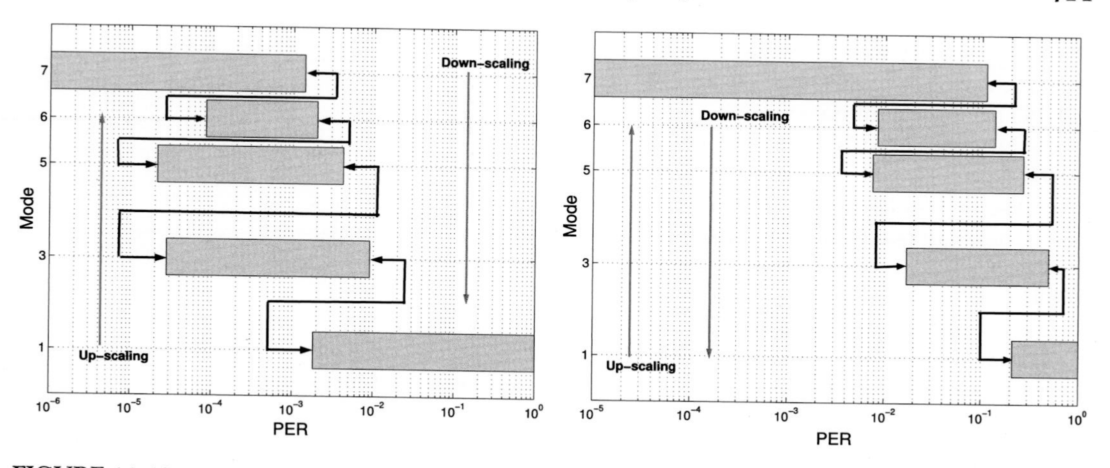

FIGURE 11.18 MCS switching characteristic comparison: video quality-based (left) vs. throughput-based (right). From [21].

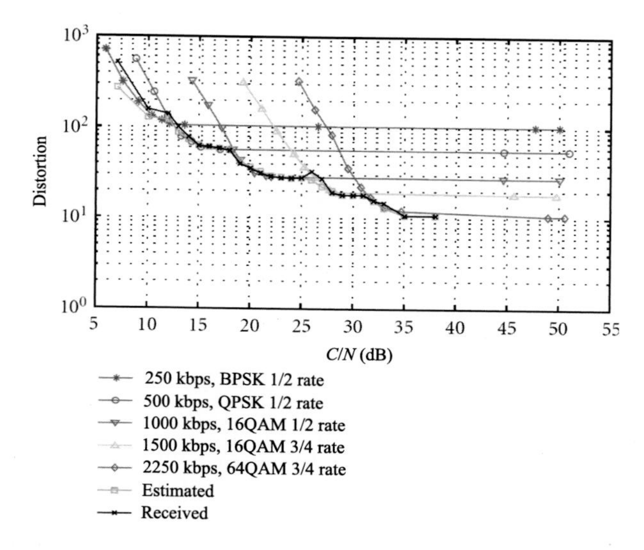

FIGURE 11.19 MCS link adaptation results from [21].

nization points without the overhead of explicit synchronization codewords. The algorithm rearranges the variable-length encoded blocks of data (for example those resulting from block-based transformation and entropy coding) into a structure with dimensions that are predefined to ensure that it has sufficient capacity.

Assume that there are N variable-length blocks, each of length $b_i : i = 1 \cdots N$, and we wish to pack these into M equal-length slots of length s. The slotted EREC data structure should therefore satisfy the following conditions:

Algorithm 11.1 EREC encoding.

1. Define offset sequence ϕ_n;
2. INPUT N blocks of data;
3. Compute slot length, s (Eq. (11.4));
4. FOR $i = 1 \cdots N$
  5. Assign block i to slot i;
  6. Compute residual length for each slot: $r_i = b_i - s$;
7. END FOR
8. FOR $n = 1 \cdots N$
  9. FOR $i = 1 \cdots N$
    10. Compute offset: $k = i + \phi_n(\text{mod}(N))$;
    11. IF $r_i > 0$ AND $r_k < 0$ THEN (pack slot k with residual from block i)
      12. IF $|r_k| \geq |r_i|$ THEN $r_k \leftarrow r_k + r_i; r_i \leftarrow 0$;
      13. IF $|r_k| < |r_i|$ THEN $r_i \leftarrow r_k + r_i; r_k \leftarrow 0$;
  14. END FOR.
  15. IF $r_i = 0 : \forall i$ THEN END.
16. END FOR.

$$\begin{cases} N \leq M, \\ T = \sum_{j=1}^{M} s_j \geq \sum_{i=1}^{N} b_i, \ s_j = s, \end{cases} \tag{11.3}$$

where T is the total data size of the slotted structure. The first condition ensures that there are enough slots while the second ensures that the EREC structure is large enough to code all the variable-length blocks of data. In most circumstances, for image or video coding, it is appropriate to set $N = M$ and we shall assume this from now on. We can thus compute the slot length for the structure as

$$s = \left\lceil \frac{1}{N} \sum_{i=1}^{N} b_i \right\rceil. \tag{11.4}$$

An N-stage algorithm is then used to pack the blocks into the slotted structure. To enable a systematic encoding process that can be reversed at the decoder, an offset sequence is used. This defines the offset used at each stage in the algorithm. At stage n, the algorithm will search slot $k = i + \phi_n(\text{mod}(N))$, where ϕ is a pseudorandom offset sequence. A pseudorandom sequence is specified since block lengths can be correlated temporally. In our simple examples however we will use the sequence $\phi = \{0, 1, 2, 3, 4 \cdots N\}$ for convenience and clarity. At each stage in the process, the EREC algorithm searches the slot at the appropriate offset to see if there is space to pack some or all of the excess data from overlong blocks – i.e., where $b_i > s$. This process is repeated until all excess data from overlong blocks is packed into the structure. The EREC encoding procedure is defined in Algorithm 11.1.

The decoding process is simply the inverse operation, progressively reconstructing, over N decoding stages, each original block until the block is fully decoded (e.g., an end of block code is detected). Thus, in the absence of channel errors, the decoder will correctly decode all the data.

A simple example of EREC encoding is shown in Fig. 11.20 for six blocks of data. Referring to Fig. 11.20, at stage 0 blocks 3, 4, and 6 are completely coded in slots 3, 4, and 6, with space left over. Blocks 1, 2, and 5 however are only partially coded and have data left to be placed in the space left in slots 3, 4, and 6. At stage 1, for an offset of $\phi = 1$, the remaining data from block 2 are coded in slot 3, and some data from block 5 are coded in slot 6. This process continues until, by the end of the sixth stage, all the data are coded. Further worked examples, illustrating both the encoding and decoding processes for a specific set of blocks, are given in Examples 11.5 and 11.6.

Example 11.5. EREC encoding

A DCT/VLC-based encoder has generated a slice of data comprising eight blocks with the following lengths.

Block i:	1	2	3	4	5	6	7	8
Length b_i	12	5	6	9	4	2	7	3

Assuming an offset sequence {0,1,2,3,4,5,6,7}, show how the EREC algorithm would encode this slice of data.

Solution. The total length of the slice is 48 bits, hence the slot length is 6 bits. The encoding process proceeds as follows.

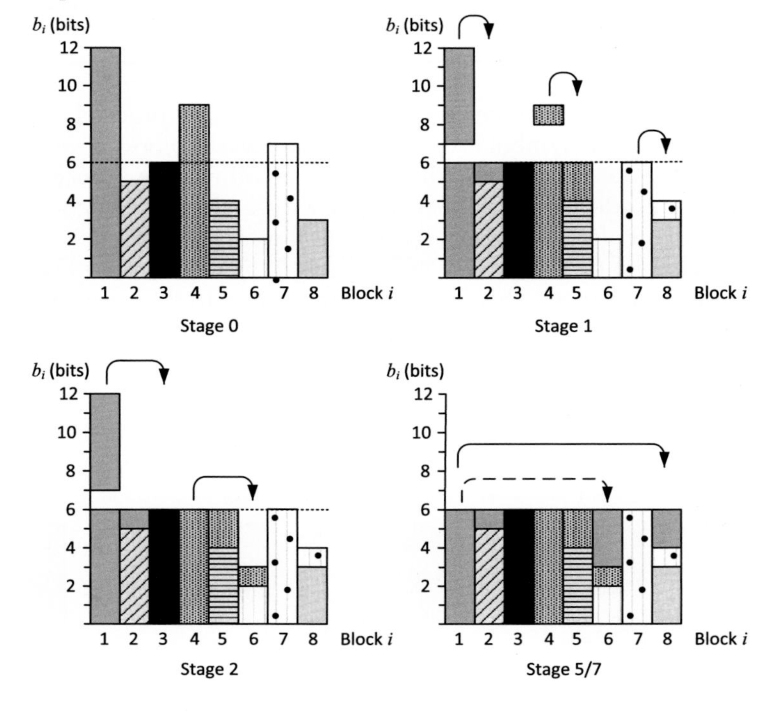

Stages 1 and 2 of the encoding process are shown above. Then, for stages 3 and 4, there is no spare space to fill in slots 4 and 5, and hence no action is taken. At stage 5, 3 bits from block 1 are packed into slot 6, and at stage 7, the remaining 2 bits from this block are packed into slot 8. These two operations are shown combined for convenience above.

Example 11.6. EREC decoding

Perform EREC decoding on the EREC encoded blocks of image data from Example 11.5 above. Assume that each block is terminated with an EOB symbol.

Solution. For convenience, here we use a more compact, tabular structure to illustrate each stage of the decoding process, where **X** indicates an incomplete block and **C** indicates a completely decoded block.

Stage	Offset	bk1	bk2	bk3	bk4	bk5	bk6	bk7	bk8
0	0	X	C	C	X	C	C	X	C
1	1	X			X			C	
2	2	X			C				
3	3	X							
4	4	X							
5	5	X							
6	6	X							
7	7	C							

At stage 0, we start decoding from the base of the first block and proceed until the end of the slot is reached or an EOB symbol is decoded. In the case of block 1, we reach the end of the slot with no EOB detected. We do the same for all other slots. Hence blocks 2, 3, 5, 6, and 8 are all fully decoded at stage 0. At stage 1 we search (for all incomplete blocks) for additional data at an offset of 1. For example, in the case of block 1 we search slot 2: we detect additional data beyond the EOB for block 2, but still no EOB is detected for block 1 so it remains incomplete. This proceeds until stage 6 in this case, when all blocks are fully decoded.

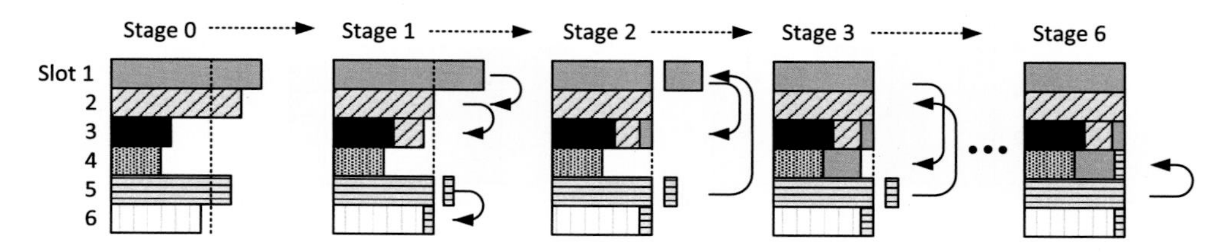

FIGURE 11.20 EREC operation.

FIGURE 11.21 EREC performance examples of reconstructed DCT/Huffman encoded images after erroneous transmission. Left: 0.1% BER. Right: 1% BER. Courtesy of D. Redmill [22].

EREC performance

In a clean channel, the EREC decoder will correctly decode all the data with negligible overhead (typically only the value of T needs to be sent as side information). When channel errors occur, their effects depend on the particular distribution of codewords in the block and how rapidly symbol resynchronization is achieved. If symbol resynchronization is not regained before the end of the block, then this can cause the EOB symbols to be missed or falsely detected. This means that the decoder will incorrectly decode following information which was packed at later stages of the algorithm. This, in turn, implies that data placed in later stages (those towards the ends of long blocks) are more susceptible to error propagation than data placed in earlier stages. For typical compression methods, these propagation effects occur in high-frequency data from active regions of the image. The important property of EREC is that, because the decoding of each block commences from a known location, block synchronization is preserved.

Examples of reconstructed images after EREC encoding for 0.1% BER and 1% BER are shown in Fig. 11.21. Comparing these with their equivalents in Fig. 11.5, the robust properties of EREC encoding can clearly be seen. Further improvements in subjective quality can be obtained if concealment is applied (see Section 11.8). An example of the relative performance of EREC versus a conventional sequential encoding is shown in Fig. 11.22 showing up to 10 dB improvement in noisy channels.

Redmill and Bull [23] have shown that EREC can perform extremely well with arithmetic coding systems as well as with conventional VLC methods. Interestingly, the paper demonstrates that without EREC, Huffman coding is significantly more robust to errors than arithmetic coding. However, with EREC, significantly improvement gains are manifest for both approaches and their performances are similar, with up to 15 dB improvement in peak signal to noise ratio (PSNR).

EREC has been considered for adoption in standards such as MPEG-4, but has never been included in the final draft. Possible reasons for this are that it requires a number of blocks to be processed before they can be packed, although these delays can be managed in a slice-structured environment. More likely, it is because the method does not fit well with conventional processing pipelines. Also, in the case of video, the use of predictive coding to

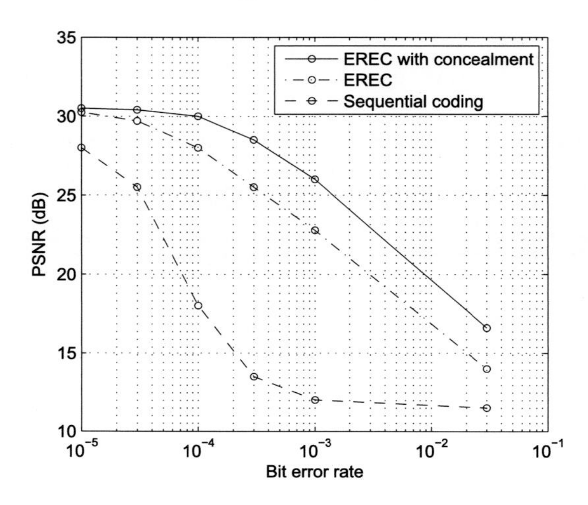

FIGURE 11.22 Graph of EREC performance for increasing BER compared to conventional DCT/VLC encoding.

remove interframe redundancy implies that an alternative approach to coding motion vectors must be employed. Swann and Kingsbury demonstrated the benefits of EREC for video coding in [24] showing significant benefits over conventional approaches in the context of MPEG-2.

11.7.2 Pyramid vector quantization (PVQ)

Principle of operation

Techniques such as EREC provide enhanced resilience through restructuring a bitstream in such a way that pseudofixed-length coding is achieved. In contrast, PVQ prevents error propagation through the use of actual FLCs. PVQ, introduced by Fischer [25], exploits the Laplacian-like probability distributions of transform coefficients to provide an efficient means of compressing them.

PVQ is a form of vector quantization optimized for i.i.d. Laplacian random variables. A vector $\mathbf{x}$ is formed by grouping L such random variables (e.g., transform coefficients). Based on the law of large numbers and the asymptotic equipartition property, it can be shown that almost all vectors will lie on a contour or region of constant probability. This region is a hyperpyramid in L-dimensional space with $2L$ vertices and $2L$ faces. Thus, for sufficiently large vector dimensions, there is negligible distortion if $\mathbf{x}$ is approximated by its nearest point on the hyperpyramid. This is the basis of the PVQ encoding algorithm.

The PVQ encoding algorithm consists of two steps. The first finds the closest vector $\hat{\mathbf{x}}$ to the input vector $\mathbf{x}$. The second step generates a bit code for $\hat{\mathbf{x}}$. Since all points on the hyperpyramid are equally probable, FLC is the optimal method for assigning bit codes. The PVQ encoded bitstream thus comprises a sequence of codebook vector indices.

Chung-How et al. [26] showed that PVQ is much more resilient to bit errors than either JPEG or H.263 and that wavelet-based PVQ outperforms DCT-based PVQ in terms of compression performance. In [27], Bokhari et al. showed that intra-mode PVQ is an effective codec

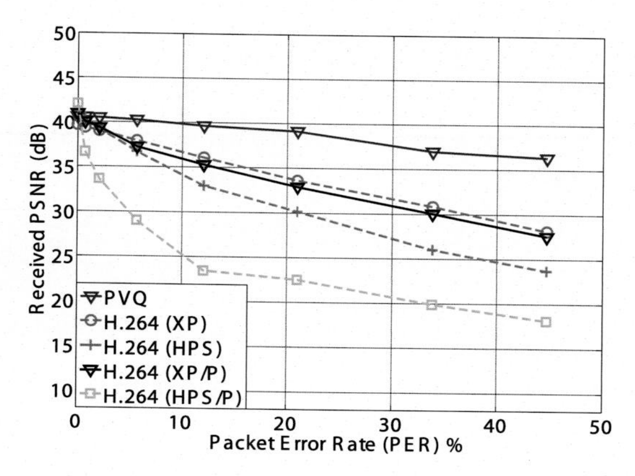

FIGURE 11.23 RD optimized PVQ performance for *tennis* in a lossy 802.11n channel with correlated errors. From [27].

for indoor wireless transmission of HD video. Intra-coding was employed since interframe encoding can lead to temporal error propagation. Although there is a loss in compression performance due to using an intra-only mode, this was more than compensated for by the increased error resilience.

Performance

Bokhari et al. [27] introduced an efficient rate-distortion (RD) algorithm for PVQ which yielded impressive compression performance, comparable to intra-H.264/AVC and Motion JPEG2000. They also evaluated the error resilience of the codec in the context of HD video using a realistic lossy wireless channel. Up to 11 dB PSNR improvement over H.264/AVC was demonstrated for lossy conditions. Intra-PVQ provides compression performance and video quality (43 dB PSNR @25 Mbps) comparable to other codecs in intra-mode. Importantly, it also obviates the need for sophisticated tools such as intra-prediction, variable-size transforms, deblocking filters, or entropy coding (CABAC). The error resilience of this approach is compared to other standardized codecs in [27] and the results of this are also shown in the graph of Fig. 11.23. PVQ offers up to 13.3 dB PSNR performance gain over the best-performing H.264/AVC counterpart in terms of error resilience. Example PVQ encoded frames are compared with H.264/AVC (High profile) in Fig. 11.24.

A major attribute of PVQ is the low variance in distortion across corrupted frames compared to H.264/AVC. This is illustrated in Fig. 11.25 (left). One of the factors that contributes to PVQ's performance is its ability to cope with corrupted data. In most wireless scenarios, the whole packet is not lost, but instead it has insufficient integrity to benefit from FEC. In such cases, the packet is usually discarded before being sent to the application layer. In PVQ and other robust formats, the codec benefits from the use of corrupted data. For VLC-based schemes, the limiting factor is the PER, whereas for FLC-based schemes it is the actual number of bit errors (in the received packet). This is defined as the residual BER (RBER). This is demonstrated in Fig. 11.25 (right), which shows that, although the PER might be high, the corresponding RBER can be much lower. In bursty or correlated channels, the instantaneous RBER varies considerably. This leads to a temporally localized psychovisual impact. In an un-

FIGURE 11.24 Frame shots for *tennis* at 25 Mb/s and 25% BER. Top: H.264/AVC (HPS). Bottom: RD optimized PVQ. From [27].

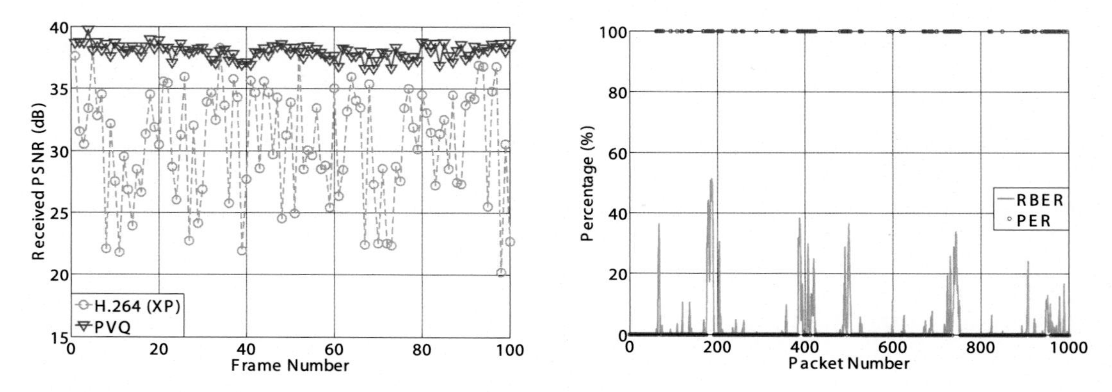

FIGURE 11.25 PVQ performance. Left: Variability in frame distortions for PVQ and H.264/AVC. Right: Packet error rate vs. residual bit error rate for an 802.11n channel [27].

correlated channel, the instantaneous RBER varies much less, giving a more consistent loss in quality throughout the sequence.

11.8 Error concealment

Error concealment is a decoder-oriented postprocessing technique which attempts to conceal the effects of errors by providing a subjectively acceptable approximation of the original data. This exploits the spatial and temporal correlations present in a video sequence to estimate the lost information from previously received data. Lost data is reconstructed using spatial or temporal interpolation in a way that reduces the perception of spatio-temporal error propagation. Error concealment methods can benefit from complimentary error-resilient encoding so that errors are as much as possible localized and have local good data upon which estimates can be based.

Concealment methods can offer solutions where other techniques, particularly interactive methods which require a feedback channel, are not possible – for example in broadcast or multicast systems. Concealment is a postprocessing operation and is not mandated in any coding standard. Many algorithms exist and some of these are described below, classified as spatial, temporal, or hybrid methods. The reader is also referred to many excellent papers and reviews on the topic, including [3,5,6,28]. Reference [29] also contains details of the performance of specific methods based on motion field interpolation and multihypothesis motion estimation. Alternative methods that show some potential are those based on texture synthesis or inpainting. These are not covered here further, but the reader is referred for example to [31].

11.8.1 Detecting missing information

Firstly, before we can apply concealment methods, we need to understand where exactly the errors are located. This can be done in a number of ways and will depend to a degree on the implementation details. Some examples are:

- **Using header information:** For example, the sequence number in a packet header can be used to detect packet loss.
- **Using forward error correction (FEC):** FEC can be used at the application layer as well as at the link layer.
- **Detection of syntax and semantic violations:** For example, illegal decoded codewords, invalid numbers of decoded units, out of range decoded parameters, etc.
- **Detection of violations to the general characteristics of video signals:** For example, blocks with highly saturated colors (pink or green), blocks where most decoded pixels need clipping, strong discontinuities at borders of blocks, etc.

None of these methods guarantee finding all errors and some can be subject to false detections. So, in practice, combinations are often employed. Let us now consider methods for concealing the errors once they have been detected.

11.8.2 Spatial error concealment (SEC)

Spatial error concealment (SEC) methods exploit the spatial correlations that exist in most video signals. They operate under the assumption of spatial continuity and smoothness. The most common method for spatial interpolation is based on the weighted averaging technique proposed in [32]. Concealment is ideally based on data that has been correctly received; however, in areas of more extensive loss, a progressive approach can be used where previously concealed blocks or pixels are used as a basis for estimation.

Spatial interpolation, based on weighted averaging, is illustrated in Fig. 11.26. After the detection of a lost block (indicated by the thicker lined boundary), each pixel, $s(x, y)$, is interpolated using the following expression:

$$s(x, y) = \frac{d_L s_R(x, y) + d_R s_L(x, y) + d_T s_B(x, y) + d_B s_T(x, y)}{d_L + d_R + d_T + d_B}, \tag{11.5}$$

where $s_R(x, y)$, etc., are the border pixels from the spatially adjacent macroblocks located to the right, left, top, and bottom of the missing block. An example of SEC applied to the *foreman*

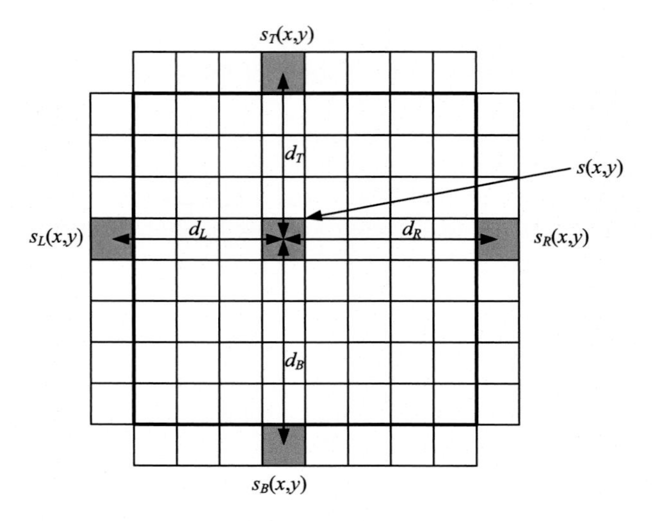

FIGURE 11.26 Spatial error concealment using nearest neighboring pixels.

FIGURE 11.27 Spatial error concealment applied to a lost row of macroblocks in the *foreman* sequence. Left: Original with error shown. Middle: Result of spatial error concealment. Right: Amplified difference signal. MSE=46.6. Courtesy of D. Agrafiotis.

sequence is shown in Fig. 11.27. The result in this case has an mean squared error (MSE) of 46.6 with respect to the original error-free frame.

There are a number of extensions to the basic SEC method that can provide additional benefit, but usually at the cost of additional complexity. For example, *edge continuity* can be enforced. This means that if an edge is detected in a neighboring block or blocks, then this edge should be preserved during concealment. Agrafiotis et al. [33] use an edge preserving method that not only preserves existing edges but also avoids introducing new strong ones. The approach switches between a directional interpolation and a conventional bilinear interpolation, based on the directional entropy of neighboring edges. The proposed strategy exploits the strengths of both methods without compromising the performance of either and offers performance improvements of over 1 dB for some cases. A similar approach was also presented by Ma et al. in [34].

Example 11.7. Spatial error concealment

Consider the missing block of data shown below with the correctly received boundary pixels indicated. By performing SEC on this block, calculate the value of the missing pixel shown encircled on the diagram.

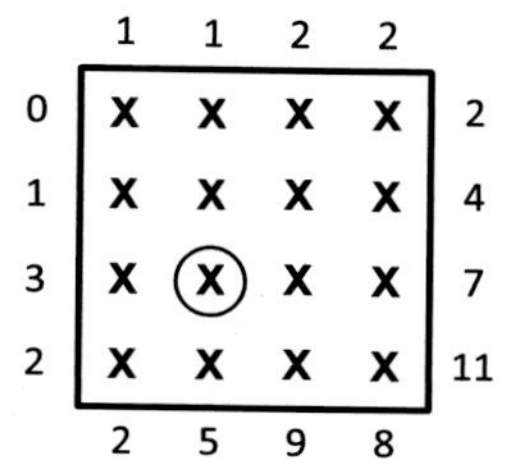

	1	1	2	2	
0	X	X	X	X	2
1	X	X	X	X	4
3	X	(X)	X	X	7
2	X	X	X	X	11
	2	5	9	8	

Solution. Using Eq. (11.5),

$$s(x, y) = \frac{2 \times 7 + 3 \times 3 + 3 \times 5 + 2 \times 1}{10} = 4.$$

All other missing pixel values in the block can be interpolated in a similar manner.

11.8.3 Temporal error concealment (TEC)

Temporal error concealment (TEC) methods operate under the assumption of temporal continuity and smoothness of the motion field [29,30]. They exploit the highly correlated nature of most natural sequences and conceal corrupted or lost pixels based on predictions from surrounding regions. They work well on inter-coded blocks, but are often less successful on intra-coded blocks. Two basic techniques are considered below.

Temporal copying (TEC_TC)

The simplest form of TEC is *temporal copying*. In this case, as shown in Fig. 11.28, the underlying model is one of zero motion. This method is extremely easy to implement, but the underlying motion model is rarely sufficient in practice. Results for a single lost packet containing one slice of the *foreman* sequence are shown in Fig. 11.29. Although this result looks reasonable, the error is actually quite large due to the person's head moving. This displacement is particularly evident if you inspect the area on the left-hand edge of his face.

Motion-compensated temporal replacement (TEC_MCTR)

As illustrated in Fig. 11.28 (bottom), this method replaces the missing block with the best candidate region from the reference frame or frames. This is significantly more complex than the previous approach as it requires producing a list of candidate motion vectors that can be evaluated in order to choose the best replacement block. The method employs a two-stage process:

1. estimation of concealment displacement,

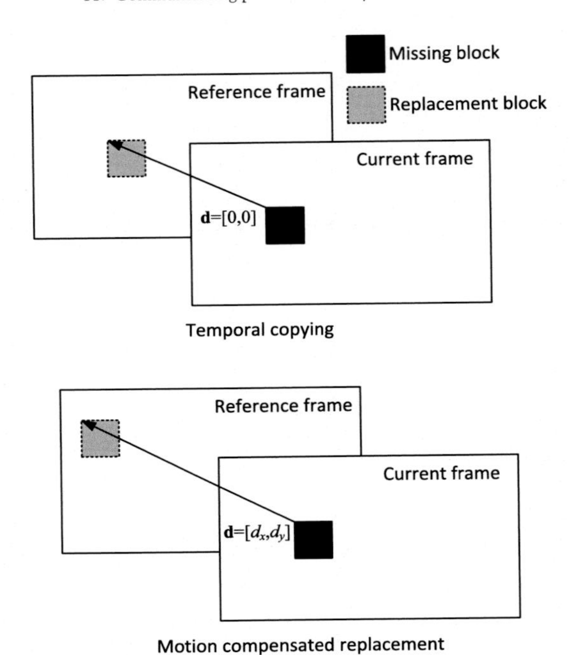

FIGURE 11.28 Temporal error concealment based on temporal copying (top) and motion-compensated prediction (bottom).

FIGURE 11.29 Temporal error concealment results for *foreman* based on temporal copying. MSE=19.86. Courtesy of D. Agrafiotis.

2. displacement evaluation and replacement selection.

The first stage raises the following question: how do we select the candidate motion vectors for TEC_MCTR? Assuming that the motion vector for the missing block is also lost, there are several likely candidates for replacement. These include:

- motion vectors from spatially neighboring blocks,
- a rank order statistic or average of the motion vectors from neighboring blocks,
- motion vectors from temporally neighboring blocks.

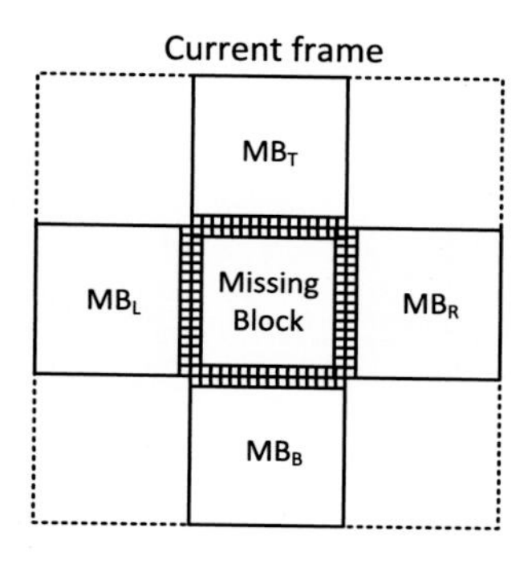

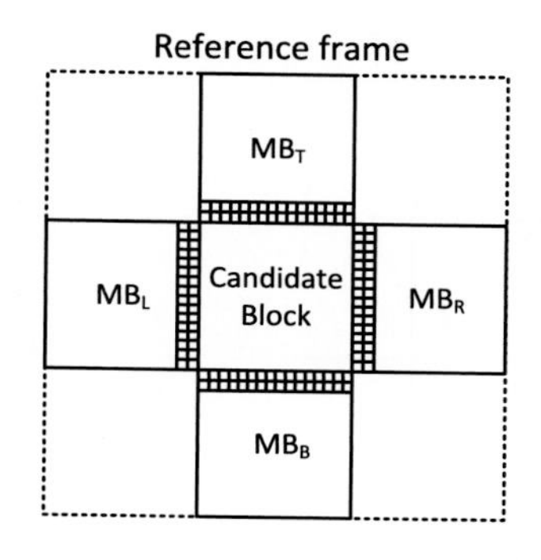

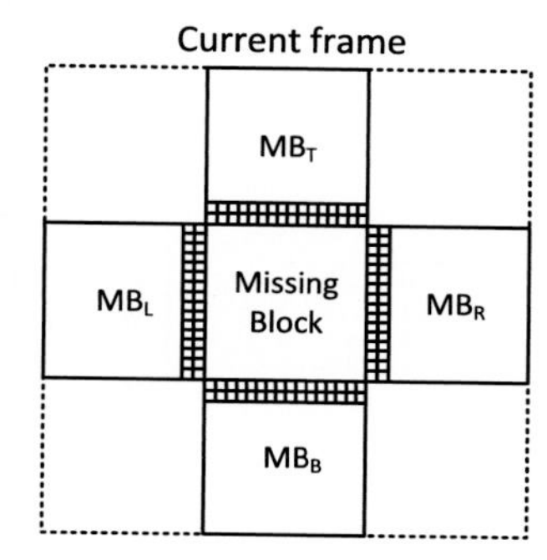

Boundary Matching

External Boundary Matching

FIGURE 11.30 Boundary matching error measures. Left: Boundary matching error (BME). Right: External boundary matching error (EBME).

We then need to evaluate these candidate motion vectors and select the one that produces the best match. So the second question is, what matching metric do we use to decide which of the candidate vectors is the best? This is normally done using a form of *boundary matching error* (BME). Options for this metric are shown in Fig. 11.30. The first method shown, referred to simply as BME, evaluates all candidate replacement blocks by computing the sum of absolute differences (SAD) value between the boundary pixels of the matched block and the adjacent pixels from the surrounding blocks. The second method, referred to as *external BME* (EBME), again computes an SAD value but this time between the row(s) of pixels surrounding the missing block in the current frame and those surrounding the candidate block in the reference frame. EBME is a superior measure because it fits with the assumption that temporal concealment preserves temporal smoothness. BME on the other hand is inconsistent with the model, as it uses a spatial continuity metric to assess temporal smoothness.

TEC results for the *foreman* sequence based on motion-compensated replacement are shown in Fig. 11.31. Because the underlying model is one of smooth translational motion, this technique generally performs significantly better that temporal copying. It can be observed from Fig. 11.31 that the residual error is lower than in Fig. 11.29 and that the artifacts at the side of the foreman's face are significantly reduced.

Example 11.8. Temporal error concealment

Consider the region of a current frame and the colocated region of a adjacent reference frame, as shown below. Assuming that the 2 × 2 block, shown in gray, has been lost during transmission, but all other data shown has been received correctly, compute the TEC_MCTR block using a BME metric.

Reference frame

8	8	8	9	7	9
8	7	7	6	7	6
6	5	8	6	4	4
7	4	7	4	7	8
5	6	8	6	7	7
4	6	7	4	4	8

Current frame

7	9	7	5	8	9
7	7	6	4	7	8
6	5	5	7	7	8
7	4	3	6	7	7
5	5	5	7	7	8
3	4	4	8	4	7

Motion vector values for adjacent blocks in the current frame are

$$\mathbf{d}_L = [0, 0],$$
$$\mathbf{d}_R = [-1, 1],$$
$$\mathbf{d}_T = [-2, 2],$$
$$\mathbf{d}_B = [2, 0].$$

Solution. We can use these four motion vectors to select candidate blocks for replacement. In each case we compute the BME value as follows:

$$d = [0, 0] : BME = 19,$$

$$d = [-1, 1] : BME = 11,$$

$$d = [-2, 2] : BME = 7,$$

$$d = [2, 0] : BME = 13.$$

The best candidate block according to the BME metric is that associated with the [−2,2] motion vector. Hence the replacement pixels in this case are

$$\begin{bmatrix} 5 & 6 \\ 4 & 6 \end{bmatrix}.$$

11.8.4 Hybrid methods with mode selection

The choice of using either SEC or TEC is not always straightforward. In most cases TEC provides the best results, and many simple systems use only a temporal copying approach, which works well if there is no motion! In practice, there are obvious situations when one method will be preferable over the other. For example, in the case of a shot cut or fade, SEC is normally superior. In the case of a static scene, temporal block copying will produce good results and in the case of camera or object motion, it is most likely that motion-compensated temporal replacement will produce the best results. However, if the scene motion does not fit well with the motion model used to find the candidate vectors, then reverting to SEC may be better.

FIGURE 11.31 Temporal error concealment results for *foreman* based on motion-compensated replacement. MSE=14.61. Courtesy of D. Agrafiotis.

Algorithm 11.2 EECMS.

---

1. INPUT video with lost block;
2. REPEAT for each corrupted block:
    3. Perform TEC: Use EBME for matching;
    4. Compute spatial activity: $SA = \mathrm{E}\left\{(s_c - \mu)^2\right\}$;
    5. Compute temporal activity: $TA = \mathrm{E}\left\{(s_c - s_r)^2\right\}$;
    6. IF ($TA<SA$) OR ($TA<$Threshold) THEN (use TEC, GOTO (8));
    7. Perform SEC; use directional interpolation;
8. UNTIL all missing blocks concealed;
9. END.

We saw in Section 11.5.5, that slice-structured modes can be beneficial in terms of error resilience. In particular, techniques such as *flexible macroblock ordering* (FMO) can help to ensure that, if a slice is lost, some good data is retained in an adjacent slice upon which an interpolation of the lost data can be based. This type of error-resilient encoding can provide significant benefits when combined with concealment. The question that arises when implementing a hybrid system is, how do we select the mode of operation – temporal or spatial? This can be done using a mode selection algorithm.

For example, Fig. 11.32 shows the architecture for the EECMS algorithm as proposed by Agrafiotis et al. [28]. This is more formally defined in Algorithm 11.2. EECMS performs mode selection based on a comparison of the motion-compensated *temporal activity* (TA) and the *spatial activity* (SA) in the vicinity of the lost block. SA is computed as the variance of the surrounding macroblocks in the current frame, whereas TA measures motion uniformity and is based on the MSE between pixels (s_c) in the macroblocks surrounding the missing block in the current frame and those (s_r) surrounding the replacement macroblock in the reference frame. SEC is invoked only if SA is lower than SA and SA is above a specified threshold. Selected results from the EECMS method are summarized in Table 11.2 and compared to the concealment algorithm used in the H.264/AVC joint model (JM). Across all sequences evaluated in [28], performance improvements up to 9 dB PSNR were reported.

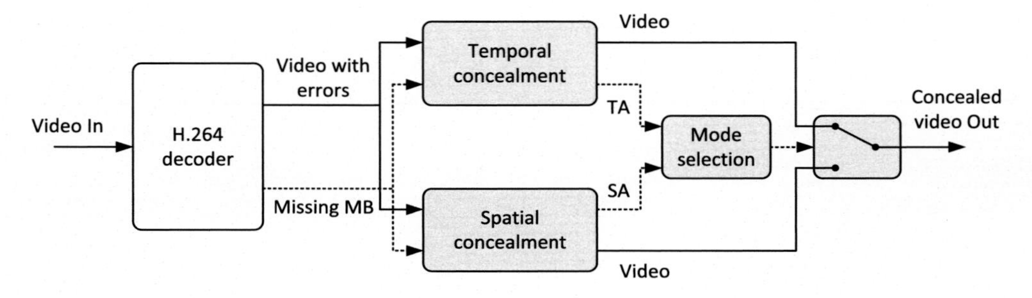

FIGURE 11.32 Enhanced error concealment with mode selection [28].

TABLE 11.2 Selected performance results for EECMS. From [28].

PSNR (dB)	0% PER	1% PER		20% PER	
		JM	EECMS	JM	EECMS
Foreman	40.14	38.11	39.44	27.33	31.10
Bus	37.20	35.08	36.33	23.79	26.99
Football	31.27	30.14	30.56	22.05	23.22

11.8.5 Intelligent error concealment

Carreira et al. [35] proposed an error concealment-aware encoding scheme based on a scalable encoding architecture and targeted at broadcast environments. The encoder signals to the decoder the preferred concealment technique based on content analysis in the context of simulated loss, followed by a Lagrangian optimization of the signaling rate–EC distortion cost. They show that this approach, using only mode signaling, provides an average PSNR gain up to 2.95 dB over conventional HEVC. When a residual signal is also encoded in the enhancement layer of SHVC, the PSNR gains increase to 3.79 dB.

As with compression in general, deep neural networks have been proposed as a basis for image interpolation, inpainting, and error concealment. Sankisa et al. [36] present an adaptable deep neural network approach to video error concealment using optical flow prediction at the decoder based on two parallel neural networks employing convolutional and long short-term memory layers. This prediction is then used to reconstruct the degraded portion of the corrupted and future video frames.

11.9 Congestion management

In cases where streams are encoded on-line, it is possible to control the rate of the video to match that of the channel. However, in many cases, channel conditions can be highly variable and detailed knowledge (especially for multicast applications) may not be available. Also, with preencoded streams, it is not possible to adapt the rate unless multiple versions of the same content are encoded. Solutions to this problem normally involve partitioning or layering the bitstream in some manner. For example, scalable encoding methods represent the bitstream in multiple hierarchical layers, whereas multiple description coding (MDC) uses multiple independent descriptions that can exploit network path diversity. Most streaming solutions however employ more explicit rate-quality partitions used in conjunction with HAS. These approaches are described below.

11.9.1 HTTP adaptive streaming (HAS)

Internet streaming is the dominant mechanism for video delivery across the globe. However, this growth would not have been possible without the introduction of mechanisms to support real-time video transmission across the internet. We have already seen that the internet does not guarantee delivery bandwidth as it was intended for best-effort, nonreal-time transmission. This is complicated further by the fact that user devices have varying communication, processing, and display capabilities.

The characteristics and requirements of internet streaming applications include:

- Higher encoder complexity can be tolerated since encoding is off-line and can be performed in parallel in the cloud.
- Decoding complexity should be kept lower to support various terminal types.
- Support for a wide range of content is needed, including high dynamic range (HDR), wide color gamut (WCG), high resolution (up to 4K), and high frame rate. Low spatial resolutions are also important to support streaming in low-bandwidth conditions.
- Coding should be cognizant of artistic processes used in content creation. For example, film grain noise is often intentionally included in movies as part of the creative process.
- Frequent random access points are needed to support frequent switching and trick modes.
- Support for multiple quality representations is needed, including scalability, simulcasting, or other adaptation mechanisms.

Typically for this application:

- Video is encoded in the cloud by software encoders.
- Source video is split into chunks, which are optimized and encoded separately at multiple quality points, in parallel.
- Closed-GOP encoding with intrapicture intervals of 2–5 seconds is used.

The class of algorithms that adjust server bit rates according to dynamic network variations is generally referred to by the term HTTP adaptive streaming (HAS). These are a critical component and used by all major streaming organizations.

HAS is a real-time adaptation mechanism driven by the requirements of a client device in the context of prevailing (dynamic) delivery conditions. The primary objective of HAS is to maximize the quality of experience (QoE) for each user of the service independently. In contrast to RTP/UDP systems, which push a predefined sequence to the client, HAS employs HTTP as the application and TCP as the transport layer protocol. The HAS approach divides the video sequence into short chunks or segments which can each be preencoded at multiple quality points and selected for transmission based on interactions between the client and the server.

As well as user preferences, device constraints, and explicit bandwidth estimation, adaptation methods can also take account of quality of experience proxies, measuring things such as playback buffer occupancy, stalls, startup delays, battery level, CPU availability, and quality variations.[2] They will also, during encoding, take account of content characteristics to determine the appropriate rate-quality profiles.

The approach most commonly used to realize HAS is MPEG-DASH [45]. DASH is the result of standardization efforts that have unified commercial offerings from Microsoft (Smooth Streaming), Apple (HLS), Adobe (HDS), and others and is now supported by most streaming organizations. Importantly, HAS standards, such as MPEG-DASH, do not specify the details of the bit rate adaptation logic. This is an area where streaming organizations and device manufacturers can innovate to differentiate performance in their players. Because HAS is HTTP-based it eases transmission through firewalls, supports the use standard server technologies, and does not require a persistent connection between server and client.

A basic HAS architecture is depicted in Fig. 11.33, and a more detailed view of the client-based adaptation process is shown in Fig. 11.34. Prior to delivery the content must first be partitioned and encoded as follows:

- **Temporal segmentation of video**: This provides the required number of segments for the sequence or movie, which may be based on simple time slices and/or include partitions at shot or subshot level.
- **Off-line encoding of segments**: To determine the appropriate range of rate-quality points for each segment and encode each segment at these rates using the codec of choice.

The basic operation of a HAS system then proceeds as described below.

- **HTTP GET request**: The client will initially request the manifest for the video sequence – metadata which is required to perform initialization of configuration.
- **Buffer filling**: Initial content requests for the first segments to fill the buffer up to an occupancy threshold when playout will commence.
- **Bandwidth estimation**: This is usually calculated as the size of the fetched segment(s) divided by the transfer time.
- **Update client state**: The state of the client is adjusted based on bandwidth and QoE estimates.
- **Request further segments**: The rate-quality point cited in subsequent GET requests is based on the current state of the client.

In this way, the client will adjust in real-time to the changing operating environment.

Most DASH systems are used in conjunction with nonscalable encoders (H.264/AVC, H.265/HEVC, VP9, and AV1). The benefits of using scalability extensions of these codecs, such as SVC and SHVC, have however been reported. This and other details are discussed in the review of adaptive streaming over HTTP presented by Bentaleb et al. [44].

Despite the advances and global impact of HAS systems, work is still required to enhance QoE estimation and to support immersive and 360-degree content. A key area of research is also how machine learning can be employed to enhance QoE estimates at the client [46]. This in turn will demand new datasets for DASH algorithm evaluation.

[2] This can present problems as there is no reference content available at the decoder.

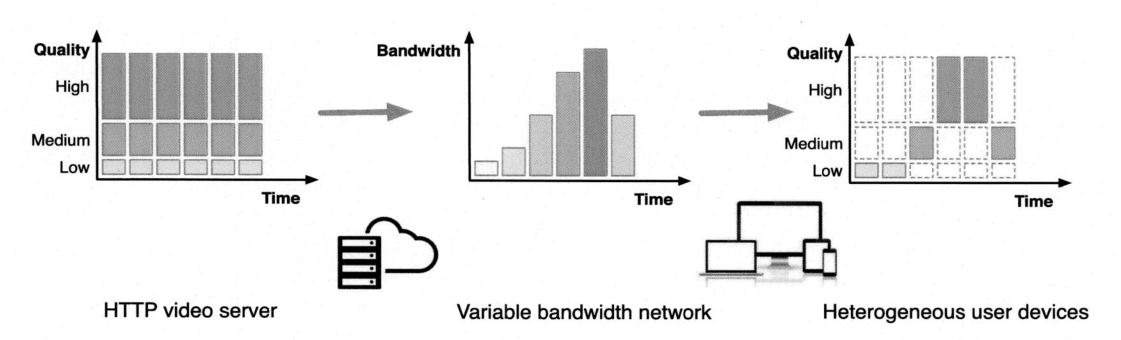

FIGURE 11.33 A generic adaptive streaming architecture.

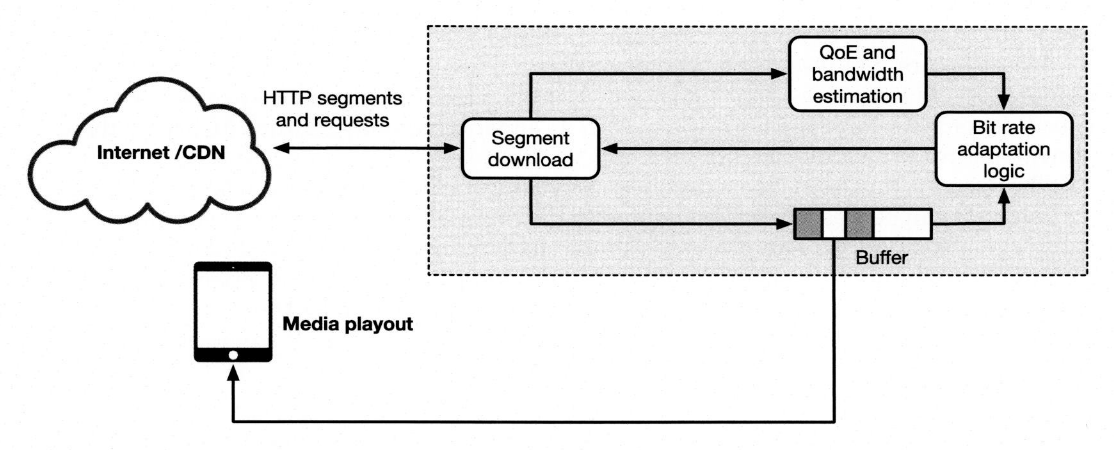

FIGURE 11.34 Client-based HAS architecture.

11.9.2 Scalable video encoding

It is common in video transmission for the signal to be encoded and transmitted without explicit knowledge of the downstream network conditions. In cases where network congestion exists in a packet switched network such as the internet, routers will discard packets that they are unable to forward due to congestion. In some cases it is possible for packets within a stream to be prioritized or embedded such that the least important information is discarded before the more important data. This mechanism lends itself to scalable or layered encoding, where a video signal is composed of a hierarchical layering in terms of spatial resolution, temporal resolution, or SNR. This enables devices to transmit and receive multilayered video streams where a base level of quality can be improved through the use of optional additional layers that enhance resolution, frame rate, and/or quality.

The first standardized codec to include optional scalable modes was MPEG-2. However, these were rarely used in practice. Since then, H.264/AVC supported a comprehensive set of scalable modes in its scalable video coding (SVC) Annex G extension. SVC has many attractive features, not least that it is compatible with the conventional H.264/AVC bitstream. In

2014 a scalability extension to H.265/HEVC was finalized [39], providing support for spatial, SNR, and color gamut scalability. As we have already seen in Chapter 6, scalable methods have also been incorporated into still image coding standards such as JPEG2000 and these have been adopted for some surveillance and video conferencing applications.

In scalable codecs, the aim is to create a hierarchical bitstream where a meaningful representation of the video can be reconstructed at the decoder even if some of the information is lost during transmission or if the terminal is incapable of decoding it. The basic architecture of a scalable video encoder and decoder pair is shown in Fig. 11.35. The figure shows the case for spatial or temporal scalability. The other case – SNR scalability – is a subset of this where the up- and downsampling operators are omitted.[3]

Consider the case of spatial scalability as shown in Fig. 11.35 and Fig. 11.36. The raw video input is first downsampled to the resolution of the base layer and is then transformed and quantized as usual. This layer is entropy coded and transmitted as the base layer. The base layer is then subjected to inverse transformation and inverse quantization before being upsampled to its original resolution. The reconstructed base layer is then subtracted from the original video to form a residual signal which is encoded as the enhancement layer. Clearly the base layer must have higher priority than its enhancement layers, thus justifying the use of unequal error protection.

At the decoder, the base layer is decoded as usual, but if the enhancement layer is available, the base layer can be upsampled and added to the decoded enhancement layer to form a full quality enhanced output. This process is similar to that for temporal and SNR scalability.

H.264/SVC is in common use for conferencing platforms offered by Zoom and Google. For further information on scalable coding the reader is referred to [4,37,38].

11.9.3 Multiple description coding (MDC)

Instead of creating a hierarchy of dependent information (as in scalable coding), where the decoding of higher layers depends on the existence of lower layers, MDC creates a number of independent encodings (descriptions). MDC uses these descriptions to exploit path diversity through a network such that, if any combination of descriptions are received correctly, then a meaningful reconstruction can be achieved at the decoder. If all descriptions are received, then the full quality reconstruction is possible. MDC attracted significant academic interest in the early 2000s and found application in live streaming and peer-to-peer distribution. It has however not been widely deployed in more recent practical systems.

The MDC approach is shown in Fig. 11.37. Many approaches have been proposed to create the independent descriptions and a review of these and their performance is provided by Apostolopoulos et al. in [40] and by Wang et al. in [41]. For example, the use of FMO slice structuring methods, as described in Section 11.5.5, has been exploited by authors such as Wang et al. [42]. Tesanovic et al. [43] extend the concept of path diversity to the case of multiple-input, multiple-output (MIMO) wireless technology based on spatial multiplexing. They propose the use of MDC as a means of emulating the spatial diversity lacking in conventional spatially multiplexed systems and providing an efficient means of mapping video

[3] For the case of SNR scalability it is normal for the encoder in Fig. 11.35 to be modified so that only the quantization and inverse quantization operations are included in the upscaling path.

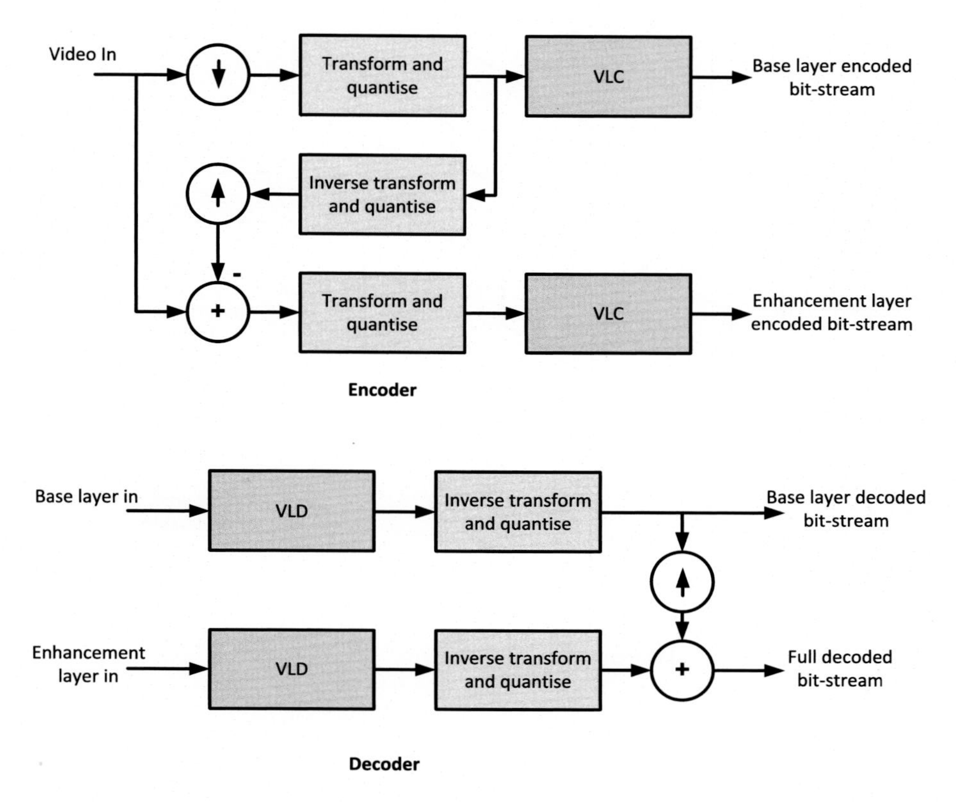

FIGURE 11.35 Scalable (spatial or temporal) video codec architecture.

content to the wireless channels. Their results indicate improvements in average PSNR of the decoded test sequences of around 5–7 dB, compared to standard, single-description video transmission.

11.10 Summary

We have examined in this chapter a number of methods that can be employed to improve the reliability of video delivery over lossy or unreliable channels. We considered why errors propagate spatially and temporally under the influence of variable-length and predictive coding regimes and saw examples which clearly demonstrate that the basic coding methods, considered earlier in this book, need to be adapted in a way that preserves rate-distortion performance while improving bitstream resilience to errors. We saw how network techniques such as FEC and ARQ can be exploited and enhanced to provide better performance, but also how the source coding itself can be modified to deliver improved error resilience with lower overhead. We saw how inherently robust encoding methods such as EREC and PVQ can

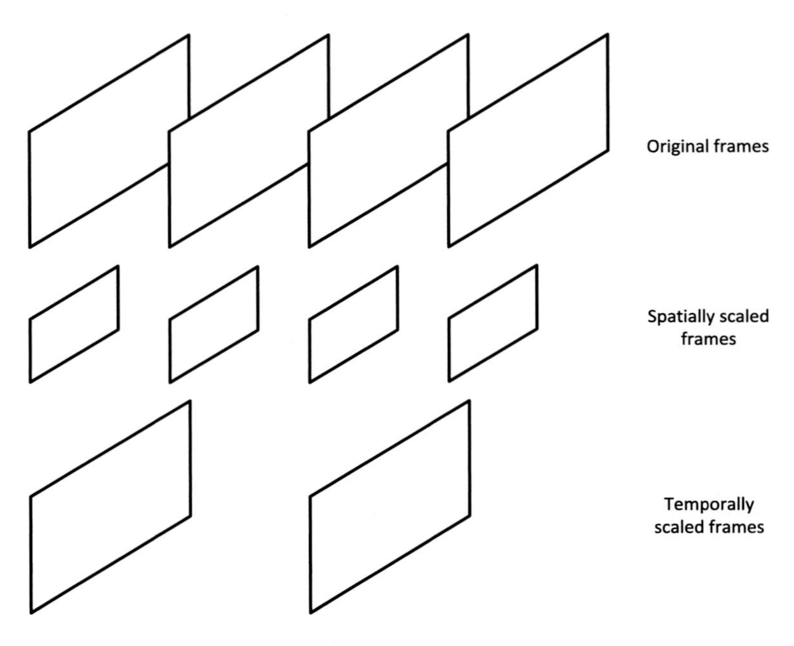

FIGURE 11.36 Frames types for scalable encoding.

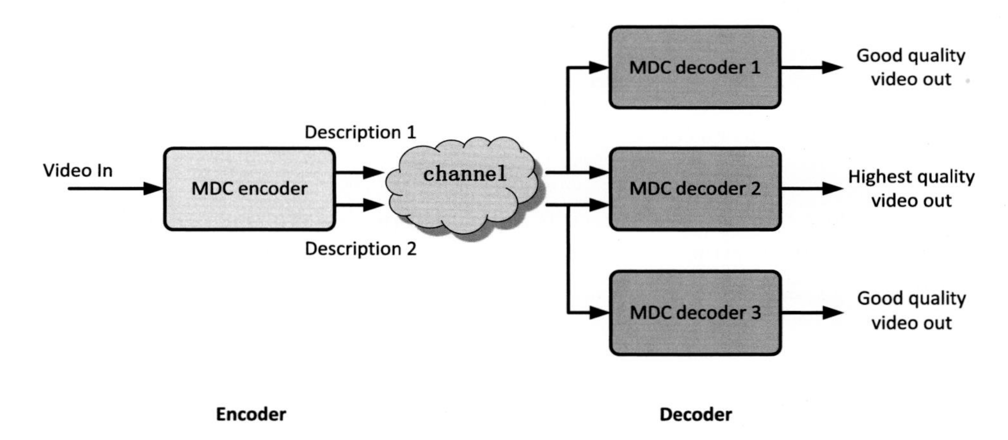

FIGURE 11.37 Multiple description codec architecture.

provide greater tolerance to corrupted data while obviating the need for complex encoding tools. We also saw how error concealment, possibly in combination with other error resilience methods, or concealment-aware encoding, can deliver significant improvements in subjective quality with little or no transmission overhead. Finally we explained the basic principles of how large-scale video streaming services are made more robust through bit rate adaption methods such as HAS and MPEG-DASH.

References

[1] C. Chong, C-M. Tan, D. Laurenson, S. McLaughlin, M. Beach, A. Nix, A new statistical wideband spatio-temporal channel model for 5-GHz band WLAN systems, IEEE J. Sel. Areas Commun. 21 (2) (2003) 139–150.
[2] G. Athanasiadou, A. Nix, J. McGeehan, A microcellular ray-tracing propagation model and evaluation of its narrow-band and wide-band predictions, IEEE J. Sel. Areas Commun. 18 (3) (2000) 322–335.
[3] T. Stockhammer, M. Hannuksela, T. Wiegand, H.264/AVC in wireless environments, IEEE Trans. Circuits Syst. Video Technol. 13 (7) (2003) 657–673.
[4] M. van der Schaar, P. Chou (Eds.), Multimedia over IP and Wireless Networks: Compression, Networking, and Systems, Academic Press, 2011.
[5] Y. Wang, S. Wenger, J. Wen, A. Katsaggelos, Review of error resilient coding techniques for real-time video communications, IEEE Signal Process. Mag. 17 (4) (2000) 61–82.
[6] Y. Wang, Q. Zhu, Error control and concealment for video communications: a review, in: Proc.IEEE, Special Issue on Multimedia Signal Processing, 1998, pp. 974–997.
[7] A. Shokrollahi, Raptor codes, IEEE Trans. Inf. Theory 52 (6) (2006) 2551–2567.
[8] M. Luby, LT-codes, in: Proc. 43rd Annual IEEE Symp. Foundations of Computer Science (FOCS), 2002, pp. 271–280.
[9] S. Lin, P. Yu, A hybrid ARQ scheme with parity retransmission for error control of satellite channels, IEEE Trans. Commun. 30 (I) (1982) 1701–1719.
[10] P. Ferre, A. Doufexi, J. Chung-How, A. Nix, Robust video transmission over wireless LANs, IEEE Trans. Veh. Technol. 57 (4) (2008) 2596–2602.
[11] J. Chung-How, D. Bull, Loss resilient H.263 video over the Internet, Signal Process. Image Commun. 16 (9) (2001) 891–908.
[12] Y. Takashima, M. Wada, H. Murakami, Reversible variable length codes, IEEE Trans. Commun. 43 (2/3/4) (1995) 158–162.
[13] A. Fraenkel, S. Klein, Bidirectional Huffman coding, Comput. J. 33 (4) (1990) 297–307.
[14] B. Girod, Bidirectionally decodable streams of prefix code-words, IEEE Commun. Lett. 3 (8) (1999) 245–247.
[15] JVT-B027.doc, Scattered Slices: a New Error Resilience Tool for H.26L, Joint Video Team of ISO/IEC MPEG and ITU-T VCEG, Feb. 2002.
[16] P. Ferre, D. Agrafiotis, D. Bull, A video error resilience redundant slices algorithm and its performance relative to other fixed redundancy schemes, Signal Process. Image Commun. 25 (3) (2010) 163–178.
[17] M. Wada, Selective recovery of video packet loss using error concealment, IEEE J. Sel. Areas Commun. 7 (5) (1989) 807–814.
[18] E. Steinbach, N. Farber, B. Girod, Standard compatible extension of H.263 for robust video transmission in mobile environments, IEEE Trans. Circuits Syst. Video Technol. 7 (6) (1997) 872–881.
[19] J. Carreira, et al., A two-stage approach for robust HEVC coding and streaming, IEEE Trans. Circuits Syst. Video Technol. 28 (8) (2018) 1960–1973.
[20] M. van der Scharr, S. Krishnamachari, S. Choi, X. Xu, Adaptive cross layer protection strategies for robust scalable multimedia video transmission, IEEE J. Selected Topics Communications 21 (10) (2003) 1752–1763.
[21] P. Ferre, J. Chung-How, A. Nix, D. Bull, Distortion-based link adaptation for wireless video transmission, EURASIP J. Adv. Signal Process. (2008) 253706.
[22] D. Redmill, N. Kingsbury, The EREC: an error resilient technique for coding variable length blocks of data, IEEE Trans. Image Process. 5 (4) (1996) 565–574.
[23] D. Redmill, D. Bull, Error resilient arithmetic coding of still images, in: IEEE Intl. Conf. on Image Processing, 1996, pp. 109–112.
[24] R. Swann, N. Kingsbury, Transcoding of MPEG-2 for enhanced resilience to transmission errors, in: IEEE Intl. Conf. on Image Processing, 1996, pp. 813–816.
[25] T. Fischer, A pyramid vector quantizer, IEEE Trans. Inf. Theory 32 (4) (1986) 568–583.
[26] J. Chung-How, D. Bull, Robust image and video coding with pyramid vector quantisation, IEEE Int. Symp. Circuits Syst. Proc. 4 (1999) 332–335.
[27] S. Bokhari, A. Nix, D. Bull, Rate-distortion-optimized video transmission using pyramid vector quantization, IEEE Trans. Image Process. 21 (8) (2012) 3560–3572.
[28] D. Agrafiotis, D. Bull, C. Canagarajah, Enhanced error concealment with mode selection, IEEE Trans. Circuits Syst. Video Technol. 16 (8) (2006) 960–973.

[29] M. Al-Mualla, C. Canagarajah, D. Bull, Video Coding for Mobile Communications, Academic Press, 2002.
[30] M. Al-Mualla, C. Canagarajah, D. Bull, Temporal error concealment using motion field interpolation, Electron. Lett. 35 (3) (1999) 215–217.
[31] H. Lakshman, M. Koppel, P. Ndjiki-Nya, T. Wiegand, Image recovery using sparse reconstruction based texture refinement, in: Proc. IEEE Intl. Conf. on Acoustics Speech and Signal Processing (ICASSP), 2010, pp. 786–789.
[32] P. Salama, E. Shroff, E. Coyle, E. Delp, Error concealment techniques for encoded video streams, IEEE Intl. Conf. Image Proces. 1 (1995) 9–12.
[33] D. Agrafiotis, D. Bull, C. Canagarajah, Spatial error concealment with edge related perceptual considerations, Signal Process. Image Commun. 21 (2) (2006) 130–142.
[34] M. Ma, O. Au, G. Chan, M-T. Sun, Edge-directed error concealment, IEEE Trans. Circuits Syst. Video Technol. 20 (3) (2010) 382–395.
[35] J. Carreira, et al., Error concealment-aware encoding for robust video transmission, IEEE Trans. Broadcast. 65 (2) (2019) 282–293.
[36] A. Sankisa, A. Punjabi, A. Katsaggelos, Video error concealment using deep neural networks, in: Proc. IEEE Intl. Conf on Image Processing, 2018, pp. 380–384.
[37] Y. Wang, J. Ostermann, Y-Q. Zhang, Video Processing and Communications, Prentice Hall, 2002.
[38] H. Schwarz, D. Marpe, T. Wiegand, Overview of the scalable video coding extension of the H.264/AVC standard, IEEE Trans. Circuits Syst. Video Technol. 17 (9) (2007) 1103–1120.
[39] J. Boyce, et al., Overview of SHVC: scalable extensions of the high efficiency video coding standard, IEEE Trans. Circuits Syst. Video Technol. 26 (1) (2015) 20–34.
[40] J. Aposolopoulos, M. Trott, W-T. Tan, Path diversity for multimedia streaming, in: M. van der Schaar, P. Chou (Eds.), Multimedia over IP and Wireless Networks: Compression, Networking, and Systems, Academic Press, 2011.
[41] Y. Wang, A. Reibmann, S. Lin, Multiple description coding for video communications, Proc. IEEE 93 (1) (2005) 57–70.
[42] D. Wang, N. Canagarajah, D. Bull, Slice group based multiple description video coding with three motion compensation loops, IEEE Intl. Symp. Circuits Syst. 2 (2005) 960–963.
[43] M. Tesanovic, D. Bull, A. Doufexi, A. Nix, Enhanced MIMO wireless video communication using multiple-description coding, Signal Process. Image Commun. 23 (4) (2008) 325–336.
[44] A. Bentaleb, et al., A Survey on Bitrate Adaptation Schemes for Streaming Media over HTTP, IEEE Communication Surveys & Tutorials, vol. 21, 2019, pp. 562–585, First Quarter.
[45] Sodagar, I., MPEG-DASH: the Standard for Multimedia Streaming over Internet, ISO/IEC JTC1/SC29/WG11 W13533, 2012.
[46] A. Balachandran, et al., Developing a predictive model of quality of experience for Internet video, ACM SIGCOMM Comput. Commun. Rev. 43 (4) (2013) 339–350.

CHAPTER

12 Video coding standards and formats

The process of standardizing video formats and compression methods has been a major influence on the universal adoption of video technology. Standards are essential for interoperability, enabling material from different sources to be processed and transmitted over a wide range of networks or stored on a wide range of devices. This interoperability provides the widest possible range of services for users. It also reduces risk for manufacturers, stimulates investment in research and development, and has created an enormous market for video equipment, with the advantages of volume manufacturing.

Each generation of video coding standard, every 7–10 years from the introduction of H.120 in 1984 through to the most recent H.265/HEVC and H.266/VVC codecs, has consistently halved the bit rate required for equivalent video quality. This chapter overviews the features that have enabled this progress to be made. It is not intended to be a definitive reference, but rather a description that enables the reader to understand how the architectures, approaches, and algorithms described in previous chapters are employed in standardized codecs that are in common use today.

12.1 The need for and role of standards

A chronology of video coding standards is represented in Fig. 12.1 (repeated for convenience from Chapter 1). Their primary features are described in the following sections. Let us first, however, revisit why we need standards and consider the process that has led to a succession of successful coding standards over the last 30 years or so.

12.1.1 The focus of video standardization

In order for a video coding standard to be successful it must clearly satisfy a demand. In particular, it must be superior in performance to any previous standards, it must not stifle competition between producers of products (i.e., it must allow innovation), and of course, it must provide interoperability through independence from specific communication networks or terminal equipment. In most cases it is advantageous if a standard is backward compatible – i.e., it should be able to decode bitstreams from prior standards. Any standardization process must therefore be mindful that future standards will need to be backward compatible with it.

Intelligent Image and Video Compression
https://doi.org/10.1016/B978-0-12-820353-8.00021-9

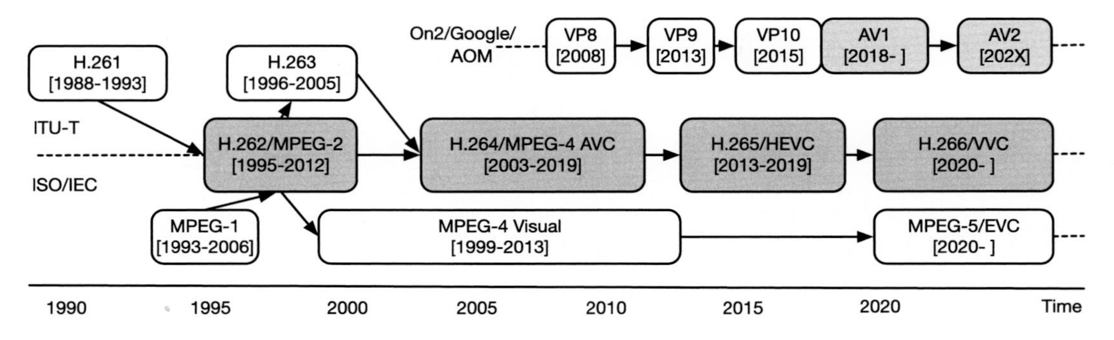

FIGURE 12.1 A chronology of video coding standards from 1984 to the present day.

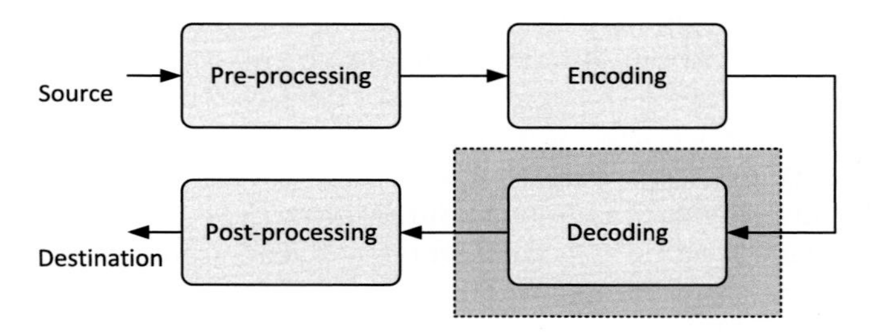

FIGURE 12.2 The scope of standardization.

Video coding standards conventionally define the bitstream format and syntax, and the decoding process, but not (for the most part) the encoding process. This is illustrated in Fig. 12.2, where the dashed box indicates the normative aspects of the standard. A standard-compliant encoder is thus one that produces a compliant bitstream and a standard-compliant decoder is one that can decode a compliant bitstream. In this context it is important to highlight the fact that the standard compliance of an encoder provides no guarantee of quality; the challenge remains for manufacturers to differentiate their products through innovative low-complexity and efficient coding solutions.

12.1.2 The standardization process

Early video coding standards were produced independently, either by the International Telecommunication Union (ITU) (previously CCITT) or by the International Standards Organization (ISO).[1] Most recent standards have however benefited from a joint approach between these two organizations.

The standardization process [4] follows a well-defined path, with the following stages:

Requirements definition: Here the scope of the standard is finalized and the requirements and goals of the standardization process are defined.

[1] Specifically the joint ISO/International Electrotechnical Commission (IEC) Technical Committee 1 (JTC 1), which addresses all computer-related activities including video compression.

Divergence: A formal call for proposals is then normally issued, enabling experts from industry and academia to present and compare their methods and results, usually against an existing standard using a predefined set of test data and conditions.

Convergence: The best algorithms contributed during the divergence phase are selected. This is normally achieved through the use of an evolving test model such as the HEVC Test Model (HM) or the VVC Test Model (VTM). This model evolves through a number of iterations using the current version as a reference for subsequent proposals in different key technical areas (KTAs, for HM) or for various core experiments (CEs, for VTM). These, if successful, are incorporated in a new improved version of the model. This process continues until the required performance specifications are met.

Verification: During verification, the resulting standard and its bitstream are validated for conformance. Conformance testing processes are then defined for compliance testing of products.

12.1.3 Intellectual property and licensing

When there is mass investment in producing products according to an international standard, the manufacturers and users rightly expect some degree of protection against patent infringement litigation. For most standardized codecs, vendors and users are required to pay royalties to the owners of associated intellectual property. This is, in the main, handled by a US organization (not affiliated with MPEG) called the MPEG Licensing Authority. The MPEG Licensing Authority administers patent licenses in connection with the patent pools for all recent standards.

Between 2005 and 2007, a dispute between two major organizations on the infringement of H.264 patents came before the US Court. The associated patents were judged to be unenforceable as they had not been disclosed to MPEG JVT prior to the H.264 standardization in 2003. This type of ruling goes some way to provide confidence in the robustness of the licensing procedures adopted for video standardization.

12.2 H.120

12.2.1 Brief history

The study group SG.XV of the CCITT commenced work on H.120 in 1980 and produced the first international digital video coding standard in 1984, followed by a second version in 1988 [5]. H.120 addressed videoconferencing applications at 2.048 Mb/s and 1.544 Mb/s for 625/50 and 525/60 TV systems, respectively. This standard was never a commercial success, partially because it was based on different coding strategies for different international regions, but mainly because its picture quality (especially temporal quality) was inadequate.

12.2.2 Primary features

H.120 implemented a conditional replenishment strategy, whereby each frame was divided into changed and unchanged regions. The changed regions were coded with intrafield differential pulse code modulation (DPCM) in parts 1 and 2 of the standard, although part 3 (for use in the USA) employed background prediction and motion-compensated interfield prediction.

12.3 H.261

12.3.1 Brief history

H.261 [6] followed H.120 in 1989 and was the first video codec that achieved widespread product adoption. It was based on a p×64 kbps (p= 1, ..., 30) model, targeted at ISDN conferencing applications. H.261 was the first block-based hybrid compression algorithm to use a combination of transformation (the discrete cosine transform [DCT]), temporal DPCM, and motion compensation. This architecture has stood the test of time as all major video coding standards since have been based on it.

12.3.2 Picture types and primary features

Macroblock, GOB, and frame format

H.261 introduced a basic hierarchy into the picture coding process that, with some modification, is still in use today. H.261 pictures, in CIF or QCIF format, are represented as a sequence of groups of blocks (GOBs), each comprising a number of macroblocks (MBs). A macroblock, as described in Chapter 4, comprises four spatial 8×8 DCT transformed luma blocks and two (subsampled) spatial 16×16 DCT transformed chroma blocks. A diagram showing the picture hierarchy is given in Fig. 12.3 for the case of a CIF frame. As can be seen, there are 33 MBs in a GOB, giving a GOB dimension of 176×48 pixels, and there are 12 GOBs in a CIF frame, giving a frame dimension of 352×288 pixels.

H.261 supports only I- and P-frames – B-frames are not supported. Motion estimation and compensation are an integral part of the coding process with a fixed block size of 16×16 and a search window of $[-15 \cdots +15]$ pixels.

Coder control

H.261 offered some flexibility in allowing switching between inter- and intraframe modes and controlling quantizer step sizes at the macroblock level. Advanced features, such as a spatial loop filter, were incorporated for removing high-frequency noise (a 3-tap filter with coefficients 0.25, 0.5, and 0.25). This allowed a reduction of prediction error by smoothing the pixels in the reference and output frames. For error resilience, an optional BCH error detection correction (511,493) scheme was also incorporated. The primary features of H.261 are summarized in Table 12.1.

12.4 MPEG-2/DVB

12.4.1 Brief history

In 1988, the Moving Picture Experts Group (MPEG) was founded, delivering, in 1992, a video coding algorithm (MPEG-1) intended for digital storage media at 1.5 Mbs/s. This

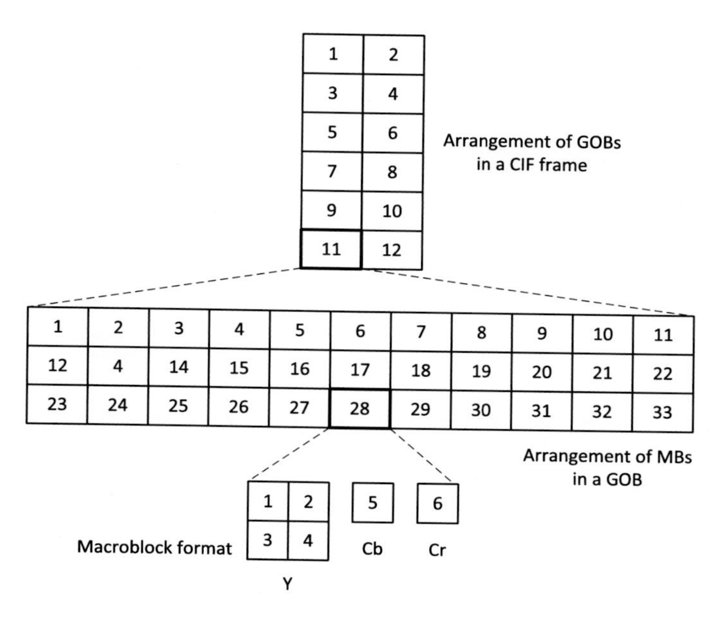

FIGURE 12.3 H.261 macroblock, GOB, and CIF frame format.

TABLE 12.1 Primary features of H.261.

Codec feature	Approach
Formats supported	CIF, QCIF
Picture format	Four layers (picture, GOB, MB, and block)
Color subsampling	4:2:0 YC_bC_r
Frame types	I, P
Intra-coding transform	Block DCT
Inter-coding transform	Block DCT
Entropy coding	VLC and Huffman coding
Quantizer	Uniform (DC) and deadzone (AC)
Motion compensation	Optional $[-15 \cdots +15]$
Coding control	Selection of inter/intra and quantizer step size
Loop filter	3-tap spatial filter
Error protection	Optional BCH (511,493) coding

was followed in 1994 by MPEG-2 [7], which specifically targeted the emerging digital video broadcasting market. MPEG-2 was instrumental, through its inclusion in all set-top boxes for more than a decade, in truly underpinning the digital broadcasting revolution.

Because of its focus on broadcasting, where there is an accepted imbalance between encoder and decoder complexity, this led to significant investment in high-cost, high-complexity encoding technology to produce high-performance studio and head-end-based encoders. An example of how the performance of MPEG-2 professional encoders improved over a decade of development is shown in Fig. 12.4, together with the innovations that led to a series of step changes in that performance.

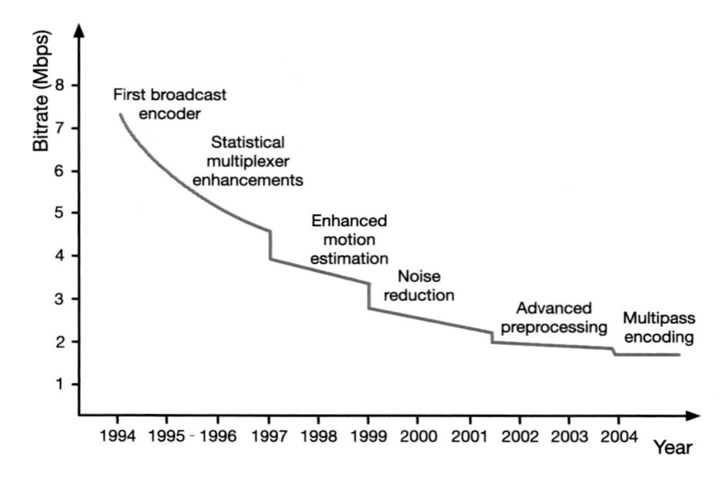

FIGURE 12.4 MPEG-2 encoder performance improvements between 1994 and 2004 (adapted from an original presented by Tandberg).

12.4.2 Picture types and primary features

MPEG-2 (ISO/IEC-13818:2000) [7], also known as H.262, is a generic audiovisual coding standard supporting a range of applications at bit rates from about 2 to 30 Mbps. The standard comprises four main parts: 13818-1: systems, −2: video, −3: audio, and −4: conformance. It employs a hybrid motion-compensated block-based DCT architecture (similar to H.261), using 8×8 blocks and 16×16 macroblocks with translational block-based motion estimation to half-pixel accuracy. Because of its focus on digital TV broadcasting, MPEG-2 necessarily supports both progressive and interlaced picture formats. The coding standard was defined primarily for error-free environments and channel coding was added according to the application scenario (e.g., terrestrial, cable, or satellite broadcasting). MPEG-2 was defined to support a broad range of applications from studio processing (capture and editing) to distribution and broadcast delivery. It is intimately coupled with the DVB-T, DVB-S, and DVB-C broadcast content delivery standards.

MPEG-2 supports three picture types: intra (I), coded without reference to other frames (least efficient); predicted (P), coded based on prediction from one previous I- or P-frame; and bidirectionally predicted (B), predicted from P- and/or I-frames but not used as a basis for further predictions (most efficient). It utilizes a group of pictures (GOP) structure as illustrated in Fig. 12.5 and was the first standard to introduce bidirectionally predicted B-frames.

12.4.3 MPEG-2 profiles and levels

To match performance against decoder capability or capacity, MPEG-2 introduced an extensive range of profiles and levels. A profile is a defined subset of the entire bitstream syntax and profiles are further partitioned into levels. Each level specifies a range of allowable values for the parameters in the bitstream.

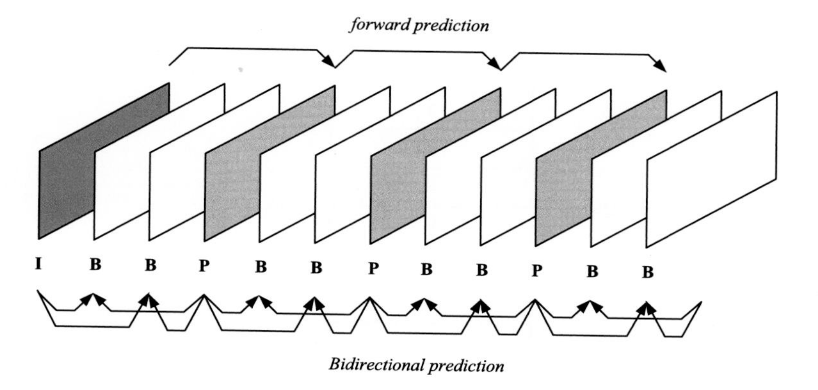

FIGURE 12.5 MPEG-2 GOP structure.

TABLE 12.2 MPEG-2 main profile @main level (MP@ML) bounds.

Parameter	Bound
samples/line	720
lines/frame	576
frames/sec	30
samples/sec	10,368,000
bit rate	15 Mb/s
buffer size	1,835,008 bits
chroma format	4:2:0
aspect ratio	4:3, 16:9 square pixels

MPEG-2 supports six profiles – Simple, Main, SNR, Spatial, High 4:2:2, and Multiview. Provisions for scalability are included in the SNR, Spatial, and High profiles, whereas the Simple and Main profiles allow only single-layer coding. It also offers four possible levels – Low, Main, High1440, and High – in each profile. The parameters for the Main level approximately correspond to normal TV resolution, the Low level corresponds to CIF resolution, and the values for High1440 and High correspond to HDTV resolution. Only two profiles were ever used in practice; the 4:2:2 profile for studio work and postproduction and the Main profile for broadcast TV delivery. Table 12.2 provides the specification for the MPEG-2 Main profile at Main level (MP@ML) as used in most delivery applications.

12.5 H.263

12.5.1 Brief history

H.263 [8] was defined by ITU-T SG15, starting in 1993 with the goal of coding at bit rates below 64 kbps. The initial application focus was on PSTN and early mobile radio applica-

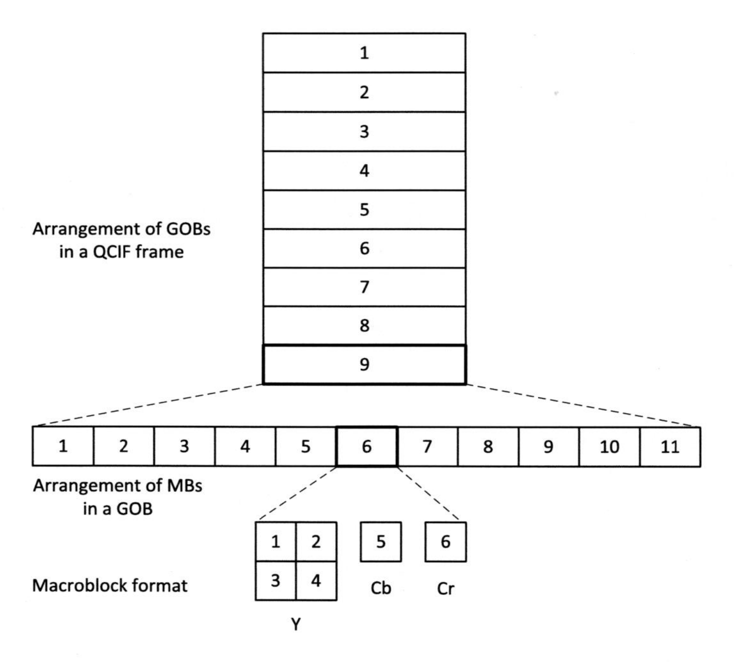

FIGURE 12.6 H.263 picture format (QCIF).

tions at bit rates between 10 and 24 kb/s. Despite mobile video being slower to take off than expected, H.263 nonetheless had a significant impact in conferencing and surveillance applications, as well as in early internet streaming. In particular, H.263 was used to encode Flash Video content for sites such as YouTube and MySpace. RealVideo was initially based on it, and it was specified in several ETSI 3GPP video services. The codec was first designed to be integrated in the H.324 framework for circuit switched applications, but has been extensively used in H.323 (RTP-based video conferencing) and other IP streaming wrappers.

12.5.2 Picture types and primary features

Macroblock, GOB, and frame format

H.263 supports the following picture formats: sub-QCIF (88 × 72), QCIF (176 × 144), CIF (352 × 288), 4CIF (704 × 576), and 16CIF (1408 × 1152), all in YC_bC_r 4:2:0 format. As with H.261, each picture is divided into a number of groups of blocks (GOBs) defined as an integer number, k, of rows of macroblocks (MBs). For SQCIF, QCIF, and CIF, k=1; for 4CIF, k=2; and for 16CIF, k=4. The structure of an H.263 QCIF picture is shown in Fig. 12.6.

Primary features of H.263

Despite using a similar coding architecture to H.261 and MPEG-2, H.263 achieved significantly enhanced performance through the incorporation of a comprehensive set of advanced features. These are summarized in Table 12.3. The standard is based on a hybrid motion-compensated DCT, using zig-zag scanning with quantization step sizes (QP=$1 \cdots 31$) changeable at each MB. 3D VLC tables are used in H.263, incorporating the events {RUN, LEVEL, LAST}, where LAST signals the last nonzero coefficient in the block. This avoids the need for an explicit end of block (EOB) symbol and provides a compact representation of the 8×8 DCT block. H.263 uses motion compensation with half-pixel precision and motion vectors are coded predictively, as shown in Fig. 12.7.

In terms of PSNR, H.263 can provide up to 3–4 dB improvement over H.261. It does this because of its half-pixel motion prediction but also by incorporating four new optional coding modes:

- **Unrestricted motion vectors**: Motion vectors are allowed to point outside of the reference picture area and the search range is extended to $[-31.5, +31.5]$.
- **Syntax-based arithmetic coding**: The conventional Huffman entropy coder is replaced with an arithmetic coder, enabling fractional word lengths per encoded symbol.
- **Advanced prediction mode**: This includes two submodes:
  - The use of 4 motion vectors per block, i.e., one motion vector for each 8×8 block rather than one per 16×16 block. This enables the encoder to better deal with multiple motions within a block.
  - Overlapped motion compensation (OMC): Here, each pixel in an 8×8 luma block is predicted as a weighted sum of three prediction values. These prediction values correspond to the vector of the current block and two of four predictions from the blocks to the top, bottom, left, and right of the current block. The two blocks selected are those closest to the pixel being computed – e.g., a pixel in the top left quadrant of the block uses the predictions from above and left. This is illustrated in Fig. 12.7 (right).
- **PB-frames mode**: In a PB-frame, two pictures are coded as a single unit – a P-frame predicted from the previous P-frame and a B-frame predicted from both adjacent P-frames using bidirectional prediction. A PB-frame macroblock comprises 12 blocks – 6 for the predicted P-block and 6 for the B-block. The benefit of a PB-frame is that motion vectors are not transmitted for the B-blocks, but are instead derived using a scaled version of that for the P-block based on the local temporal activity relative to the corresponding P-block in the previous P-frame. This process enables an effective increase in frame rate without a significant increase in bit rate.

H.263 extensions (H.263+ and H.263++)

H.263 Version 2, also known as H.263+, was standardized in 1998 and extends H.263 with many new modes and features that further improve compression efficiency. H.263++ (Version 3) provided still further enhancements in 2000. The 12 new modes in H.263+ improve coding gain, improve error resilience, enable scalable bitstreams, introduce flexibility in picture size and clock frequency, and provide supplemental display capabilities. Many of these new modes were pulled through into the later H.264/AVC standard.

Firstly, several new modes to support error resilience were introduced and these are summarized below. We have covered many of these already in Chapter 11 and they are also reviewed by Wenger et al. in [9].

TABLE 12.3 Primary features of H.263.

Codec feature	Approach
Formats supported	SQCIF, QCIF, CIF, 4CIF, 16CIF
Picture format	Four layers (picture, GOB, MB, and block)
Color subsampling	4:2:0 YC_bC_r
Frame types	I, P, B, (PB)
Intra-coding transform	Block 8 × 8 DCT
Inter-coding transform	Block 8 × 8 DCT
Quantizer	Uniform (DC) and deadzone (AC) QP=1···31
Motion compensation	Half-pixel, unrestricted vectors, overlapped
Entropy coding	3D VLC, Huffman and arithmetic coding
Loop filter	None
Coding control	Inter/intra, quantizer step size, and Transmit/Skip at MB level
Error protection	Provided by transport layer or BCH-FEC

- **Slice-Structured Mode (Annex K)**: In this mode, the GOB structure is replaced by a set of slices. All macroblocks in one slice can be decoded independently since prediction dependencies are not permitted across slice boundaries.
- **Independent Segment Decoding Mode (Annex R)**: A segment boundary acts in the same way as a picture boundary. A segment can be a slice, a GOB, or a collection of GOBs, and the shape of a segment must be identical from frame to frame. Independent segments support error resilience as error propagation is eliminated between defined parts of the picture. They also enable special effects in a similar manner to the object planes in MPEG-4.
- **Reference Picture Selection Mode (Annex N)**: As described in Chapter 11, this allows flexibility in the choice of reference picture. It is also possible to apply the reference picture selection mode to individual segments rather than to full pictures. This mode supports error resilience if a feedback channel exists, but is also a precursor to the multiple reference frame methods used in H.264 and HEVC.
- **Temporal, SNR, and Spatial Scalability Mode (Annex O)**: This mode introduced layering into the H.263 standard to support flexibility in delivery to terminals of different capabilities and in congestion management. As discussed in Chapter 11, a scalable bitstream comprises a base layer and enhancement layers, where the base layer provides an acceptable level of quality which can be further enhanced by the other layers (in terms of improved signal to noise ratio [SNR], improved temporal resolution, or improved spatial resolution) if they are available.

Other H.263 modes that support enhanced coding gain are:

- Advanced Intra-coding Mode (Annex I),
- Modified Unrestricted Motion Vectors Mode (Annex D),
- Improved PB-frames Mode (Annex M),
- Deblocking Filter Mode (Annex J),
- Reference Picture Resampling Mode (Annex P),
- Reduced-Resolution Update Mode (Annex Q),
- Alternative Inter VLC Mode (Annex S),
- Modified Quantization Mode (Annex T).

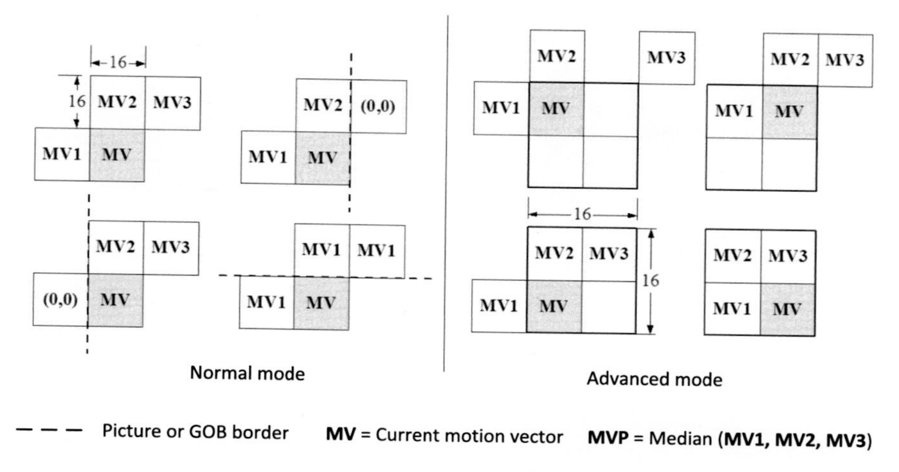

FIGURE 12.7 Motion vector coding in H.263.

Finally one mode was introduced to provide additional bitstream information:

- **Supplemental Enhancement Information Mode (Annex L)**: This enables the incorporation of additional information that may or may not be decodable by a specific decoder, such as chroma keying, time segments, picture freezing, and picture resizing.

H.263++ provided further enhancements to H.263 in terms of introducing a range of nine profiles and seven levels similar to those in MPEG-2. The profiles enabled formalization of specific combinations of modes to support different application scenarios (e.g., Baseline and Wireless). The levels enabled matching of coder performance to environmental constraints (e.g., maximum bit rate, spatial resolution, etc.). It also introduced some new modes and revised others – a Data Partitioned Slice Mode (Annex V), Additional Supplemental Enhancement Information (Annex W), and an Enhanced Reference Picture Selection Mode (Annex U).

12.6 MPEG-4

12.6.1 Brief history

MPEG-4 [10] was a very ambitious project that sought to introduce new approaches using object-based as well as waveform-based methods. The MPEG-4 standard was ratified in 2000, as ISO/IEC 14496, potentially offering significant encoding improvements over and above MPEG-2. It was however, in general, found to be too complex. It is interesting to compare the standards documents for H.261 and MPEG-4. The former was 25 pages long while MPEG-4 comprised 500 pages for the visual part alone. Only the Advanced Simple Profile (ASP) was used in practice, and this gained some traction forming the basis for the emerging digital camera technology of the time. The early MPEG-4 part 2 offered little practical advantage

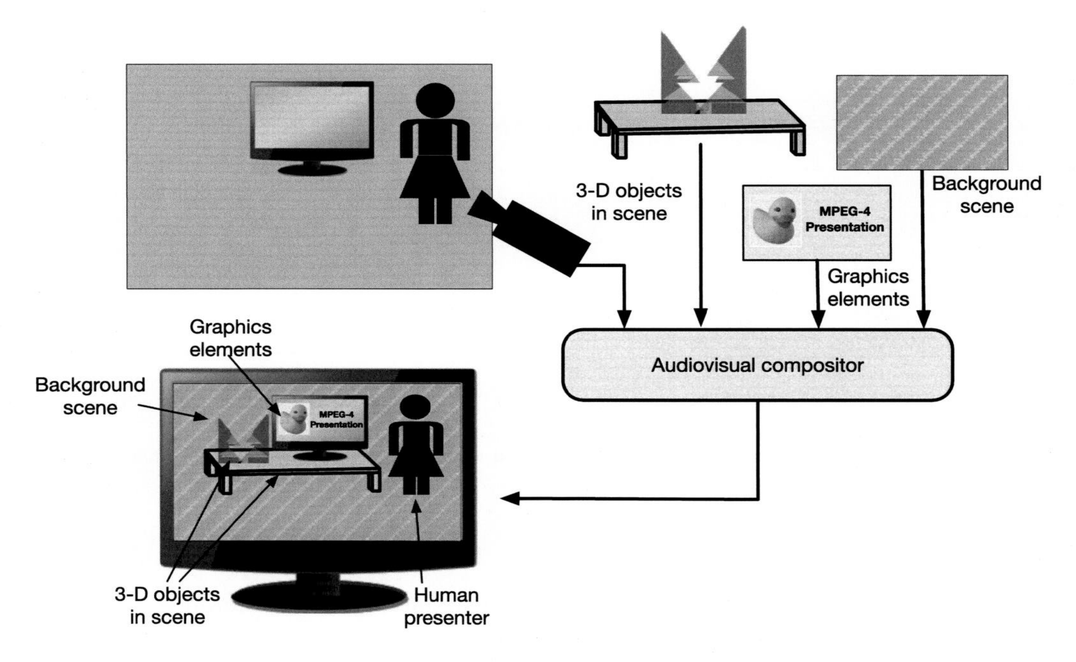

FIGURE 12.8 An example of an MPEG-4 audiovisual scene.

over MPEG-2 or H.263 and was rapidly overtaken by MPEG-4 part 10/AVC, now known as H.264/AVC.

12.6.2 Picture types and primary features

MPEG-4 was aimed at encoding video at lower bit rates and higher video qualities than MPEG-2, effectively creating a multimedia transmission and storage framework. MPEG-4 (part 2) was designed to be suitable for a wide range of video encoding scenarios ranging from the studio and movies to video applications on mobile phones. In this context it focused on object-based as well as waveform-based coding.

MPEG-4 provided a set of coding tools for audiovisual scenes, supporting the coding of arbitrarily shaped objects, as shown in Fig. 12.8. The scene in this figure represents a 2D background, a video playing on the screen, a presenter, and 3D objects such as the table and the tree. MPEG-4 was aimed at the independent processing of such audiovisual objects as well as their compositing and allowed modification of both natural and synthetic (computer-generated) content.

As with MPEG-2 and H.263++, MPEG-4 initially defined a number of conformance points using a Simple profile, a Core profile, and a Main profile. The Simple and Core profiles address scene sizes of QCIF and CIF at 64, 128, and 384 kbps and 2 Mbps. The Main profile is targeted at CIF, ITU-R 601, and HD sizes, with bit rates at 2, 15, and 38.4 Mbps. The standard also incorporated support for sprites, scalability, and error resilience.

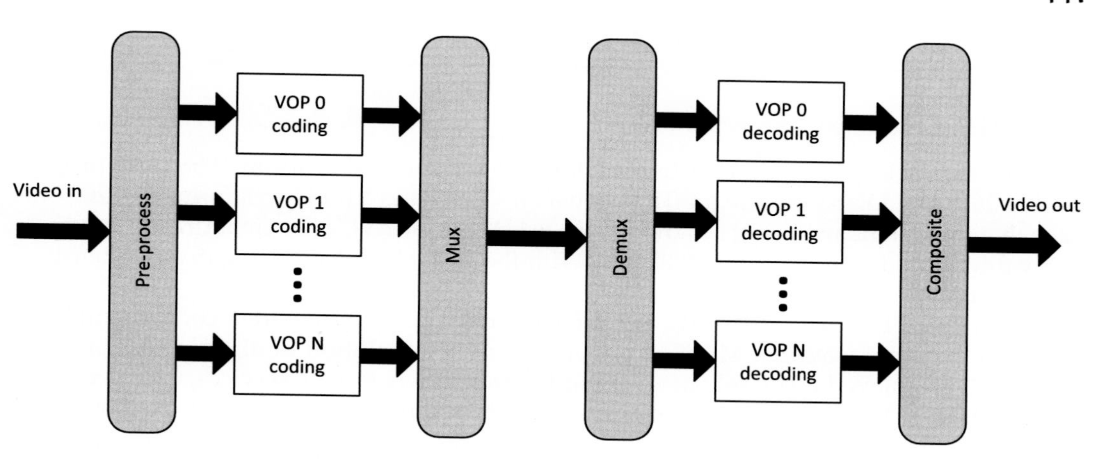

FIGURE 12.9 Generic representation of MPEG-4 video object plane (VOP) coding.

MPEG-4 also provided, for the first time, support for integration with the MPEG-7 content description scheme to facilitate the use of metadata. An excellent overview of MPEG-4 natural video coding is provided by Ebrahimi and Horne in [11].

Coding framework

Although we will not devote a lot of time to describing the details of MPEG-4, it is worth briefly putting its aims into context.

MPEG-4 supported the composition or decomposition of video content into visual objects, allowing these to be processed independently as a number of video object planes (VOPs) (Fig. 12.9). In order to describe these object planes it was necessary to code their textures, shapes, and motions. Shape coding is done using a binary mask, or a gray-scale alpha channel, allowing transparency. Both motion compensation and DCT-based texture coding have to take account of object boundaries. Error resilience is provided by resynchronization markers, data partitioning, header extension codes, and reversible variable-length codes. Scalability is provided for both spatial and temporal resolution enhancement. MPEG-4 also provided scalability on an object basis, with the restriction that the object shape has to be rectangular.

MPEG-4 part 2 advanced simple profile (ASP)

The Advanced Simple profile was the most commonly used MPEG-4 profile, offering support for interlaced video, B-pictures, quarter-pixel motion compensation, and global motion compensation. The quarter-pixel accuracy was subsequently adopted in H.264/AVC whereas the global motion compensation feature was not generally supported in most implementations. As the reader will notice, ASP does not offer a lot that is not present in H.263; in fact the uptake of H.263 was far more widespread than that of ASP.

Soon after the introduction of ASP, MPEG-4 visual work refocused on part 10 and this became a joint activity with ITU-T under the H.264/AVC banner. This major step forward is the topic of the next section.

12.7 H.264/AVC

12.7.1 Brief history

ITU-T/SG16/Q6 (Video Coding Experts Group) commenced work in 1998 on a project called H.26L. The Joint Video Team (JVT) between VCEG and MPEG was formed in 2001 to establish a joint standard project known as H.264/MPEG-4 AVC. The final draft for formal approval was produced in March 2003 [12], with the scalable video coding (SVC) extension finalized in 2007.

MPEG-4 (part 10), or H.264/AVC, is by far the most ubiquitous video coding standard to date. In the same way that MPEG-2 underpinned the revolution in digital broadcasting, H.264/AVC has played a key role in enabling Internet video, mobile services, OTT services, IPTV, and HDTV. H.264/AVC is a mandatory format for Blu-ray players and is used by most internet streaming sites, including Vimeo, YouTube, and iTunes. It is used in Adobe Flash Player and Microsoft Silverlight and it has also been adopted for HDTV cable, satellite, and terrestrial broadcasting.

ITU-T, in partnership with ISO/IEC, delivered H.264/AVC in 2004. We have already examined many of the features that contribute to the success and performance of H.264 in previous chapters and we summarize these below.

12.7.2 Primary features

The aim of the H.264/AVC project was to provide:

- **Improved coding efficiency**: A targeted average bit rate reduction of 50% given fixed fidelity, compared to any other video standard.
- **Error robustness**: Tools to deal with packet loss and congestion management.
- **Network friendliness**: Major targets were mobile networks and the Internet – separation of coding layers through network abstraction.
- **Adaptation to delay constraints**: To provide low delay modes.
- **Simple syntax specification**: Avoiding an excessive quantity of optional features or profile configurations.

These are all sensible objectives, but the last one is interesting inasmuch as it acknowledges the overcomplex nature of its predecessor.

Let us now consider some of these attributes in more detail. A summary is provided in Table 12.4; although not all the details of a complex codec such as H.264 can be captured in a table of this type, it can be seen that many of the features are similar to those in the extensions to H.263. For further details, the reader is referred to the standards documents [12], or overviews such as that by Wiegand et al. [13]. A very accessible introduction to the standard is provided by Richardson in [14].

12.7.3 Network abstraction and bitstream syntax

Fig. 12.10 shows the layered structure of H.264/AVC coding. This ensures that the coding layers are independent of the network, relying on protocols as shown to provide mapping to

TABLE 12.4 Primary features of H.264/AVC.

Codec feature	Approach
Formats supported	SQCIF to UHDTV (4k)
Picture format	Four layers (picture, GOB, MB, and block)
Color subsampling	4:0:0, 4:2:0, 4:2:2, 4:4:4 YC_bC_r
Partitions	Flexible slice structuring
Frame types	I, P, B, SP, SI
Intra-coding transform	Block 8×8 or 4×4 integer transform
Inter-coding transform	Block 8×8 or 4×4 integer transform
Intra-prediction	Multidirection, multipattern
Quantizer	Scalar with logarithmic control, QP=$0 \cdots 51$
Motion compensation	Quarter-pixel, unrestricted vectors, multiple reference frames
ME block size	16×16, 16×8, 8×16, 8×8, 8×4, 4×8, 4×4
Entropy coding	CAVLC or CABAC
Coding control	Many sophisticated coding control modes – RDO
Loop filter	In loop nonlinear deblocking filter
Error resilience	Slices, FMO, SI, SP frames, multiple reference frames

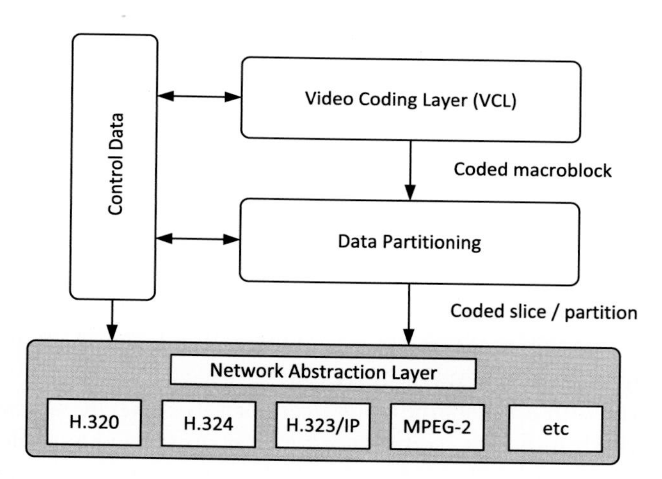

FIGURE 12.10 H.264/AVC layer structure.

the appropriate video transport mechanism. Video coding layer (VCL) data can thus easily be mapped to transport layers such as RTP/IP for real-time wired or wireless Internet, MP4 for storage and MMS, H.32X for wireline and wireless conversational services, or MPEG-2 systems for broadcasting.

In H.264/AVC, the bitstream is organized hierarchically, comprising the network abstraction layer (NAL) and the VCL. When coded, H.264 information is represented as a series of NAL units (NALUs). The NALU header indicates the type of NALU which may be a sequence parameter set (SPS), a picture parameter set (PPS), VCL coded slice data, or supplemental enhancement information (SEI). The SPS contains information about the whole sequence such

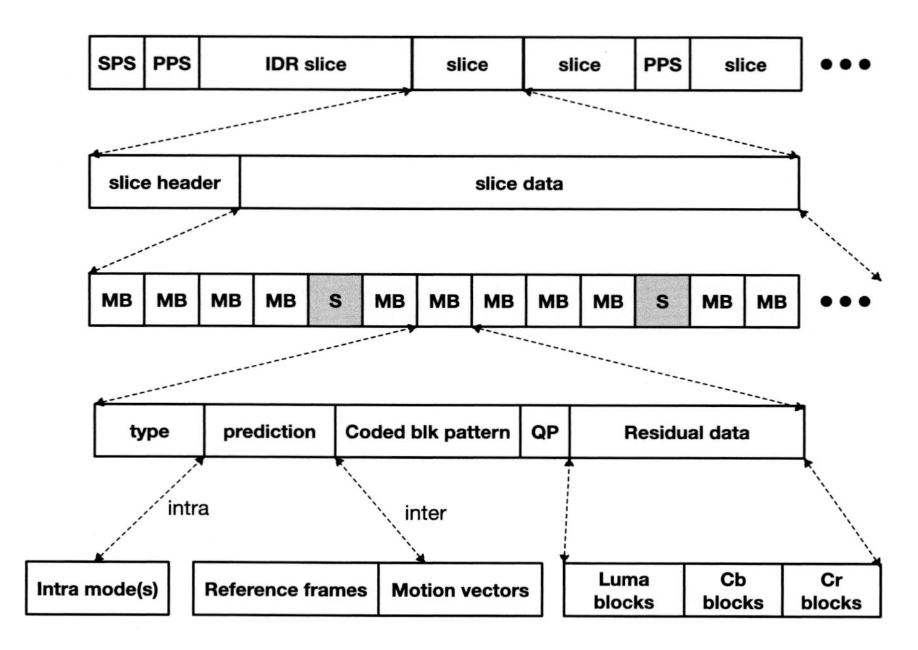

FIGURE 12.11 H.264 syntax.

as profile, level, frame size, and other characteristics important to the decoder. A PPS contains more localized information, relevant to a subset of frames, such as the number of slice groups (SGs), entropy coding mode, whether weighted prediction is used, and other initialization parameters. SEI relates to information that is not essential for decoding but may assist in, for example, buffer management or other nonnormative tasks. A diagram showing the hierarchical H.264 syntax is shown in Fig. 12.11.

Each sequence starts with an instantaneous decoder refresh (IDR) access unit. This is an intra-coded picture which indicates to the decoder that no subsequent picture will require reference to pictures prior to it. The video slice data corresponds to a sequence of coded macroblocks, each of which contains information relevant to decoding plus the actual encoded residual data. The following describes the MB level information, with reference to Fig. 12.11:

- **Type:** I, P, B.
- **Prediction:** I-block mode (16×16, 8×8 or 4×4), P-block partition (16×16, 16×8, 8×16, or 8×8 [contains subblock partitions]), B-block partition (as for P-block), choice of reference frames, and motion vector data. Note: P- and B-blocks may indicate a SKIP condition to signal that no residual data is sent.
- **Coded block pattern:** This indicates for which blocks coefficients are present.
- **Quantization parameter:** QP value for the MB.
- **Residual data:** Coded residual data where indicated by the CBP parameter.

12.7.4 Pictures and partitions

Picture types

H.264/AVC supports the processing of both progressive and interlaced formats in a consistent manner as a single unit. The fact that the frame is progressive or interlaced is indicated in the PPS and adaptive field-frame encoding can be used (either to encode a frame as a single unit or as two fields) but this has no impact on the way the picture is decoded via the decoder buffer.

H.264/AVC supports I-frames, P-frames, and B-frames. B-frames are referred to as generalized B-frames in H.264 in that they can be predicted from one or from two reference frames. H.264/AVC supports a very wide range of picture formats form QCIF to 4K UHDTV.

Slices and slice groups

As discussed in Chapter 11, slices can be arranged into SGs, where each SG can contain one or more slices. The number of slices is signaled by a PPS parameter. In cases where there is more than one SG, the mapping arrangement of slices must be signaled using a value in the range of 0 to 6. As examples, SG type 0 is interleaved, SG type 1 refers to the dispersed allocation, and SG type 2 represents the foreground and leftover case. Depending on the SG map type, additional syntax elements are required to enable the decoder remap the received macroblocks to the slice groups used at the encoder. We introduced the various types of slice used in H.264/AVC in Section 12.7.3, and selected SG patterns are illustrated in Fig. 12.12.

SGs in H.264 support independent decoding, for example as required in multiple description coding and error concealment (both described in Chapter 11). Switching slices are also included in H.264. SI and SP slices direct a decoder to jump to a different point in the stream or to a different stream. These features are useful for video streaming where operations such as fast-forward are used.

Blocks

The macroblock in H.264 is based on a 16×16 sample arrangement, comprising luminance and chrominance residual data. The difference with H.264/AVC is that this structure can be subdivided in various ways according to the available prediction modes. This is discussed in more detail below.

12.7.5 The video coding layer

Intra-coding

- **I_PCM mode:** This is the most basic H.264 mode and switches off all of the normal prediction and transform modes associated with the standard. This does not offer any compression but can be useful in high-quality regimes where the PCM approach may offer better rate-distortion performance than a prediction–transform approach.
- **Intra-prediction:** H.264/AVC has benefited from the introduction of a range of intra-prediction modes. These offer significant benefits when there is a high level of local detail, particularly with oriented features such as edges. H.264 supports intra-prediction for 4×4, 8×8 (High profiles only), and 16×16 blocks. Coding of a given block is performed with

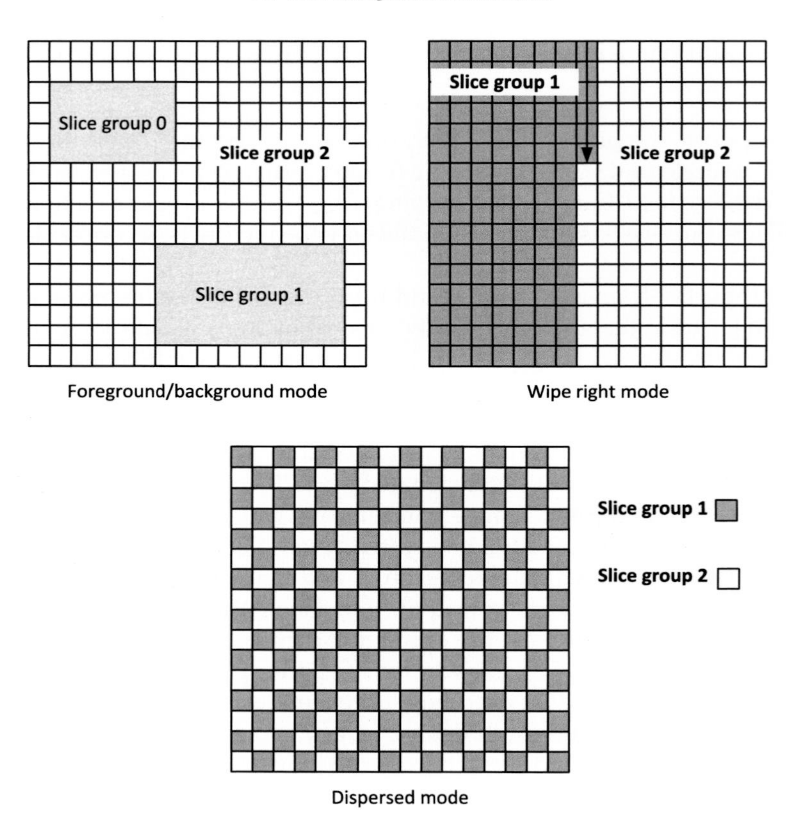

FIGURE 12.12 Example slice structures.

reference to a subset of samples from previously coded blocks that lie to the left of and above the current block. These modes were discussed in detail in Chapter 9.

- **Transforms:** In the case of intra_4 × 4 modes, the residual signal after prediction is transformed using the core 4 × 4 integer transform as described in Chapter 9. For the case of intra_16 × 16 modes, each of the 16 4 × 4 residual blocks is transformed as shown in Fig. 12.13 and each of the 16 DC coefficients are further transformed using a 4 × 4 Hadamard transform. A similar approach is taken for the DC coefficients of the transformed chroma residuals, but this time a 2 × 2 Hadamard transform is applied.

Inter-coding

- **Variable block sizes:** We saw in Chapter 9 how the use of multiple block sizes for predictive coding of frames can provide significant benefits in dealing with complex textures and motions, offering up to 15% bit rate saving over the use of fixed block sizes. The range of block sizes available in H.264/AVC are shown again for convenience in Fig. 12.14 and an example of a typical block decomposition is given in Fig. 12.15. In order to make the most

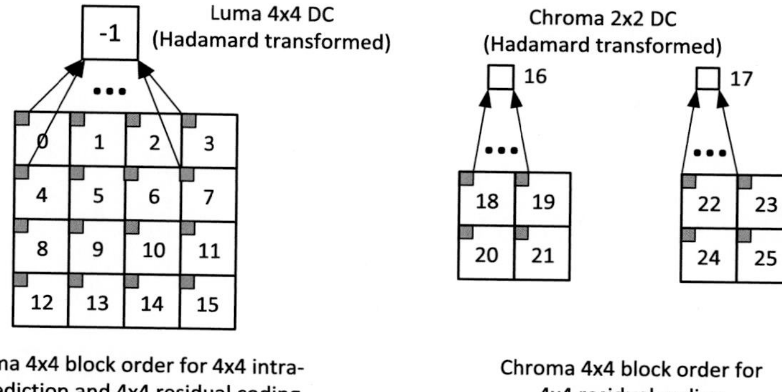

FIGURE 12.13 Hadamard transformation of prediction residuals and coded block ordering.

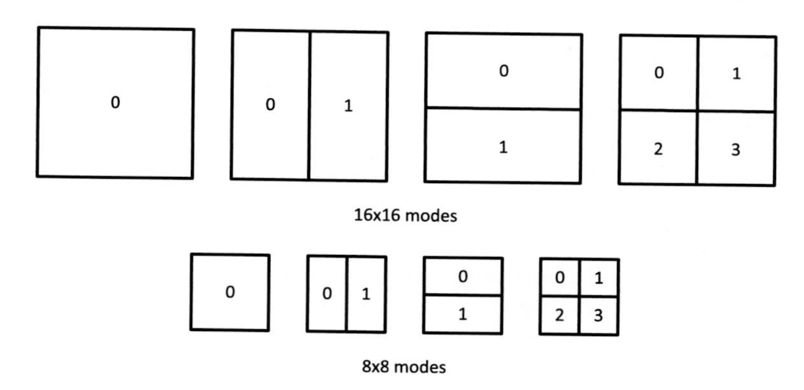

FIGURE 12.14 Variable block sizes supported by H.264/AVC. Top: 16 × 16 modes 1 to 4. Bottom: 8 × 8 modes 1 to 4.

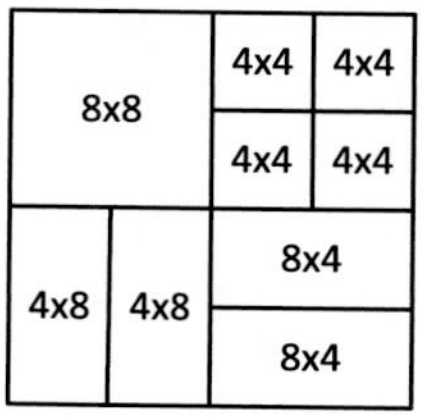

FIGURE 12.15 Example H.264/AVC macroblock partition for inter-coding.

benefit of this flexibility however, significant complexity must be added to the encoding process in terms of rate-distortion optimization (see Chapter 10).

- **Transforms:** H.264/AVC adopts the 4 × 4 DCT-like integer transform (described in Chapters 5 and 9) for residual coding in its Baseline and Main profiles. This is complemented by the use of a 2 × 2 Hadamard transform on the DC values for the case of chroma signals.

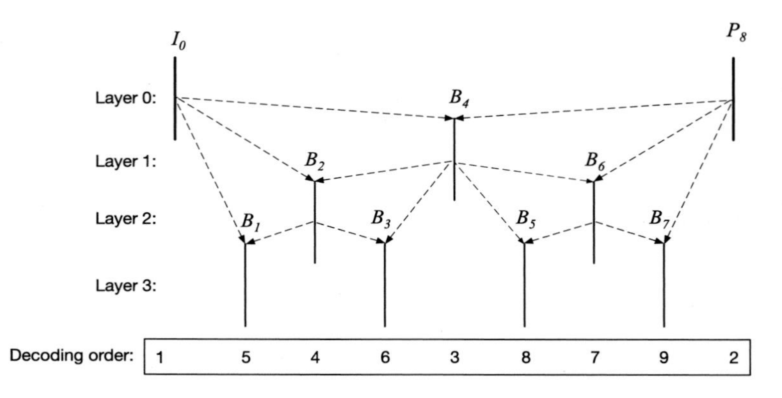

FIGURE 12.16 Hierarchical B-pictures in H.264.

For the High profiles, the standard also includes an 8×8 transform as this is often better suited to content with higher spatial resolutions.

- **Multiple reference frames**: Up to 16 frames or 32 fields (depending on profile) can be buffered in the H.264 reference buffer. The reference buffer does not have to contain the 16 most recent frames; frames can be updated according to their utility, provided that the encoder and decoder remain synchronized. For further details on MRF-ME, the reader is referred back to Chapter 9.
- **Prediction structures:** The H.264 coding regime enables a much more flexible approach to temporal prediction. For example, in the case of low delay, low memory constraints, an IPPP$\cdots$I structure is most suitable, as offered by the baseline profile. If some delay is tolerable, along with the availability of B-frames and multiple reference frames, then improved compression performance can be achieved. An example of this is shown in Fig. 12.16, where a hierarchical prediction structure is imposed. Using this approach it has been demonstrated [15] that, if the quantization parameter (QP) is increased (defined in terms of a QP offset relative to the sequence QP) in a controlled fashion, with layer, then more efficient compression performance can be achieved. Fig. 12.17 illustrates the distribution of bits and PSNR variations when a hierarchical B-frame structure is employed with a QP offset of +1 (per layer). In this case, we can observe significant variations in frame level bit allocation and reconstruction quality are evident. This temporal inconsistency of frame quality can result in the reduction of overall perceived video quality.
- **Weighted prediction**: This is a useful tool, introduced in H.264, that enables the weighting/offsetting of prediction sample values in B-frames. This provides coding gains in the case of accelerating motion or during scene cuts.
- **Quarter-pixel prediction:** H.264 benefits from the availability of subpixel motion compensation to $\frac{1}{4}$-pixel accuracy. This has been estimated to provide up to 20% saving over the case of integer-pixel accuracy. A comprehensive description of the H.264/AVC subpixel motion estimation and compensation process was provided in Chapter 9. H.264/AVC uses a 5-tap filter for $\frac{1}{2}$-pixel interpolation and a simple 2-tap filter for local refinement to $\frac{1}{4}$-pixel accuracy.

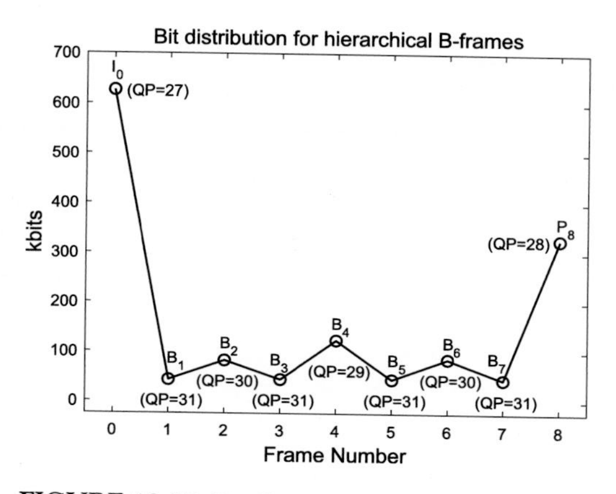

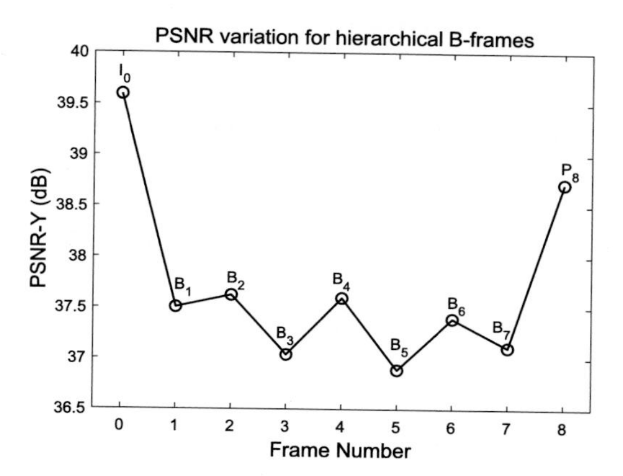

FIGURE 12.17 Bit allocation and PSNR variation for hierarchical B-frames (first nine frames of the *BasketballDrive* sequence using the H.264/AVC Joint Model [JM] 19.0, QP=27, seven hierarchical B-frames with a QP offset of +1).

Deblocking operations

The H.264 deblocking filter (DBF) has been assessed to reduce the coded bit rate for the same subjective quality by up to 10%. It improves subjective visual and objective quality of the decoded picture. This highly content-adaptive nonlinear filter removes blocking artifacts and does not unnecessarily blur the visual content. Its structure and operation were described in Chapter 9.

Variable-length coding

- **Exp-Golomb coding:** This is a highly structured and simple variable-length coding method, used almost universally in H.264 for all symbols apart from transform coefficients.
- **CAVLC:** Context-adaptive VLC (CAVLC) is a relatively low-complexity, but effective method, used as the Baseline and Extended profiles for entropy coding of transform coefficients. Local contexts are used at the encoder to select between different VLC tables according to the local statistics, in this case the number of nonzero coefficients in neighboring blocks.
- **CABAC:** CABAC is a more complex, arithmetic coding-based method that is only supported in the H.264 Main and High profiles. It uses local contexts to adjust its conditioning probabilities. CABAC has been found to reduce bit rate by between 10% and 20% compared to CAVLC, dependent on content type and quantization level.

Further details on all of the above entropy coding methods can be found in Chapter 7.

Coder control

Coder control is a nonnormative part of H.264/AVC. It is however a key element in achieving the optimal performance from the standard, since without good rate-distortion optimization (RDO), many of the advanced features (e.g., block size selections) included will not deliver major savings. The goal of coder control is to select what parts of the video signal should be encoded using which methods and with what parameter settings. Often

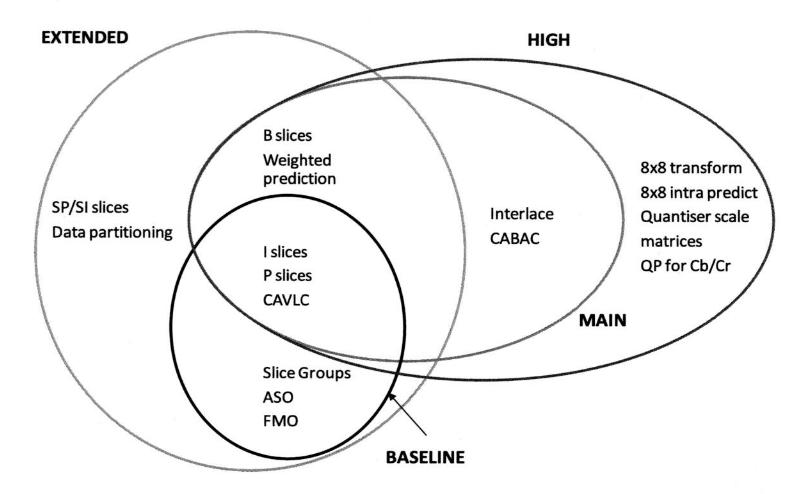

FIGURE 12.18 Selected H.264 profiles.

a Lagrangian optimization approach is adopted where the distortion measure, D, is based on sum of squared difference and the rate, R, includes all bits associated with the decision including header, motion, side information, and transform data. Further details on the RDO methods used in H.264/AVC can be found in Chapter 10.

12.7.6 Profiles and levels

H.264/AVC started with a relatively small number of profiles and levels. The number has however grown considerably since its introduction, to reflect its popularity in a widening range of applications and the demand for higher-quality profiles to deal with emerging formats. The standard now supports 21 profiles ranging from the simplest Constrained Baseline profile (CBP) and Baseline profile (BP) to the High 4:2:2 profile (Hi422P – used for studio work) and the High 4:4:4 Predictive profile (Hi444PP). The Main profile (MP) is typically used for standard-definition (4:2:0) DVB broadcasting, The Extended profile (XP) is used in streaming and the High profile (HiP) is used in Blu-ray players and for DVB HDTV delivery.

H.264 also supports a range of decoder constraints on memory and computational power in terms of its 17 levels, specifying parameters such as decoded bit rate, decoded picture buffer size, picture size, and processing speed in terms of numbers of macroblocks per second. A selection of the H.264 profiles are shown in Fig. 12.18 and a good summary of all profiles and levels is provided in [16].

12.7.7 Performance

The performance of H.264/AVC relative to previous standards is summarized in [17]. An example RD profile is shown in Fig. 12.19 for the case of an entertainment application. This shows approximately 50% saving in bit rate, for the same PSNR, for H.264/AVC over

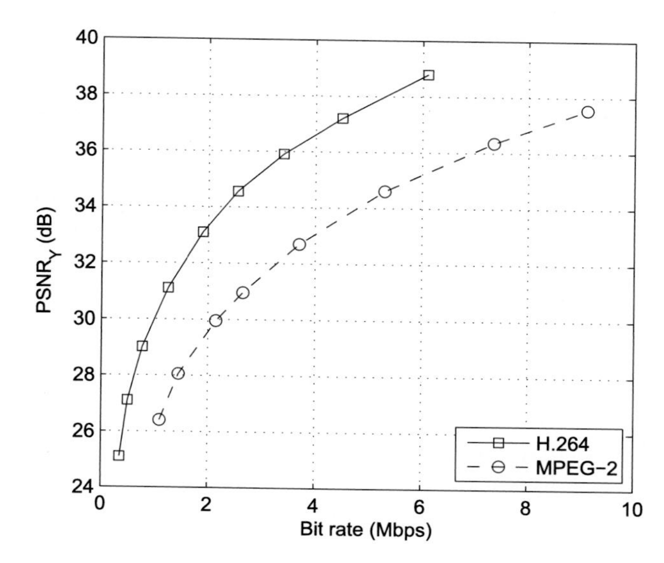

FIGURE 12.19 Typical H.264/AVC (MP) performance relative to MPEG-2 for standard-definition entertainment content.

MPEG-2. A more recent comparison including HEVC is provided by Ohm et al. in [18] (see Table 12.11).

12.7.8 Scalable extensions

A scalable extension to H.264/AVC (SVC) was approved in 2007. As described in Chapter 11, bitstream scalability is useful for providing graceful degradation of performance in the context of dynamic channel conditions. SVC contains five extra profiles, offering a layered approach to scalability. This creates an embedded bitstream providing SNR, spatial, and/or temporal scalability, where packets can be discarded in an orderly fashion to manage congestion or terminal limitations. As can be observed from Fig. 12.16, the use of hierarchical B-frames is a very useful tool in providing temporal scalability. The reader is referred to [19] for more details on SVC.

12.7.9 Multiview extensions

Finally, to cater for the predicted boom in 3D stereoscopic video content and broadcasting, a multiview extension (MVC) was approved in 2009 comprising three additional H.264 profiles. As its name suggests, MVC supports the coding of content acquired from multiple cameras. Its primary focus is stereoscopic (3D) video, but it also supports free-viewpoint and multiview 3D applications, such as television and multiview 3D television. The Stereo High profile is used in 3D Blu-ray devices. For further information on MVC, the reader is referred to [20,22].

12.8 H.265/HEVC

12.8.1 Brief background

HEVC, formally known as ISO/IEC MPEG-H Part 2 and ITU-T H.265, was developed collaboratively between ISO/IEC MPEG and ITU-T VCEG. The standard was approved in April 2013 [23] and offers the potential for up to 50% compression efficiency improvement over AVC and targets solutions for mobile video, broadcast, and streaming services with increased quality demands and more immersive experiences.

HEVC has a specific focus on bit rate reduction for increased video resolutions and on support for parallel processing and ease of integration with appropriate transport mechanisms. An excellent introduction to the range of HEVC features is provided by the papers in a special issue of the IEEE Transactions on Circuits and Systems for Video Technology published in December 2012, in particular the overview by Sullivan et al. [24] and the performance comparisons by Ohm et al. [18]. For full details, the reader should consult the standards documentation [23] and the excellent book by Mathias Wien [27].

12.8.2 Primary features

HEVC reduces bit rate requirements roughly by half compared to its predecessor for the same image quality. It does this using the proven hybrid approach that has served us well since H.261, but with the addition of several new and important features. While some of these add complexity to the codec, many actually offer simplicity compared to H.264/AVC and facilitate parallel processing, compatible with recent hardware developments. HEVC, like previous standards, can trade off computational complexity, compression rate, robustness to errors, and processing delay time, according to application requirements, and this range of operating points is defined by its profile and level structure. A major focus for HEVC is the next generation of HDTV displays and acquisition systems which feature progressively scanned frames, higher frame rates, and resolutions up to UHDTV.

In the following sections, we examine more closely the video coding and syntactic structure of HEVC. The reader will see that a large amount of HEVC technology has been pulled through from H.264/AVC, with some H.264 features omitted and other new and modified features added where appropriate to improve performance. The primary new features in HEVC relate to (i) its block structure, where supermacroblocks, known as coding tree units (CTUs), are used, where the size of a CTU can range up to 64×64 samples; (ii) the way in which these CTUs are partitioned into variable-sized coding, transform, and prediction blocks (PBs) using a quadtree structure; (iii) significantly enhanced intra-prediction, supporting 35 modes; and (iv) the use of sample-adaptive offset (SAO) in conjunction with the DBF in the inter-prediction loop to reduce banding and ringing artifacts. In the HEVC test model (HM) [1], three default coding modes have been defined for different evaluation scenarios: all intra, low delay, and random access, although customized configurations can also be designed by users:

- **All intra mode**: All frames are encoded as intra (I)-pictures.

TABLE 12.5 Primary new coding tools employed in H.265/HEVC.

Codec feature	Approach
Picture partitions	New coding tree units (CTUs) with sizes up to 64 × 64 for the luma component. A quadtree structure used for the partitions. Maximum transform size for luma blocks of 32 × 32.
Intra-prediction	Enhanced intra-prediction modes supporting 35 modes. Improved reference and boundary sample smoothing.
Inter-prediction	Redesigned subpixel interpolation filters. Advanced motion vector prediction (AMVP) mode.
Transform and quantization	Transform block sizes supported from 4 × 4 to 32 × 32. Alternative DST integer transform for encoding intra residuals. Frame and CTU level QP variations. Improved coefficient scanning approaches.
Entropy coding	Enhanced CABAC algorithm for entropy coding.
In-loop filter	New sample-adaptive offset (SAO) approach alongside the deblocking filter.

- **Low delay mode**: All frames are encoded as either I-pictures or inter-predicted frames. The encoding frame order is identical to the temporal order.
- **Random access mode**: A hierarchical structure is employed with 15 consecutive B-pictures in a GOP (this can be configured). Large QP offsets are applied at different layer levels.

The QP configuration and RDO parameter calculations for these three modes have been described in Example 10.2 of Chapter 10. The architecture of the HEVC codec, illustrating many of its features, is shown in Fig. 12.20. These features are summarized in Table 12.5 and described in more detail in the following sections.

12.8.3 Network abstraction and high-level syntax

NAL units

The high-level syntax of HEVC retains a number of the elements used in H.264/AVC to provide network abstraction and the basis for error resilience. NALUs are still employed as the basic interface to the transport mechanism and HEVC supports 63 NALU (VCL and non-VCL) types [23,24].

Slice structures

Slices enable independent decoding and assist with resynchronization and error robustness. The slice structure in HEVC has been largely pulled through from H.264/AVC and the reader is referred to Chapters 9 and 11 for a more detailed description.

Parameter sets

The same parameter sets as in H.264/AVC are employed, supplemented by a new video parameter set (VPS). VPS can include metadata to describe the general characteristics of coded layers, such as could be useful in future scalable extensions. Information in the NALU header also supports the identification of temporal layers.

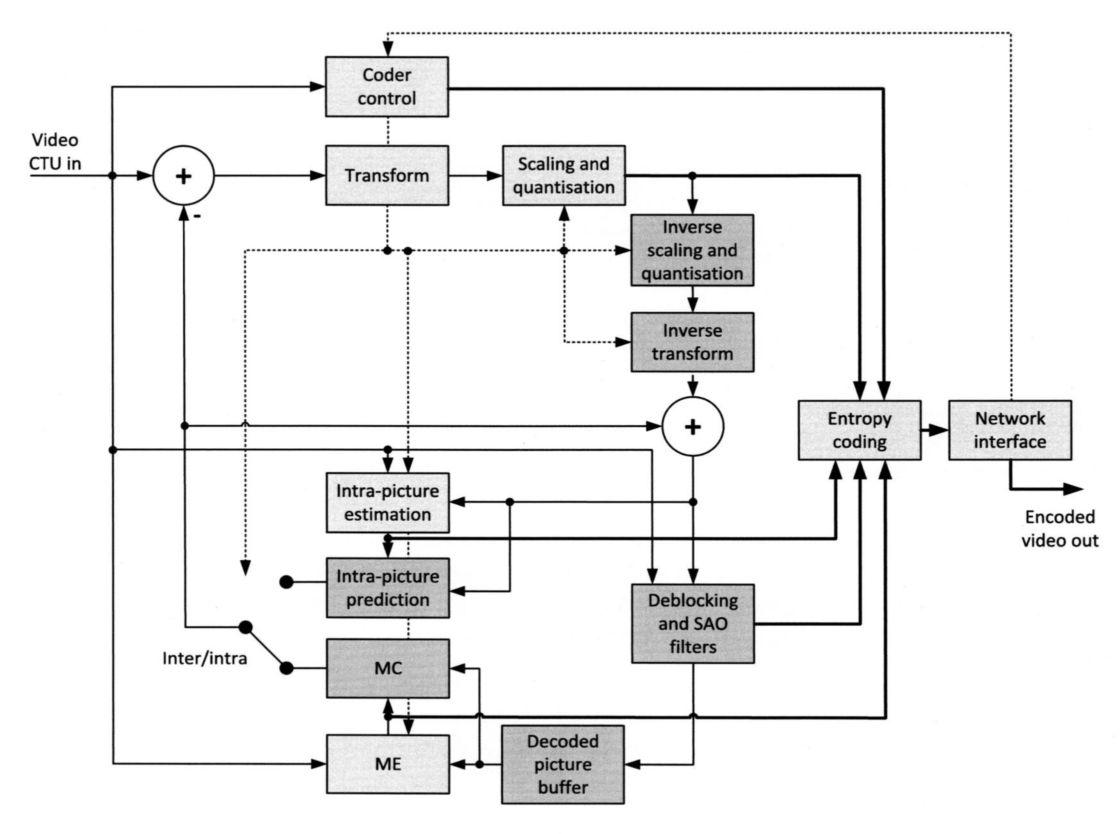

FIGURE 12.20 HEVC video encoder architecture.

Reference picture sets and reference picture lists

It is important to manage the content of the decoded picture buffer (DPB) to ensure synchronization between encoder and decoder. The retained set of pictures is referred to as a reference picture set (RPS). HEVC supports two reference picture lists, RPL0 and RPL1. In the case of unidirectional prediction, a reference picture can be selected from either list and in the case of bidirectional prediction, one picture must come from each list. Some improvements have been made in HEVC to make the identification of the RPS more robust than in H.264/AVC.

12.8.4 Pictures and partitions

HEVC supports a very wide range of picture sizes and frame rates, as illustrated in the level examples in Table 12.6. The basic picture types supported in HEVC are identical to those in H.264/AVC, with the small exception of a modified version of the IDR frame known now as a clean random access (CRA) picture.

TABLE 12.6 Example HEVC levels (* indicates High Tier specification, others are Main Tier).

Level	Max. luma sample rate	Max. luma picture rate	Max bit rate (kbps)	Example resolution
1	552,960	36,864	350	176×144@15
3	16,588,800	552,960	6,000	720×576@30
5	267,386,880	8,912,896	*100,000	3,840×2160@60
6.2	4,278,190,080	35,651,584	*800,000	7680×4320@120

An HEVC coding unit (CU) can contain up to three blocks (arrays of luma and chroma samples) and any information required to decode these blocks. Each CU is subdivided into prediction units (PUs) and these are coded using either intra- or inter-prediction. The appropriate prediction parameters (inter- or intra-) are signaled within each PU. A residual quadtree (RQT) is then used to divide the CU into transform units (TUs) within which the residual information is coded. These units all contain coding blocks (CBs), PBs, and transform blocks (TBs) for the constituent luma and chroma components, as described above. This structure is described in more detail in [28]. The decomposition of the HEVC picture is described in more detail below.

Coding tree units (CTUs) and coding tree blocks (CTBs)

HEVC replaces the macroblock of earlier standards with an L×L CTU where L=16, 32, 64. This provides more efficient coding for higher-resolution formats. The CTU is made up of luma and chroma coding tree blocks (CTBs).

Quadtree CTU structure

CTBs can be partitioned into CUs comprising luma and chroma CBs with smaller block sizes using a quadtree decomposition. The size of the largest CU (LCU) is 64 × 64. The manner in which the CTU is decomposed is not specified in the standard (it is nonnormative) but nonetheless forms an important part of the encoder's RDO processing. An example of CTU partitioning is given in Fig. 12.21. Each CU is further partitioned into TUs and PUs. This approach provides the encoder with much more flexibility to optimize block partitions according to the video content.

Prediction units (PUs)

A much larger variety of block decompositions for intra- and inter-coding is supported in HEVC, from 4 × 4 to 64 × 64.

Transform units (TUs)

Prediction residuals, whether from intra- or inter-coding, are transformed using integer transforms similar to the DCT, with sizes from 4 × 4, 8 × 8, 16 × 16, and 32 × 32. For the case of intra-prediction residuals, a 4 × 4 integer sine transform is also available as this provides better performance on less correlated residuals.

Random access points (RAPs) and clean random access (CRA) pictures

The concept of CRA pictures is the same as that for IDR frames in H.264/AVC. CRAs generalize this by embodying independently coded pictures, where following frames have

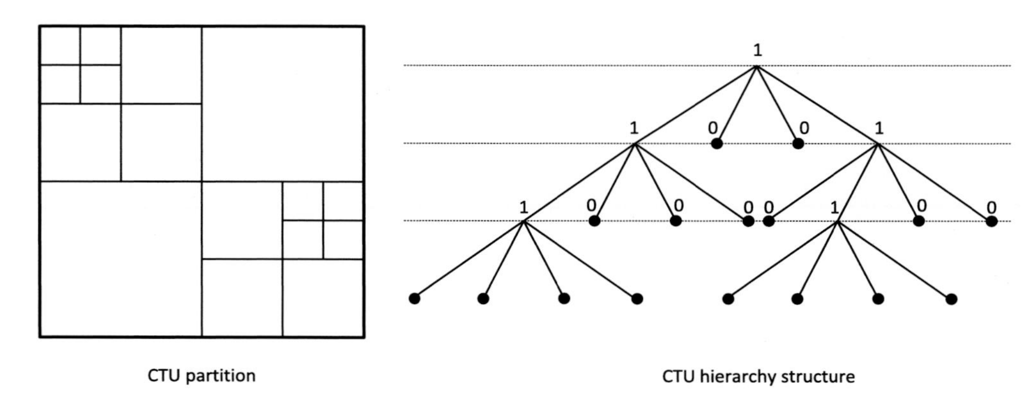

FIGURE 12.21 CTU partitioning in HEVC – an example.

no dependence on frames received before the CRA picture. CRA pictures can be placed at locations where random access is required, known as random access points (RAPs).

12.8.5 The video coding layer (VCL)

Similar to H.264, HEVC uses a hybrid coding structure combining motion compensation with block transform and entropy coding. Several innovations have been incorporated into HEVC that provide significant gains in performance and these are described below.

Intra-coding

In a similar manner to H.264/AVC, the boundary pixel values from adjacent decoded blocks can be used as a basis of spatial prediction. Whereas H.264/AVC supports only eight prediction modes, this has been significantly enhanced in HEVC with 35 modes including 33 directional modes (Fig. 12.22) plus surface fitting and DC prediction. Further details on the computation of the predictions for these modes can be found in [23]. The amplitude of the planar surface in mode 0 is computed based on horizontal and vertical slopes derived from the reference pixel values in the surrounding blocks. The DC mode (1) is computed in the same way as for H.264/AVC – based on the average of the block boundary reference values.

HEVC also extends the concept of reference and boundary sample smoothing. These operations help to reduce residual and reconstruction edges by filtering either the reference samples (from surrounding blocks) or the boundary samples (within the predicted block) prior to prediction.

When intra-modes are used the PB size is the same as that for the CB, except for the smallest CB size where a further partition into four PB quadrants of 4×4 is permitted. After prediction, residuals are coded using a forward integer transform, quantized, and entropy coded.

Inter-coding

The approach to inter-prediction and coding is similar to H.264/AVC. Multiple reference frames are stored in the decoded picture buffer and are used in a similar way to H.264/AVC,

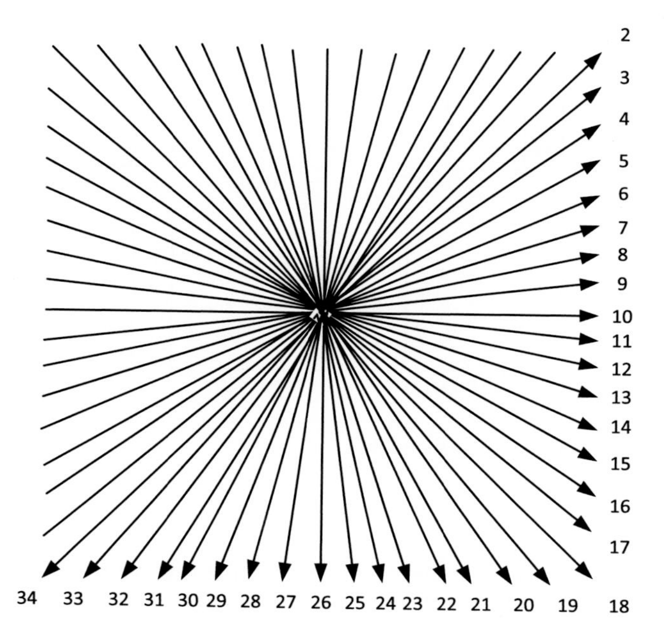

FIGURE 12.22 HEVC intra-prediction modes (mode 0 =planar and mode 1 =DC).

supporting both unipredictive and bipredictive modes. As with H.264/AVC, weighted prediction can be employed. Some modifications included in HEVC are as follows:

- **Subpixel motion compensation:** Quarter-pixel precision is used for estimating motion during temporal prediction. Local luma interpolation to half-pixel resolution is provided by an 8-tap filter, rather than the 6-tap filter in H.264/AVC. A 7-tap filter is then used for quarter-sample interpolation rather than the 2-tap filter used in H.264/AVC. These are applied separably and (unlike H.264) without intermediate rounding. This improves precision and simplifies the prediction architecture. The filter coefficients for interpolation are given in Eqs. (12.1) and (12.2). Further details on the use of these can be found in [23,24]. We have

$$h_h[i] = \{-1, 4, -11, 40, 40, -11, 4, -1\}/64, \tag{12.1}$$

$$h_q[i] = \{-1, 4, -10, 58, 17, -5, 1\}/64. \tag{12.2}$$

- **Motion vector prediction:** An advanced motion vector prediction (AMVP) mode is used for initialization and derivation of likely motion vector candidates. Also, improved skip modes are included along with a capability of merging motion vectors between adjacent blocks.
- **CB partitioning into prediction blocks:** In the case of inter-prediction, luma and chroma CBs can be divided into one, two, or four PBs (again the latter only for the smallest CB size). These partitioning options are illustrated in Fig. 12.23.

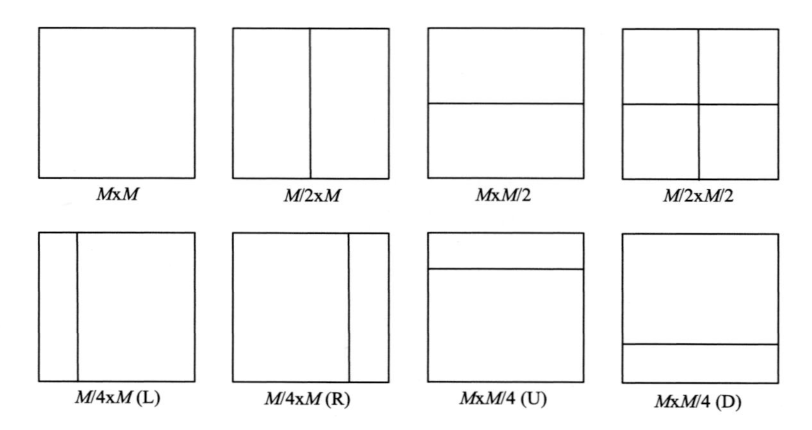

FIGURE 12.23 CB partitions for inter-prediction PBs.

Transforms in HEVC

Residual data after prediction is coded with a separable integer transform whose structure depends on prediction mode and block size. The transform block sizes supported in HEVC are 4×4, 8×8, 16×16, and 32×32. As for the case of H.264/AVC, the transform basis functions are derived from the DCT. The transform is defined as a 32×32 matrix of basis functions and the smaller transform sizes are obtained by subsampling this larger matrix. For example the 16×16 matrix is derived from the 32×32 matrix by taking the first 16 values from rows 0, 2, 4, etc. A similar subsampling is used to produce 8×8 and 4×4 transform matrices. As an example, the 8×8 transform is given as

$$\mathbf{A}_{8\times8}=\begin{bmatrix} 64 & 64 & 64 & 64 & 64 & 64 & 64 & 64 \\ 89 & 75 & 50 & 18 & -18 & -50 & -75 & -89 \\ 83 & 36 & -36 & -83 & -83 & -36 & 36 & 83 \\ 75 & -18 & -89 & -50 & 50 & 89 & 18 & -75 \\ 64 & -64 & -64 & 64 & 64 & -64 & -64 & 64 \\ 50 & -89 & 18 & 75 & -75 & -18 & 89 & -50 \\ 36 & -83 & 83 & -36 & -36 & 83 & -83 & 36 \\ 18 & -50 & 75 & -89 & 89 & -75 & 50 & -18 \end{bmatrix}. \tag{12.3}$$

In the case of intra-coded 4×4 residuals, an alternative integer transform is used – derived from the DST. It has been observed that the statistics of intra residuals better fit the basis functions of the DST since the error values increase with distance from the boundary samples used to predict them. The 4×4 transform matrix is thus

$$\mathbf{A}_{4\times4\mathrm{I}}=\begin{bmatrix} 29 & 55 & 74 & 84 \\ 74 & 74 & 0 & -74 \\ 84 & -29 & -74 & 55 \\ 55 & -84 & 74 & -29 \end{bmatrix}. \tag{12.4}$$

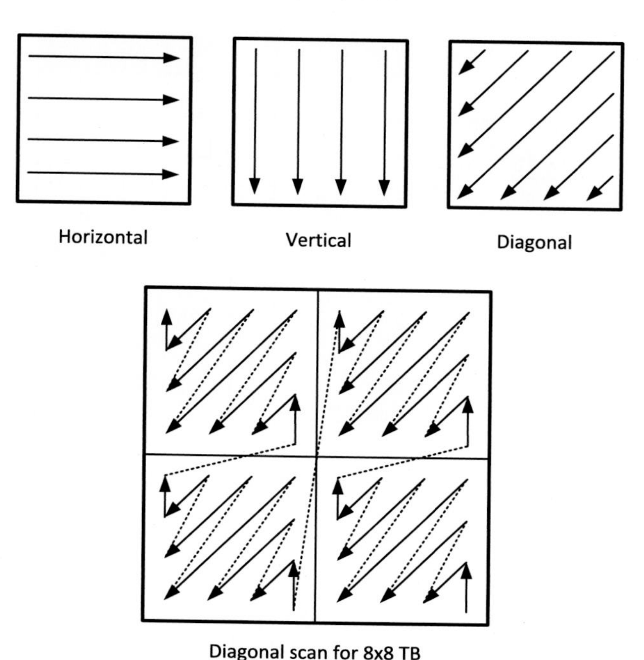

FIGURE 12.24 Coefficient scanning in HEVC.

Quantization in HEVC

HEVC employs a similar quantization method for transform coefficients to that used in H.264. It uses uniform quantization with a range of scaling matrices available for each block size. For the design of perceptual quantization matrices in HEVC, readers are referred to Chapter 5 (Section 5.6). HEVC also supports frame and CTU level QP variations (using different QP offsets and adaptive QPs). Examples of frame level QP configurations were given in Example 10.2 and in Fig. 12.17.

Coefficient scanning

The scanning of coefficients in an HEVC TB [30] is based on the use of multiples of 4×4 blocks for all TB sizes. The three coefficient scanning options available to represent each coefficient group (CG; a group of 16 consecutive coefficients) are shown in Fig. 12.24. In the case of intra-blocks, a new method referred to as mode-dependent coefficient scanning (MDCS) is used, where the mode closest to the orientation of the intra-prediction mode is selected. For inter-coded blocks, only the diagonal mode is used. For larger inter-blocks, coefficients are scanned from the least significant to the most significant as shown in Fig. 12.24 for the case of an 8 × 8 TB. In contrast to the zig-zag scanning used in H.264/AVC and previous standards, this simplified approach enables better matching of scanning to coefficient contexts. In terms of coding efficiency, improvements of around 3.5% compared to H.264/AVC have been reported [30].

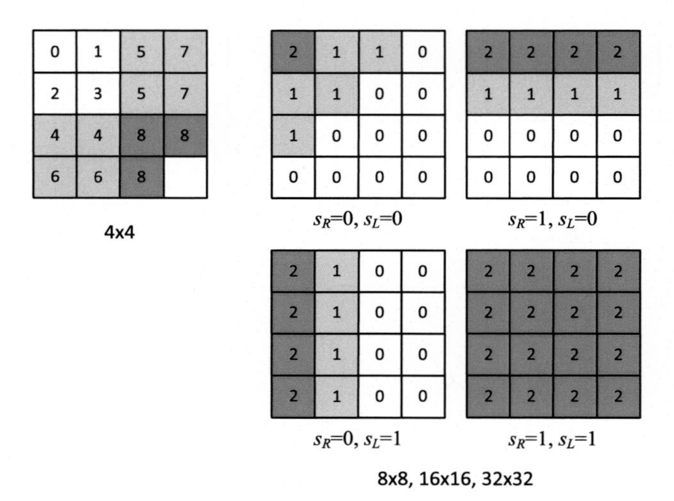

FIGURE 12.25 Significance flag contexts for 4 × 4 and 8 × 8 HEVC transform blocks.

A further modification in HEVC relative to AVC is related to the signaling of the last significant coefficient in a scan. As we have seen, signaling the last significant coefficient reduces the number of coded bins by eliminating the need to explicitly code runs of trailing zeros. In HEVC the position of the last significant coefficient is coded explicitly followed by the flags to indicate significant coefficients.

Context and significance

As with H.264/AVC, HEVC signals the significance of a TB using a coded block flag (CBF) which indicates whether the TB contains any nonzero coefficients. The context of a CBF depends on the level of the block in the quadtree decomposition. The significance maps within HEVC are coded in a way that exploits the sparsity of the significance map, using two passes. Firstly the significance of coefficient groups and then the significance of the coefficients within them is encoded. The coefficient significance contexts in HEVC for 4 × 4 and 8 × 8 blocks are shown in Fig. 12.25. The significance flags are coded for each coefficient between the last and the DC based on a context model that depends its location and the significance of its neighbors. In the case of 4 × 4 TBs, coefficients are grouped according to frequency and the distribution, where coefficients with similar distributions are grouped together.

For the case of 8 × 8, 16 × 16, and 32 × 32 TBs in HEVC, context modeling is again based on position but according to the templates shown in Fig. 12.25 (right). One of the four templates is selected according to the values of s_R and s_L, which correspond to the coded subblock flag for the right-hand and lower subblocks, respectively.

Entropy coding

CABAC, as described in Chapter 7, is a powerful arithmetic coding method that was supported in the H.264/AVC Main and High profiles. Because of its performance advantages, CABAC has also been adopted for entropy coding in HEVC, and is the only variable-length

TABLE 12.7 SAO edge classification.

EdgeIdx	Condition	Case
0	Exceptions to cases (1...4)	Monotonic
1	$p < n_0$ and $p < n_1$	Local min.
2	$p < n_0$ and $p = n_1$ or $p < n_1$ and $p = n_0$	Edge
3	$p > n_0$ and $p = n_1$ or $p > n_1$ and $p = n_0$	Edge
4	$p > n_0$ and $p > n_1$	Local max.

coding method employed for block level information. As with H.264/AVC, CABAC in HEVC uses local contexts to adjust its conditioning probabilities, but with the following improvements:

- **Better context modeling**: Based on exploitation of HEVC's tree-structured approach.
- **Reduced number of context coded bins**: Although the number of contexts is lower than in H.264/AVC, it achieves better performance with reduced memory requirements.
- **An enhanced bypass mode**: More extensive use of bypass is made to reduce the amount of data that needs to be processed using CABAC contexts, resulting in a throughput increase.

In-loop filters

HEVC includes two stages of in-loop filtering prior to updating its decoded picture buffer: a DBF and an SAO process.

- **Deblocking filter:** The deblocking operation in HEVC is very similar to that for H.264/AVC, which was described in Chapter 9. It is slightly simplified to make it more amenable to parallel processing. DBF is applied to all TU and PU boundaries, except when they coincide with picture or slice boundaries. In the case of 4×4 blocks the DBF is simplified by only using it when block boundaries coincide with 8 × 8 boundaries. This simplifies processing generally with no noticeable degradation in performance. In the case of HEVC, only three deblocking strengths are available rather than the five with H.264/AVC.
- **Sample-adaptive offset:** SAO is a new feature in HEVC that is invoked after DBF, in order to increase picture quality through a reduction in banding and ringing artifacts. It is a nonlinear adjustment that adds offsets to samples based on a look-up table created at the encoder (and transmitted to the decoder) using histogram analysis of signal amplitudes. Two SAO modes are available: edge offset mode and band offset mode. The first operates on a CTB basis using one of four gradient operators to compare a sample with adjacent values in a given orientation. Five conditions can result according to the relative value of the sample compared to its neighbors. Consider for example the case of a horizontal orientation; if a sample p has a left neighbor n_0 and a right neighbor n_1, then the regions are classified according to Table 12.7. In the second case, band offsetting, the offset is applied directly based on the sample amplitude. The full amplitude range is divided into 32 bands and the values of samples in 4 consecutive bands are modified using transmitted offsets. This can help to reduce banding artifacts in smooth areas.

Coding tools for screen content, extended range, and 3D videos

HEVC has, since its fourth version, improved coding performance for still or moving computer rendered graphics, text, and animation content (screen content). Primary coding tools developed for screen content coding (SCC) [29] include:

- **Adaptive color transform**: This feature employs an alternative color space conversion (RGB to YC_oC_g) function (Eq. (12.5)) to reduce the redundancy among color components. This has been shown to be more efficient compared to conventional RGB to YC_bC_r conversion for the case of screen content with highly saturated color samples. We have

$$\begin{bmatrix} Y \\ C_o \\ C_g \end{bmatrix} = \begin{bmatrix} 0.25 & 0.5 & 0.25 \\ 0.5 & 0 & -0.5 \\ -0.25 & 0.5 & -0.25 \end{bmatrix} \begin{bmatrix} R \\ G \\ B \end{bmatrix}. \tag{12.5}$$

- **Intra-block copying**: This copies reconstructed block samples from the current frame. Initially developed for H.264/AVC, it did not achieve significant performance gains for natural video content. It has however proved beneficial for screen content due to the likelihood of repeated textures and structures. This has also been adopted in the AOMedia Video 1 (AV1) codec (see Section 12.10.2).
- **Adaptive motion vector resolution**: The trajectory of movement for computer rendered content is typically aligned with integer sample positions, which produces integer motion vectors. For SCC, HEVC has supported an adaptive motion vector resolution feature, flagging the use of full-pixel motion estimation in the current slice.
- **Palette mode**: A look-up table (palette) is created to represent color (in RGB or YC_bC_r color space) values. Color indices for samples (based on this table) are transmitted alongside the palette table. This is particularly useful for blocks containing a relatively small number of color values.

New coding tools have also been adopted in HEVC for 3D/multiview videos and content with extended dynamic range. These are summarized in Table 12.8. More details can be found in overview articles such as [25,26].

12.8.6 Profiles and levels

Version 1 of HEVC specified three profiles, Main, Main 10, and Main Still Picture.

Main profile

The Main profile supports a bit depth of 8 bits per sample and 4:2:0 (and 4:0:0) chroma sampling, employing the features described in the previous subsections.

Main 10 profile

The Main 10 profile supports bit depths up to 10 bits and up to 4:2:2 chroma sampling. In recent tests, it has been demonstrated that the Main 10 profile outperforms the Main profile with a bit rate reduction of approximately 5% for identical PSNR values.

TABLE 12.8 Primary new coding tools employed in H.265/HEVC range and 3D/multiview extensions.

Extension	Coding tools
Range	Cross-component prediction Adaptive chroma QP offset Residual DPCM
3D and Multiview	Neighboring block disparity vector (NBDV) Extended temporal motion vector prediction for merge Inter-view motion prediction Disparity information merge candidate NBDV-based residual prediction Illumination compensation NBDV-based depth refinement View synthesis prediction Depth-based block partitioning Inter-view motion prediction Intra-wedge mode Depth look-up table Quadtree limitation Texture merge candidate Intra contour mode

Main still picture profile

HEVC also includes, like H.264, a still image coding profile, which is a subset of the Main profile. Performance comparisons based on MOS and PSNR scores have provided indications that HEVC still image coding outperforms JPEG2000 in terms of bit rate reduction by approximately 20% based on PSNR and 30% for MOS scores. Likewise it has been shown to outperform JPEG by 40% and 60%.

Levels

In the same way as previous standards, HEVC also supports a range of levels that impose parameter limits that constrain decoder performance. There are currently 13 levels associated with HEVC. Some examples are shown in Table 12.6.

12.8.7 Extensions

At the time of writing, a number of HEVC range extensions and 21 additional profiles have been, or are in the process of being, developed. These include:

- the Monochrome, Monochrome 10, Monochrome 12, and Monochrome 16 profiles;
- the Main 12 profile;
- the Main 4:2:2 10 and Main 4:2:2 12 profiles;
- the Main 4:4:4, Main 4:4:4 10, and Main 4:4:4 12 profiles;
- the Main Intra, Main 10 Intra, Main 12 Intra, Main 4:2:2 10 Intra, Main 4:2:2 12 Intra, Main 4:4:4 Intra, Main 4:4:4 10 Intra, Main 4:4:4 12 Intra, and Main 4:4:4 16 Intra profiles;

TABLE 12.9 Comparison of video coding standards for entertainment applications. Average bit rate savings are shown for equal objective quality values measured using PSNR.

Video standard	Relative bit rate savings (%)		
	H.264/AVC	MPEG-4	MPEG-2
HEVC MP	35.4	63.7	70.8
H.264/AVC HP	X	44.5	55.4
MPEG-4 ASP	X	X	19.7
H.263 HLP	X	X	16.2

- the Main 4:4:4 Still Picture and Main 4:4:4 16 Still Picture profiles.

The extended ranges in terms of color sampling and bit depths are obvious from their titles. Alongside these extensions, HEVC provides another 19 profiles for 3D and multiview video, screen content, and high-throughput and scalable coding applications. Their definitions can be found in the HEVC standard document [23].

12.8.8 Performance

A comprehensive comparison between HEVC and previous standards is reported by Ohm et al. in [18]. We reproduce their results here in Table 12.9, noting that even greater improvements were reported for interactive applications. As a rule of thumb, we have stated previously that video coding standards have, since the introduction of H.120 in 1984, delivered a halving of bit rate for the equivalent video quality every 7–10 years. This is evidenced by the results in Table 12.9.

12.9 H.266/VVC

12.9.1 Brief background

In order to provide better support for immersive video formats and enhance compression capabilities beyond those offered by current standards, a new video coding standard, H.266/Versatile Video Coding (VVC), has been developed by a joint collaborative team of ITU-T and ISO/IEC experts (the Joint Video Experts Team [JVET]) and was finally approved in July 2020. The formal call for VVC proposals was issued in 2017 with the aim of significantly improving coding efficiency over HEVC. Twenty-one proponent groups from 32 industrial and academic organizations submitted 46 solutions: 22 for standard dynamic range (SDR) content and 12 each for high dynamic range (HDR) and 360-degree video content. Various CEs were then designed to investigate the contribution of the proposed tools at different coding stages. As a result, around 40 new features were adopted. The most recent version of the VVC Test Model (VTM) can achieve over 30% coding gain compared to the HEVC reference encoder (HM) on SDR content, with even greater performance gains evident for HDR and 360-degree content.

TABLE 12.10 Primary new coding tools employed in H.266/VVC.

Codec feature	Approach
Picture partitions	A quadtree structure with nested multitype partitions using binary and ternary segmentation
	Maximum CTU size for luma blocks is 128×128
	Maximum transform size for luma blocks is 64×64
Intra-prediction	65 directional intra-modes, plus surface fitting and DC prediction
	Mode-dependent intra-smoothing
	Multireference line intra-prediction
Inter-prediction	Affine motion model-based inter-prediction
	$\frac{1}{16}$-pixel motion vector precision
	Combined intra- and inter-prediction
	Adaptive motion vector resolution
Transform and quantization	Multiple primary transform selections
	Dependent quantization with max QP increased to 63
Entropy coding	CABAC engine with an adaptive double probability update model
In-loop filter	Adaptive loop filter
	Luma mapping with chroma scaling
360-degree video coding	Horizontal wrap-around motion compensation

In the following subsections, we will summarize some of the key innovations in VVC and report performance comparisons between the test models of VVC, HEVC, and AVC based on objective quality assessment. For more details of VVC, readers are referred to the standard description in [2,3].

12.9.2 Primary features

One of the primary aims of VVC was to improve compression efficiency for UHD, HDR, and wide color gamut (WCG) content, and for encoding immersive video formats such as 360-degree. Although VVC still employs a traditional block-based hybrid coding architecture, it includes numerous new features related to picture partitioning, intra- and inter-prediction, transformation, quantization, entropy coding, and in-loop filtering. Table 12.10 summarizes primary new coding tools adopted in VVC.

12.9.3 High-level syntax

NAL units

Compared to HEVC, VVC reduces the number of NALU types (NUTs) from 63 to 32, including 13 NUTs for the VLC, 7 for parameter sets and picture headers, and 12 for other non-VLC NUTs.

Parameter sets

Three new parameter sets have been employed in addition to the existing ones in HEVC, including decoding capability information (DCI), an adaptation parameter set (APS), and

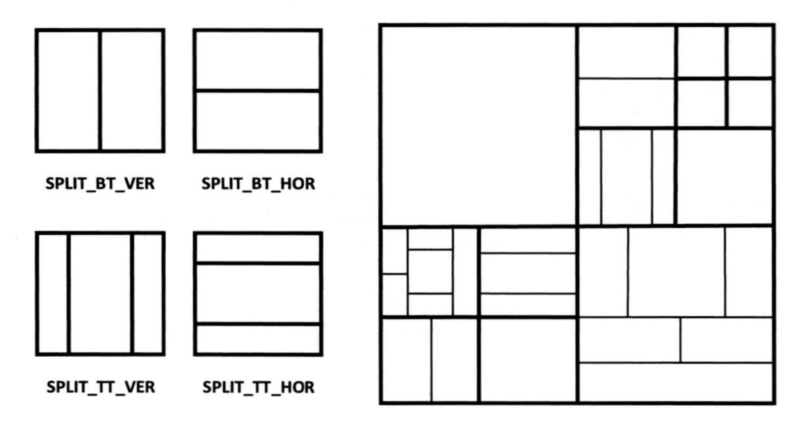

FIGURE 12.26 An example of CTU partitions in VVC.

a picture header (PH). DCI includes profile, level, and subprofile information, while APS contains adaptive parameters such as adaptive loop filter (ALF) coefficients. PH contains PPSs and SPSs that apply to all slices in a picture.

12.9.4 Picture partitioning

In order to better deal with complex motions and object boundaries, VVC supports increased flexibility in the way pictures can be partitioned. It employs the same CTU concept as in HEVC, but with its maximum luma block size extended to 128 × 128. Each CTU can be further divided into a nested multitype tree through binary and ternary segmentation. VVC has also simplified the multiple partition unit types (CU, PU, and TU) defined in HEVC, and only enables them when a CU is larger than the maximum transform size. In addition to the quadtree segmentation in HEVC with square CU shapes, VVC also supports various rectangular CU shapes through vertical binary splitting (SPLIT_BT_VER), horizontal binary splitting (SPLIT_BT_HOR), vertical ternary splitting (SPLIT_TT_VER), and horizontal ternary splitting (SPLIT_TT_HOR). An example of CTU partitions is illustrated in Fig. 12.26. The maximum transform size supported in VVC has increased to 64 × 64 for luma components and 32×32 for chroma channels.

12.9.5 Intra-prediction

Extended intra-coding modes

Following the trend in HEVC, VVC has further refined intra-coding with 65 directional modes (extended from 33 in HEVC) alongside the existing surface fitting and DC prediction modes. These are shown in Fig. 12.27. Since VVC allows blocks with a rectangular shape, only the average value of neighboring pixels on the longer side is calculated for DC prediction.

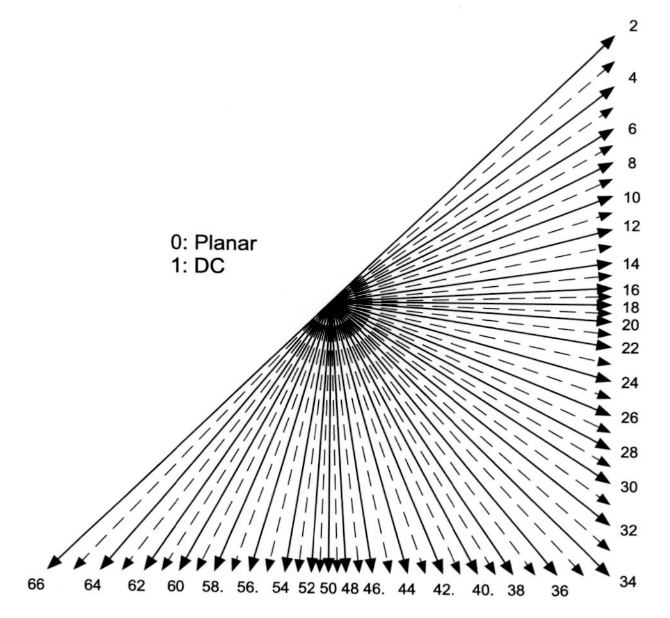

FIGURE 12.27 VVC intra-prediction modes. Extended directional modes compared to HEVC are highlighted with dashes.

Mode-dependent intra-smoothing

Instead of the 2-tap filter in HEVC, a series of 4-tap Gaussian interpolation filters are employed in VVC to smooth directionally intra-predicted content. The choice of filter is based on the prediction orientation in each mode.

Multiple reference lines

In addition to extra orientations, VVC also supports multiple reference lines (MRL) for intra-prediction (Fig. 12.28). Instead of just using the nearest reference line (reference line 0) to the block being analyzed, VVC can base intra-prediction on any one of multiple neighboring reference lines (reference lines 0, 1, and 3) to achieve enhanced prediction accuracy. However, additional bits are needed to signal the reference line in use.

12.9.6 Inter-prediction

Affine motion inter-prediction

VVC has introduced a new inter-prediction mode based on block-based affine transform motion compensation. This provides improved prediction performance for content exhibiting complex temporal motions, such as zooms, rotations, and perspective motions. For coding this type of information, two affine motion models (four-parameter and six-parameter) are employed. Based on Eqs. (12.6) and (12.7), the motion vector field of the current block can be obtained from the motion vectors of its neighbors at the top left, top right, and bottom left

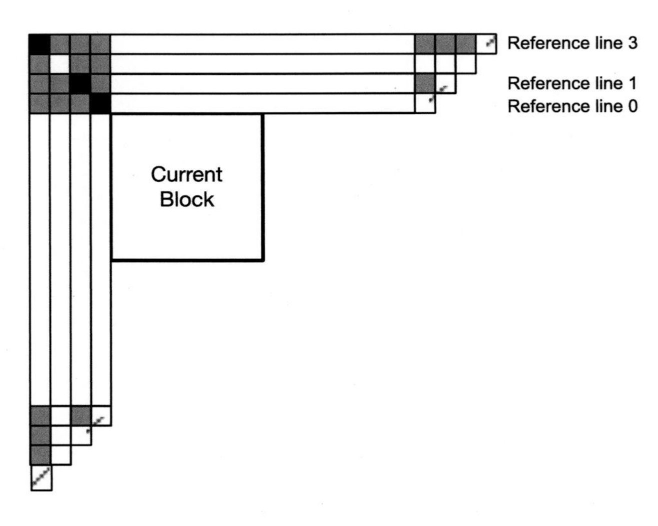

FIGURE 12.28 An example of multiple reference lines used in VVC intra-prediction.

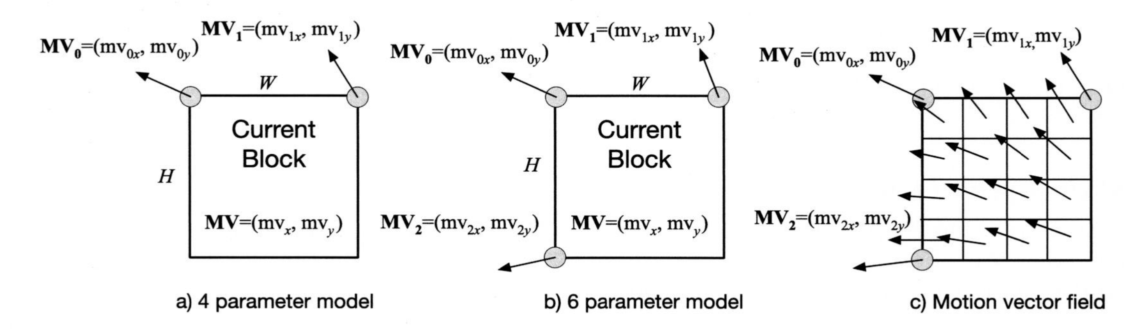

FIGURE 12.29 Affine motion inter-prediction in VVC.

(for six-parameter affine model-only) corners (as illustrated in Fig. 12.29 (a and b)). During motion compensation, the motion vector at the central location is calculated with 1/16-pixel precision for each 4×4 subblock, as shown in Fig. 12.29 (c). We have

$$\begin{cases} \mathrm{mv}_x(i,j) = \frac{\mathrm{mv}_{1x} - \mathrm{mv}_{0x}}{W} i - \frac{\mathrm{mv}_{1y} - \mathrm{mv}_{0y}}{W} j + \mathrm{mv}_{0x}, \\ \mathrm{mv}_y(i,j) = \frac{\mathrm{mv}_{1y} - \mathrm{mv}_{0y}}{W} i + \frac{\mathrm{mv}_{1x} - \mathrm{mv}_{0x}}{W} j + \mathrm{mv}_{0y}, \end{cases} \tag{12.6}$$

$$\begin{cases} \mathrm{mv}_x(i,j) = \frac{\mathrm{mv}_{1x} - \mathrm{mv}_{0x}}{W} i + \frac{\mathrm{mv}_{2x} - \mathrm{mv}_{0x}}{H} j + \mathrm{mv}_{0x}, \\ \mathrm{mv}_y(i,j) = \frac{\mathrm{mv}_{1y} - \mathrm{mv}_{0y}}{W} i + \frac{\mathrm{mv}_{2y} - \mathrm{mv}_{0y}}{H} j + \mathrm{mv}_{0y}, \end{cases} \tag{12.7}$$

where $(\mathrm{mv}_x, \mathrm{mv}_y)$ represents the motion vector field of the current block and (i, j) are the pixel coordinates.

Adaptive motion vector resolution

To signal motion vectors, VVC employs CU-based adaptive motion vector resolution (AMVR). This encodes motion vector differences (between the actual and predicted motion vectors) with adaptive precision. Two AMVP modes are available: normal AMVP and affine AMVP. The former is employed for conventional motion compensation, supporting quarter-pixel, half-pixel, integer-pixel, and four-pixel luma samples, while the latter is designed for affine motion prediction which only supports quarter-pixel, integer-pixel, and 1/16-pixel luma samples.

Combined inter- and intra-prediction mode

A new mode in VVC supports a combined inter- and intra-prediction (CIIP) mode for CUs containing more than 64 luma samples where the height and width of the block are both less than 128 pixels. CIIP combines regular merge mode-based inter-predicted (P_{inter}) results with those from the planar intra-prediction mode (P_{intra}) through the following equation:

$$P_{\text{CIIP}} = ((4 - w) \cdot P_{\text{inter}} + w \cdot P_{\text{intra}} + 2) \gg 2. \tag{12.8}$$

The value of weight w is determined according to the coding modes of neighboring blocks, and equals either 1, 2, or 3.

12.9.7 Transformation and quantization

Larger maximum transform block size

In order to achieve improved coding performance on high-spatial resolution content, VVC supports transforms with block sizes up to 64×64 (HEVC supports up to 32 × 32). For transform blocks with sizes larger than 32 × 32, the high-frequency transform coefficients outside the top left 32 × 32 area are set to zero in order to reduce computational complexity.

Multiple transform selection

New DST and DCT transforms (DST-VII and DCT-VIII) have also been introduced in VVC in addition to the DCT-II used in HEVC. Their N-point basis functions are given below. VVC also supports a multiple transform selection (MTS) method [21] for coding both intra and inter luma blocks. The horizontal and vertical transform methods are selected separately among those given below, according to coding mode and block size:

$$C_{\text{DST-VII}}(m,n) = \frac{2}{\sqrt{2N+1}} \cdot \sin\left(\frac{\pi \cdot (2m+1) \cdot (n+1)}{2N+1}\right), \; m,n = 0,1,\cdots,N-1, \tag{12.9}$$

$$C_{\text{DCT-VIII}}(m,n) = \frac{2}{\sqrt{2N+1}} \cdot \cos\left(\frac{\pi \cdot (2m+1) \cdot (2n+1)}{4N+1}\right), \; m,n = 0,1,\cdots,N-1, \tag{12.10}$$

$$\begin{aligned} C_{\text{DCT-II}}(m,n) &= a(n) \cdot \sqrt{\tfrac{2}{N}} \cdot \cos\left(\tfrac{\pi \cdot (2m+1) \cdot n}{2N}\right), \; m,n = 0,1,\cdots,N-1, \\ \text{where } a(n) &= \begin{cases} \sqrt{\tfrac{2}{N}}, & n = 0, \\ 1, & n = 1,2,\cdots,N. \end{cases} \end{aligned} \tag{12.11}$$

Quantization in VVC

VVC has extended the quantization parameter (QP) range in HEVC from 0–51 to 0–63. It has also introduced a dependent scalar quantization approach to replace the independent scalar quantization used in HEVC. This selects between two scalar quantizers based on state transition rules to achieve denser packing of transform coefficients. By doing this, the average distortion between an input and its reconstructed coefficient vector can be reduced.

12.9.8 Entropy coding

VVC has further enhanced CABAC entropy by enabling context models and probabilities to be updated using a multihypothesis approach. This independently updates two probability estimates associated with each context model with pretrained adaptation rates. This replaces the look-up tables used in HEVC.

VVC also uses a different residual coding structure for transform coefficients to achieve more effective energy compaction. This is particularly useful for coding screen content. VVC also supports context modeling for coefficient coding, which selects probability models based on their absolute values and partially reconstructed absolute values from the neighbors of the current block.

12.9.9 In-loop filters

VVC employs three in-loop filters, i.e., a DBF, SAO, and an ALF. The DBF and SAO are similar to those described for HEVC. ALF is however a new innovation that is invoked after deblocking filtering and SAO. Fig. 12.30 illustrates the complete in-loop filtering process in VVC.

The ALF operation includes luma ALF, chroma ALF, and cross-component ALF (CC-ALF). CC-ALF is applied to refine chroma components by combining an adaptively filtered luma signal with ALF processed chroma components. All the adaptive loop filters are designed to have linear diamond shapes, shown in Fig. 12.31. For luma components, each 4×4 block is classified into 25 classes based on its local gradients. Blocks in each class are applied on one of 25 ALF filters to achieve optimal refinement. A single set of filter coefficients are employed for chroma and cross-component ALFs.

12.9.10 Coding tools for 360-degree video

Horizontal wrap around motion compensation

To achieved enhanced compression performance for 360-degree videos, VVC has introduced new coding tools to support this, including horizontal wrap around motion compensation. This is applied when a reference block is partially outside the boundaries (to the left or right) of the reference picture. Rather than repetitive padding of the nearest neighbors (as in conventional motion compensation) the spherically neighboring pixels within the reference frame are wrapped around to form the boundary extension. This is not used for the top and bottom boundaries, where repetitive padding is still employed.

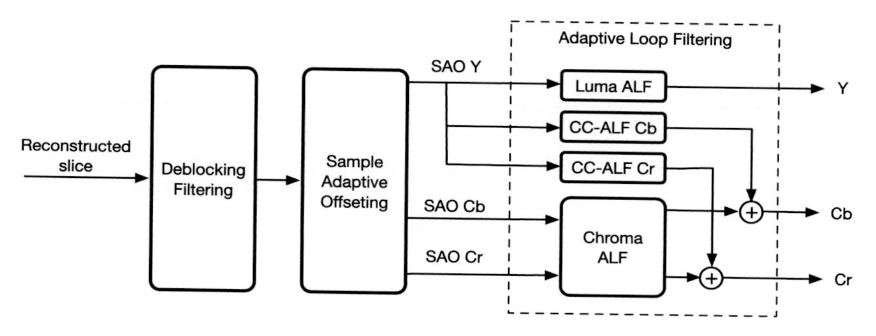

FIGURE 12.30 In-loop filtering in VVC [2].

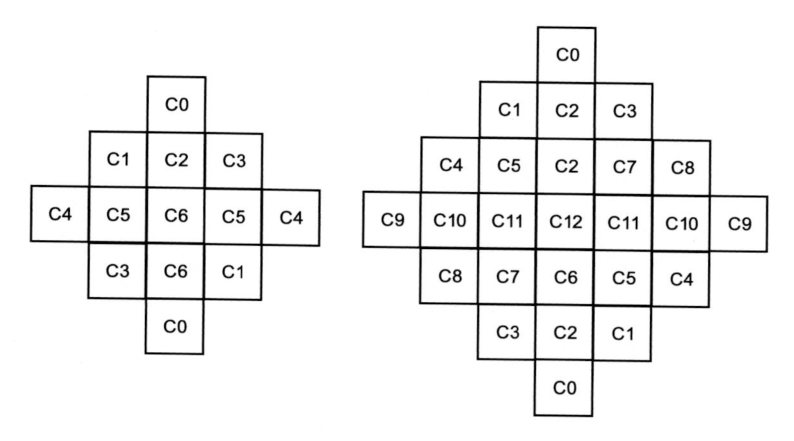

FIGURE 12.31 ALF filter shapes for luma (7×7) and chroma (5×5) components.

12.9.11 Profiles, tiers, and levels

The first version of VVC, launched in July 2020, defines four profiles. These are: Main 10, Main 10 Still Picture, Main 4:4:4 10, and Main 4:4:4 10 Still Picture. Main 10 and Main 10 Still Picture profiles support formats with bit depths up to 10 bits and 4:2:0 chroma sampling. Main 4:4:4 10 and Main 4:4:4 10 Still Picture support color formats up to 4:4:4 with bit depths up to 10 bits. The definitions of levels and tiers are similar to those in HEVC. Version 1 of VVC has included the same number of levels (13) as in the current version of HEVC standard.

12.9.12 Performance gains for VVC over recent standards

As part of the standardization process, the VTM test model performance has been extensively compared with previous video coding standards. Table 12.11 summarizes comparison results between one of recent VTM versions (7.1) and the test models of HEVC (HM 16.20) and H.264/AVC (JM 19.0) based on the JVET common test conditions (in random access mode) using SDR test video content. It can be seen that VVC achieves an average bit rate

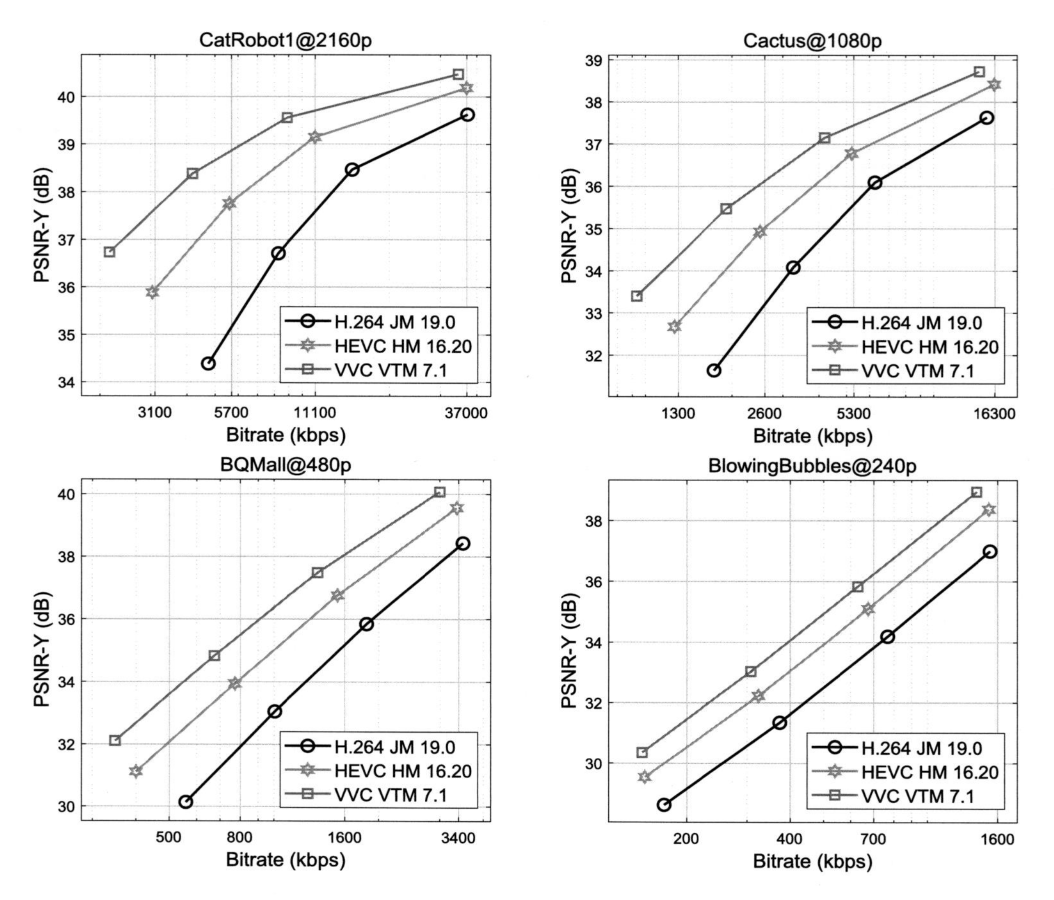

FIGURE 12.32 Example rate distortion curves for H.264/AVC, H.265/HEVC, and H.266/VVC test models on four test sequences with various spatial resolutions. The bit rates here are with logarithmic scale.

saving of 32% over HEVC for these conditions. Coding gains are more significant on higher-resolution (2160p and 1080p) content. Fig. 12.32 shows example rate-distortion curves for three test codecs.

12.10 The alliance for open media (AOM)

The Alliance for Open Media (AOM) was founded in 2015 to address the perceived financial uncertainties and licensing conditions associated with MPEG standards, especially for streaming organizations. Its members include companies such as Amazon, Apple, ARM, Cisco, Facebook, Google, IBM, Intel, Microsoft, Mozilla, Netflix, Nvidia, Samsung Electronics, and Tencent. AOM is a nonprofit consortium focused on the development of open, royalty-

TABLE 12.11 Comparison of video coding standards based on JVET common test conditions (CTC) using SDR test sequences. Average bit rate savings are shown for equal objective quality values measured using PSNR.

Video standard	Relative bit rate savings (%) against	
	H.264/AVC	**HEVC MP**
Class A (2160p)	−70.0%	−39.0%
Class B (1080p)	−65.1%	−35.3%
Class C (480p)	−53.5%	−27.9%
Class D (240p)	−50.0%	−26.0%
Overall	−59.7%	−32.1%

free coding solutions and its first published standard was AV1, a successor to VP9 and competitive with HEVC.

12.10.1 VP9 and VP10

Google's VP9 is a successor to VP8 and is open-source and royalty-free. VP8 was created by On2 Technologies, A company acquired by Google in 2010. YouTube and other streaming companies still use VP9, alongside AV1 and MPEG codecs. Shortly after the development of V10, a successor to VP9, Google led the formation of AOM and work on VP10 was merged with the development of AV1.

12.10.2 AV1

The first-generation AOM codec, AV1 (AOMedia Video 1), was released in 2018 and has been progressively enhanced since then. Similar to VP9, AV1 is designed to be open-source and royalty-free, and is the primary competitor of MPEG standards. While its basic architecture is based on the conventional hybrid block-based paradigm, various unique coding tools have been developed and adopted in AV1 that differ from those in HEVC or VVC [31]:

- **A 10-way partition tree structure**: With 4:1/1:4 rectangular CB sizes (shown in Fig. 12.33).
- **New nondirectional intra-modes**: Extended from VP9. Three of these provide quadratic interpolation in vertical and/or horizontal directions. The fourth mode supports the prediction of direction for larger gradient values.
- **Intra-block copy**: This is useful for encoding screen content containing repeated patterns and textures.
- **Overlapped block motion compensation**: Combines block-based prediction with a secondary inter-prediction across edges to reduce possible errors due to motion mismatches.
- **Asymmetric discrete sine transform (ADST)**: To provide increased flexibility for decorrelating a wider range of content types.
- **Frame superresolution**: Based on a Wiener filter, this enables the codec to encode video content at lower spatial resolutions and superresolve it within the coding loop to recover its original resolution.

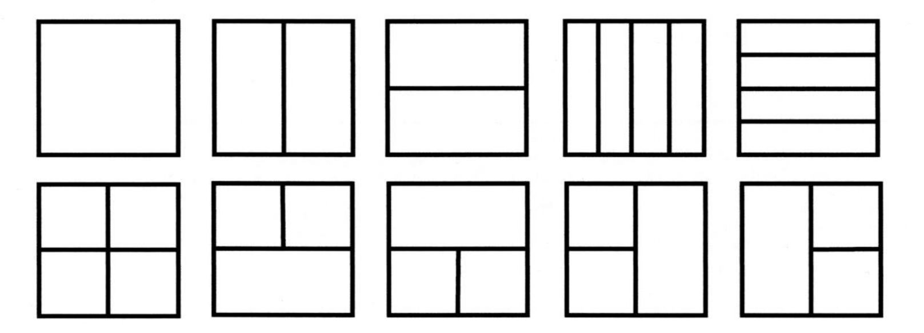

FIGURE 12.33 The 10-way partition tree employed in AV1. [31].

- **Film grain synthesis**: This is a pre/postprocessing tool applied outside the coding loop, which estimates and eliminates the grain noise before encoding and synthesizes it at the decoder based on an autoregressive (AR) model and transmitted noise parameters.
- **Nonbinary multisymbol arithmetic coding**: The entropy coding in AV1 is based on an adaptive multisymbol arithmetic entropy coder with symbol-adaptive coding, in which multiple symbols are encoded with up to 15-bit probabilities and combined into nonbinary symbols. This has been reported to offer comparable performance with binary coders such as CABAC [32].
- **Two-pass coding**: Similar to VP9, AV1 supports a 2-pass coding mode. The first pass collects the statistical information from the given content. This feature is primarily designed for adaptive streaming applications which do not require real-time encoding. More recent versions of AV1 have significantly enhanced 1-pass coding, which now provide similar performance to the 2-pass mode.

Since the first stable version of AV1 was published in 2018, its performance has been evaluated and compared to MPEG and other standard video codecs. One of the latest studies (by AV1 developers) has reported significant improvements achieved by AV1 relative to HEVC. They also report similar performance to VVC, with AV1 outperforming VVC by 5% [31].

Since the first stable version of AV1 was published in 2018, its performance has been evaluated and compared to MPEG and other standard video codecs. One of the latest studies (by AV1 developers) has reported significant improvements achieved by AV1 relative to HEVC. They also report similar performances for AV1 and VVC, with AV1 reported to be some 5% better than VVC [31].

12.11 Other standardized and proprietary codecs

A number of other video codecs exist. While these are worthy of mention, we will not cover them in detail here.

12.11.1 VC-1

The Microsoft WMV 9 codec was a proprietary codec, adopted in 2006 as international standard SMPTE 421M or VC-1. VC-1 remains a recognized format for DVDs and Blu-ray devices.

12.11.2 Dirac or VC-2

Dirac was developed by BBC Research and Development as a royalty-free alternative to the mainstream standards. Dirac employs wavelet compression rather than block transforms such as the DCT, and offers good performance for HDTV formats and beyond. An I-frame only version, Dirac Pro, has been used by BBC internally for studio and outside broadcast applications and was standardized by SMPTE as VC-2 in 2010. A portable version, known as Schrödinger, also exists.

12.11.3 RealVideo

RealVideo comprises a series of proprietary video compression formats developed by RealNetworks for their players. The current version, rv60, is believed to be based on HEVC. RealNetworks themselves claim higher compression performance than HEVC speed gains over x265 and VP9.

12.12 Codec comparisons

As mentioned in Chapter 4, video coding algorithms can be compared based on their rate-distortion (RD) or rate-quality (RQ) performance across a range of test sequences. As discussed in Chapter 10, the selection of test content is important and should provide a diverse and representative coverage of the video parameter space. Objective quality metrics or subjective opinion measurements are normally employed to assess compressed video quality, and the overall RD or RQ performance difference between codecs can then be calculated using Bjøntegaard measurements (as described in Chapter 4) or SCENIC [33] (for subjective assessments). There have been several recent reports on comparisons between MPEG codecs (H.264/AVC, HEVC, and VVC) and royalty-free (VP9 and AV1) codecs [34–36]. However, these results are often inconsistent, mainly due to the different coding configurations and test data employed.

Table 12.12 summarizes comparisons between the test models for four major video coding standards (HEVC, VVC, VP9, and AV1) based on the assessment of PSNR. The configurations employed for these codecs were similar to the random access mode in the JVET common test conditions. The test dataset contains 18 source sequences at UHD and HD resolutions [36], which covers various texture types (static and dynamic) and camera motions. To achieve a fair comparison, none of these sequences are from JVET or AOM test datasets. We can observe that AV1 achieves an average bit rate saving of 38.3% (measured by BD-rate) against VP9. The new MPEG standard, VVC, also provides significant coding gains (27.8%) over its predecessor, HEVC. The difference between VVC and AV1 is much smaller – only 9.4% benchmarked on AV1. We also note that the performance results here differ from those re-

TABLE 12.12 Compression performance comparison of video coding standards on selected UHD and HD content [36]. Average bit rate savings are shown for equal objective quality values measured using PSNR (luma component only). For AV1, 1-pass encoding is used (rather than 2-pass) for fair comparison.

Video codec	Relative bit rate savings (%)			
	VP9	**HEVC**	**AV1**	**VVC**
VP9 (1.7.0-426-g19222548a)	X	X	X	X
HEVC HM 16.18	25.1	X	X	X
AV1 (1.0.0-5ec3e8c)	38.3	18.3	X	X
VVC VTM 4.0.1	45.4	27.8	9.4	X

TABLE 12.13 Encoder computational complexity (benchmarked to VP9) for four video codecs. These were executed on the CPU nodes of a shared cluster, Blue Crystal Phase 3, based at the University of Bristol.

Video codec	VP9	HEVC	AV1	VVC
Relative encoder complexity	1×	7×	39×	57×

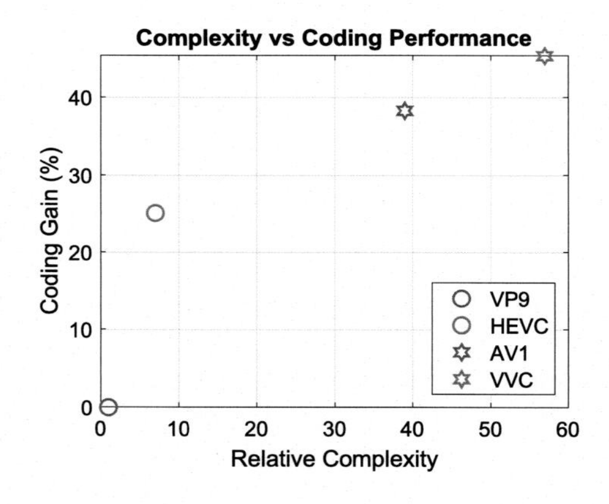

FIGURE 12.34 The relationship between the relative codec complexity and encoding performance (in terms of average bit rate savings benchmarked on VP9).

ported in [31] (Section 12.10.2). As mentioned above, this is likely to be due to the different coding configurations and test sequences employed in these two experiments.

The average complexity figures for encoding these 18 UHD and HD sequences are presented in Table 12.13. As can be seen, encoder complexity is consistent with coding performance. The VVC encoder has the highest average computational complexity, 57 times that of VP9. AV1 also has relatively high complexity, 39 times of that of VP9. The relationship between the relative complexity and coding performance is illustrated in Fig. 12.34, where VP9 is employed as the benchmark for both figures. It should be noted that these comparisons are all based on the standard test models cited in Table 12.12, so relative performances will inevitably change over time as the test models evolve.

12.13 Summary

We have examined in this chapter the primary features of all major video compression standards from H.120 through H.261, H.263, and MPEG-2 H.264/AVC, to H.265/HEVC and the new H.266/VVC codec. Although these have been largely built on the same hybrid motion-compensated transform architecture, continued innovations over the past three decades or so have delivered, on average, a halving of bit rate for the equivalent video quality every 7–10 years. While this is impressive progress, the demand for high volumes of high-quality video content keeps growing and new approaches will continue to be needed to fulfill future requirements. Some of the potential candidate solutions are considered in the next chapter.

References

[1] C. Rosewarne, K. Sharman, R. Sjöberg, G.J. Sullivan, High Efficiency video coding (HEVC) test model 16 (HM 16) encoder description update 12, in: The JCT-VC Meeting, No. JCTVC-AK1002. ITU-T and ISO/IEC, 2019.

[2] J. Chen, Y. Ye, S.H. Kim, Algorithm description for versatile video coding and test model 9 (VTM 9), in: The JVET Meeting, No JVET-R2002. ITU-T and ISO/IEC, 2020.

[3] B. Benjamin, J. Chen, S. Liu, Y.-K. Wang, Versatile video coding (draft 10), in: The JVET Meeting, No JVET-S2001. ITU-T and ISO/IEC, 2020.

[4] S. Okubu, Reference model methodology-a tool for the collaborative creation of video coding standards, Proc. IEEE 83 (2) (1995) 139–150.

[5] CCITT/SG XV. Codecs for videoconferencing using primary group transmission. Rec. H.120, CCITT (now ITU-T), 1989.

[6] Int. Telecommun. Union-Telecommun. (ITU-T), Recommendation H.261, Video codec for audiovisual services at px64 kbit/s, version 1, 1990; version 2, 1993.

[7] ITU-T and ISO/IEC JTC 1, Generic Coding of Moving Pictures and Associated Audio Information—Part 2: Video, ITU-T Rec. H.262 and ISO/IEC 13818-2 (MPEG-2 Video), version 1, 1994.

[8] ITU-T, Video Coding for Low Bitrate Communication, ITU-T Rec. H.263, version 1, 1995, version 2, 1998, version 3, 2000.

[9] S. Wenger, et al., Error resilience support in H.263+, IEEE Trans. Circuits Syst. Video Technol. 8 (7) (1998) 867–877.

[10] ISO/IEC JTC 1, Coding of Audio-Visual Objects—Part 2: Visual, ISO/IEC 14496-2 (MPEG-4 Visual), version 1, 1999, version 2, 2000, version 3, 2004.

[11] T. Ebrahimi, C. Horne, MPEG-4 natural video coding: an overview, Signal Process. Image Commun. 15 (4–5) (2000) 365–385.

[12] ITU-T and ISO/IEC JTC 1, Advanced Video Coding for Generic Audiovisual Services, ITU-T Rec. H.264 and ISO/IEC 14496-10 (AVC), version 1, 2003, version 2, 2004, versions 3, 4, 2005, versions 5, 6, 2006, versions 7, 8, 2007, versions 9, 10, 11, 2009, versions 12, 13, 2010, versions 14, 15, 2011, version 16, 2012.

[13] T. Wiegand, et al., Overview of the H.264/AVC video coding standard, IEEE Trans. Circuits Syst. Video Technol. 13 (7) (2003) 560–576.

[14] I. Richardson, The H.264 Advanced Video Compression Standard, 2e, Wiley, 2010.

[15] H. Schwartz, D. Marpe, T. Wiegand, Analysis of hierarchical B-pictures and MCTF, in: IEEE Intl. Conf. on Multimedia and Expo, 2006, pp. 1929–1932.

[16] https://en.wikipedia.org/wiki/Advanced_Video_Coding. (Accessed August 2020).
[17] J. Ostermann, et al., Video coding with H.264/AVC: tools, performance, and complexity, IEEE Circuits Syst. Mag. 4 (1) (2004) 7–28.
[18] J-R. Ohm, G. Sullivan, H. Schwartz, T. Tan, T. Wiegand, Comparison of the coding efficiency of video coding standards—including high efficiency video coding (HEVC), IEEE Trans. Circuits Syst. Video Technol. 22 (12) (2012) 1669–1684.
[19] H. Schwarz, D. Marpe, T. Wiegand, Overview of the scalable video coding extension of the H.264/AVC standard, IEEE Trans. Circuits Syst. Video Technol. 17 (9) (2007) 1103–1120.
[20] P. Merkle, A. Smolic, K. Müller, T. Wiegand, Efficient prediction structures for multiview video coding, IEEE Trans. Circuits Syst. Video Technol. 17 (11) (2007) 1461–1473.
[21] J. Lainema, CE6: 2-mode MTS (CE6-2.1), in the JVET meeting, no. JVET-N0053, ITU-T and ISO/IEC, 2019.
[22] A. Vetro, T. Wiegand, G.J. Sullivan, Overview of the stereo and multiview video coding extensions of the H.264/MPEG-4 AVC standard, Proc. IEEE 99 (4) (2011) 626–642.
[23] Joint Collaborative Team on Video Coding (JCT-VC) of ITU-T SG 16 WP 3 and ISO/IEC JTC 1/SC 29/WG 11 ISO/IEC 23008-2 and ITU-T Recommendation H.265, High Efficiency Video Coding (HEVC), November 2019.
[24] G. Sullivan, J-R. Ohm, W. Han, T. Wiegand, Overview of the high efficiency video coding (HEVC) standard, IEEE Trans. Circuits Syst. Video Technol. 22 (12) (2012) 1648–1667.
[25] G. Tech, Y. Chen, K. Müller, J.R. Ohm, A. Vetro, Y.K. Wang, Overview of the multiview and 3D extensions of high efficiency video coding, IEEE Trans. Circuits Syst. Video Technol. 26 (1) (2015) 35–49.
[26] D. Flynn, et al., Overview of the range extensions for the HEVC standard: tools, profiles, and performance, IEEE Trans. Circuits Syst. Video Technol. 26 (1) (2015) 4–19.
[27] M. Wien, High Efficiency Video Coding: Coding Tools and Specification, Springer, 2015.
[28] I. Kim, J. Min, T. Lee, W. Han, J. Park, Block partitioning in the HEVC standard, IEEE Trans. Circuits Syst. Video Technol. 22 (12) (2012) 1697–1706.
[29] J. Xu, R. Joshi, R. Cohen, Overview of the emerging HEVC screen content coding extension, IEEE Trans. Circuits Syst. Video Technol. 26 (1) (2015) 50–62.
[30] J. Sole, et al., Transform coefficient coding in HEVC, IEEE Trans. Circuits Syst. Video Technol. 22 (12) (2012) 1765–1777.
[31] Y. Chen, et al., An overview of coding tools in AV1: the first video codec from the alliance for open media, APSIPA Trans. Signal Inf. Proces. 9 (2020).
[32] T. Laude, et al., A comparison of JEM and AV1 with HEVC: coding tools, coding efficiency and complexity, in: Picture Coding Symposium (PCS), IEEE, 2018, pp. 36–40.
[33] P. Hanhart, T. Ebrahimi, Calculation of average coding efficiency based on subjective quality scores, J. Vis. Commun. Image Represent. 25 (3) (2014) 555–564.
[34] P. Akyazi, T. Ebrahimi, Comparison of compression efficiency between HEVC/H.265, VP9 and AV1 based on subjective quality assessments, in: 10Th Intl. Conf. on Quality of Multimedia Experience (QoMEX), IEEE, 2018, pp. 1–6.
[35] D. Grois, T. Nguyen, D. Marpe, Coding efficiency comparison of AV1, VP9, H.265/MPEG-HEVC, and H. 264/MPEGAVC encoders, in: Picture Coding Symposium (PCS), IEEE, 2016, pp. 1–5.
[36] F. Zhang, A. Katsenou, M. Afonso, G. Dimitrov, D. Bull, Comparing VVC, HEVC and AV1 using objective and subjective assessments, arXiv preprint, arXiv:2003.10282, 2020.

CHAPTER

13

Communicating pictures – the future

In this chapter we briefly review the demands of future content types in the context of an extended video parameter space, to examine how parameters such as spatial resolution, temporal resolution, dynamic range, and color gamut affect our viewing experience, both individually and collectively. The delivery requirements associated with this extended parameter space are considered, with a focus on how compression might affect and influence video quality.

Emphasizing the impact of, and the interactions between, the video parameters and also their content dependence, the justification is made for an increased use of perceptual models in an analysis–synthesis framework. It is argued that perception-based compression techniques, particularly those exploiting deep learning methods, can improve video quality, providing the basis for a new rate-quality optimization (RQO) approach for effective coding of the immersive formats of the future.

13.1 The motivation: more immersive experiences

The moving image industry is a trillion-dollar business worldwide, across film, broadcast, and streaming. According to the USA Bureau of Labor Statistics, we spend one-fifth of our waking lives watching movies, television, or some other form of edited moving image. Visual experiences are thus important drivers for technology development. Cisco predicts that video will account for 79% of all mobile internet traffic by 2022 (up from 59% in 2017), with total annual IP traffic rising to 4.8 zettabytes over the same period (1 zettabyte = 10^{21} bytes = 1000 exabytes) [1]. Despite these impressive numbers, other areas are less buoyant; cinema attendance and revenues across much of the developed world are flat, despite the significant increase in the number of digital 3D screens, and revenues from 3D movies have actually decreased. It is clear that the industry emphasis is shifting towards higher spatial resolutions and high dynamic range formats with larger and brighter screens.

Another area that has grown over recent years is the use of augmented and virtual reality (AR and VR) technologies. VR content is normally simulated or synthetic and finds applications in games, simulation/training, and short film productions, whereas AR overlays synthetic content on the user's view of the real world and has found applications in collaborative design and tourism. A hybrid version, mixed reality (MR), creates environments where real and virtual objects coexist and interact. VR, AR, and MR have the potential for major

Intelligent Image and Video Compression
https://doi.org/10.1016/B978-0-12-820353-8.00022-0

growth and, although past predictions have not been reached in reality, developments in displays, interactive equipment, mobile networks, edge computing, and compression are likely to facilitate these in the coming years.

So there is a significant demand for new, more immersive content: from users who want to experience it, from producers who want to add value to their content, and from service providers who want to expand the use of their networks. So in this context, the questions we need to answer are: what are the best formats for such content, how do we assess its quality, and how do we preserve its value during acquisition, delivery, and display?

13.2 New formats and extended video parameter spaces

The drive for increased immersion coupled with greater mobility means that compression efficiency is a priority for the media, communications, and entertainment industries. Two particular recent events – the explosion of video consumption on portable and personal devices and the increased investment in UHD content and services – demand significant increases in compression efficiency.

However, to produce and deliver more immersive content, it is not sufficient just to increase spatial resolution or screen size. Although these parameters play an important role, other factors such as frame rate, dynamic range, and color gamut are also key to delivering high-value experiences. We will examine some of these briefly below.

13.2.1 Influences

There are many things that can influence our perception of video quality and affect the immersiveness of the viewing experience, not least our interest in the subject matter. However, putting the narrative to one side, the main factors are:

- **Resolution**: Temporal, spatial, and bit depth.
- **Motion characteristics**: Blurring and masking effects.
- **Colorimetry**: Gamut bit depth, tone mapping, and grading.
- **Display**: Size, aspect ratio, dynamic range, and viewing environment.
- **Views**: Monoscopic, stereoscopic, and multiview.
- **Cinematography**: Including set design, lighting, shot length, scene geometry, motion, and color.
- **Other psychovisual characteristics**: Such as peripheral vision, saliency, depth cues, and image noise.
- **Audio**: Ambisonics and wavefields are both exciting developments that complement the visual to enhance the experience.

Complex interactions between these exist and they will, of course, be confounded by strong content dependence. Their impact is however highly moderated by the method and degree of compression employed.

TABLE 13.1 Parameter set for UHDTV / ITU-R BT.2020.

Parameter	Values
Picture aspect ratio	16×9
Pixel count ($H \times V$)	7680×4320, 3840×2160
Sampling lattice	orthogonal
Pixel aspect ratio	1:1 (square pixels)
Frame frequency (Hz)	120, 60, 60/1.001, 50, 30, 30/1.001, 25, 24, 24/1.001
Color bit depth	10 or 12 bits per component
Scan mode	progressive
Viewing distance	$0.75H$
Viewing angle	100 degrees

13.2.2 Spatial resolution

Why spatial detail is important

Spatial resolution is important as it influences how sharply we see objects (particularly when they do not move). The key parameter is not simply the number of pixels in each row or column of the display, but the angle subtended, θ, by each of these pixels on the viewer's retina. Thought of it in this way, increased spatial resolution is particularly important as screen sizes become larger and viewing distances closer. Also, the *effective spatial resolution* will depend on how the compression system interacts with the displayed content. Contrary to popular advertising, there is no such thing as "HD quality." Let us look at this a little closer.

UHDTV and ITU-R BT.2020

The ultrahigh-definition television (UHDTV) standard ITU-R BT.2020 [2] has been a major influence on compression methods in recent years. It is intended to deliver the highest-quality immersive viewing experience to the home, business, or mobile user.

Table 13.1 shows the basic parameters for UHDTV and some interesting points can be observed from this table. Firstly, the assumed viewing distance places the viewer very close to the picture at 0.75 times the screen height, compared to $3H$ for HDTV (ITU-R BT.709). Obviously large screens are assumed as the viewing angle is stated as 100 degrees, in contrast to that for conventional HDTV of around 33 degrees. This will mean significant changes in the way we interact (psychovisually) with the content through increased head and eye movements. In practice this author suspects that the viewing distance will be greater than this for most people in many environments. Nonetheless, different cinematographic methods will be required. Secondly, the maximum frame rate is 120 Hz, rather than the 25, 30, 50, or 60 Hz we are familiar with. This was introduced in response to pressure from organizations such as BBC and NHK, who gained significant experience of spatio-temporal interactions via the Super HiVision project [3]. We will consider this aspect more in Section 13.2.3.

Spatial resolution and visual quality

Since higher spatial resolutions have been supported in recent formats, cameras, displays, and video coding standards, it is important to investigate how much benefit can be achieved from these. Li et al. [4] explored the impact of different downsampling filters on viewer preference scores using a small video dataset (eight source sequences). Their results indicate that

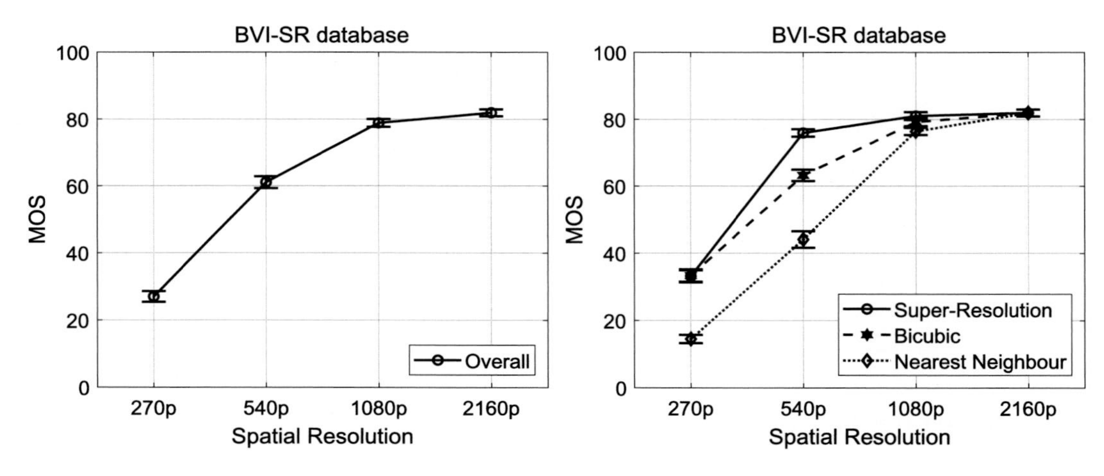

FIGURE 13.1 Results from the subjective experiment [6] showing the overall relationship between visual quality with spatial resolution (left) and that for each upsampling method tested. Error bars represent the standard error among source sequences.

the (then) state-of-the-art filters could not achieve similar perceptual quality to the 2160p reference resolution when upsampling from 1080p and 720p resolutions. Van Wallendael et al. [5] compared 2160p and 1080p resolutions on a larger video dataset (31 source sequences), and found a positive sharpness difference between the reference 2160p and upscaled HD resolutions with a Lanczos-3 filter (although content dependence was reported). Mackin et al. [6] conducted a more comprehensive subjective experiment investigating the relationship between visual quality and spatial resolution when using conventional resampling filters and a convolutional neural network (CNN)-based superresolution approach. The video database employed is publicly available and contains 24 unique video sequences at a range of spatial resolution from 270p to 2160p. The subjective results are shown in Fig. 13.1, which demonstrates a clear relationship between perceived visual quality and spatial resolution. We can observe that the overall visual quality loss (among all test sequences) from UHD to upsampled HD is much less significant than that from upsampled HD to upsampled 540p (or from upscaled 540p to upscaled 270p). The quality reduction also depends on the upsampling approach employed – the CNN-based superresolution offers superior resolution reconstruction performance compared to bicubic and nearest neighbor filters.

Compression performance

Firstly let us take a look at the general interactions between spatial resolution and compression performance and dispel the myth that UHD is always better than HD or HD better than SD. Fig. 13.2 shows how the R-D curves for different formats overlap as the bit rate increases, clearly indicating threshold bit rates where it is better to switch format in order to ensure the best quality. The problem is that it is highly content-dependent and hence content-adaptive techniques are needed to identify these cross-over points. The overall implication is that there is no point increasing spatial resolution unless the compression system can preserve the perceptual benefits associated with that resolution.

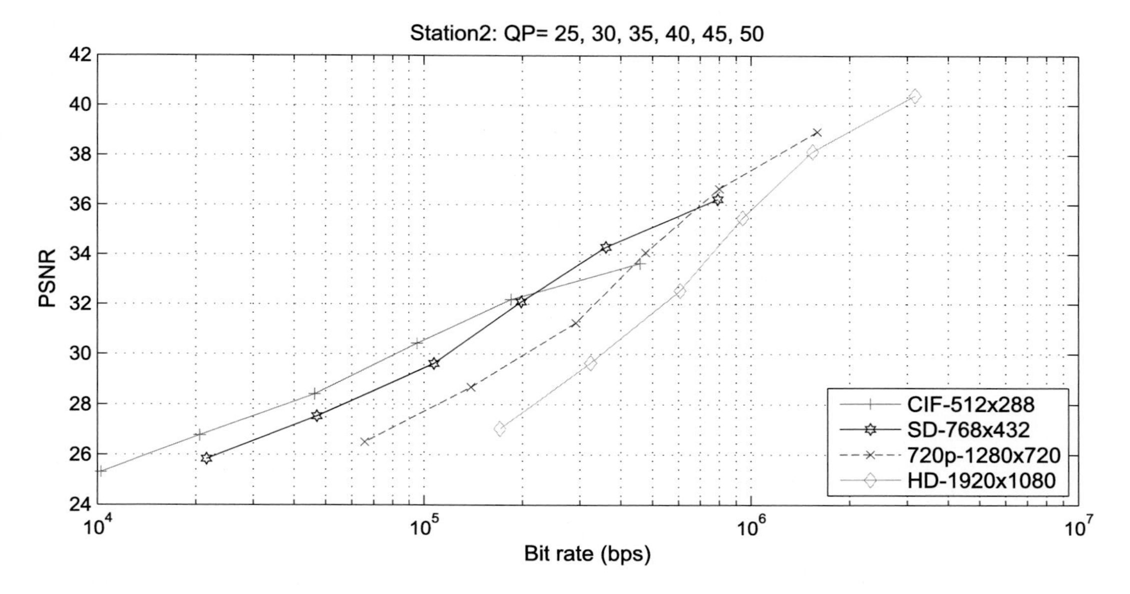

FIGURE 13.2 Does increased resolution mean increased quality?

13.2.3 Temporal resolution

Why rendition of motion is important

We saw in Chapter 2 that motion is important. Our retina is highly tuned to motion, even in the periphery, and our visual system is designed to track motion through retinal and head movements. In terms of video content, motion is conveyed through a combination of frame rate and shutter angle. Frame rates for film were standardized at 24 fps in the 1920s and TV frame rates have been fixed at 25 or 30 fps (using 50 or 60 fields) since the BBC adopted Marconi's 405-line system in 1937. These rates were appropriate at the time, were compatible with mains frequencies, and provided an excellent trade-off between bandwidth, perceived smoothness of motion, and the elimination of flicker. However, as screen sizes grow, along with spatial resolutions, a significant mismatch has developed between frame rate and spatial resolution that few people appreciate [4]. In an attempt to mitigate this, TV manufacturers have introduced sophisticated upsampling at the receiver to smooth motion at 300 or 600 Hz. While this approach can reduce peripheral flicker for larger screens, it does little or nothing to improve motion blur.

The portrayal of motion for a given frame rate is a trade-off between long shutters that create motion blur and short shutters that, in cases of lower frame rates, can cause jerky motion and aliasing. We explore this further below.

Frame rates and shutter angles – static and dynamic resolutions

Fig. 13.4 shows, for a fixed frame rate, the effect that shutter angle has on motion blur over a sequence of three frames. The difference between high frame rate capture and conventional

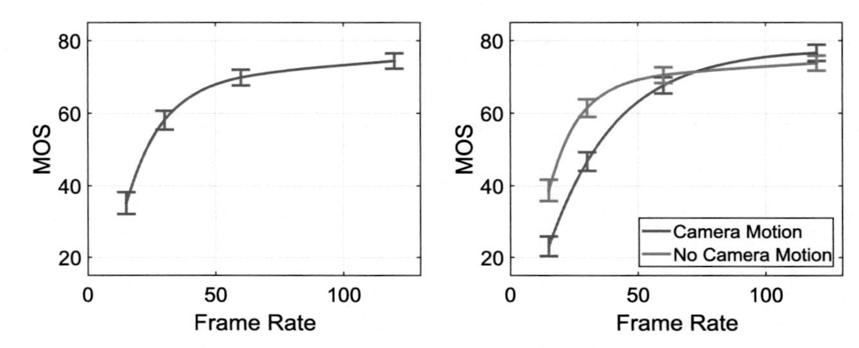

FIGURE 13.3 (Left) The measured relationship between frame rate and average MOS values for all source sequences reported in [10]. (Right) The influence of camera motion. Error bars represent standard error of the mean.

rates is also demonstrated by Fig. 13.5, where the effects of motion blur on the hands and spokes can clearly be seen. To reduce motion blur, we could shorten the shutter time. However, this leads to aliasing effects and jitter, especially in areas that are not being tracked by the viewer.

Let us define dynamic resolution as the effective spatial resolution of a format in the presence of motion. The relationship between resolution and frame rate for the case of TV systems is shown in Fig. 13.6 [7]. This illustrates an important relationship between static and dynamic resolution. The top dashed line shows the frame rate needed to preserve dynamic resolution, extrapolated from a baseline of 50-Hz SDTV, as the spatial resolution is increased. It can be seen that, even at HDTV spatial resolutions, there is a significant reduction in dynamic resolution. Taking this further, to the 8k UHDTV image format, this extrapolation indicates that the corresponding frame rate should be something approaching 600 fps! In contrast, UHDTV has been standardized with a maximum frame rate of 120 Hz, which is clearly well below this line (see Example 13.1).

Higher frame rates hold the potential to significantly improve the rendition of motion, especially for large screens where viewers will engage much more in wide angle tracking. They can reduce motion judder and the occurrence of background distractors caused by motion artifacts and flicker; they also provide an increased perception of depth in the image.

Frame rates and visual quality

The relationship between frame rate and perceived quality has been investigated through psychophysical experiments. However, few studies using natural video content have investigated frame rates beyond 60 fps. Emoto et al. [8] considered frame rates up to 240 Hz, although their source sequences are not publicly available. Sugawara et al. [9] investigated the perception of motion blur, which is of key significance when analyzing frame rates. Mackin et al. reported experimental results based on a large video database with various frame rates up to 120 fps [10]. This subjective evaluation demonstrated a significant relationship between frame rates and perceived quality, and showed the relationship to be highly content-dependent. Fig. 13.3 shows this relationship between frame rate and average MOS among 22 source sequences, and also demonstrates the influence of camera motion on this relationship.

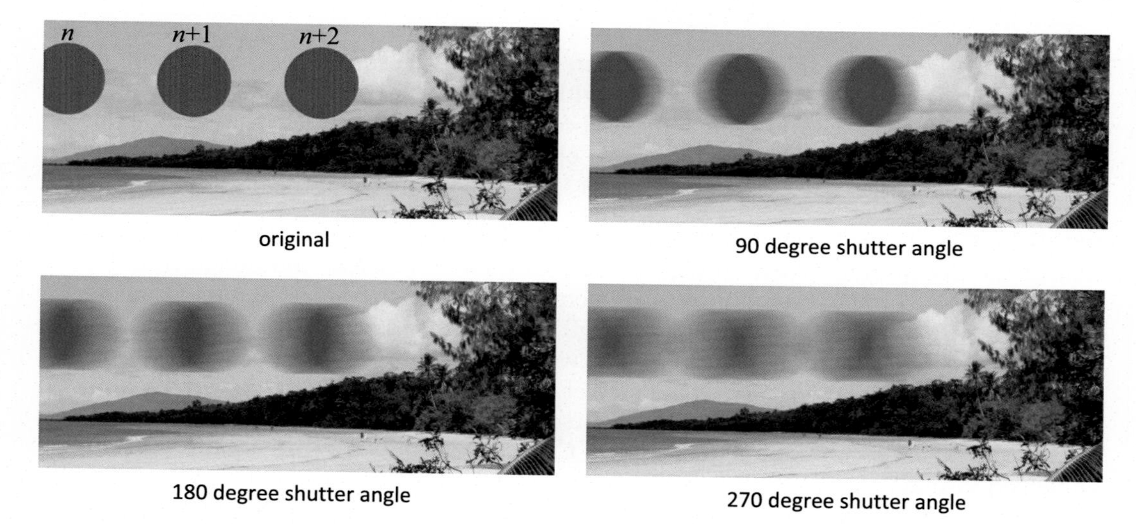

FIGURE 13.4 Frame rate and shutter angle. The figure shows the simulated effect of a ball thrown from left to right, captured at various shutter angles.

FIGURE 13.5 The influence of frame rate (*outdoor* sequence). Left: 25 fps. Right: 600 fps.

Compression methods and performance

So how does frame rate interact with compression performance? A number of factors come into play here. Firstly, the temporal correlation between adjacent frames increases with frame rate. This produces smaller motion vector magnitudes and more correlated motion vectors. Secondly, with reduced frame periods, motion will more closely fit the translational model assumed in most codecs. Thus residual energy will be reduced and fewer bits will be required to code it. Higher frame rates, however, can introduce more high-frequency spatial detail, which can be harder to code in some cases. We can see these effects in Figs. 13.7 and 13.8.

We would thus expect that the bit rate increase would be lower than the fame rate increase. Unpublished results from the Visual Information Laboratory at the University of Bristol indicate approximately a 2:1 ratio between the frame rate increase and the bit rate increase using H.264/AVC or HEVC. This is however highly content-dependent.

Example 13.1. Static and dynamic resolution

Consider the case of an SD video format with 720 horizontal pixels, compared with an 8k UHD system with 7680 horizontal pixels, both at 30 fps. What is the effective dynamic resolution of both formats for the case of an object traversing the width of the frame in 4 s?

Solution. Clearly the horizontal static resolution of the UHDTV format is over 10 times that of the SD format. When an object tracks horizontally across the image frame in 4 seconds, this corresponds to a blurring across 6 pixels per frame for the SD image, compared with 64 pixels per frame for the UHD image. If the screens are of the same size, then the dynamic resolution of both formats will be the same (i.e., 120 pixels). However, if the UHD screen is larger and/or the viewer is closer to it, then its perceived dynamic resolution will actually be significantly worse than that for SD, due to the increased angle subtended at the retina by each pixel. Clearly this is not a great situation for our enhanced format.

Let us say, for example, that the viewing angle for the UHDTV screen is 100 degrees and that for the SD screen is 20 degrees. The dynamic resolution for each case is then given by

$$\text{SD: } 120/20=6 \text{ pixels per degree,}$$

$$\text{UHD: } 120/100=1.2 \text{ pixels per degree.}$$

The expression for the viewer screen normalized dynamic resolution in pixels per degree is thus

$$r_\theta = \frac{f}{2v_o \tan^{-1}\left(\frac{W}{2D}\right)},$$

where v_o is the object velocity in terms of frame widths per second, f is the frame rate, W is the width of the screen, and D is the viewing distance.

13.2.4 Dynamic range

Why dynamic range is important

It is known that increasing the dynamic range of content when displayed, both in terms of bit depth and screen bright-to-dark range, can make video appear more lifelike, increasing the perceived depth in the image and even imparting a sense of 3D. The human visual system (HVS) can cover between 10 and 14 stops without adaptation and with adaption it can accommodate 20 stops for photopic vision, and more for scotopic vision.

A typical modern flat panel TV, however, has a dynamic range capability of only around 8–10 stops. So in order to fully exploit dynamic range, methods such as black-stretch and pregamma mappings are used. The grading process in production also lifts detail in darker areas, and compresses highlights into the normal displayed range. New professional displays

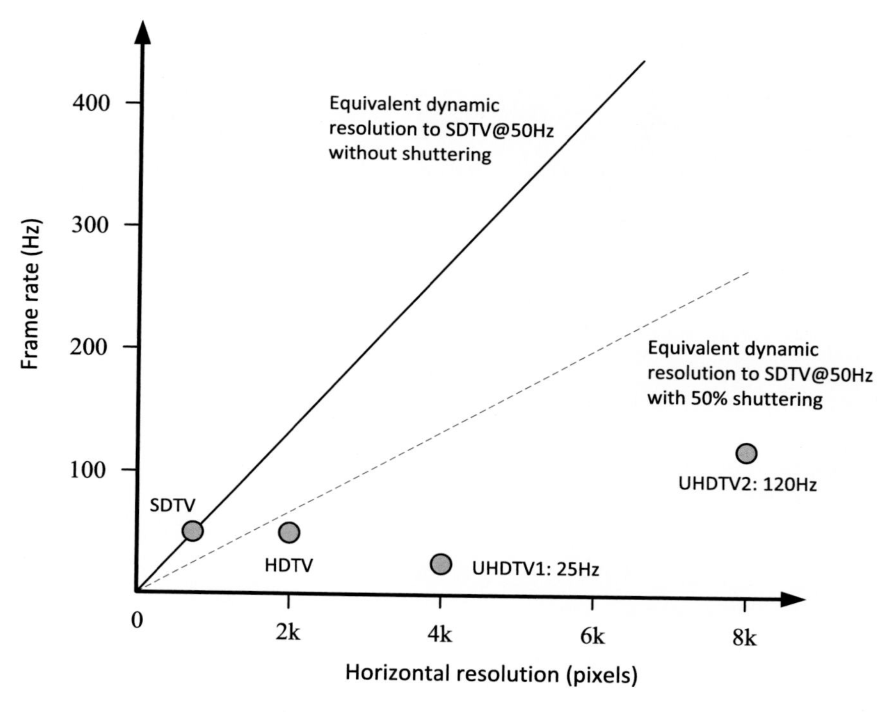

FIGURE 13.6 Static vs. dynamic resolution – the relationship between frame rate and spatial resolution (adapted from an original by Richard Salmon, BBC).

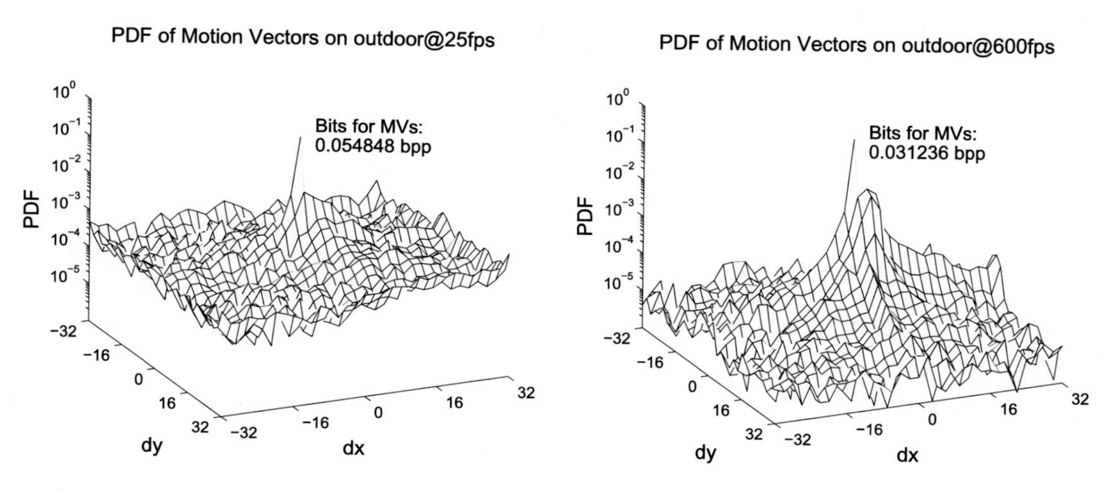

FIGURE 13.7 Motion vector pdf comparisons for 25 fps and 600 fps.

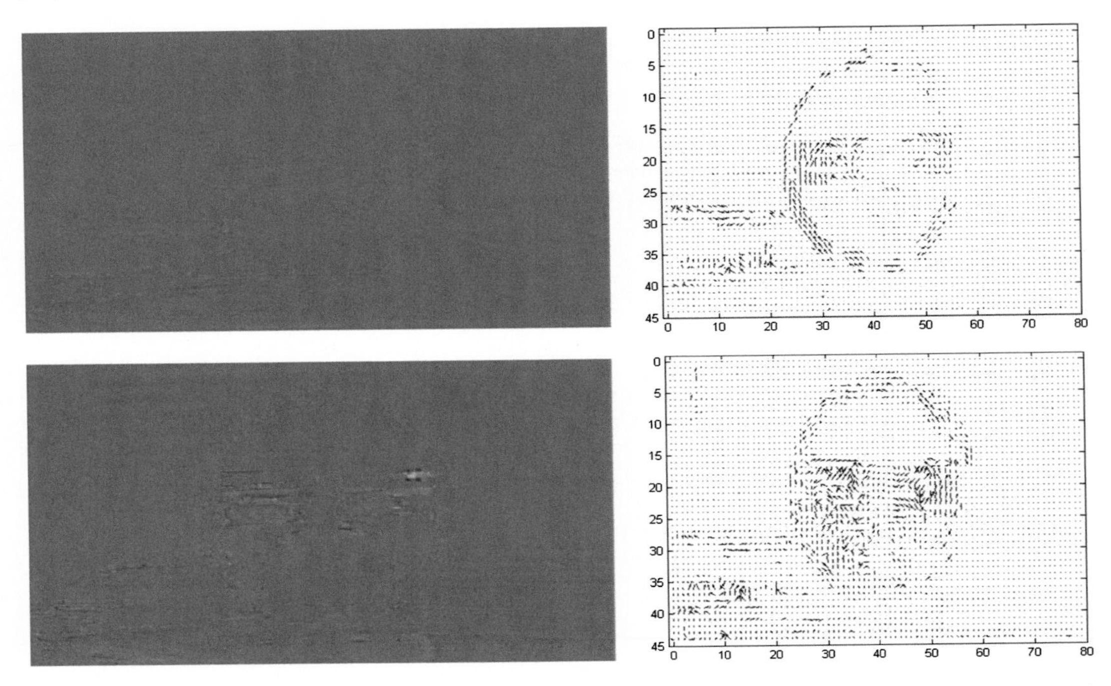

FIGURE 13.8 Motion vectors and residual content for high-frame rate content. Top: 400 fps. Bottom: 25 fps. Left: DFD signal. Right: Spatial distribution of motion vectors.

from companies such as such as SIM2 (e.g., the HDR47E) offer the prospect of getting much closer to the capabilities of the HVS – delivering from approximately 0.4 to 4000 cd/m^2. This is achieved using a high dynamic range (HDR) (but low-resolution) back-light LED array, in conjunction with a conventional LCD panel. Professional camera sensors are also emerging with impressive HDR capability. The RED Monstro 8kVV is a 46-mm sensor with 35.4 megapixels offering 17 stops of dynamic range and, integrated into bodies such as the DSMC2, can support 60 fps of full-format 8k video and 32k at 240 fps. These cameras also support complementary, dual exposures for each frame, enabling the possibility for even greater dynamic range.

Technological challenges of understanding and realizing HDR video still remain. We must, for example, understand the masking effects of this type of content as well as the impact of the viewing environment. We also need to better understand the impact of compression on the immersive properties of such content.

High dynamic range formats

Recommendation ITU-R BT.2100 [11] specifies the parameters for HDR content production and transmission based on two methods, perceptual quantization (PQ) and Hybrid Log-Gamma (HLG). The PQ method employs a nonlinear transfer function based on the characteristics of the HVS, which can achieve a relatively wide brightness range for a given bit depth. Commonly used HDR formats, such as HDR10 and HDR10+, have adopted PQ

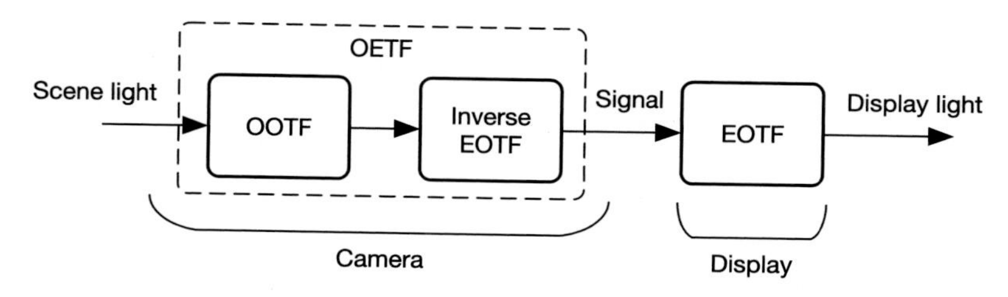

FIGURE 13.9 The PQ HDR system [11].

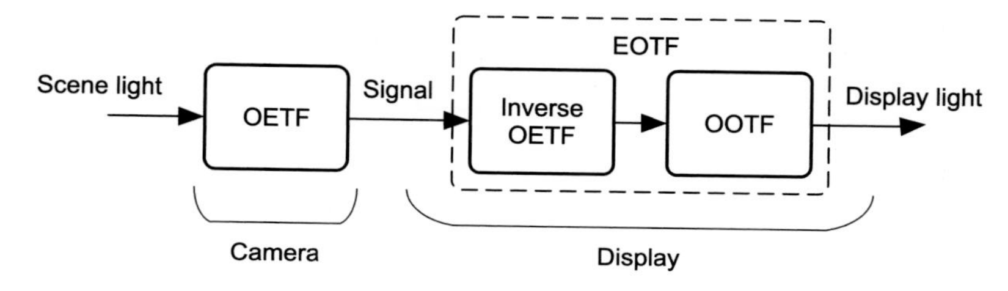

FIGURE 13.10 The HLG HDR system [11].

transfer functions. On the other hand, HLG provides better compatibility with conventional (non-HDR) displays, and is therefore more suitable for generic broadcasting applications.

The high-level system structures for PQ and HLG are illustrated in Figs. 13.9 and 13.10. The opto-electronic transfer function (OETF) is employed within a camera to convert scene light into a video signal. The electro-optical transfer function (EOTF) converts the video signal to linear screen light on the display. The opto-optical transfer function (OOTF) is invoked during the rendering process, converting scene light to display luminance based on viewing environment and display parameters. The primary difference between PQ and HLG is the location of the transformation between optical and electronic signals. This defines the method to be scene-orientated (HLG) or display-orientated (PQ). Both are supported by the current MPEG standards, HEVC and VVC. Detailed transfer function parameters of these two systems can be found in their related standard documents [11,12].

Coding tools for HDR content

Since the first version of HEVC was published, its coding performance on HDR content has been refined by the joint development team JCT-VC, and also investigated within the wider research community [13]. Improvements have been achieved by applying intelligent format conversion to (i) solve chroma leakage problems and (ii) adjust quantization parameters (QPs) based on the characteristics of HDR content. Among the responses to the Joint Call for Proposals on Versatile Video Coding, there were eight contributions specifically focused on HDR compression. Some of these have been adopted in the current version of VVC, including luma mapping with chroma scaling (LMCS) and luma-adaptive deblocking [14].

Other approaches have been reported that exploit human perception during HDR coding. Naccari and Pereira [27] proposed a perceptual video coding architecture incorporating a

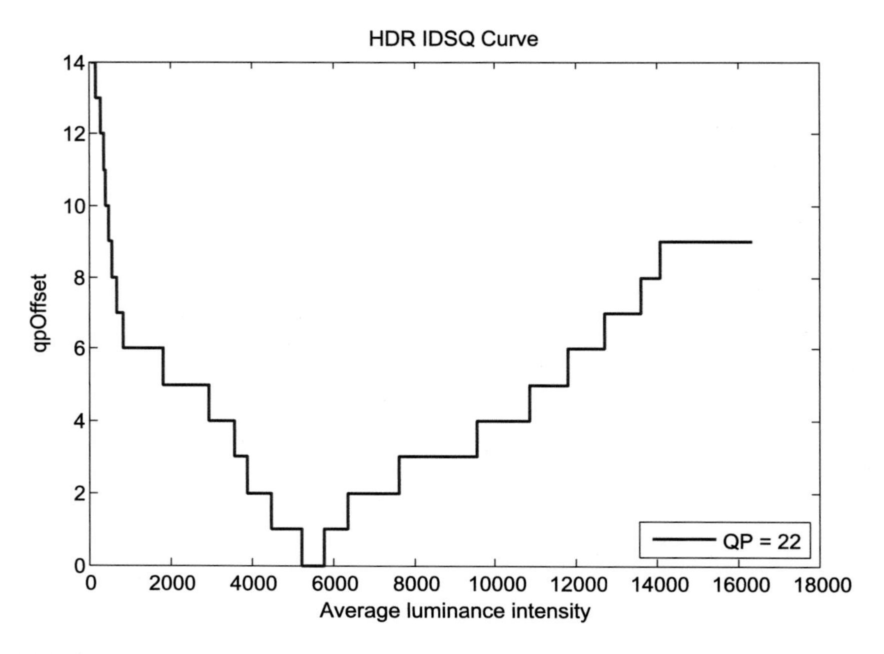

FIGURE 13.11 Intensity-dependent quantization for HDR extensions to HEVC.

JND model based on spatio-temporal masking used for rate allocation and RQO. Zhang et al. proposed a perception-based quantization method that exploits luminance masking [29] to enhance the performance of HEVC. Extending the idea in [28] a profile scaling is proposed, based on a tone-mapping curve computed for each HDR frame. The quantization step is then perceptually tuned on a transform unit basis. The proposed method has been integrated into the HEVC HM reference codec and its performance was assessed using the HDR-VDP-2 image quality metric reporting a bit rate reduction of 9%. An example of this quantization curve is shown in Fig. 13.11.

13.2.5 360-Degree video

What is 360-degree video?

Alongside the extension of the video parameter space with higher spatial resolution, higher frame rate, wider dynamic range, and wider color gamuts, new and more immersive video formats have emerged which enable the viewer to experience three degrees of freedom (3DoF) – i.e., they can look around from a fixed position while watching the video. These are referred to as omnidirectional media, VR, or 360-degree video. This experience can be further enhanced with extension to six degrees of freedom (6DoF), where the user is allowed to move round within the physical environment. These can form the basis of a VR, AR, or MR system with content being wholly synthesized animation as in a game, or natural video content acquired using a 360-degree camera. The definitions of different terminology related to 360-degree video are given below [30].

FIGURE 13.12 An example of a 360-degree video frame. Example frame from the *harbor* sequence in the JVET 360-degree video test set [31].

- **Virtual reality (VR)**: Reconstructs a complete virtual environment to create an immersive experience, typically via a head-mounted display.
- **Augmented reality (AR)**: Implies an interactive experience (often using a smartphone or tablet) with computer-generated objects residing in the (enhanced) real environment.
- **Mixed reality (MR)**: Combines the real and virtual world in a new immersive environment, where physical and digital objects/background coexist and interact.
- **Extended reality (XR)**: A generic term that includes all virtual and physical immersive experiences and the interactions between the user and the system. It is a superset containing VR, AR, and MR.

Compression of 360-degree video

360-degree videos are formatted by mapping or projecting a spherical field of view to 2D images or video frames for representation, processing, communication, and display. Commonly used projection methods include the equirectangular projection (ERP), the Lambert cylindrical equal-area projection (EAP), and the cubemap projection (CMP). An example of a projected 360-degree video frame is shown in Fig. 13.12. Projected videos can then be compressed by conventional video codecs, such as HEVC and VVC. Specific coding tools have also been proposed for encoding 360-degree video content [32], including:

- adaptive quantization to compensate the sampling distortions due to projection,
- disabling inefficient "frame packed neighbors" and employing spherical neighbors during intra-prediction, motion estimation, and in-loop filtering,
- alternative boundary padding approaches in motion compensation (see Section 12.9.10).

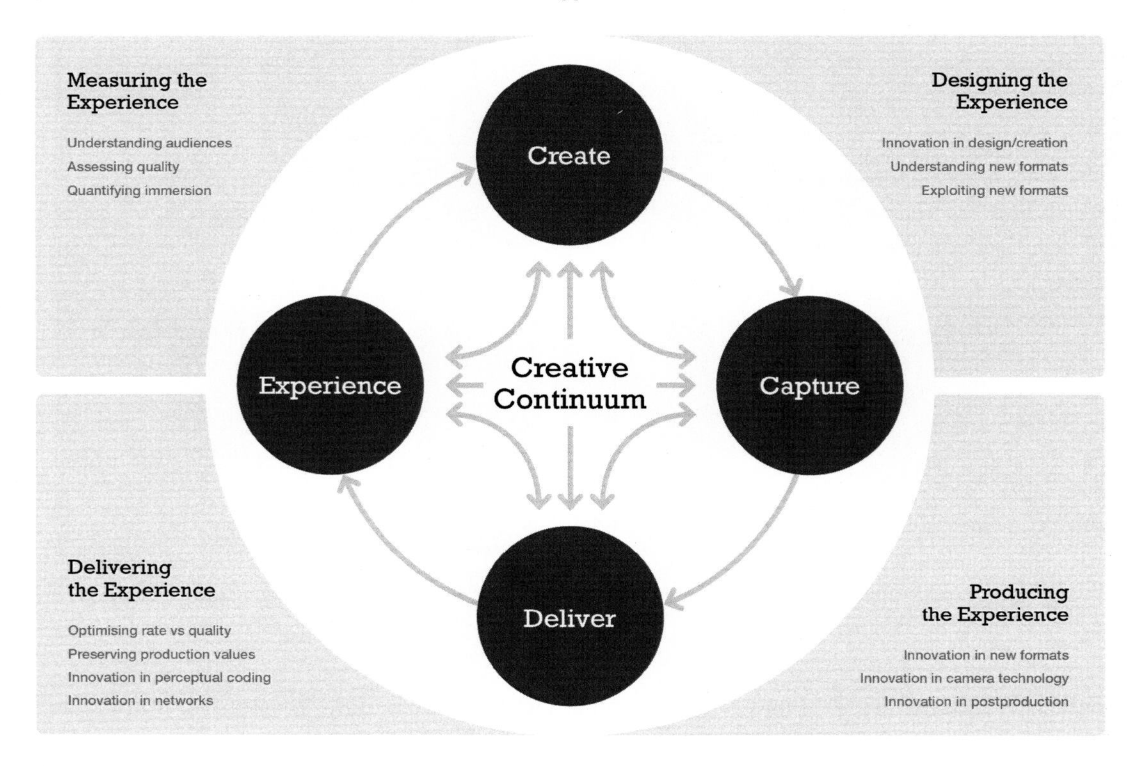

FIGURE 13.13 The Creative Continuum.

13.2.6 Parameter interactions and the creative continuum

So what is required to enable us to create an optimal viewing experience? In essence, we need:

- an understanding of the influence of, and interactions within, the extended visual parameter space,
- generalized psychovisual measures of quality, separating narrative from format/medium, and environment,
- use of these measures to characterize the impact of distortion, dynamic range, color palette, and spatial and temporal resolution,
- representations, formats, acquisition, display, and delivery processes that preserve or enhance immersive properties,
- parameter optimization and adaptation methods that take account of the influences of content and context,
- compression methods that minimize bit rate while preserving immersive properties,
- an understanding of the impact of cinematographic methods, e.g., shot length, framing, camera placement, and camera and subject motion,
- new acquisition, delivery, and display technologies.

Because of these content-dependent parameter interactions, a question that naturally arises is, should we maintain the idea of a fixed parameter set at all, or should we be looking more into formats that move away from the notion of fixed frame rates and resolutions?

Perhaps the most important message is that we cannot consider acquisition, production, compression, display, and assessment as independent processes, but should instead view these as a creative continuum. This intimately links production, cinematography, and acquisition to delivery, display, consumption, and quality assessment. Rather than an end-to-end delivery system we should instead consider a set of continuous relationships within an extended parameter space, as indicated in Fig. 13.13, where:

- **The experience** must maximize engagement with the displayed content. Measuring it is essential if we are to fully understand the influences of the delivery and display processes.
- **The delivery** processes must ensure that the content is delivered in a manner that preserves the immersive properties of the format.
- **The capture** processes must employ formats and parameters that enable immersive experiences.
- **The creation** processes must be matched to the acquisition formats to ensure optimal exploitation of the format in terms of sets, lighting, etc.

Importantly, an ability to measure the quality of the experience, not simply in terms of video quality, but in terms of how engaging it is, is critical in optimizing other parameters and processes in the delivery chain.

13.3 Intelligent video compression

13.3.1 Challenges for compression: understanding content and context

While the demand for new video services will, to some extent, be addressed through efficiency improvements in network and physical layer technology, the role of video compression remains of central importance in ensuring that content is delivered at an acceptable quality while remaining matched to the available bandwidth and variable nature of the channel. In the context of these growing demands for more content, at higher qualities and in more immersive formats, there is a need for transformational solutions to the video compression problem, and these must go well beyond the capabilities of existing standards.

All major video coding standards since H.261 have been based on incremental improvements to the hybrid motion-compensated block transform coding model. While this approach has produced impressive rate-distortion improvements, more disruptive techniques offer the potential to provide substantial additional gains. As we have seen, H.264/AVC is based on the picture-wise processing and waveform-based coding of video signals. HEVC and VVC are a generalization of this approach offering gains through improved intra-prediction, larger block sizes, more flexible ways of decomposing blocks for inter- and intra-coding, and better exploitation of long-term correlations and picture dependencies.

Example 13.2. The delivery challenge

Let us for now put the above discussion in context by considering the effect that the immersive parameter set might have on bit rate. Let us assume that, for given content, optimal immersion is provided by the following fixed parameter values:

- frame rate: 200 fps,
- spatial resolution: UHDTV resolution at 7680×4320 pixels,
- dynamic range: requiring 16 bits of dynamic range in R, G, and B.

Calculate the uncompressed bit rate for this format and compare its compression requirements with those for existing HDTV broadcasting and internet streaming systems.

Solution. Assuming no color subsampling, the overall uncompressed bit rate would be around 3×10^{11} bps. This is approximately 100 times greater than HDTV at 1080p50, 50,000 times greater than a typical current broadcast compressed bit rate, and 100,000 times greater than high-quality internet HD delivery.

Consider now Example 13.2. This clearly illustrates that, despite video compression advances providing a 50% reduction in bit rate every 7–10 years, such immersive parameter sets demand far greater bandwidths than are currently available. We thus not only need to discover the optimal video parameters, but we also need to understand the perceptual implications of compression in order to specify a compression ratio that, while exploiting psychovisual redundancy, exploits masking effects to minimize bit rate. It is clear that conventional compression techniques are unlikely to meet these requirements and hence new, perceptually driven methods will be necessary.

New approaches should be predicated on the assumption that, in most cases, the target of video compression is to provide good subjective quality rather than to minimize the error between the original and coded pictures. It is thus possible to conceive of a compression scheme where an analysis–synthesis framework replaces the conventional energy minimization approach. Such a scheme could offer substantially lower bit rates through region-based parameterization and reduced residual and motion vector coding. New and alternative frameworks of this type are beginning to emerge, where prediction and signal representations are based on a parametric or data-driven model of scene content. These often invoke a combination of waveform coding and texture replacement, either using explicit computer graphic models or trained neural networks to synthesize or interpolate target textures at the decoder. These approaches can also benefit from the use of higher order motion models for texture warping and mosaicing, from the use of contextual scene knowledge and from embedded perceptually driven quality metrics (see Chapter 10).

Work to date has demonstrated the potential for significant rate-quality improvements with these methods. The choice of texture analysis and synthesis models, alongside meaningful quality metrics and the exploitation of long-term picture dependencies, will be key if an effective and reliable system is to result with acceptable complexity. Furthermore, in order to rigorously evaluate these new methods, challenging and representative test data will be required. The hypothesis that underpins this approach is the following.

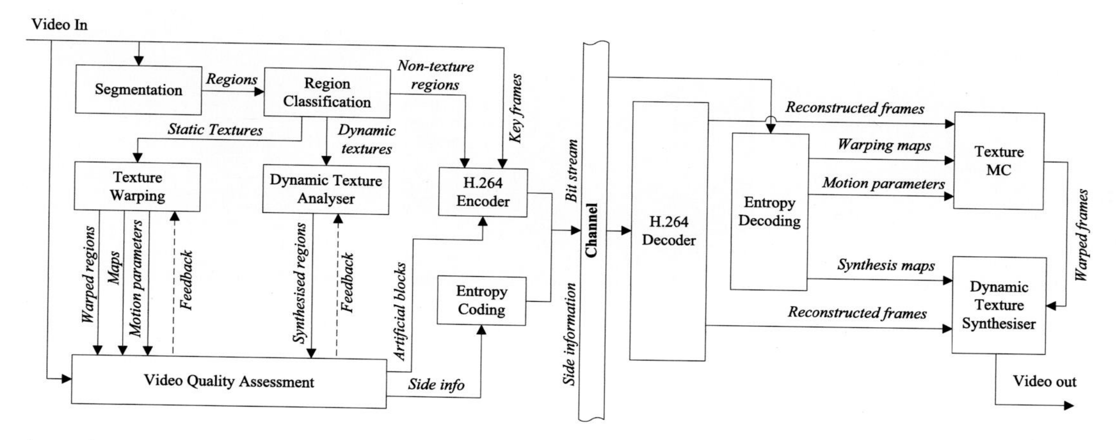

FIGURE 13.14 Parametric video compression architecture [20].

If consistent, representative spatio-temporal descriptions can be obtained for static and dynamic textures and a valid perceptual distortion measure can be defined, then these descriptions can be employed within a rate-quality optimized parametric framework to dramatically improve video compression performance.

13.3.2 Parametric video compression

Prior to recent advances in machine learning, there was a clear understanding that a better understanding of perceptual processes and their link to the content being viewed was essential in order to enhance coding performance [21]. Early approaches, based on parameterization of content texture types, employed an analysis–synthesis framework rather than the conventional energy minimization approach. Parametric methods were employed to describe texture warping and/or synthesis by Ndjiki-Nya et al. [15,16], Bosch et al. [17], Bryne et al. [18,19], Stojanovic et al. [25] Zhu et al. [25,26], and Zhang and Bull [20].

The work of Zhang and Bull [20] combined explicit dynamic and static texture synthesis within a framework of robust region segmentation, classification, and quality assessment, hosted as a mode within a conventional block-based codec. A spatial texture-based image segmentation algorithm [23] was employed to obtain and classify textured regions. Static textures were coded using a perspective model, while dynamic textures are coded using a modified version of the synthesis method of Doretto et al. [24].

An ongoing challenge for synthesis-based coding is to create a reliable in-loop quality assessment measure as a proxy for subjective quality. Existing distortion-based metrics either do not well with this type of compression, introduce too much delay, or are too complex. A simple metric (AVM) was developed by Zhang and Bull to provide a monotonic quality estimate for synthesized textures.

The overall architecture of this scheme is shown in Fig. 13.14, while Fig. 13.15 gives an example of a synthesized frame compared with H.264/AVC. The results presented in [20] indicated savings up to 60% over and above those offered by the host codec (H.264/AVC) alone for the same subjective quality.

 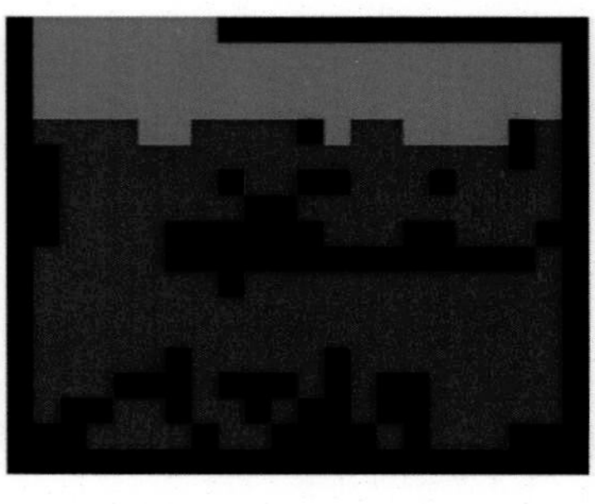 

FIGURE 13.15 Coding results from parametric video coding. Left: H.264 frame. Right: proposed method. Center: Coding mode employed [20].

13.3.3 Context-based video compression

Other unconventional approaches have exploited spatio-temporal relationships based on prior knowledge of a scene. For example, certain applications, such as sports broadcasting, are highly demanding due to their activity levels and the perceptual quality levels required. Such content is however often captured in a closed and well-defined environment such as a sports arena. The context-based coding method of Vigars et al. [22] exploits prior knowledge about the geometry of a scene. It applies a planar-perspective motion model to rigid planar and quasiplanar regions of the video. This enables independent planar regions to be coded accurately without the need for registration, blending, or residual coding.

Fig. 13.16 shows the matching process across two frames based on feature point matching. Firstly prior knowledge of the environment in which the video is captured is encapsulated into a scene model. Feature matching is used to detect salient points in each frame, and to compose descriptors of neighborhoods for matching and tracking between images. This is then used to track known planar structures in the video. Planar structures in each frame of the video are thus located, facilitating perspective motion estimation between them. Foreground segmentation is then used to separate regions of the video which do not conform to the planar model; these are processed by the host codec. Vigars et al. report savings up to 50% for this method, when applied to suitable content, compared with conventional H.264/AVC encoding.

13.4 Deep video compression

As introduced in Chapter 3, deep neural networks (DNNs) are gaining popularity in the image and video compression research community due to their ability to achieve consistently greater coding gain than conventional approaches. They provide an important framework for generalizing, extending, and enhancing the context-based synthesis methods described in Section 13.3.

Deep compression methods are also now being considered in mainstream video coding standardization bodies such as MPEG (VVC) and AOM (AV2), and JVET has recently es-

FIGURE 13.16 Sped-up robust features (SURF)-based feature matching for context-based coding of planar regions. Left: Reference frame. Right: Current frame. Circles represent RANSAC inliers used for planar motion modeling.

tablished a working group on this topic. In this section, we review some of the most recent approaches, including both individual coding tools and new end-to-end architectures. We also emphasize the need for diverse and representative training data and how this can be supported by data synthesis and augmentation. For further information, a recent review of deep video compression methods can be found in [34] and a broader review of artificial intelligence in the creative industries in [33].

13.4.1 The need for data – training datasets for compression

Coverage, generalization, and bias

A deep learning framework comprises a computational architecture, a learning strategy, and a data environment. The composition of the training database is thus key to the success of a deep learning system. A well-designed training database with appropriate size and coverage can help significantly with model generalization and avoiding problems of overfitting. Good datasets will contain large numbers of examples with a statistical distribution matched to the problem domain. This enables the network to estimate gradients in weight-error surface which in turn enable it to converge to an optimal solution with robust decision boundaries.

Datasets should comprise:

1. data that are statistically similar to the inputs when the models are used in real situations and
2. ground truth annotations or target content that inform the machine what the desired outputs are.

For example, in segmentation applications, the dataset would comprise the images and the corresponding segmentation maps indicating homogeneous or semantically meaningful regions in each image. Similarly, for object recognition, the dataset would also include the original images while the ground truth would be the object categories, e.g., car, house, human, type of animals, etc. For compression, the target outputs would typically be the uncompressed content.

A number of labeled datasets exist for public use,[1] but these are limited, especially in cases where data is difficult to collect and label. One of the largest, ImageNet, contains over 14 million images labeled into 22,000 classes. Care must be taken when collecting or using data to avoid imbalance and bias – skewed class distributions can cause machine learning algorithms to develop a bias towards classes with a greater number of instances, preferentially predicting majority class data. Features of minority classes are treated as noise and are often ignored.

Numerous approaches have been introduced to address the issue of imbalanced distributions. For example, Zhang et al. [52] reweight the loss of each pixel based on the pixel color rarity. Recently, Lehtinen et al. have introduced an innovative approach to learning via their Noise2Noise network [53] which demonstrates that it is possible to train a network without clean data if the corrupted data complies with certain statistical assumptions. Typical data manipulation techniques include downsampling majority classes, oversampling minority classes, or both.

Data synthesis and augmentation

Data augmentation techniques are frequently used to increase the volume and diversity of a training dataset without the need to collect new data. In order to increase the representation of minority classes and avoid overfitting, existing data is subjected to transformations such as cropping, flipping, translating, rotating, and scaling [54]. Generative adversarial networks (GANs) (see Section 3.7) have recently been employed with success to enlarge training sets; for example, CycleGAN was improved in [55] with a structure-aware network to augment training data for vehicle detection. IBM Research introduced a balancing GAN [56], where the model learns useful features from majority classes and uses these to generate images for minority classes that avoid features close to those of majority cases. An extensive survey of data augmentation techniques can be found in [57].

Scientific or parametric models can be exploited to generate synthetic data in those applications where it is difficult to collect real data, and where data augmentation techniques cannot increase variety in the dataset. Synthetic data are often created by degrading the clean data. For example, in [58], synthetic motion blur is applied on sharp video frames to train the deblurring model.

Training datasets for compression

Learning-based compression demands volumes of training material much greater than typically used for conventional compression. These should include diverse content covering different formats and video texture types. However, most databases developed for computer vision applications do not provide sufficient content coverage and diversity for coding applications. As a consequence, the generalization performance of networks cannot be ensured. Research by Ma et al. [51] at the University of Bristol has demonstrated the importance of large and diverse datasets when developing CNN-based coding tools. Their BVI-DVC

[1] https://en.wikipedia.org/wiki/List_of_datasets_for_machine-learning_research, https://ieee-dataport.org/.

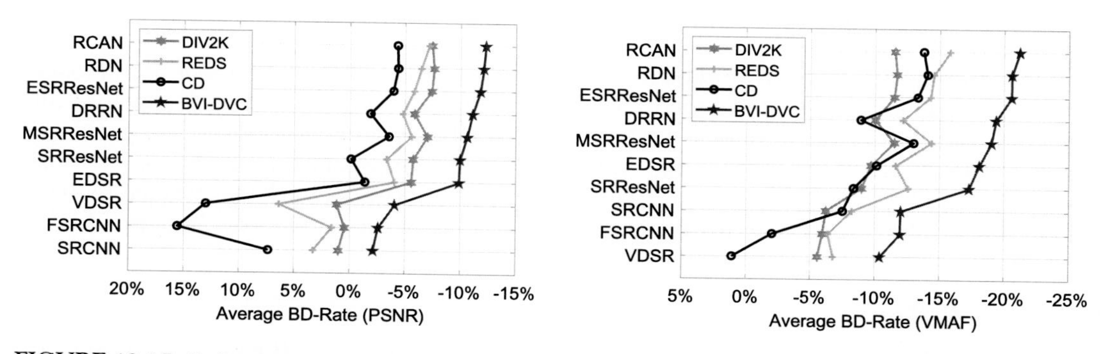

FIGURE 13.17 Coding gains from BVI-DVC compared to other training databases, DIV2K [59], REDS [60], and CD [61]. Results are for 10 popular network architectures trained for various coding modules in HEVC. Negative BD-rate values indicate bit rate savings or coding gains.

database is publicly available and produces significant improvements across a wide range of deep learning networks for a variety of coding tools, including loop filtering and post-decoder enhancement. Fig. 13.17 summarizes the coding gains achieved by using BVI-DVC compared to three other training databases used in deep video compression, for the same network architectures and various coding modules in the context of HEVC. BD-rate is used to evaluate the coding gains based on two quality metrics, PSNR and VMAF, benchmarked against an HEVC anchor.

13.4.2 Deep optimization of compression tools

Deep learning has been successfully applied in research laboratories to optimize a range of coding tools for use within conventional codec architectures. These methods have been applied to optimize decorrelating transforms, intra-prediction, motion prediction, loop filters, and entropy coding. They have also shown significant benefits as an intelligent post-processing tool. Among the responses to the call for proposals for VVC, there were four proposals containing coding tools based on deep learning. Recent innovations are briefly reviewed below.

Transforms and quantization

A conventional discrete cosine transform (DCT) can be replaced by a pretrained CNN model, as demonstrated by Liu et al. in [41], where a CNN-based transform was reported to achieve 9% bit rate savings over the DCT. Deep learning can also be employed to enhance conventional quantization schemes. Alam et al. [42] proposed a new adaptive quantization strategy, which employs a fast CNN model (with only three layers) to predict artifact visibility. Local QP values (e.g., for each CTU) were adjusted based on the CNN-estimated visibility thresholds and then used for compression. This approach was reported to provide 11% coding gains over HEVC based on an assessment using the structural similarity image metric (SSIM).

Intra-prediction

Deep learning has also been applied to enhance intra-prediction for both lossless and lossy compression. Schiopu et al. developed a CNN-based prediction model, which replaced all 33 angular intra directions in HEVC [39]. Consistent bit rate savings (5% on average) were reported against HEVC lossless coding. Similar approaches have been applied to lossy compression, including the work of Li et al., where a fully connected network was trained to map from multiple reference lines to the current block [43]. When this model was integrated into HEVC, coding gains of approximately 3.4% were claimed.

Motion prediction

For inter-prediction, deep learning has been employed to improve fractional motion estimation. Liu et al. [47] utilize a grouped variation CNN to interpolate subpixels at different locations. This achieved promising bit rate savings of 2.2% integrated into HEVC using its low delay mode. Zhao et al. presented a different strategy, using a CNN-based approach to enhance bidirectional prediction in HEVC [40]. This CNN-based synthesis model outperformed the original biprediction module in HEVC, delivering a 3.0% coding gain. This type of CNN-based synthesis has also been used to generate virtual reference frames to improve inter-prediction performance [46].

Entropy coding

Entropy coding operations can also be enhanced using deep learning. Song et al. developed an arithmetic coding approach to predict the probability distribution of intra-prediction modes. This method was integrated in HEVC to replace the binarization and context models for intra-coding, offering up to 9.9% bit rate savings compared to the original CABAC [48].

Postprocessing and loop filtering

Postprocessing is commonly used following video decoding to reduce the visual impact of coding artifacts and to enhance the overall quality of reconstructed frames. Due to its powerful enhancement capability, CNN-based postprocessing approaches have the potential to provide more significant quality improvements compared to conventional filter-based methods [44]. A notable example was proposed by Zhang et al. [35], who modified a residual block-based CNN architecture targeted at postprocessing compressed content. This method was evaluated on the VVC test model, showing consistent improvements for wide QP and bit rate ranges. This work has been further extended and integrated into both HEVC and VVC encoders as an additional in-loop filter [36].

Deep resampling and the ViSTRA architecture

Spatial resolution adaptation has also been integrated into video compression to improve coding efficiency, especially for high-spatial resolution content. These methods encode downsampled versions of the original video frames and reconstruct the original resolution during the decoding process. This approach was originally adopted for relatively low bit rate applications due to inconsistent reconstruction performance when conventional upsampling filters were used. Inspired by recent advances in CNN-based superresolution processing, spatial resolution adaptation has been applied across extended bit rate ranges and has delivered significantly enhanced reconstruction performance.

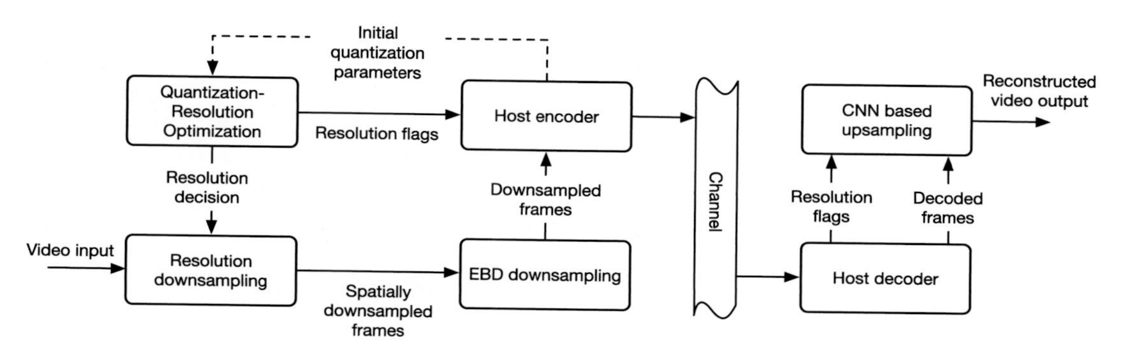

FIGURE 13.18 Coding framework of the intelligent resampling method ViSTRA [38].

This adaptation process can be applied at block (CTU) or frame level. A typical example of frame-level resolution adaptation is the ViSTRA codec proposed by Afonso et al. [62], which employs perceptual criteria during the downsampling stage and reconstructs the full-resolution video using a deep CNN. ViSTRA has been further enhanced with effective bit depth (EBD) adaptation [38], and the original resolution and bit depth can be reconstructed using a single CNN model. The structure of the ViSTRA coding architecture is illustrated in Fig. 13.18. ViSTRA has been integrated into both HEVC and VVC, and achieved consistent and significant compression gains against these test models, with average BD-rate savings of 12.6% over HM and 5.5% over VTM. Fig. 13.19 shows example rate-PSNR curves for four JVET test sequences with various spatial resolutions. Example frame shots are shown in Fig. 13.20.

Perceptual loss functions

In most of the deep learning-based coding tools introduced above, mean absolute difference (MAD) (ℓ_1) and mean squared error (MSE) (ℓ_2) have been used as the loss functions for training in order to minimize pixel-wise distortions. As we introduced in Chapter 10, these measurements do not always correlate well with perceived visual quality. Using a GAN architecture, Ma et al. modified the CNN training strategy by incorporating perceptual loss functions based on SSIM and multiscale SSIM (MS-SSIM), and retrained the CNN models for spatial resolution and bit depth upsampling [37,50]. The experimental results demonstrate that the models trained using perceptual loss functions provide significant coding gains compared to ℓ_1 or ℓ_2 trained CNNs based on subjective perceptual comparisons. Comparative results are shown in Table 13.2, where the coding gains for CNN-based spatial resolution adaptation using ℓ_1 and perceptual loss functions are presented for four UHD JVET test sequences. We can observe that, for the perceptual quality metric VMAF, the rate quality performance is improved when ℓ_1 loss is replaced by perceptual loss functions in the training process.

The complexity issues of deep video compression

Although the deep learning-based coding tools discussed above offer powerful solutions compared to conventional coding algorithms, they have the disadvantage of increased com-

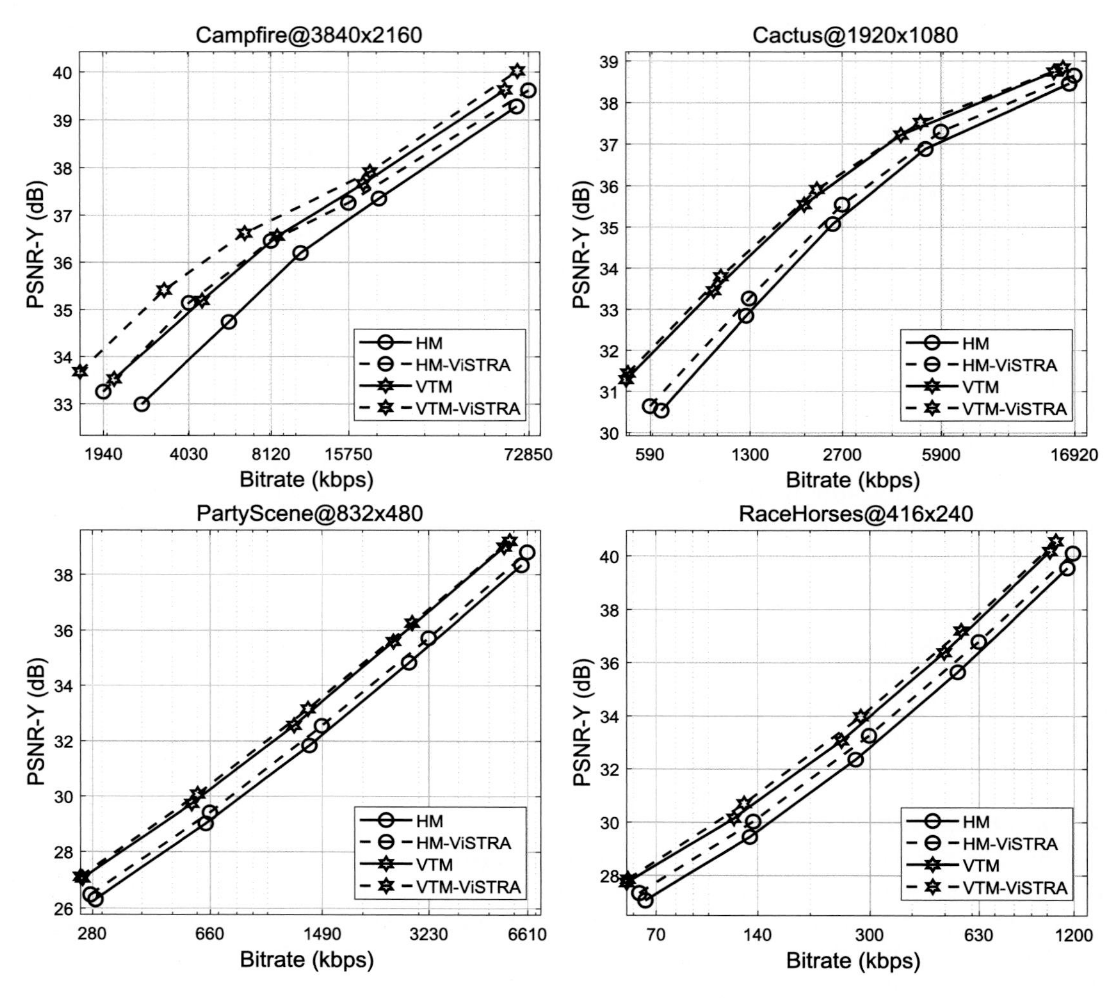

FIGURE 13.19 Rate-PSNR curves for HEVC HM and VVC VTM and their corresponding ViSTRA codecs on JVET test sequences with various spatial resolutions.

TABLE 13.2 Coding gains for CNN-based spatial resolution adaptation using ℓ_1 and perceptual loss functions for model training. HEVC HM is used as a benchmark (negative BD-rate values indicate coding gains).

Sequence	ℓ_1 loss		Perceptual loss	
	BD-rate (PSNR)	BD-rate (VMAF)	BD-rate (PSNR)	BD-rate (VMAF)
Campfire	−24.5%	−42.0%	−21.4%	−46.2%
Tango2	−16.4%	−23.0%	−13.8%	−27.6%
CatRobot1	−5.0%	−22.5%	−0.2%	−33.5%
ParkRunning3	−24.4%	−34.7%	−23.2%	−44.0%

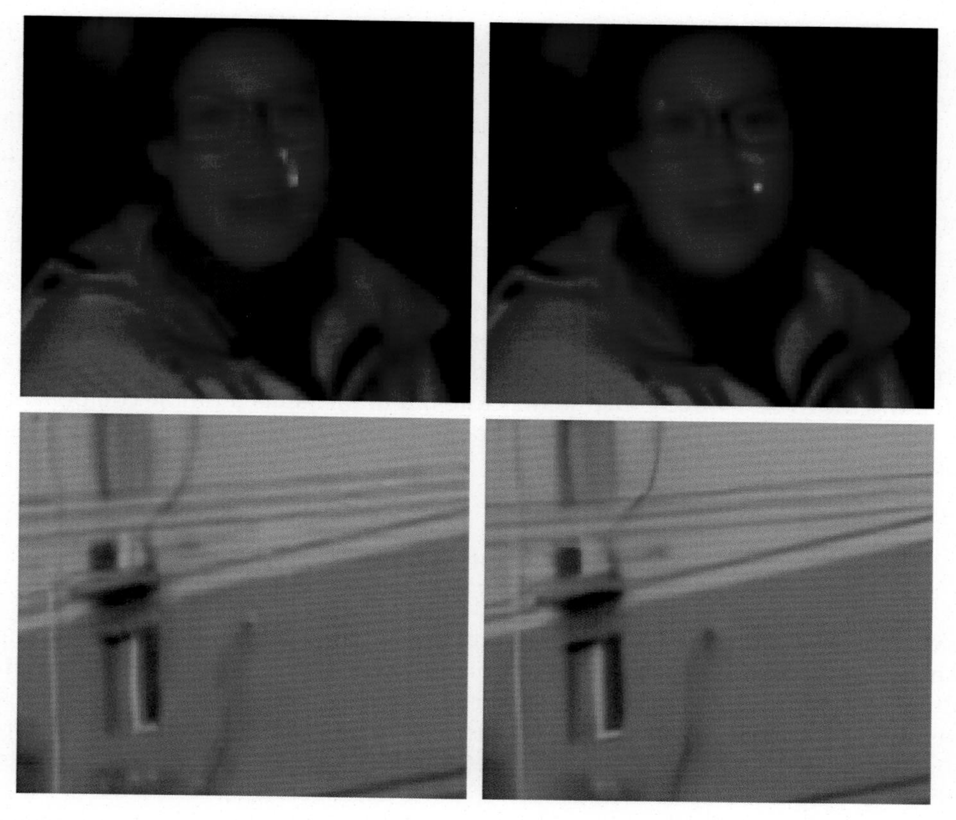

FIGURE 13.20 Perceptual comparisons between HEVC HM 16.20 (left) and HM with CNN-based spatial resolution adaptation using ℓ_1 (right) for model training. Each pair of patches are extracted from encoded test sequences with similar bit rates, 5.7 Mb/s for *campfire* (top) and 2.9Mb/s for *DaylightRoad* (bottom).

putational complexity, in particular when CNN-based methods are located at the decoder. As a consequence, standards such as VVC have not adopted any coding tools based on deep learning in its first version. However, a machine learning extension is currently under investigation. The complexity issue has been recently addressed by Ma et al., who modified the spatial resolution adaptation framework in [38] to support flexible allocation of complexity between the encoder and the decoder. This approach employs a CNN model for video downsampling at the encoder and uses a Lanczos3 filter to reconstruct full resolution at the decoder, which can reduce the computational complexity at both ends (29% for encoder and 10% for decoder) yet still achieve significant bit rate savings (more than 10%) over conventional codecs.

13.4.3 End-to-end architectures for deep image compression

As an alternative to the enhancement of specific coding tools within a conventional compression framework, several authors have investigated deep network architectures for end-to-end training and optimization. This type of approach presents a radical departure from conventional coding strategies and, while it is not yet competitive with state-of-the-art conventional video codecs, it holds significant promise for the future. Ballé et al. presented a general framework for rate-distortion optimized image compression [63] based on nonlinear transforms, which consists of multiple convolutional filters and nonlinear activation functions. All the network parameters employed in the transforms (in both encoder and decoder) are jointly trained using stochastic gradient descent to achieve optimal overall rate-distortion performance. This work was reported to offer improvements over JPEG and JPEG2000 based on PSNR and MS-SSIM, and with further enhancement to provide coding gains against HEVC intra-coding (BPG) [65]. This framework has been recently extended to video compression based on optical flow [64,66] and was reported to outperform HEVC fast codec (x265) based on MS-SSIM. A further recent contribution was proposed by Lu et al. [45,49], which follows the hybrid video codec architecture but implements multiple components using (deep) neural networks. Their results show performance improvements over x265 (in very fast mode) of up to 0.6 dB.

13.5 Summary

This chapter has summarized the likely future demands and challenges for video compression. It has been postulated that the increased bit rates demanded will, for many applications, outstrip those provided by advances in network technology and conventional compression methods. The hybrid codec, which has served us well for the past 30 years, will no doubt continue to dominate for the foreseeable future, but could be enhanced through the exploitation of intelligent perception-based methods such as those described here.

We have seen that the HVS exhibits nonlinear sensitivities to the distortions introduced by lossy image and video coding. There are several factors that contribute to this, including luminance masking, contrast masking, and spatial and temporal frequency masking. Coding methods in the future must attempt to exploit these phenomena to a greater extent. New approaches to compression, using deep neural networks that implicitly learn context and analysis–synthesis models, could provide the next step toward delivering content at the required quality for future immersive applications. There are however significant issues of complexity and compatible perceptual loss functions that require further investigation.

References

[1] Cisco visual networking index: global mobile data traffic forecast update, 2017–2022, https://www.cisco.com/c/en/us/solutions/executive-perspectives/annual-internet-report/index.html, 2019.

[2] Recommendation ITU-R BT.2020 (10/2015), Parameter values for ultra-high definition television systems for production and international programme exchange, ITU-R, 2015.

[3] S. Sakaida, N. Nakajima, A. Ichigaya, M. Kurozumi, K. Iguchi, Y. Nishida, E. Nakasu, S. Gohshi, The super HiVision codec, in: Proc. IEEE Intl. Conf. on Image Processing, 2007, pp. 121–124.

[4] J. Li, et al., Comparing upscaling algorithms from HD to ultra HD by evaluating preference of experience, in: Sixth International Workshop on Quality of Multimedia Experience (QoMEX), IEEE, 2014, pp. 208–213.

[5] G. Van Wallendael, et al., Perceptual quality of 4K-resolution video content compared to HD, in: Eighth Intl. Conf. on Quality of Multimedia Experience (QoMEX), IEEE, 2016, pp. 1–6.
[6] A. Mackin, M. Afonso, F. Zhang, D. Bull, A study of subjective video quality at various spatial resolutions, in: Proc. IEEE Intl. Conf. on Image Processing, 2018, pp. 2830–2834.
[7] Salmon, R., Armstrong, M., and Jolly S., Higher frame rates for more immersive video and television, BBC White Paper WHP209, BBC, 2009.
[8] M. Emoto, Y. Kusakabe, M. Sugawara, High-frame-rate motion picture quality and its independence of viewing distance, J. Disp. Technol. 10 (8) (2014) 635–641.
[9] M. Sugawara, et al., Temporal sampling parameters and motion portrayal of television, in: Society for Information Display Intl. Symposium, vol. 40(1), 2009, pp. 1200–1203.
[10] A. Mackin, F. Zhang, D. Bull, A study of high frame rate video formats, IEEE Trans. Multimed. 21 (6) (2019) 1499–1512.
[11] Recommendation ITU-R BT.2100-2, Image parameter values for high dynamic range television for use in production and international programme exchange, ITU-R, 2018.
[12] Recommendation ITU-R BT.2390-8, High dynamic range television for production and international programme exchange, ITU-R, 2020.
[13] E. François, et al., High dynamic range video coding technology in response to the joint call for proposals on video compression with capability beyond HEVC, IEEE Trans. Circuits Syst. Video Technol. 30 (5) (2020) 1253–1266.
[14] J. Chen, Y. Ye, S.H. Kim, Algorithm description for versatile video coding and test model 9 (VTM 9), in: The JVET Meeting, No JVET-R2002. ITU-T and ISO/IEC, 2020.
[15] P. Ndjiki-Nya, C. Stuber, T. Wiegand, Texture synthesis method for generic video sequences, IEEE Intl. Conf. Image Processing 3 (2007) 397–400.
[16] P. Ndjiki-Nya, T. Hinz, C. Stuber, T. Wiegand, A content-based video coding approach for rigid and non-rigid textures, IEEE Intl. Conf. Image Processing (2006) 3169–3172.
[17] M. Bosch, M. Zhu, E. Delp, Spatial texture models for video compression, IEEE Intl. Conf. Image Processing (2007) 93–96.
[18] J. Byrne, S. Ierodiaconou, D.R. Bull, D. Redmill, P. Hill, Unsupervised image compression-by-synthesis within a JPEG framework, in: Proc. IEEE Int Conf. Image Process, 2008, pp. 2892–2895.
[19] S. Ierodiaconou, J. Byrne, D. Bull, D. Redmill, P. Hill, Unsupervised image compression using graphcut texture synthesis, IEEE Intl. Conf. Image Processing (2009) 2289–2292.
[20] F. Zhang, D. Bull, A parametric framework for video compression using region-based texture models, IEEE J. Sel. Top. Signal Process. 6 (7) (2011) 1378–1392.
[21] J. Lee, T. Ebrahimi, Perceptual video compression: a survey, IEEE J. Sel. Top. Signal Process. 6 (2012) 684–697.
[22] R. Vigars, A. Calway, D. Bull, Context-based video coding, in: Proc IEEE Intl. Conf. on Image Processing, 2013, pp. 1953–1957.
[23] R. O'Callaghan, D. Bull, Combined morphological-spectral unsupervised image segmentation, IEEE Trans. Image Process. 14 (1) (2005) 49–62.
[24] G. Doretto, A. Chiuso, Y. Wu, S. Soatto, Dynamic textures, Int. J. Comput. Vis. 51 (2) (2003) 91–109.
[25] A. Stojanovic, M. Wien, J-R. Ohm, Dynamic texture synthesis for H.264/AVC inter coding, IEEE Intl. Conf. Image Processing (2009) 1608–1611.
[26] C. Zhu, X. Sun, F. Wu, H. Li, Video coding with spatio-temporal texture synthesis and edge-based inpainting, Proc. ICME (2008) 813–816.
[27] M. Naccari, F. Pereira, Advanced H.264/AVC-based perceptual video coding: architecture, tools, and assessment, IEEE Trans. Circuits Syst. Video Technol. 21 (6) (2011) 766–782.
[28] M. Naccari, M. Mrak, D. Flynn, A. Gabriellini, Improving HEVC compression efficiency by intensity dependent spatial quantisation, in: JCTVC-J0076, 10th Meeting, Stockholm, 2012.
[29] Y. Zhang, M. Naccari, D. Agrafiotis, M. Mrak, D. Bull, High dynamic range video compression exploiting luminance masking, IEEE Trans. Circuits Syst. Video Technol. 26 (5) (2016) 950–964.
[30] Z. He, et al., Progress in virtual reality and augmented reality based on holographic display, Appl. Opt. 58 (5) (2019) A74–A81.
[31] Segall, a. et al., Joint call for proposals on video compression with capability beyond HEVC, JVET meeting, no. JVET-H1002. ITU-T and ISO/IEC, 2017.

[32] Y. Yan, J. Boyce, P. Hanhart, Omnidirectional 360° video coding technology in responses to the joint call for proposals on video compression with capability beyond HEVC, IEEE Trans. Circuits Syst. Video Technol. 30 (5) (2019) 1241–1252.
[33] N. Anantrasirichai, D. Bull, Artificial intelligence in the creative industries: a review, arXiv:2007.12391v2, 2020.
[34] S. Ma, et al., Image and video compression with neural networks: a review, IEEE Trans. Circuits Syst. Video Technol. 30 (6) (2020) 1883–1898.
[35] F. Zhang, F. Chen, D. Bull, Enhancing VVC through CNN-based post-processing, in: IEEE Intl. Conference on Multimedia and Expo, 2020.
[36] D. Ma, F. Zhang, D. Bull, MFRNet: a new CNN architecture for post-processing and in-loop filtering, IEEE J. Sel. Top. Signal Process. (2020).
[37] D. Ma, M. Afonso, F. Zhang, D. Bull, Perceptually-inspired super-resolution of compressed videos, in: Proc. SPIE 11137, Applications of Digital Image Processing XLII, 2019, p. 1113717.
[38] F. Zhang, M. Afonso, D. Bull, ViSTRA2: video coding using spatial resolution and effective bit depth adaptation, arXiv:1911.02833, 2019.
[39] I. Schiopu, H. Huang, A. Munteanu, CNN-based intra-prediction for lossless HEVC, IEEE Trans. Circuits Syst. Video Technol. 30 (7) (2020) 1816–1828.
[40] S. Zhao, et al., Enhanced bi-prediction with convolutional neural network for high-efficiency video coding, IEEE Trans. Circuits Syst. Video Technol. 29 (11) (2019) 3291–3301.
[41] D. Liu, et al., CNN-based DCT-like transform for image compression, MultiMedia Modeling (2018) 61–72.
[42] M. Alam, et al., A Perceptual Quantization Strategy for HEVC Based on a Convolutional Neural Network Trained on Natural Images, Applications of Digital Image Processing, vol. XXXVIII, SPIE, 2015, p. 9599.
[43] J. Li, et al., Fully connected network-based intra prediction for image coding, IEEE Trans. Image Process. 27 (7) (2018) 3236–3247.
[44] Y. Xue, J. Su, Attention based image compression post-processing convolutional neural network, in: IEEE Conf. Computer Vision and Pattern Recognition Workshop (CVPRW), 2019.
[45] G. Lu, et al., An end-to-end learning framework for video compression, IEEE Trans. Pattern Anal. Mach. Intell. (2020).
[46] L. Zhao, et al., Enhanced motion compensated video coding with deep virtual reference frame generation, IEEE Trans. Image Process. 28 (10) (2019) 4832–4844.
[47] J. Liu, et al., One-for-all: grouped variation network based fractional interpolation in video coding, IEEE Trans. Image Process. 28 (5) (2019) 2140–2151.
[48] R. Song, D. Liu, H. Li, F. Wu, Neural network-based arithmetic coding of intra prediction modes in HEVC, in: IEEE Visual Communications and Image Processing Conf. (VCIP), 2017, pp. 1–4.
[49] G. Lu, et al., DVC: an end-to-end deep video compression framework, in: IEEE/CVF Conf. Computer Vision and Pattern Recognition (CVPR), 2019, pp. 10998–11007.
[50] D. Ma, F. Zhang, D. Bull, GAN-based effective bit depth adaptation for perceptual video compression, in: IEEE Intl. Conf. Multimedia and Expo (ICME), 2020, pp. 1–6.
[51] D. Ma, F. Zhang, D. Bull, BVI-DVC: A training database for deep video compression, arXiv:2003.13552, 2020.
[52] R. Zhang, P. Isola, A. Efros, Colorful image colorization, in: European Conference on Computer Vision (ECCV), 2016, pp. 649–666.
[53] J. Lehtinen, et al., Noise2Noise: learning image restoration without clean data, in: Proc 35th International Conference on Machine Learning, vol. 80, 2018, pp. 2965–2974.
[54] A. Krizhevsky, I. Sutskever, G. Hinton, ImageNet classification with deep convolutional neural networks, in: Proc. 25th Intl. Conf. on Neural Information Processing Systems, vol. 1, 2012, pp. 1097–1105.
[55] S. Huang, et al., AugGAN: cross domain adaptation with GAN-based data augmentation, in: European Conference on Computer Vision (ECCV), 2018.
[56] G. Mariani, et al., BAGAN: data augmentation with balancing GAN, arXiv:1803.09655v2, 2018.
[57] C. Shorten, T. Khoshgoftaar, A survey on image data augmentation for deep learning, J. Big Data 6 (60) (2019).
[58] S. Su, et al., Deep video deblurring for hand-held cameras, in: IEEE Conf. Computer Vision and Pattern Recognition (CVPR), 2017, pp. 237–246.
[59] E. Agustsson, R. Timofte, NTIRE 2017 challenge on single image superresolution: dataset and study, in: IEEE Conf. Computer Vision and Pattern Recognition (CVPR) Workshops, 2017.
[60] S. Nah, et al., NTIRE 2019 challenge on video deblurring and super-resolution: dataset and study, in: Proc. IEEE Conf. on Computer Vision and Pattern Recognition Workshops, 2019.

[61] D. Liu, et al., Robust video super-resolution with learned temporal dynamics, in: Proc. IEEE Intl. Conf. on Computer Vision, 2017, pp. 2507–2515.
[62] M. Afonso, F. Zhang, D. Bull, Video compression based on spatio-temporal resolution adaptation, IEEE Trans. Circuits Syst. Video Technol. 29 (1) (2019) 275–280.
[63] J. Balle, V. Laparra, E. Simoncelli, End-to-end optimized image compression, in: 5th Intl. Conf. on Learning Representations (ICLR), 2017.
[64] E. Agustsson, D. Minnen, N. Johnston, J. Balle, S.J. Hwang, G. Toderici, Scale-space flow for end-to-end optimized video compression. proc.IEEE conf, in: Computer Vision and Pattern Recognition (CVPR), 2020, pp. 8503–8512.
[65] D. Minnen, J. Ballé, G.D. Toderici, Joint autoregressive and hierarchical priors for learned image compression, in: Advances in Neural Information Processing Systems, 2020, pp. 10771–10780.
[66] O. Rippel, S. Nair, C. Lew, S. Branson, A.G. Anderson, L. Bourdev, Learned video compression, in: Proc. IEEE Intl. Conf. Computer Vision (ICCV), 2019, pp. 3454–3463.

APPENDIX

A

Glossary of terms

1D	One-dimensional
2D	Two-dimensional
3D	Three-dimensional
4PMV+0	Four predicted motion vectors plus zero vector
3DTV	Three-dimensional television
AC	Alternating current. Used to denote all transform coefficients except the zero frequency coefficient
ACR	Absolute category rating
ACK	Acknowledge
A-D	Analog to digital
ADSL	Asymmetric digital subscriber line
ADST	Asymmetric discrete sine transform
AFD	Active Format Description
AI	Artificial intelligence or all intra
ALF	Adaptive loop filter
AMVP	Advanced motion vector prediction
AMVR	Adaptive motion vector resolution
ANN	Artificial neural network
AOM	Alliance for Open Media (or AOMedia)
APS	Adaptation parameter set
AR	Augmented reality or autoregressive
ARQ	Automatic repeat request
ASP	Advanced simple profile (of MPEG-4)
AV	Audiovisual
AV1	AOMedia Video 1
AV2	AOMedia Video 1
AVC	Advanced video codec (H.264)
AVM	Artifact-based video metric
B	Bipredicted picture
BDM	Block distortion measure
BER	Bit error rate
BMA	Block matching algorithm
BMME	Block matching motion estimation

BPG	Better portable graphics
bpp	Bits per pixel
bps	Bits per second
BPSK	Binary phase shift keying
BVI	Bristol Vision Institute
CABAC	Context-based adaptive binary arithmetic coding
CAVLC	Context-adaptive variable-length coding
CB	Coding block
CBF	Coded block flag
CBP	Coded block pattern
CBR	Constant bit rate
CCD	Charge coupled device
CCIR	International radio consultative committee (now ITU)
CD	Compact disc
CDMA	Code division multiple access
CE	Core experiment
CFF	Critical flicker frequency
CIE	Commission Internationale de L'Eclairage (the International Color Science Commission)
CIF	Common intermediate format
CIIP	Combined inter- and intra-prediction
CMP	Cubemap projection
CMY	Color primaries cyan, magenta, and yellow
CNN	Convolutional neural network
codec	Encoder and decoder
CPB	Coded picture buffer
cpd	Cycles per degree
CRA	Clean random access
CRT	Cathode ray tube
CSF	Contrast sensitivity function
CTB	Coding tree block
CTC	Common test conditions
CTU	Coding tree unit
CU	Coding unit
CVS	Coded video sequence
CW-SSIM	Complex wavelet SSIM
DAB	Digital audio broadcasting
DASH	Dynamic adaptive streaming over HTTP
DC	Direct current. Refers to zero-frequency transform coefficient
DCI	Decoding capability information
DCR	Degradation category rating
DCT	Discrete cosine transform
DFD	Displaced frame difference
DFT	Discrete Fourier transform

DoG	Difference of Gaussians
DL	Deep learning
DLM	Detail loss metric
DMOS	Difference mean opinion score
DNN	Deep neural network
DPB	Decoded picture buffer
DPCM	Differential pulse code modulation
DS	Diamond search
DSCQS	Double-stimulus continuous quality scale
DSIS	Double-stimulus impairment scale
DSP	Digital signal processor
DST	Discrete sine transform
DT-CWT	Dual tree continuous wavelet transform
DVB	Digital video broadcasting
DVC	Deep video compression
DVD	Digital versatile disc
DWHT	Discrete Walsh–Hadamard transform
DWT	Discrete wavelet transform
EAP	Equal-area projection
EBCOT	Embedded block coding with optimized truncation
EBD	Effective bit depth
EBMA	External boundary matching algorithm
EBME	External block matching error
EBU	European Broadcasting Union
EECMS	Enhanced error concealment with mode selection
EEG	Electroencephalography
EG	Exp-Golomb probability distribution and entropy coder
EOB	End of block
EOTF	Electro-optical transfer function
EREC	Error-resilient entropy coding
ERP	Equirectangular projection
EZW	Embedded zero-tree wavelet
FD	Frame difference
FEC	Forward error correction
FEF	Frontal eye field
FFA	Fusiform face area
FFT	Fast Fourier transform
FGS	Fine granularity scalability
FIR	Finite impulse response (filter)
FLC	Fixed-length coding
FMO	Flexible macroblock ordering
fMRI	Functional magnetic resonance imaging
fps	Frames per second
FR	Full reference

FRTV	Full reference TV VQEG database
FS	Full search
GAN	Generative adversarial network
GIF	Graphics interchange format
GOB	Group of blocks
GOP	Group of pictures
GPU	Graphics processing unit
HARQ	Hybrid ARQ
HAS	HTTP adaptive streaming
HDR	High dynamic range
HDTV	High-definition television
HEVC	High-efficiency video codec
HFR	High frame rate
HRD	Hypothetical reference decoder
HTTP	Hypertext transfer protocol
HVS	Human visual system
HEXBS	Hexagon-based search method
HM	HEVC test model
I-picture	Intra-coded picture
IAR	Image aspect ratio
IDR	Instantaneous decoder refresh
IEC	International Electrotechical Commission
IEEE	Institute of Electrical and Electronic Engineers
i.i.d.	Independent and identically distributed
IIR	Infinite impulse response (filter)
IP	Internet Protocol
ISDN	Integrated services digital network
ISO	International Standards Organization
IT	Inferior temporal cortex
ITU	International Telecommunications Union. -R Radio; -T Telecommunications
JCT-VC	Joint collaborative team on video coding
JM	H.264 Joint reference model
JND	Just noticeable difference
JPEG	Joint Photographic Experts Group
JPSEC	JPEG2000 security extension
JPWL	JPEG2000 wireless extension
JVC	Joint video team
JVET	Joint video experts team
kbps	Kilobits per second
KLT	Karhunen–Loeve transform
KTA	Key technical area
LAN	Local area network

LCC	Linear correlation coefficient
LCD	Liquid crystal display
LCU	Largest coding unit
LD	Low delay
LDPC	Low density parity check (codes)
LGN	Lateral
LIVE	Laboratory for Image and Video Engineering
LMCS	Luma mapping with chroma scaling
LPS	Least probable symbol
LSI	Linear shift-invariant
LTE	Long term evolution (4G mobile radio technology)
LTI	Linear time-invariant
LZW	Lempel–Ziv–Welch
MAC	Multiply and accumulate operation
MAC	Medium access control
MAD	Mean absolute difference
MAD	Most apparent distortion
MB	Macroblock
Mbps	Megabits per second
MC	Motion compensation
MCP	Motion-compensated prediction
MCS	Modulation and coding mode selection
MCTF	Motion-compensated temporal filtering
MDC	Multiple description coding
MDCS	Mode-dependent coefficient scanning
ME	Motion estimation
MEC	Motion estimation and compensation
MIMO	Multiple-input, multiple-output
ML	Machine learning
MLP	Multilayer perceptron
MOS	Mean opinion score
MOVIE	Video quality metric from LIVE
MPEG	Motion Picture Experts Group
MPS	Most probable symbol
MR	Mixed reality
MRF	Multiple reference frames
MRL	Multiple reference lines
MSB	Most significant bit
MSE	Mean squared error
MS-SSIM	Multiscale SSIM
MST	Medial superior temporal
MT	Medial temporal
MTS	Multiple transform selection
MV	Motion vector
MVC	Multiview video coding

MVD	Motion vector difference
NACK	Negative acknowledge
NAL	Network abstraction layer
NALU	Network abstraction layer unit
NCCF	Normalized cross-correlation coefficient
NR	No reference (metric)
NSS	N-step search
NUT	NAL unit type
OBMC	Overlapped block motion compensation (or just OMC)
OETF	Opto-electronic transfer function
OOTF	Opto-optical transfer function
OR	Outlier ratio
P	Predicted picture
PAL	Phase alternating line
PB	Prediction block
PCA	Principal component analysis
PCM	Pulse code modulation
pdf	Probability density function
PER	Packet error rate
PH	Picture header
PHY	Physical layer
PMR	Private mobile radio
PNG	Portable network graphics
PRF	Periodic reference frame
PPS	Picture parameter set
PSD	Power spectral density
PSNR	Peak signal to noise ratio
PU	Prediction unit
PVM	Perception inspired video metric
PVQ	Pyramid vector quantization
QAM	Quadrature amplitude modulation
QCIF	Quarter CIF resolution
QMF	Quadrature mirror filter
QPSK	Quadrature phase shift keying
QoE	Quality of experience
QoS	Quality of service
QP	Quantization parameter
RA	Random access
RAP	Random access point
RBER	Residual bit error rate
RDO	Rate-distortion optimization
RF	Radio frequency
RGB	Red, green, and blue color primaries

RLC	Run-length coding
RPS	Reference picture selection
RPS	Reference picture set (HEVC)
RQO	Rate-quality optimization
RQT	Residual quadtree
RR	Reduced-reference (metric)
RS	Redundant slice
RSE	Reed–Solomon erasure code
RSSI	Residual signal strength indication
RTCP	Real-time control protocol
RTP	Real-time transmission protocol
RTSP	Real-time streaming protocol
RVLC	Reversible variable-length coding
SAD	Sum of absolute differences
SAMVIQ	Subjective Assessment Methodology for Video Quality
SAO	Sample-adaptive offset
SATD	Sum of absolute transform differences
SCC	Screen content coding
SDR	Standard dynamic range
SDTV	Standard-definition television
SE	Syntax element
SEC	Spatial error concealment
SEI	Supplemental enhancement information
SG	Study group (of ITU)
SG	Slice group
SI	Spatial information
SIFT	Scale-invariant feature transform
SMPTE	Society of Motion Picture and Television Engineers
SNR	Signal to noise ratio
SPIHT	Set partitioning into hierarchical trees
SPS	Sequence parameter set
SROCC	Spearman rank order correlation coefficient
SSD	Sum of squared differences
SSIM	Structural similarity index
SSCQS	Single-stimulus continuous quality scale
SSIS	Single-stimulus impairment scale
ST-MAD	Spatio-temporal most apparent distortion
SURF	Sped-up robust features
SVC	Scalable video coding
SVD	Singular value decomposition
SVM	Support vector machine
TB	Transform block
TCP	Transmission control protocol
TDL	Two-dimensional logarithmic

TEC	Temporal error concealment
TI	Temporal information
TS	Transport stream
TSCES	Triple-stimulus continuous evaluation scale
TSS	Three-step search
TU	Transform unit or truncated unary
TV	Television
TZS	Test zone search
UDP	User datagram protocol
UHDTV	Ultrahigh-definition television
UHF	Ultrahigh frequency
URL	Universal resource locator
UMTS	Universal mobile telecommunications system
V1	Region of visual cortex (also V2–V5)
VAE	Variational autoencoder
VBR	Variable bit rate
VCEG	Video coding experts group
VCL	Video coding layer
VDP	Visible difference predictor
VDSL	Very high-bit rate digital subscriber line
VIF	Visual information fidelity
ViSTRA	Video compression based on spatio-temporal resolution adaptation
VLC	Variable-length coding
VLD	Variable-length decoding
VMAF	Video multimethod assessment fusion
VOP	Video object plane
VPS	Video parameter set
VQ	Vector quantization
VQEG	Video Quality Experts Group
VQM	Video quality metric
VR	Virtual reality
VSNR	Visual signal to noise ratio
VSTM	Visual short-term memory
VTM	VVC test model
VVC	Versatile video coding
WBA	Warping-based algorithm (motion estimation)
WCG	Wide color gamut
WSS	Wide sense stationary
XR	Extended reality
YC_bC_r	Color coordinate system comprising luminance, Y, and two chrominance channels, C_b and C_r
YUV	Color coordinate system comprising luminance, Y, and two chrominance channels, U and V

APPENDIX

B

Tutorial problems

Further tutorial questions and outline solutions can be found on the website which accompanies this book.

Chapter 1: Introduction

Q1.1

What is the primary purpose of standardization in video compression? List two other advantages of standardization.

Q1.2

Using the example of a DVB-T2 terrestrial broadcast system transmitting HDTV video content to the home, explain why digital video compression is needed.

Q1.3

Consider the case of 4K UHDTV, with the original video in 4:2:2 (a luma signal of 3840×2160 and two chroma signals of 1920×2160) format at 10 bits and a frame rate of 50 fps. Calculate the compression ratio if this video is to be transmitted over a DVB-T2 link with an average bandwidth of 15 Mb/s.

Chapter 2: The human visual system

Q2.1

Assuming that the field of view within the fovea is 2 degrees, compute the number of pixels that fall horizontally within the foveated visual field. Assume a 1 m wide screen with 1920 horizontal pixels viewed at a distance of 3H, where H is the screen height.

Q2.2

The following table lists a number of important features of the human visual system (HVS). Complete the table by describing the influence each feature has on the design of a digital video compression system.

Q2.3

Calculate the normalized contrast sensitivity of the HVS for a luminance-only stimulus at a spatial frequency of 10 cycles per degree.

HVS characteristic	Implication for compression
HVS more sensitive to high-contrast image regions than low-contrast regions.	?
HVS is more sensitive to luminance than chrominance information.	?
HVS is more sensitive to low spatial frequencies than high spatial frequencies.	?
In order to achieve a smooth appearance of motion, the HVS must be presented with image frames above a certain minimum rate (and this rate depends on ambient light levels).	?
HVS responses vary from individual to individual.	?

Chapter 3: Signal processing and information theory fundamentals

Q3.1

Plot the sinusoidal signal $x(t) = \cos(20\pi t)$. Compute the frequency spectrum of this sinusoid and plot its magnitude spectrum.

Q3.2

Assume that the sinusoidal signal in Q3.1 is sampled with $T = 0.1$ s. Sketch the magnitude spectrum of the sampled signal and comment on any aliasing issues.

Q3.3

If an HDTV (full HD) screen with aspect ratio 16:9 has a width of 1.5 m and is viewed at a distance of 4H, what is the angle subtended by each pixel at the retina?

Q3.4

The impulse response for the LeGall high-pass analysis filter is $h_1[n] = \{0.25 \quad -0.5 \quad 0.25\}$. Compute the output from this filter for the input sequence $x[n] = \{\ 1 \quad 2 \quad 3 \quad 0 \quad 0 \quad \cdots\ \}$.

Q3.5

Compute the z plane pole zero plots and the frequency responses for the following filter pair: $H_0(z) = 1 + z^{-1}$ and $H_1(z) = 1 - z^{-1}$.

Q3.6

Perform median filtering on the following input sequence using a 5-tap median filter and comment on the result:

$$x[n] = \{1, 3, 5, 13, 9, 11, 6, 15, 17, 19, 29\}.$$

Q3.7

Compute the basis functions for the two-point DFT.

Q3.8

Consider the two 1D digital filters:

$$\mathbf{h}_1 = [\,1 \quad 2 \quad 2 \quad 1]^{\mathrm{T}},$$

$$\mathbf{h}_2 = [\,1 \quad -3 \quad -3 \quad 1]^{\mathrm{T}}.$$

Compute the equivalent 2D digital filter where $\mathbf{h}_1$ performs horizontal filtering and $\mathbf{h}_2$ performs vertical filtering.

Q3.9

Compute biased and unbiased correlation estimates for the following sequence:

$$x[n] = \{1, 2, 5, -1, 3, 6, -4, -1\}.$$

Q3.10

Plot the autocorrelation function $r_v(k)$ for a white noise sequence $v[n]$ with variance σ_v^2. Form the autocorrelation matrix for this sequence for lags up to ± 3. What is the inverse of this autocorrelation matrix?

Q3.11

A feedback-based linear predictor with quantization uses a predictor $P(z) = (z^{-1} + z^{-2})/2$. Assume an input sequence as

$$x[n] = \{1, 3, 4, 3, 5, 6 \cdots\}$$

and assume that quantization is performed as follows, with rounding of 0.5 values toward zero:

$$e_Q[n] = \mathrm{rnd}\left(\frac{e[n]}{2}\right), \quad e_R[n] = 2e_Q[n].$$

Compute the predictor output sequence $e_Q[n]$ and the reconstructed output signal $y[n]$ from the decoder. Comment on your results in terms of numerical precision.

Q3.12

Compute the entropies of the sequences $y[n]$ and $e_Q[n]$ from Q3.11. Comment on your result.

Chapter 4: Digital picture formats and representations

Q4.1

In video coding schemes it is usual to code the color components in the form Y, C_b, C_r rather than R, G, B. Explain why this approach is justified, its benefits in terms of compression, and how it is exploited in image sampling. Explain how pictures can be efficiently stored or transmitted using a 4:2:0 format.

Q4.2

Compute YUV and Y C_b C_r vectors for the following RGB vectors (assume Rec. 601 format):

$$\text{a)}\quad [R, G, B] = [\ 128 \quad 128 \quad 128\ ],$$
$$\text{b)}\quad [R, G, B] = [\ 255 \quad 255 \quad 255\ ],$$
$$\text{c)}\quad [R, G, B] = [\ 100 \quad 0 \quad 0\ ].$$

Q4.3

If a color movie of 100 minutes duration is represented using ITU-R.601 (720 × 576, 25 fps @8 bits, 4:2:0 format):

a. What hard disk capacity would be required to store the whole movie?
b. If the movie is encoded at a compression ratio CR=40:1 and transmitted over a satellite link with 50% channel coding overhead, what is the total bit rate required for the video signal?

Q4.4

Given a video sequence with a spatial resolution of 1920 × 1080 at 50 fps using 10-bit color sampling, compute the (uncompressed) bit rates of 4:4:4, 4:2:0, and 4:2:2 systems.

Q4.5

If, for a given 1920 × 1080 I-frame in 4:2:0 format, the mean entropy of each coded 16 × 16 luminance block is 1.3 bits/sample and that for each corresponding chrominance block is 0.6 bits/sample, then estimate the total number of bits required to code this frame.

Q4.6

Calculate the MAD between the following original image block **X** and its encoded and decoded version, $\tilde{\mathbf{X}}$:

$$\mathbf{X} = \begin{bmatrix} 2 & 4 & 4 & 6 \\ 3 & 6 & 6 & 6 \\ 5 & 7 & 7 & 8 \\ 3 & 7 & 7 & 8 \end{bmatrix}, \quad \tilde{\mathbf{X}} = \begin{bmatrix} 2 & 5 & 5 & 5 \\ 4 & 4 & 6 & 6 \\ 5 & 5 & 7 & 8 \\ 4 & 5 & 7 & 7 \end{bmatrix}.$$

Q4.7

Assuming a 5-bit digital image block, X, the reconstruction after image compression is given by Y. Calculate the PSNR for the reconstructed signal and provide an interpretation of the results in terms of error visibility. We have

$$\mathbf{X}=\begin{bmatrix} 3 & 8 & 1 & 8 \\ 7 & 0 & 5 & 0 \\ 2 & 6 & 0 & 5 \\ 4 & 1 & 10 & 2 \end{bmatrix}, \quad \mathbf{Y}=\begin{bmatrix} 4 & 10 & 1 & 9 \\ 7 & 0 & 6 & 1 \\ 3 & 6 & 0 & 4 \\ 5 & 1 & 12 & 3 \end{bmatrix}.$$

Q4.8

Assuming an 8-bit digital image, **X**, the reconstructions due to two alternative coding schemes are given by $\mathbf{Y}_1$ and $\mathbf{Y}_2$ below. Calculate the peak signal to noise ratio (PSNR) for each of the reconstructions and give an interpretation of the results in terms of error visibility. We have

$$\mathbf{X}=\begin{bmatrix} 20 & 17 & 18 \\ 15 & 14 & 15 \\ 19 & 13 & 14 \end{bmatrix}, \quad \mathbf{Y}_1=\begin{bmatrix} 19 & 18 & 17 \\ 16 & 15 & 14 \\ 18 & 14 & 13 \end{bmatrix}, \quad \mathbf{Y}_2=\begin{bmatrix} 20 & 17 & 18 \\ 15 & 23 & 15 \\ 19 & 13 & 14 \end{bmatrix}.$$

Q4.9

Describe a typical GOP structure used in MPEG-2 television broadcasting, explaining the different types of frame coding employed, the predictive relationships between all frames in the GOP, and the transmission order of the frames. If a transmission error affects the second P-frame in the GOP, how many pictures are likely to be affected due to error propagation?

Q4.10

An HDTV satellite operator allocates 10 Mb/s to each program in the DVB-S multiplex. State what color subsampling format will be used for this and calculate the required compression ratio for a 1080i25 system.

Q4.11

Compute the gamma corrected version of the following image block for the case where $\gamma = 0.45$ assuming an 8-bit word length:

$$\mathbf{X}=\begin{bmatrix} 20 & 17 & 18 \\ 15 & 14 & 15 \\ 19 & 13 & 14 \end{bmatrix}.$$

Q4.12

Bayer filtering (demosaicing) is commonly used to interpolate color planes for single sensor cameras.

a. Draw a diagram of a 5-by-5 color filter array.
b. Derive the demosaicing kernels for this array for the R, G, and B color planes. Assume that bilinear interpolation is used for demosaicing.
c. Given the following acquired pixel values for the array in (a) in [R G B] format, calculate missing values for location [x,y]=[1,1] using the interpolation kernel in (b).

[x x 10]	[x 10 x]	[x x 2]	[x 3 x]	[x x 1]
[x 10 x]	[20 x x]	[x 5 x]	[4 x x]	[x 0 x]
[x x 20]	[x 10 x]	[x x 4]	[x 2 x]	[x x 0]
[x 10 x]	[20 x x]	[x 5 x]	[4 x x]	[x 1 x]
[x x 20]	[x 10 x]	[x x 3]	[x 2 x]	[x x 1]

Chapter 5: Transforms for image and video coding

Q5.1

Derive the 1D two-point KLT for a stationary real-valued process, x, that has the following autocorrelation matrix:

$$\mathbf{R}_x = \begin{bmatrix} r_{xx}(0) & r_{xx}(1) \\ r_{xx}(1) & r_{xx}(0) \end{bmatrix}.$$

Q5.2

Prove that the following vectors are an orthonormal pair:

$$\mathbf{a}_0 = \begin{bmatrix} \frac{\sqrt{2}}{2} & \frac{\sqrt{2}}{2} \end{bmatrix}^{\mathrm{T}}, \quad \mathbf{a}_0 = \begin{bmatrix} \frac{\sqrt{2}}{2} & -\frac{\sqrt{2}}{2} \end{bmatrix}^{\mathrm{T}}.$$

Q5.3

Prove that the four-point DWHT is a unitary transform.

Q5.4

Compute the basis functions for the eight-point 1D DWHT. Compute the first four basis functions for the four-point 2D DWHT.

Q5.5

Use the DWHT to transform the following 2D image block, **S**. Assuming that all data and coefficients are represented as 8-bit numbers and that compression is achieved in the transform domain by selecting only the most dominant coefficient for transmission, compute the decoded data matrix and its PSNR. We have

$$\mathbf{S} = \begin{bmatrix} 10 & 10 & 10 & 10 \\ 10 & 10 & 10 & 10 \\ 10 & 10 & 9 & 9 \\ 10 & 10 & 9 & 9 \end{bmatrix}.$$

Q5.6

Given the four-point DWHT basis functions, if the coefficients after transformation are $c(0) = 1$, $c(1) = \frac{1}{4}$, $c(2) = -\frac{1}{2}$, $c(3) = 3$, plot the weighted basis functions and hence reconstruct the original signal waveform.

Q5.7

The 1D discrete cosine transform is given by

$$C(k) = \sqrt{\frac{2}{N}}\varepsilon_k \sum_{n=0}^{N-1} x[n] \cos\left(\frac{\pi (n+0.5) k}{N}\right), \quad 0 \le n, k \le N-1,$$

where

$$\varepsilon_k = \begin{cases} 1/\sqrt{2}, & k = 0, \\ 1, & \text{otherwise.} \end{cases}$$

Calculate the numerical values of the basis functions for a 1D 4×4 DCT.

Q5.8

The DCT is an orthonormal transform. Explain the term orthonormal and describe what it means in practice for the transform. Write down the formula and the basis function matrix for the 1D four-point DCT. Using this, show that, for the 1D DCT,

$$\epsilon_k = \begin{cases} 1/\sqrt{2}, & k = 0, \\ 1, & \text{otherwise.} \end{cases}$$

Q5.9

Compute the DCT transform coefficient vector for an input sequence, $\mathbf{x} = [1001]^T$.

Q5.10

Calculate the DCT of the following 2×2 image block:

$$\mathbf{X} = \begin{bmatrix} 21 & 19 \\ 15 & 20 \end{bmatrix}.$$

Q5.11

Quantize the result from Q5.10 using the following quantization matrix:

$$\mathbf{Q} = \begin{bmatrix} 4 & 8 \\ 8 & 8 \end{bmatrix}.$$

Q5.12

Perform inverse quantization and an inverse DCT on the output from Q5.11.

Q5.13

Compute the 2D DCT of the following (4×4) image block:

$$\mathbf{X} = \begin{bmatrix} 1 & 0 & 0 & 0 \\ 0 & 0 & 0 & 0 \\ 0 & 0 & 0 & 0 \\ 1 & 0 & 0 & 0 \end{bmatrix}.$$

Q5.14

Given the following block of DCT coefficients and the associated quantization matrix, compute the block of quantized coefficients. Perform zig-zag scanning to form a string of {run/value} symbols (where "run" is the number of zeros preceding a nonzero value) appropriate for entropy coding. We have

$$\mathbf{C} = \begin{bmatrix} 128 & 50 & -20 & 22 & 12 & 27 & -5 & 7 \\ 40 & -25 & 26 & 20 & -34 & -2 & 13 & -5 \\ -10 & 22 & 12 & 12 & 26 & 12 & 3 & 8 \\ 12 & -2 & 16 & -7 & 9 & 3 & 17 & 17 \\ -32 & 6 & 21 & 9 & 18 & 5 & 4 & 7 \\ -10 & -7 & -14 & 3 & -2 & 13 & 18 & 18 \\ 11 & -9 & -9 & 4 & 8 & 13 & 6 & 9 \\ -7 & 19 & 15 & 8 & 6 & -6 & 18 & 33 \end{bmatrix},$$

$$\mathbf{Q} = \begin{bmatrix} 8 & 16 & 19 & 22 & 26 & 27 & 29 & 34 \\ 16 & 16 & 22 & 24 & 27 & 29 & 34 & 37 \\ 19 & 22 & 26 & 27 & 29 & 34 & 34 & 38 \\ 22 & 22 & 26 & 27 & 29 & 34 & 37 & 40 \\ 22 & 26 & 27 & 29 & 32 & 35 & 40 & 48 \\ 26 & 27 & 29 & 32 & 35 & 40 & 48 & 58 \\ 26 & 27 & 29 & 34 & 38 & 46 & 56 & 69 \\ 27 & 29 & 35 & 38 & 46 & 56 & 69 & 83 \end{bmatrix}.$$

Q5.15

Calculate the number of multiply and accumulate (MAC) operations required to compute a conventional (4×4)-point 2D DCT. Assume that the separability property of the 2D DCT is exploited.

Q5.16

The complexity of the DCT can be reduced using "fast" methods such as McGovern's algorithm. Derive McGovern's algorithm for a four-point 1D DCT. Compare its complexity (again assuming exploitation of separability) with that of the conventional approach for the case of a (4×4)-point 2D DCT.

Q5.17

Compute the sum of absolute transformed differences (SATD) for the image blocks **X** and **Y** in Q4.7.

Chapter 6: Filter-banks and wavelet compression

Q6.1

For a filter-bank downsampler, show that the frequency domain behavior of the output signal $x_d[n]$ is related to that of the input x[n] by

$$X_d(\Omega) = 0.5\left[X\left(e^{j\Omega/2}\right) + X\left(e^{-j\Omega/2}\right)\right].$$

Q6.2

The figure below shows a simple two-band analysis–synthesis filter-bank and a representative input spectrum.

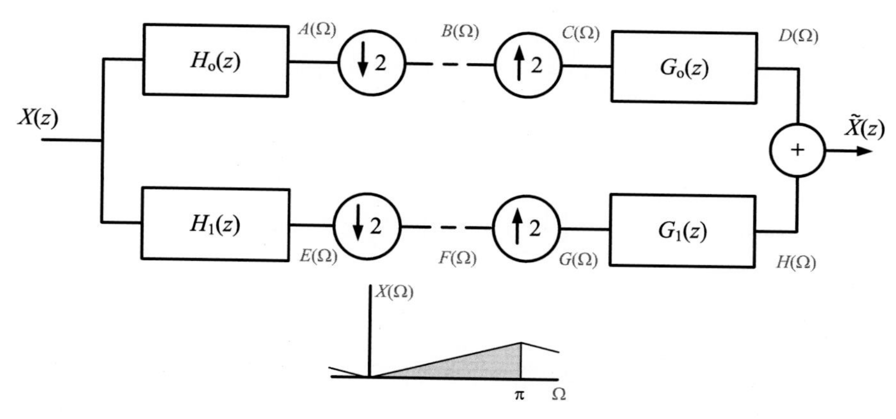

It can be shown that the output of this system is given by

$$\tilde{X}(z) = \frac{1}{2}X(z)\left[G_0(z)H_0(z) + G_1(z)H_1(z)\right] + \frac{1}{2}X(-z)\left[G_0(z)H_0(-z) + G_1(z)H_1(-z)\right],$$

where the upsampler and downsampler relationships are given by

$$C\left(e^{j\Omega}\right) = B\left(e^{j2\Omega}\right),$$

$$B\left(e^{j\Omega}\right) = 0.5\left[A\left(e^{j\Omega/2}\right) + A\left(-e^{j\Omega/2}\right)\right].$$

Using the upsampler and downsampler relationships given above, compute and sketch the spectra at points A to H. Assume that all the filters have ideal brickwall responses. Hence demonstrate graphically that the system is capable of perfect reconstruction.

Q6.3

Given the subband filters

$$H_0(z) = \frac{1}{\sqrt{2}}\left(1 + z^{-1}\right), \qquad H_1(z) = \frac{1}{\sqrt{2}}\left(1 - z^{-1}\right),$$
$$G_0(z) = \frac{1}{\sqrt{2}}\left(1 + z^{-1}\right), \qquad G_1(z) = \frac{1}{\sqrt{2}}\left(-1 + z^{-1}\right),$$

show that the corresponding two-band filter-bank exhibits perfect reconstruction.

Q6.4

Demonstrate that the following filter relationships produce a filter-bank that is alias-free and offers perfect reconstruction:

$$H_1(z) = zG_0(-z), \qquad G_1(z) = z^{-1}H_0(-z).$$

Assuming that $G_0(z) = H_0(z) = z^{-1}$, what is the output of this filter-bank given an input sequence {1,1,0}?

Q6.5

Referring to the figure below, a signal, x[n], is downsampled by 2 and then upsampled by 2. Show that, in the z domain, the input–output relationship is given by

$$\tilde{X}(z) = 0.5\left[X(z) + X(-z)\right].$$

$x[n]$ → (↓2) → (↑2) → $\tilde{x}[n]$

Q6.6

Demonstrate that the following quadrature mirror filter relationships produce a filter-bank that is alias-free and offers perfect reconstruction:

$$H_0(z) = z^{-2} + z^{-3}, \qquad G_0(z) = H_1(-z),$$
$$H_1(z) = H_0(-z), \qquad G_1(z) = -H_0(-z).$$

What are the limitations of the above filter-bank? How, in practice, are more useful QMF filter-banks designed?

Q6.7

A two-channel single-stage 1D QMF filter-bank is constructed using the low-pass prototype filter $H_0(z) = (1 + z^{-1})$. Derive the other filters needed for this system and compute the output signal for an input $x[n] = \{2, 3, 6, 4\}$.

Q6.8

A 1D wavelet filter-bank comprises two stages of decomposition and uses the same filters as defined in Q6.7 (but factored by $1/\sqrt{2}$ to ensure exact reconstruction). The input sequence is

$x[n] = \{1, 1, 1, 1, 1, 1, 1, 1\}$. Using boundary extension and assuming critical sampling between analysis and synthesis banks, compute the signal values at all internal nodes in this filter bank, and hence demonstrate that the output sequence is identical to the input sequence.

Q6.9

Repeat Q6.8, but this time assume that the bit allocation strategy employed preserves the low-frequency subband but completely discards the high-pass subband. Assuming a word length of 4 bits, what is the PSNR of the reconstructed signal after decoding?

Q6.10

Draw the diagram for a 2D, three-stage wavelet filter-bank. Show how this decomposition tiles the 2D spatial frequency plane and compute the frequency range of each subband. Assuming that the input is of dimensions 256 × 256 pixels, how many samples are contained in each subband?

Q6.11

Prove that the following two diagrams are equivalent.

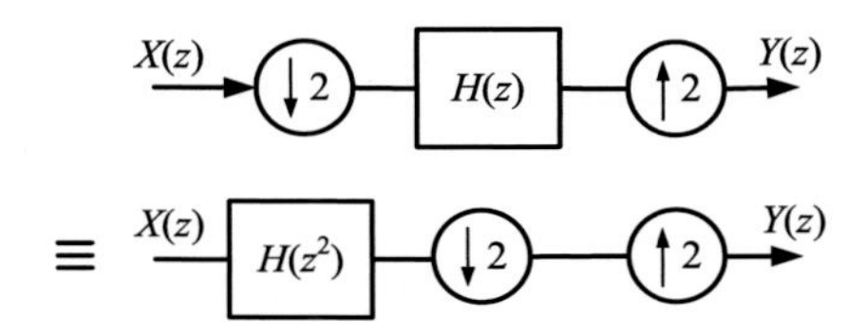

Chapter 7: Lossless compression methods

Q7.1

Consider the following codewords for the given set of symbols.

Symbol	C1	C2	C3
a_1	0	0	01
a_2	10	01	10
a_3	110	110	11
a_4	1110	1000	001
a_5	1111	1111	010

Identify which are prefix codes.

Q7.2

Derive the set of Huffman codewords for the symbol set with the following probabilities:

$P(s_0) = 0.06,\quad P(s_1) = 0.23,\quad P(s_2) = 0.30,\quad P(s_3) = 0.15,$

$P(s_4) = 0.08,\quad P(s_5) = 0.06,\quad P(s_6) = 0.06,\quad P(s_7) = 0.06.$

What is the transmitted binary sequence corresponding to the symbol pattern s_0, s_4, s_6? What symbol sequence corresponds to the code sequence 000101110? Calculate the first order entropy of this symbol set, the average codeword length, and the coding redundancy.

Q7.3

A quantized image is to be encoded using the symbols $I \in \{I_0 \cdots I_{11}\}$. From simulation studies it has been estimated that the relative frequencies for these symbols are as follows: I_0: 0.2; $I_1..I_3$: 0.1; $I_4..I_7$: 0.05; $I_8..I_{11}$: 0.075. Construct the Huffman tree for these symbols and list the resultant codewords in each case.

Q7.4

If the minimum-variance set of Huffman codes for an alphabet A is as shown in the table below, determine the efficiency of the corresponding Huffman encoder.

Symbol	Probability	Huffman code
a_1	0.5	0
a_2	0.25	10
a_3	0.125	110
a_4	0.0625	1110
a_5	0.0625	1111

Q7.5

Consider the following 4 × 4 matrix, **C**, of DCT coefficients, produced from a block-based image coder. Using zig-zag scanning and run-length coding (assume a {run, value} model where value is the integer value of a nonzero coefficient and run is the number of zeros preceding it), determine the transmitted bitstream after entropy coding. We have

$$\mathbf{C} = \begin{bmatrix} 2 & 0 & 3 & 0 \\ 0 & 3 & 1 & 0 \\ 1 & 0 & 0 & 0 \\ 0 & 0 & 0 & 0 \end{bmatrix}.$$

Use the following symbol to codeword mappings:

Symbol	Code	Symbol	Code
{0,1}	00	{1,3}	0111
{0,2}	10	{2,1}	0110
{0,3}	110	{2,2}	0101
{1,1}	1111	{2,3}	01001
{1,2}	1110	EOB	01000

Q7.6

Assuming that the DCT coefficient matrix and quantization matrix in Q5.14 form part of a JPEG baseline codec, derive the Huffman coded sequence that would be produced by the codec for the AC coefficients.

Q7.7

What is the Exp-Golomb code for the symbol index 132_{10}?

Q7.8

Show how the Exp-Golomb codeword 000010101 would be decoded and compute the value of the corresponding symbol index. What is the corresponding Golomb–Rice code for this index (assume m=4)?

Q7.9

Given the symbols from an alphabet, $A = \{a_1, a_2, a_3, a_4, a_5\}$, and their associated probabilities of occurrence in the table below, determine the shortest arithmetic code which represents the sequence $\{a_1, a_1, a_2\}$.

Symbol	Probability
a_1	0.5
a_2	0.25
a_3	0.125
a_4	0.0625
a_5	0.0625

Q7.10

Given the following symbols and their associated probabilities of occurrence, determine the binary arithmetic code which corresponds to the sequence $\{a_1, a_2, a_3, a_4\}$ and demonstrate how an arithmetic decoder matched to the encoder would decode the bitstream 010110111.

Symbol	Probability
a_1	0.5
a_2	0.25
a_3	0.125
a_4 (EOB)	0.125

Q7.11

Given the following symbols and their associated probabilities of occurrence, determine the arithmetic code which corresponds to the sequence a_1, a_2, a_3 (where a_3 represents the EOB symbol).

Symbol	Probability
a_1	0.375
a_2	0.375
a_3	0.125
a_4	0.125

Show how the bitstream produced above would be decoded to produce the original input symbols.
Repeat this question using Huffman encoding rather than arithmetic coding. Compare your results in terms of coding efficiency.

Q7.12

Derive the arithmetic codeword for the sequence $s_1, s_1, s_2, s_2, s_5, s_4, s_6$, given a symbol set with the following probabilities:

$$P(s_0) = 0.065, \quad P(s_1) = 0.20, \quad P(s_2) = 0.10, \quad P(s_3) = 0.05,$$
$$P(s_4) = 0.30, \quad P(s_5) = 0.20, \quad P(s_6) = 0.10 = \text{EOB}.$$

Q7.13

Consider an alphabet, $A = \{a_1, a_2, a_3, a_4\}$, where $P(a_1) = 0.6$, $P(a_2) = 0.2$, $P(a_3) = 0.1$, $P(a_4) = 0.1$. Compute the sequence, S, of three symbols which corresponds to the arithmetic code 0.58310.

Q7.14

Given an alphabet of two symbols A={a,b}, where $P(a) = 0.25$, $P(b) = 0.75$, draw a diagram showing coding and probability intervals for the associated arithmetic coder. Derive a binary arithmetic code for the sequence S=baa.

Q7.15

Assuming the symbol probabilities in the table below, compute the arithmetic codeword for the sequence $s2 \quad s2 \quad s7 \quad s8$. Use interval scaling in your solution.

Symbol	Prob	Symbol	Prob
s1{0,1}	0.0625	s5{2,1}	0.0625
s2{0,2}	0.125	s6{2,2}	0.125
s3{1,1}	0.0625	s7{3,1}	0.25
s4{1,2}	0.0625	s8{EOB}	0.125
		s9{Other}	0.125

Chapter 8: Coding moving pictures: motion prediction

Q8.1

Implement the full search BBME algorithm on the 6×6 search window, **S**, using the current-frame template, **M**, given as follows:

$$\mathbf{S} = \begin{bmatrix} 1 & 5 & 4 & 9 & 6 & 1 \\ 6 & 1 & 3 & 8 & 5 & 1 \\ 5 & 7 & 1 & 3 & 4 & 1 \\ 2 & 4 & 1 & 7 & 6 & 1 \\ 2 & 4 & 1 & 7 & 8 & 1 \\ 1 & 1 & 1 & 1 & 1 & 1 \end{bmatrix}, \quad \mathbf{M} = \begin{bmatrix} 7 & 7 \\ 7 & 7 \end{bmatrix}.$$

Q8.2

Given the following reference window and current block, show how an N-step search algorithm would locate the best match (assume an SAD optimization criterion). Determine the motion vector for this block. Does this produce the same result as an exhaustive search? We have

$$\begin{bmatrix} 1 & 4 & 3 & 2 & 1 & 2 & 3 & 2 & 2 \\ 0 & 1 & 2 & 0 & 2 & 3 & 0 & 2 & 1 \\ 0 & 2 & 0 & 0 & 1 & 2 & 0 & 1 & 0 \\ 0 & 0 & 0 & 0 & 1 & 1 & 1 & 1 & 0 \\ 1 & 2 & 1 & 1 & 4 & 4 & 0 & 0 & 1 \\ 0 & 3 & 2 & 3 & 0 & 1 & 2 & 2 & 3 \\ 0 & 3 & 3 & 1 & 0 & 1 & 1 & 2 & 2 \\ 0 & 4 & 2 & 3 & 0 & 2 & 1 & 2 & 1 \\ 0 & 1 & 2 & 1 & 2 & 3 & 2 & 2 & 0 \end{bmatrix} \begin{bmatrix} 1 & 2 & 3 \\ 1 & 2 & 0 \\ 2 & 2 & 0 \end{bmatrix}.$$

Q8.3

Use bidirectional exhaustive search motion estimation, such as that employed in MPEG-2, to produce the best match for the following current block and two reference frames:

$$\begin{bmatrix} 1 & 2 & 3 & 4 \\ 1 & 2 & 2 & 3 \\ 1 & 1 & 2 & 2 \\ 2 & 2 & 2 & 1 \end{bmatrix} \begin{bmatrix} 1 & 1 & 3 & 4 \\ 1 & 1 & 0 & 3 \\ 1 & 1 & 2 & 0 \\ 2 & 1 & 0 & 1 \end{bmatrix} \begin{bmatrix} 1 & 2 \\ 1 & 2 \end{bmatrix}.$$

Q8.4

Explain how the two-dimensional logarithmic (TDL) search method improves the search speed of block-based motion estimation. Illustrate this method for the case of a $\pm 6 \times \pm 6$ search window where the resultant motion vector is [2,5]. Quantify the savings for this particular example over a full search. What is the main advantage and disadvantage of this method?

Q8.5

Implement a hexagonal search block matching algorithm on the search window, **S**, using the current-frame template, **M**, as given below. Determine the motion vector for this search. Assume that any search points where the hexagon goes outside of the reference frame are invalid. We have

$$\mathbf{S} = \begin{bmatrix} 0 & 1 & 2 & 3 & 22 & 19 & 18 & 23 \\ 7 & 3 & 6 & 5 & 33 & 31 & 13 & 22 \\ 4 & 6 & 3 & 7 & 23 & 23 & 15 & 26 \\ 8 & 4 & 1 & 3 & 11 & 22 & 29 & 19 \\ 2 & 8 & 9 & 7 & 8 & 14 & 16 & 18 \\ 5 & 0 & 7 & 3 & 7 & 15 & 12 & 13 \\ 7 & 4 & 6 & 6 & 9 & 8 & 8 & 12 \\ 1 & 2 & 3 & 9 & 10 & 9 & 8 & 12 \end{bmatrix}, \quad \mathbf{M} = \begin{bmatrix} 1 & 4 \\ 10 & 7 \end{bmatrix}.$$

Q8.6

Given the following current block P and its set of three adjacent neighbors A, B, C, D with motion vectors as indicated, use motion vector prediction to initialize the search for the best motion vector for this block.

$\mathbf{d}_B$=[1,1]	$\mathbf{d}_C$=[1,0]	$\mathbf{d}_D$=[1,0]	
$\mathbf{d}_A$=[1,2]	P		

Q8.7

Motion vectors for six adjacent blocks in two partial rows of a frame are shown below. Using the prediction scheme $\hat{\mathbf{d}}_P = \text{med}(\mathbf{d}_A, \mathbf{d}_C, \mathbf{d}_D)$, compute the predicted motion vectors for each of these blocks together with their coding residuals. State any assumptions that you make regarding the prediction of motion vectors for blocks located at picture boundaries.

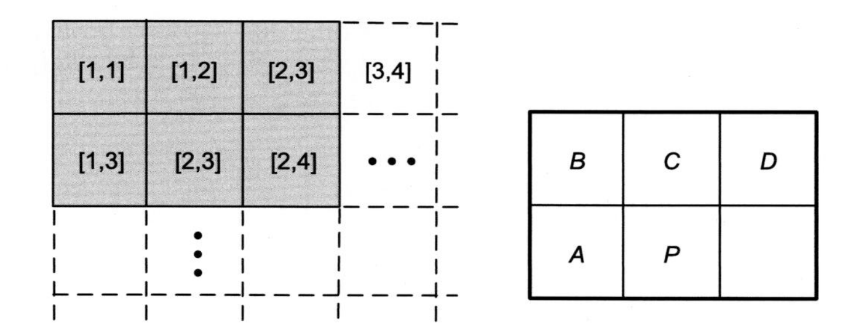

Chapter 9: The block-based hybrid video codec

Q9.1

The figure below shows a block diagram of a basic hybrid (block transform, motion-compensated) video coding system.

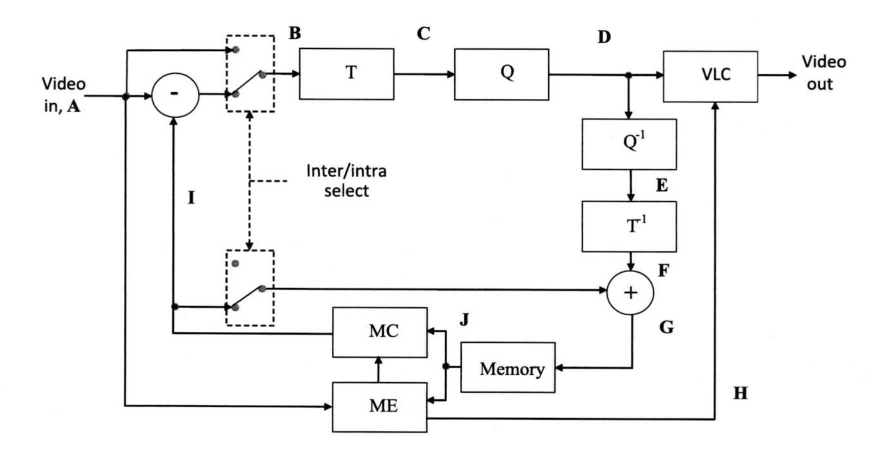

Consider the following assumptions and data at time n:

Image frame size: 12×12 pixels

Macroblock size: 4×4 pixels

Operating mode: Interframe

Motion estimation: Linear translational model (4×4 blocks)

Transform matrix: $$\mathbf{T}=\frac{1}{2}\begin{bmatrix}1 & 1 & 1 & 1\\ 1 & 1 & -1 & -1\\ 1 & -1 & -1 & 1\\ 1 & -1 & 1 & -1\end{bmatrix}$$

Quantization matrix: $$\mathbf{Q}=\begin{bmatrix}1 & 1 & 4 & 4\\ 1 & 2 & 4 & 4\\ 4 & 4 & 4 & 4\\ 4 & 4 & 4 & 4\end{bmatrix}$$

Current input block: $$\mathbf{A}=\begin{bmatrix}2 & 2 & 0 & 0\\ 2 & 2 & 0 & 0\\ 0 & 0 & 0 & 0\\ 0 & 0 & 0 & 0\end{bmatrix}$$; assume exact center of frame.

Reference memory: $$\mathbf{J}=\begin{bmatrix}0 & 0 & 2 & 2 & 2 & 1 & 1 & 1 & 1 & 1 & 1 & 1\\ 4 & 10 & 4 & 7 & 4 & 7 & 4 & 4 & 4 & 3 & 2 & 1\\ 8 & 4 & 7 & 6 & 7 & 1 & 1 & 0 & 0 & 0 & 0 & 0\\ 6 & 4 & 6 & 5 & 4 & 1 & 1 & 0 & 0 & 0 & 0 & 0\\ 3 & 2 & 10 & 6 & 8 & 0 & 0 & 0 & 0 & 0 & 0 & 0\\ 0 & 4 & 14 & 4 & 4 & 0 & 0 & 0 & 0 & 0 & 0 & 0\\ 0 & 0 & 5 & 6 & 6 & 6 & 6 & 0 & 0 & 7 & 0 & 0\\ 0 & 0 & 0 & 12 & 0 & 0 & 0 & 3 & 4 & 2 & 3 & 4\\ 0 & 0 & 0 & 0 & 0 & 0 & 0 & 4 & 4 & 5 & 8 & 0\\ 7 & 7 & 8 & 9 & 0 & 0 & 0 & 0 & 6 & 7 & 0 & 0\\ 0 & 0 & 0 & 10 & 3 & 0 & 0 & 0 & 0 & 7 & 0 & 0\\ 0 & 0 & 2 & 2 & 2 & 0 & 0 & 0 & 0 & 0 & 0 & 0\end{bmatrix}$$

Compute:

1. The current motion vector, **H**, at time n.
2. The motion-compensated output, **I**, at time n.
3. The DFD for the current input frame, **B**.
4. The transformed and quantized DFD output, **D**.

Q9.2

If the encoded bitstream from Q9.1 is decoded by a compliant decoder, compute the decoder output at time n that corresponds to the input block, **A**.

Q9.3

Compute the transformed and quantized DFD output in Q9.1, but this time for the case of intraframe coding.

Q9.4

Using horizontal, vertical, and DC modes only, produce the H.264/AVC intra-prediction for the following highlighted 4×4 luminance block:

1	2	3	4	2	3	5	4	7	8	9	5
2	4	4	6	7	8	4	6	4	4	3	2
4	3	5	6	3	3	3	4	4	4	3	3
4	4	5	6	3	3	3	4	4	4	3	3
1	1	4	6	**2**	**3**	**6**	**3**	x	x	x	x
4	3	3	5	**3**	**2**	**5**	**3**	x	x	x	x
2	2	3	5	**3**	**3**	**4**	**4**	x	x	x	x
1	3	5	3	**1**	**2**	**4**	**4**	x	x	x	x

Q9.5

Following the full search result from Q8.1, refine the motion vector for this search to ½-pixel accuracy using the 2-tap interpolation filter $(s_i + s_j)/2$ (where s_i and s_j are the horizontal or vertical whole-pixel locations adjacent to the selected ½ pixel location).

Q9.6

Use the H.264/AVC subpixel interpolation filter to generate the half-pixel values for the shaded locations in the following search window.

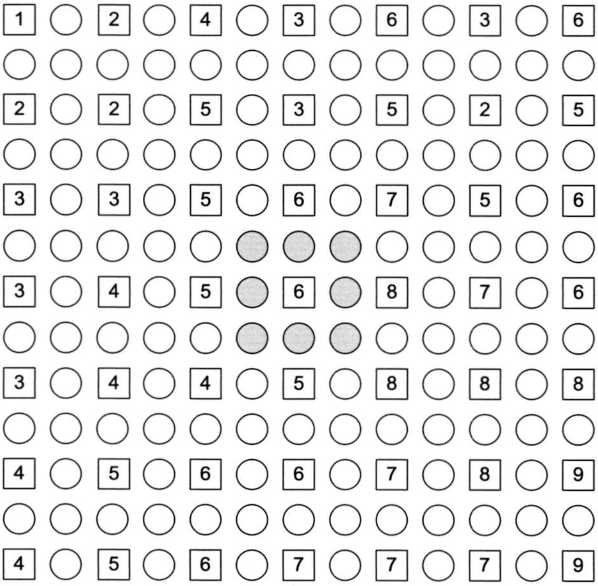

1	○	2	○	4	○	3	○	6	○	3	○	6
○	○	○	○	○	○	○	○	○	○	○	○	○
2	○	2	○	5	○	3	○	5	○	2	○	5
○	○	○	○	○	○	○	○	○	○	○	○	○
3	○	3	○	5	○	6	○	7	○	5	○	6
○	○	○	○	○	●	●	●	○	○	○	○	○
3	○	4	○	5	●	6	●	8	○	7	○	6
○	○	○	○	○	●	●	●	○	○	○	○	○
3	○	4	○	4	○	5	○	8	○	8	○	8
○	○	○	○	○	○	○	○	○	○	○	○	○
4	○	5	○	6	○	6	○	7	○	8	○	9
○	○	○	○	○	○	○	○	○	○	○	○	○
4	○	5	○	6	○	7	○	7	○	7	○	9

Q9.7

H.264 employs a 4×4 integer transform instead of the 8×8 DCT used in previous coding standards. Prove that the four-point integer approximation to the 1D DCT transform matrix,

A, is given by

$$\mathbf{A} = \begin{bmatrix} 1 & 1 & 1 & 1 \\ 2 & 1 & -1 & -2 \\ 1 & -1 & -1 & 1 \\ 1 & -2 & 2 & -1 \end{bmatrix} \otimes \mathbf{E}_f,$$

where $\mathbf{E}_f$ is a 4×4 scaling matrix. State any assumptions made during your derivation.

Chapter 10: Measuring and managing picture quality

Q10.1

List the primary factors that should be controlled and recorded during subjective video assessment trials.

Q10.2

List the primary attributes that a good subjective database should possess.

Q10.3

Given the following three consecutive frames, compute the temporal activity (TI) for this sequence:

$$\mathbf{S}_1 = \begin{bmatrix} 1 & 2 & 3 & 4 \\ 2 & 3 & 4 & 5 \\ 3 & 4 & 5 & 6 \\ 4 & 5 & 6 & 7 \end{bmatrix}, \quad \mathbf{S}_2 = \begin{bmatrix} 1 & 2 & 3 & 5 \\ 2 & 3 & 4 & 6 \\ 3 & 4 & 5 & 7 \\ 4 & 5 & 6 & 8 \end{bmatrix}, \quad \mathbf{S}_3 = \begin{bmatrix} 1 & 2 & 4 & 6 \\ 2 & 3 & 5 & 7 \\ 3 & 4 & 6 & 8 \\ 4 & 5 & 7 & 9 \end{bmatrix}.$$

Q10.4

Consider the operation of a constant bit rate video transmission system at 25 fps with the following parameters:

$$R_o = 500 \text{ kbps}, \ B = 150 \text{ kb}, \ F_i = 100 \text{ kb}, \ \text{GOP} = 6 \text{ frames}.$$

If the pictures transmitted have the following sizes, compute the occupancy of the decoder buffer over time and determine whether underflow or overflow occurs.

Frame no.	Picture size (kbits)
1	20
2–6	10
7	40
8–12	10

Q10.5

The following rate (bits) and distortion (SSD) results are for three mode candidates when encoding a 64 × 64 coding tree unit (CTU) of an inter-predicted frame of the *BasketballDrill* (843 × 480, 10 bit) sequence using an HEVC codec. Following the Lagrangian multiplier approach for rate-distortion optimization, evaluate which of the three modes offers the best RD performance. Assume that the value of the Lagrange parameter λ for mode selection is 232.

Mode	Rate (bits)	SSD
Skip	1	127368
Intra	358	136823
Inter	34	95984

Chapter 11: Communicating pictures: delivery across networks

Q11.1

Using the Huffman codes in the following table, decode the following encoded bitstream: 0 0 1 0 1 0 0 1 1 1 1 1 1 1 0 1 0 0.

Symbol	Probability	Huffman code
a_1	0.5	0
a_2	0.25	10
a_3	0.125	110
a_4	0.0625	1110
a_5	0.0625	1111

Assuming codeword 110 represents an EOB signal, what would be the effect of a single error in bit position 2 of the above sequence?

Q11.2

An image encoder uses VLC and Huffman coding of transform coefficients, based on a set of four symbols with the following mappings:

$$A \leftrightarrow 0, \quad B \leftrightarrow 10, \quad C \leftrightarrow 110, \quad D \leftrightarrow 111.$$

Assuming that the sequence transmitted is ABCDAC, determine the received symbol sequences for the following scenarios:

1. An error occurring in the third bit position.
2. An error occurring in the first bit position.
3. An error occurring in the fourth bit position.
4. An error occurring in the eighth bit position.

For each scenario, assuming that B represents the EOB symbol, comment on the impact that the error has on the final state of the decoder and on the reconstructed transform coefficients.

Q11.3

VLC codes for symbols {a, b, c} are a = 0, b = 11, c = 101, where P(a) = 0.5, P(b) = 0.25, P(c) = 0.25. Comment on any specific property these codewords exhibit and on its benefits in a lossy transmission environment. Compare the efficiency of these codewords with that of a conventional set of Huffman codes for the same alphabet.

Q11.4

The sequence of VLC codewords produced in Q7.5 is sent over a channel which introduces a single error in the 15th bit transmitted. Comment on the impact that this error has on the reconstructed transform coefficients and any subsequent blocks of data.

Q11.5

Use EREC to code four blocks of data with lengths $b_1 = 6$, $b_2 = 5$, $b_3 = 2$, $b_4 = 3$. Assuming that an error occurs in the middle of block 3 and that this causes the EOB code to be missed for this block, state which blocks in the frame are corrupted and to what extent.

Q11.6

A video coder uses VLC and Huffman coding based on a set of four symbols with probabilities 0.4, 0.3, 0.15, and 0.15. Calculate the average bit length of the resulting Huffman codewords. If this video coder is to employ reversible codes (RVLC) to improve error resilience, suggest appropriate codewords and calculate the bit rate overhead compared to conventional Huffman coding.

Q11.7

Assume that $\mathbf{S_1}$ and $\mathbf{S_2}$ below represent corresponding regions in two temporally adjacent video frames. Due to transmission errors, the central 4×4 block (pixels marked as "×") in the received current frame, $\mathbf{S}'_2$, is lost. Assume that the codec operates on the basis of 2×2 macroblocks.

a) Perform temporal error concealment using frame copying to provide an estimate of the lost block.

b) Perform motion-compensated temporal error concealment, based on the BME measure. Assume that the candidate motion vectors obtained from the four adjacent blocks are either [0,1], [1,0], or [1,1]. We have

$$\mathbf{S_1} = \begin{bmatrix} 1 & 1 & 1 & 2 & 2 & 2 & 3 & 3 \\ 2 & 2 & 2 & 3 & 3 & 3 & 4 & 4 \\ 3 & 3 & 3 & 4 & 4 & 4 & 4 & 4 \\ 3 & 3 & 3 & 4 & 4 & 4 & 5 & 5 \\ 4 & 4 & 4 & 4 & 4 & 4 & 4 & 4 \\ 5 & 5 & 5 & 6 & 6 & 7 & 6 & 6 \\ 6 & 6 & 6 & 7 & 6 & 6 & 6 & 6 \\ 7 & 7 & 7 & 7 & 8 & 8 & 8 & 8 \end{bmatrix}, \quad \mathbf{S}'_2 = \begin{bmatrix} 2 & 2 & 3 & 3 & 3 & 4 & 4 & 3 \\ 3 & 3 & 4 & 4 & 4 & 4 & 5 & 5 \\ 3 & 3 & 4 & 4 & 4 & 4 & 4 & 4 \\ 4 & 4 & 5 & \times & \times & 4 & 4 & 6 \\ 5 & 5 & 6 & \times & \times & 6 & 6 & 6 \\ 7 & 6 & 7 & 6 & 6 & 6 & 6 & 7 \\ 6 & 7 & 7 & 6 & 6 & 8 & 8 & 8 \\ 9 & 9 & 9 & 7 & 8 & 7 & 8 & 8 \end{bmatrix}.$$

Q11.8

Assuming again that the central 2 × 2 block (pixels marked as "x") in the received current frame, $\mathbf{S}_2'$ (above), is lost, calculate the missing elements using spatial error concealment.

Q11.9

A block-based image coder generates a slice comprising five DCT blocks of data as follows.

Block number	Block data
1	01011011101111
2	10111001011101111
3	1100101111
4	101101101111
5	0001111

Here the symbols and entropy codes used to generate this data are as follows (assume E is the end of block symbol).

Symbol	Huffman code
A	0
B	10
C	110
D	1110
E	1111

Code this data using error-resilient entropy coding (EREC). Choose an appropriate EREC frame size and offset sequence for this slice and show all the encoding stages of the algorithm.

Q11.10

Perform EREC decoding on the bitstream generated from Q11.9.

If the slice data given in the above question is corrupted during transmission such that the values of the last 2 bits in each of block 2 and block 3 are complemented, how many blocks are corrupted after EREC decoding? Compute how many blocks would contain errors if the slice were transmitted as a conventional stream of entropy codewords without EREC.

Chapter 12: Video coding standards and formats

Q12.1

Assuming a CIF format luminance-only H.261 sequence with synchronization codewords at the end of each GOB, calculate the (likely) percentage of corrupted blocks in a frame if bit errors occur in block 1 of macroblock 7 in GOB 2 and block 1 of macroblock 30 of GOB 10.

Q12.2

What parts of a video codec are normally subject to standardization in terms of compliance testing? Why is this?

Q12.3

Discuss the concepts of profiles and levels in MPEG-2 and later standards, indicating how these have enabled a family of compatible algorithms covering a range of applications and bit rates to be defined.

Q12.4

Compare the features of H.264/AVC and MPEG-2 and highlight those features that have enabled H.264 to approximately halve the bit rate of MPEG-2 for equivalent picture quality.

Q12.5

Compare the features of H.265/HEVC and H.264/AVC and highlight those features that have enabled HEVC to approximately halve the bit rate of H.264 for equivalent picture quality.

Q12.6

Why has H.265/HEVC increased its maximum block size to 64×64 pixels and VVC increased this further to 128×128?

Q12.7

Compare the features of H.265/HEVC and H.266/VVC and highlight those features that have enabled VVC to approximately halve the bit rate of HEVC for equivalent picture quality.

Chapter 13: Communicating pictures – the future

Q13.1

What are the primary challenges and demands for video compression in the future?

Q13.2

Consider the case of an 8K resolution video signal in 4:2:2 format with 14 bits dynamic range and a frame rate of 300 fps. Calculate the total bit rate needed for this video.

Q13.3

How might video compression algorithms develop in the future in order to cope with the demands of formats such as that described in Q13.2?

Index

D

F

M

N

O

P

Q

R

U

V

W

Z

Printed in the United States
by Baker & Taylor Publisher Services